Hurlel Elliott, Heinz Bloch

Compressor Technology Advances

Also of interest

Fluid Machinery.
Life Extension of Pumps, Gas Compressors and Drivers
Bloch, 2020
ISBN 978-3-11-067413-2, e-ISBN 978-3-11-067415-6

Process Safety.
An Engineering Discipline
Hoorelbeke, 2021
ISBN 978-3-11-063205-7, e-ISBN 978-3-11-063213-2

Process Engineering.
Addressing the Gap between Study and Chemical Industry
Kleiber, 2020
ISBN 978-3-11-065764-7, e-ISBN 978-3-11-065768-5

Engineering Risk Management.
Meyer, Reniers, 2021
ISBN 978-3-11-066531-4, e-ISBN 978-3-11-066533-8

Process Intensification.
Breakthrough in Design, Industrial Innovation Practices, and Education
Harmsen, Verkerk, 2020
ISBN 978-3-11-065734-0, e-ISBN 978-3-11-065735-7

Hurlel Elliott, Heinz Bloch

Compressor Technology Advances

Beyond 2020

DE GRUYTER

Authors
Hurlel G. Elliott
2951 Marina Bay Drive
Suite 130-275
League City, TX 77573
USA
hurlge5@gmail.com

Heinz P. Bloch, P. E.
267 Sunnyvale East
Montgomery, TX 77356
USA
heinzpbloch@gmail.com

ISBN 978-3-11-067873-4
e-ISBN (PDF) 978-3-11-067876-5
e-ISBN (EPUB) 978-3-11-067883-3

Library of Congress Control Number: 2020947735

Bibliographic information published by the Deutsche Nationalbibliothek
The Deutsche Nationalbibliothek lists this publication in the Deutsche Nationalbibliografie; detailed bibliographic data are available on the Internet at http://dnb.dnb.de.

Cover image: Sergey Ryzhov/Centrifugal compressor/stock.adobe.com
Typesetting: Integra Software Services Pvt. Ltd.
Printing and binding: CPI books GmbH, Leck

www.degruyter.com

We have fond memories of many of our mentors at the Esso/Exxon Research and Engineering, later at Exxon Chemicals and ExxonMobil. Their willingness and patience in sharing technical knowledge and practical wisdom always proved encouraging and motivating.

Exxon Chemical's first two chief engineers, Ralph James and Siegfried Zierau, are remembered for their remarkable practice of fairness and sensitivity in human relations. They placed these valuable attributes ahead of even their own superb professionalism and competence. Their insightful priority was evident; it built character in coworkers who appreciated their approachability and cooperative style. Success in telling truth to power made these two men tower over the rest, which is why this book is dedicated to them.

Preface

As coauthors with a combined professional experience (in 2016) of well over 100 years, we are keenly aware of the profusion of compressor texts available to the reader. In our personal reference libraries, we probably have more outdated compressor books than we care to enumerate. However, access to these personal library books allows us to make our main point: Changing technology made some books outdated. Educational approaches have certainly changed over the years, and a new set of needs has emerged in the many decades since most of these texts were written. We decided that the existing books on compressor technology did not at all, or did not sufficiently, delve into the details of compression technology of interest to us in 2020.

While compressor basics and their underlying thermodynamics and physics are immutable and have not changed, some compressor books and compression-related texts cater to a relatively narrow range of interests. Conversely, other texts have attempted to address too wide a spectrum of readers and/or are simply too voluminous in either size or scope. This then prompted our decision to approach De Gruyter with a solid proposal for this unique book.

Much of the input was developed for us by subject matter experts. We asked for their input on the distinct topic of new approaches to a very mature technology. Today, these new approaches are primarily pursued by a handful of safety and reliability-focused users and process operating companies. They then commission the most experienced engineering–procurement–construction companies to implement plants making the fullest use of known best-available or fully optimized technology approaches. The best and most profitable owner–operators judiciously superimpose their own experts to ascertain that these approaches are consistently and efficiently implemented.

This book explains and describes in detail where, why, and how new approaches to compression technology are of value to all process plants. We refer to existing facilities contemplating upgrades as well as facilities that have merely reached cost estimating or preengineering and definition-of-scope status.

Our contents speak to the need and delineate where we are going with this book. It was our goal to keep it to 460 or so pages. We endeavored to steer clear of preexisting or widely disseminated prior books; these materials are referenced – if necessary – but will not be reused as a filler material for *Compressor Technology Advances*.

Heinz P. Bloch and Hurlel G. Elliott

https://doi.org/10.1515/9783110678765-202

Acknowledgements

We are indebted to individual engineers and manufacturers who kindly responded to our requests for compiling, and also granting permission, to include material in this book. Their able assistance made it possible for us to move the book from concept to reality. We wish to also thank many of these contributors for spending considerable personal time compiling highly relevant input.

Our listing follows the order in which their input appears in the book

- Masayuki Kita, and Hiroaki Osaki–Mitsubishi Heavy Industries Compressor Corporation (MHI), Hiroshima, Japan. We are grateful for receiving illustrations for Chapter 2; the material relating to effective wash nozzle arrangements (Section 3.2.1.2); modern design and manufacturing considerations (Section 3.2.2); re-thinking driver selection for Ethylene Refrigeration Trains (Section 3.2.5); Discussion of Executing a Mechanical Run Test Per API 617 (Section 5.2).
- C. Hunter Cloud – BRG Machinery, Charlottesville, Virginia: Thanks for Chapter 7; figures for APPENDIX C1, and reviewing of final document for accuracy
- Robert Whitney, and Pat McCormack-Indikon/Riverhawk Company, New Hartford, New York. Both contributed write-ups to Chapter 6, Chapter 12, and APPENDIX C2 – showing how torsional vibration problems can be mitigated using new design viscous rotational damper
- Craig Holmes, Torquemeters Ltd., Ravensthorpe, Northampton, England. Our thanks are due for his solid contributions to Chapter 6
- Cameron Compressors, Buffalo, NY. Their engineering staff graciously contributed Chapter 10
- Kenneth Atkins, Engineering Dynamics Inc. (EDI), San Antonio, Texas. A well-respected former Exxon colleague and (later) value adder at EDI, he contributed Section 4.2.1 – Understanding various industrial interpretations of rod loading
- Asst. Prof. Jorge Tito-Izquierdo, University of Houston, Houston, Texas: Graciously provided the excellent write-up on Chapter 11, "Design Basics of Tabletop Foundations for Machinery Engineers;" also APPENDIX D – MQA of Tabletop Foundations
- Serge Staroselsky, Compressor Controls Corporation (CCC), Des Moines, Iowa. He provided Chapter 13 – "Turbomachinery Control: Architecture and functional details."
- Dr. Christian Tümmers, and Reinaldo Bermudez – MAN-ES, Germany. Their engineering offices in Augsburg/Germany and Houston/Texas contributed many of the illustrations in Chapter 2; their insights on "Optimizing of flow path including blade profiles to improve performance for Axial Compressors" (Section 3.1) are, and will continue to be, greatly valued.
- Kenji Fujimatsu – Kobelco, Japan. Contributed relevant illustrations and key points for Chapter 2; also the write-up on "Using Innovative New Rotor Profile to Maximize Performance" (for Section 4.1.1)

https://doi.org/10.1515/9783110678765-203

- Kyle Carpenter, Christiana Bash, and Derrick Bauer–Elliott Turbomachinery, Jeanette, Pennsylvania, USA. These professionals were instrumental in providing material for Section 3.2.1.1: "Surface Treatment Technologies; " also on "mitigating fouling in Centrifugal Compressors."
- Pierre Noack–Machinenfabrik, Aerzen, Germany: Important updates and illustrations on modern twin-screw rotary compressors. Their expertise covers wet screw compressors operating not just with an oil fill, but with a succession of liquids ranging from water to the truly exotic.
- Glen Schmidt–EagleBurgmann, Houston, Texas, USA. Contributed much to the discussions on Dry Gas Seals (DGS, Chapter 9)
- "Zee" Zahroof – Zahroof Valves Inc., Houston, Texas, USA. Zee contributed the easily understood technical input and final illustrations relating to highly important modular reed valves in our Section 4.2.2

Contents

Chapter 1
Thermodynamics refresher

1.1 Applicable laws

The intent of this chapter is not to rehash the fundamentals of classical thermodynamics but to use those fundamental principles and processes. All are well known in the academic arena as they apply to compression technology. Our assumption is that the reader has studied classical thermodynamics and can understand, and use, these fundamental laws and equations as we apply them in designing gas compression machinery.

Fluids can exist in any of the three states, namely solid, liquid, or gas (superheated vapor). Figure 1.1 illustrates the states of a typical fluid as the fluid changes state, say from solid A to liquid B, then vapor C, and finally, beyond C it becomes gas making D a superheated vapor. B to C is the latent heat of vaporization where temperature as a property remains constant.

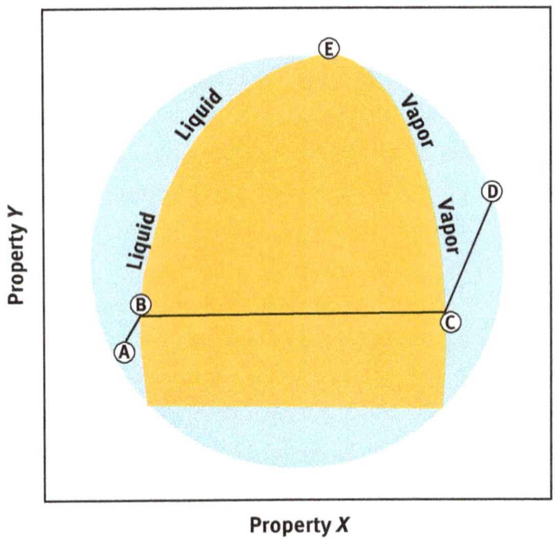

Figure 1.1: Changes in the state of typical fluids as a function of change in properties.

Point E is regarded as the critical point of the fluid where the liquid and vapor phases are indistinguishable and conflate into a single identical density with no latent heat of vaporization. The pressure and temperature associated with this point in a fluid are described as critical pressure and critical temperature.

https://doi.org/10.1515/9783110678765-001

As shown in Figure 1.1, by knowing the intercept between property Y and property X, we can define the state point and hence the state of the fluid. So how are we defining properties of fluid? The properties of a fluid are specific thermodynamic characteristics whose values depend on the state of the fluid. Common thermodynamic properties are temperature, pressure, volume, internal energy, entropy, and enthalpy. Knowing the value of any two independent properties will pinpoint the state of the fluid.

Compression processes occur when the pressure of a system increases and its volume decreases. Later, as we superimposed different properties on Figure 1.1, it will be shown that the compression process takes place in the region C–D. Various methods and types of machines are used to achieve this process, but they all follow basic gas laws established centuries ago.

The laws associated with a gas compression process are best explained with the ideal gas equation. Though many gases such as air and oxygen are sometimes treated as ideal gases, ideal gas laws only give an approximation since there are no perfect gases.

1.1.1 Boyle's law

For a gas at constant temperature, the pressure will vary inversely to its volume. In other words, for the pressure of a gas to increase, its volume has to decrease proportionately at constant temperature:

$$P \propto \frac{1}{v} \text{ or } Pv = \text{constant} \tag{1.1}$$

where P is the absolute pressure and v is the specific volume.

When a gas is compressed from state 1 to state 2, eq. (1.1) can be rewritten as follows:

$$P_1v_1 = P_2v_2 \tag{1.2}$$

1.1.2 Charles' law

Varying the volume of a gas at constant pressure will vary its temperature proportionately:

$$\frac{v}{T} = \text{constant} \tag{1.3}$$

where T is the absolute temperature.

1.1.3 Equation of state for ideal gases

By combining both Boyle's and Charles' laws and denoting the constant as R, we arrive at the equation of state (EOS) for ideal gases:

$$Pv = RT \tag{1.4}$$

where R is the *characteristic gas constant* for a specific gas.

The units of R are derived as follows:

$$R = \frac{Pv}{T} = \frac{\text{energy units}}{\text{mass} \times \text{absolute temperature}}$$

For SI units

$$= \frac{\text{Newton} - \text{meter}}{\text{kg} \times {}^{\circ}\text{K}}$$

$$= \frac{\text{Joule}}{\text{kg} - \text{K}} \text{ or } \frac{\text{kJ}}{\text{kg} - \text{K}}$$

For US customary units

$$= \frac{\text{ft} - \text{lbf}}{\text{lb} - {}^{\circ}\text{R}}$$

If "m" is the amount of mass being compressed, then multiplying both sides of eq. (1.4), the equation becomes

$$PV = mRT \tag{1.5}$$

where is m the mass of gas, R is the characteristic gas constant, and V is the total volume.

Typical units of mass "m" are pound-mass (lbm), kilogram (kg), and gram (g).

All gases exist as *molecules*, meaning that over two or more atoms are bonded together chemically. Examples are oxygen (O_2), hydrogen (H_2), and methane (CH_4).

The *molecular weight* of any gas is determined by first multiplying the number of atoms in each element existing in the molecule, by their respective atomic mass, and then adding the total to find the *molecular weight*.

Using the three gases mentioned as examples

Molecular weight of oxygen (O_2) = 2 × atomic mass = 2 × 16 = 32 mass/mol

Molecular weight of hydrogen (H_2) = 2 × atomic mass = 2 × 1 = 2 mass/mol

Molecular weight of methane (CH_4) = 1 × atomic mass of carbon + 4 × atomic mass of hydrogen = 1 × 12 + 4 × 1 = 16 mass/mol

1.1.4 Mole

A mole is defined as the mass of a substance numerically equal to its molecular weight.

Denoting the molecular weight of any gas as Mw, and its mass as "m," we define the number of moles in the gas as follows:

$$n = \frac{m}{\text{Mw}} \tag{1.6}$$

1.1.5 Avogadro's law

Equal volumes of gases at the same temperature and pressure contain the same number of molecules.

Consider 1 mol of each of the gases mentioned previously: O_2, H_2, and CH_4. Working backward we can determine if they each have the same volume at constant temperature and pressure. Assuming the temperature of each gas to be 32 °F (0 °C) and pressure to be atmospheric (14.7 lb/in^2 or 101 kPa), calculations are done for 1 lb mol and 1 kg mol. Table 1.1 shows the results in US customary units and SI units.

Volume V in Table 1.1 is determined by rearranging eq. (1.5).

Table 1.1: Volume calculation for 1 mol of gas at standard temperature and pressure (STP).

Volume calculation for 1 mol of gas – US customary units					
	Variables		**Gas molecules**		
			O_2	**H_2**	**CH_4**
	Input				
Pressure	lb f/in^2	P =	Atmospheric	Atmospheric	Atmospheric
Temperature	°F	T =	32	32	32
	Output				
Number of moles		n =	1	1	1
Mass of gas	– lbm	m =	32	2	16
Pressure (lb/ft^2)	– (0 + 14.679 lb/in^2) × 144	P =	2,116.8	2,116.8	2,116.8

Table 1.1 (continued)

Volume calculation for 1 mol of gas – US customary units					
	Variables		**Gas molecules**		
			O_2	**H_2**	**CH_4**
	Input				
Temperature (°R)	– $T + 460$ °R	$T =$	492	492	492
Individual gas constant	– ft-lb/lb-°R	$R =$	48.294	766.541	96.329
Volume of gas	– Cubic feet	$V =$	**359.19**	**356.33**	**358.23**
Volume calculation for 1 mol of gas – SI units					
	Variables		**Gas molecules**		
			O_2	**H_2**	**CH_4**
	Input				
Pressure	Pa	$P =$	Atmospheric	Atmospheric	Atmospheric
Temperature	°C	$T =$	0	0	0
	Output				
Number of moles		$n =$	1	1	1
Mass of gas	– kg m	$m =$	32	2	16
Pressure (Pa)	– (0 + 101.325 ka) × 1,000	$P =$	101,325	101,325	101,325
Temperature (K)	– $T + 273$ K	$T =$	273	273	273
Individual gas constant	– kJ/kg-K	$R =$	0.2598	4.1242	0.5183
Volume of gas	– Cubic meter	$V =$	**0.022399**	**0.022224**	**0.022343**

See Table A1 (Appendix A) for individual gas constants for some typical gases.

1.1.6 Universal gas constant

If two gases A and B have the same volume at the same temperature and pressure, then using eq. (1.6) we arrive at the following equations:

$$\frac{m_{\mathrm{A}}}{m_{\mathrm{B}}} = \frac{\mathrm{Mw}_{\mathrm{A}}}{\mathrm{Mw}_{\mathrm{B}}} \tag{1.7}$$

where m_A and m_B are the respective masses for each gas and Mw_A and Mw_B are the respective molecular weights of each gas.

From eq. (1.7), we derive the following relationships:

$$\frac{m_A}{m_B} = \frac{R_B}{R_A} = \frac{Mw_A}{Mw_B}$$

where R_A and R_B are respective characteristic gas constants for each gas.

Therefore,

$$Mw_A R_A = Mw_B R_B = R_u \tag{1.8}$$

where R_u is known as the **universal gas constant** and is the product of the molecular weight of a gas and its characteristic gas constant.

Values for R_u = 1,545 ft-lb/lb mol °R
= 8.314 kJ/kg mol K
= 8.314 Nm/g mol °R

Equation (1.5) can now be rewritten as

$$PV = nR_u T \tag{1.9}$$

1.1.7 Dalton's law of partial pressure

The pressure of a mixture of perfect gases is equal to the sum of the pressures each gas would exert if it existed alone in the same volume at the same temperature:

$$P = p_1 + p_2 + p_3 + \cdots + p_j \tag{1.10}$$

where P is the total pressure of the mixture and P_1 through P_j are the partial pressures of each gas.

From eq. (1.9)

$$p_1 = \frac{n_1}{V} R_u T, \quad p_2 = \frac{n_2}{V} R_u T, \text{etc.}$$

$$P = \Sigma(n_1 + n_2 + n_3 + \cdots)\,\frac{R_u T}{V} \tag{1.11}$$

1.1.8 Amagat–Leduc law of additive volumes

This law states that the total volume of a gas mixture of ideal gases is the sum of the component volumes if the components existed individually at the pressure and temperature of the mixture. The formula for total volume is derived in a manner similar to what was done in eq. (1.9)

$$V = V_1 + V_2 + V_3 + \cdots + V_j \tag{1.12}$$

1.2 Gas mixtures

Many practical applications in thermodynamics involve multicomponent gases or mixtures. A variety of compressed gases used in hydrocarbon and chemical manufacturing processing are multicomponent gases or mixtures. It is therefore important that we understand how the properties of these mixtures are derived.

Consider a number of gases and/or vapors mixed together in a container having volume V and temperature T; the total mass m of the mixture is given by eq. (1.12)

$$m = m_1 + m_2 + m_3 + \cdots + m_j \tag{1.13}$$

or

$$m_m = \sum_{i=1}^{j} m_i \tag{1.14}$$

The *mass fraction* is the ratio of the mass of one component divided by the total mass m_m.

Mass fraction of the ith component:

$$X_j = \frac{m_i}{m_m} \tag{1.15}$$

Therefore

$$\sum_{i=1}^{j} X_i = 1 \tag{1.16}$$

Similarly, the *molar fraction* of the ith component Y_i is

$$Y_i = \frac{n_i}{n} \tag{1.17}$$

or

$$\sum_{i=1}^{j} Y_i = 1 \tag{1.18}$$

If we designate Mw_m as the molecular weight of the mixture, n_m as the total number of moles in the mixture, and m_m as the mass of the mixture, then

$$Mw_m = \frac{m_m}{n_m} = \frac{\sum m_i}{n_m} = \frac{\sum n_i Mw_i}{n_m} = \sum_{i=1}^{j} Y_i M w_i \tag{1.19}$$

The characteristic gas constant of the mixture $R_m = R_u/Mw_m$ where $R_u = 8.313$ kJ/kmol °K

For a gas mixture at pressure "p" and temperature "T" and with total volume V_m

$$pV_i = n_i R_u T \text{ for the } i\text{th component} \tag{1.20}$$

For the mixture

$$V_m = n_m R_u T \tag{1.21}$$

Dividing eq. (1.20) by eq. (1.21)

$$\frac{V_i}{V_m} = \frac{n_i}{n_m} = Y_i \tag{1.22}$$

1.2.1 Relationship between molar fraction and mass fraction

Since

$$n_i = \frac{m_i}{Mw_i} \text{ and } Y_i = \frac{n_i}{n_m} = \frac{m_i}{Mw_i\, n_m} = \frac{m_i}{Mw_i} \times \frac{Mw_m}{m_m}$$
$$= \frac{Mw_m}{Mw_i} \times \frac{m_i}{m_m}$$
$$= X_i \frac{Mw_m}{Mw_i}$$

Example 1: A vessel contains a mixture of gas consisting of 5 kg of O_2, 3 kg of H_2, and 10 kg of CH_4. Determine (a) the mass fraction of each component, (b) the mole fraction of each component, and (c) the molecular weight of the mixture and gas constant of the mixture.

(a) The total mass of the mixture $m_m = (5 + 3 + 10) = 18$ kg

The *mass fraction* of each component:

$$O_2 = \frac{5}{18} = 0.28$$

$$H_2 = \frac{3}{18} = 0.17$$

$$CH_4 = \frac{10}{18} = 0.56$$

(b) To find the mole fractions, we must first find the total amount of moles in the mixture.

$$\text{Number of moles in } O_2 = \frac{5\,\text{kg}}{32\,\text{kg/kmol}} = 0.156\,\text{kmol}$$

$$\text{Number of moles in } H_2 = \frac{3\text{kg}}{2\text{kg/kmol}} = 1.5\,\text{kmol}$$

$$\text{Number of moles in } CH_4 = \frac{10\text{kg}}{16\text{kg/kmol}} = 0.625\,\text{kmol}$$

$$\text{Total number of moles in mixture} = 0.156 + 1.5 + 0.625 = 2.281\,\text{mol}$$

The mole fraction for each component:

$$O_2 = \frac{0.156\,\text{kmol}}{2.281\,\text{kmol}} = 0.0684$$

$$H_2 = \frac{1.5\,\text{kmol}}{2.281\,\text{kmol}} = 0.6576$$

$$CH_4 = \frac{0.625\,\text{kmol}}{2.281\,\text{kmol}} = 0.2740$$

$$Y(O_2) + Y(H_2) + Y(CH_4) = 0.0684 + 0.6576 + 0.2740 = 1.00$$

(c) The molecular weight of the mixture Mw_m

$$= \frac{18\,\text{kg}}{2.281\,\text{kmol}} = 7.891\,\text{kg/kmol}$$

The characteristic gas constant R_m of the mixture

$$= \frac{8.314\,\text{kJ/kmol} - \text{K}}{7.891\,\text{kg/kmol}} = 1.0536\,\text{kJ/kg} - \text{K}$$

1.3 Ideal gas equation of state

Gases are highly superheated vapors with properties that can be reduced to simple relationships. The perfect gas would be one that obeys the relationship between pressure, volume, and temperature in the ideal gas equation of state shown in eq. (1.9). Figure 1.2 reminds us of the three gas laws that are combined to give eq. (1.9), namely *Boyle's law, Charles's law,* and *Avogadro's law.*

If we transpose eq. (1.9) to account for the molar volume of the ideal gas, the equation becomes

$$\frac{V}{n} = \frac{R_u}{P}\,T \tag{1.23}$$

This suggests a proportionality between molar volume and temperature for a gas at constant pressure.

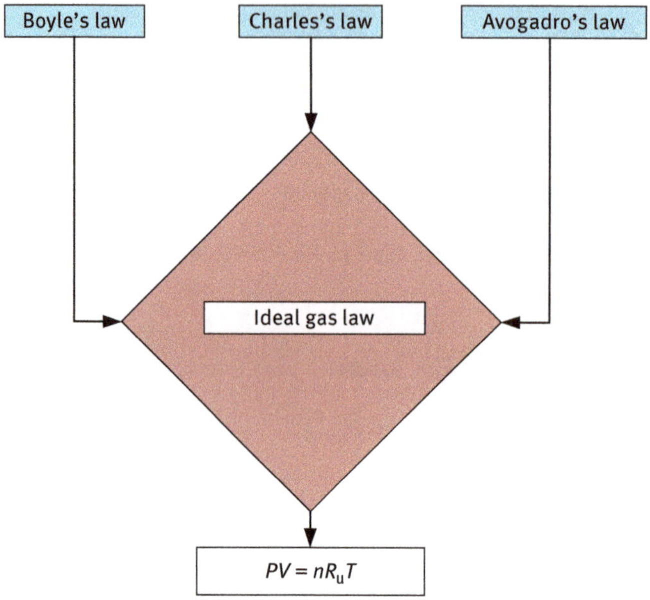

Figure 1.2: Illustration showing three gas laws combined to produce the ideal gas law.

Figure 1.3 graphically represents (not to scale) the proportionality of molar volume with respect to temperature for a hypothetical gas following the ideal gas EOS at three different pressures. If gases followed this equation, then all the straight lines should merge at point "A," the absolute zero temperature

$$T = (t\ ^\circ\mathrm{F} + 460)\ \text{degrees Rankine}$$
$$T = (t\ ^\circ\mathrm{C} + 273)\ \text{Kelvin}$$

One of the inconsistencies of the ideal gas EOS is that the volume at point "A" in Figure 1.3 becomes zero, and the gas remains a gas. We know real gases will condense to a liquid as the temperature decreases, and if the temperature continues to decrease, the liquid will become a solid.

Many gases, such as oxygen, hydrogen, carbon dioxide, and air, can be treated as perfect gases following the ideal gas EOS as a first approximation. While this is a good approximation, it has several limitations.

There is less than 1% error using the ideal gas EOS when calculating molar volume based on the following conditions:

For diatomic gases (e.g. N2, H2, O2 etc.) if $\frac{V}{n} = \frac{Ru}{P} T > 80\text{ft}^3$ per lb-mole

For all other gases

$$\frac{V}{n} = \frac{R_u}{P} T > 320 \text{ ft}^3 \text{ per lb-mol}$$

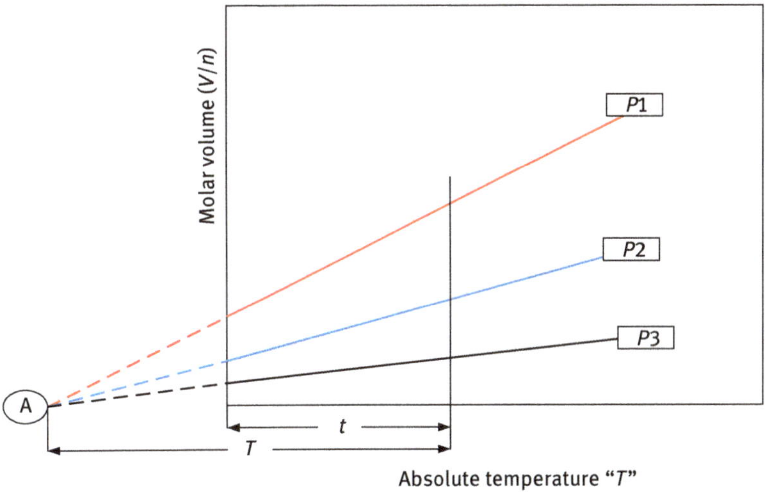

Figure 1.3: The proportionality of molar volume with respect to absolute temperature.

In summary, the following are some of the basic limitations of the ideal gas EOS:
1. This equation holds well as long as the density is kept low.
2. Works well at low pressures and high temperatures. The limitations are, therefore, high pressures and low temperatures.
3. Most gases do not behave ideally above 1 atm pressure.
4. Does not work well near the condensation conditions of the gas.

Dividing eq. (1.4) by RT yields $Pv/RT = 1$ for an ideal gas. However, if the gas is not ideal or the conditions of pressure and temperature limit the gas to follow the ideal gas equation, then $Pv/RT \neq 1$.

1.4 Compressibility factors

All gases that exist in nature are known as real gases. Figure 1.4 shows how a real gas behaves when pressure and temperature are varied; it is then compared with an ideal gas. The relationship between the properties of a real gas approaches those of

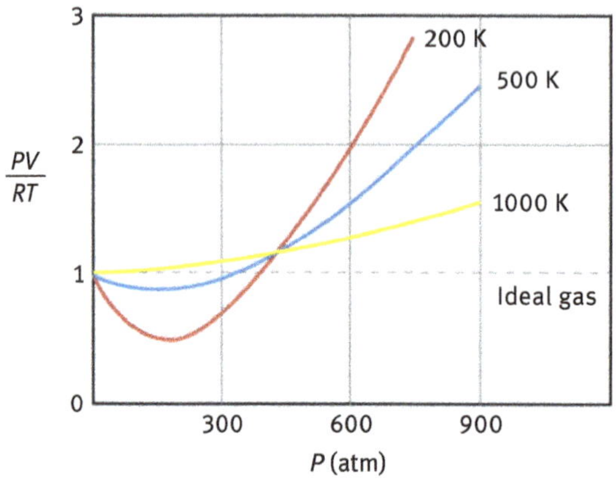

Figure 1.4: The departure of a real gas from the ideal gas equation of state when subjected to varying pressures and temperatures.

an ideal gas only at certain conditions of pressure and temperature. However, for real gases, eq. (1.4) can be modified in the following manner:

$$\frac{Pv}{RT} = Z$$

where "Z" is called the compressibility factor.

The compressibility factor is equal to 1 for an ideal gas, and for real gases it varies with pressure and temperature. This factor is empirically determined for each gas as a function of pressure "P" and temperature "T." While charts of "Z" as a function of P and T have been published for different gases, a single *generalized compressibility factor* chart was published by Nelson and Obert. Although slightly less accurate than individual gas charts, it gives reasonable results.

The *generalized compressibility factor* chart developed by Nelson and Obert is based on the fact that the critical pressure, temperature, and specific volume are unique for each gas at the critical point "E" previously mentioned in conjunction with Figure 1.1.

By using a reduced EOS, the reduced coordinates are as follows:

Reduced pressure $P_r = P/P_c$, where P_c is the critical pressure in absolute units.

Reduced temperature $T_r = T/T_c$, where T_c is the critical temperature in absolute units.

Reduced specific volume $v_r = v/v_c$.

(Reference is made here to Table A2 of Appendix A, which shows critical properties for some common gases. Charts A1 through A3 of Appendix A show a few generalized compressibility factor.)

Example 2: Determine the compressibility factor and specific volume for butane at 2,000 kPa and temperature at 170 °C.

Pressure $P = 2{,}101.35$ kPa-a
Temperature $T = 443.15$ K
From Table A2
Critical temperature $T_c = 152$ °C (425.15 K),
Critical pressure $P_c = 550.4$ psia (3,794.874 kPa-a)
Pr = 2,101.35/3794.874 = 0.554
Tr = 443.15/425.15 = 1.04
From Chart A1, $Z = 0.81$
From Table A1, $R = 0.1431$ kJ/kg. K
Specific volume

$$v = Z\frac{RT}{P}$$

$$= (0.81)\,(0.1431)\,(443.15)/2{,}101.35 = 0.0244\ \text{m}^3/\text{kg}$$

1.5 Real gas equations of state

Many EOS have been developed over the years to compensate for the shortcomings of the ideal gas EOS. Refinement of the EOS for ideal gases started with J. D. van der Waals in 1873 and continues even today. The EOS for various fluids including hydrocarbons or methods to compute compressibility were developed in a sequential manner and are based on work done by

- Van der Waals (circa 1873)
- Redlich–Kwong (circa 1949)
- Benedict–Webb–Rubin (circa 1951)
- Redlich–Kwong–Soave (circa 1972)
- Lee–Kessler (circa 1975)
- Peng–Robinson ("PR") circa 1976

An EOS is basically an analytical means of relating *pressure*, *volume*, and *temperature* in one equation. The EOS should be able to accurately predict the volumetric behavior of a fluid for both the gas phase and the vapor/liquid equilibrium (VLE) phase.

Numerous EOS have been proposed since van der Waals introduced his expression in 1873. However, because many of these EOS are very complex, discussions of each method are outside the scope of this book.

1.5.1 Van der Waals' EOS

Although J. D. van der Waals' EOS has been around since 1873, engineers still laud his achievements as a critical step in revolutionizing the EOS. In today's world, the van der Waals EOS does not have any practical applications. However, it is still viewed as the foundation from which most of the EOS in present use were derived.

The basic van der Waals EOS is

$$\left(P + \frac{a}{v^2}\right)(v - b) = R_u T \tag{1.24}$$

where a and b are constants calculated at the critical pressure and critical temperature, P_c is the critical pressure = $a/(27b^2)$, T_c is the critical temperature = $(8a)/(27R_u b)$, P is the absolute pressure, v is the molar volume, T is the absolute temperature, and R_u is the universal gas constant.

Equation (1.24) laid the foundations for modern cubic EOS. The modern EOS increased in complexity to be able to cover a wide range of states and applications. The next to follow the van der Walls EOS was the Redlich–Kwong EOS in 1949.

1.5.2 Redlich–Kwong's EOS

The Redlich–Kwong EOS (RK EOS) was considered a great improvement over the van der Waals EOS at the time. It is still used in many applications due to its simplicity.

The RK EOS is

$$P = \frac{R_u T}{v - b} - \frac{a}{v(v - b)T^{0.5}} \tag{1.25}$$

where a and b are constants and different from those in the van der Waals EOS:

$$a = 0.427\, R_u{}^2 \frac{T_c{}^2}{P_c} \text{ and } b = 0.0866\, R_u \frac{T_c}{P_c}$$

The RK EOS is adequate for calculation of gas phase properties when the ratio of the pressure to the critical pressure, that is, the *reduced pressure*, is less than one-half of the ratio of the temperature to the critical temperature, that is, the *reduced temperature*:

$$\frac{P}{P_c} < \frac{T}{2T_c}$$

The RK EOS performs poorly with respect to the liquid phase of most fluids and, therefore, should not be used for calculating VLE with a high degree of accuracy.

1.5.3 Soave–Redlich–Kwong's EOS

In 1972, Soave improved the RK EOS by adding a third constant "*m*." The new equation is now known as the Redlich–Kwong–Soave EOS (SRK EOS):

$$P = \frac{R_u T}{v-b} - \frac{a}{v(v-b)T^{0.5}} \times \left\{ 1 + m\left[1 - \left(\frac{T}{T_c}\right)^{0.5}\right] \right\}^2 \tag{1.26}$$

$$m = 0.48 + 1.574\omega - 0.17\omega^2$$

where ω is called the acentric factor. Table A6 shows this acentric factor "ω" for various gases, including hydrocarbons.

The SRK EOS modification is a significant improvement over the RK EOS; the SRK EOS is well suited for the prediction of vapor-liquid-equilibria. However, it still does not provide reliable values of liquid density.

1.5.4 Benedict–Webb–Rubin

The Benedict–Webb–Rubin EOS (BWR EOS) was developed in 1951 specifically for light hydrocarbons and mixtures of light hydrocarbons.

It is an eight constant EOS

$$P = \frac{RT}{v} + \left(B_0 RT - A_0 - \frac{C_0}{T^2}\right)\frac{1}{v^2} + (bRT - a)\frac{1}{v^3} + \frac{a\alpha}{v^6} + \frac{c}{v^3 T^2}\left(1 + \frac{\gamma}{v^2}\right)e^{\frac{-\gamma}{v^2}} \tag{1.27}$$

Constants for the BWR EOS are found in Table A4 of Appendix A.

1.5.5 Peng–Robinson's EOS

In 1976, Peng and Robinson introduced a variation of the van der Waals EOS. The key objective of the Peng–Robinson EOS (PR EOS) was to improve density predictions.

The PR EOS introduced only two constants "*a*" and "*b*":

$$P = \frac{R_u T}{v-b} - \frac{a(T)}{v(v+b) + b(v-b)}$$

where

$$a(T_c) = 0.45724\frac{(R_u T_c)^2}{P_c} \quad \text{and} \quad b = 0.07780\frac{(R_u T_c)}{P_c} \tag{1.28}$$

Hydrocarbons range from a mere few to ones containing well over 100 carbon atoms. This could mean that users of just about any of the available EOS must make minor adjustments to arrive at accurate results.

Comparisons have shown that the prediction of vapor pressure of several substances using PR EOS and SRK EOS compare very well with each other. They also show very good results generating enthalpy values but break down at around C10–C11 and heavier compounds.

There are many other EOS, too numerous to mention, that are used in the process industry. Generally, an EOS is tailored for use in specific applications. Each equation has limitations when used outside its applicable range or niche.

1.6 Applied laws of thermodynamics

1.6.1 First law of thermodynamics

In practical terms, there are many *processes* that result in work because of heat being transferred. In a cyclic *process*, where heat and work are being transferred, the algebraic sum of the work transfers is proportional to the algebraic sum of the heat transfers.

Simply put, this is the *first law of thermodynamics*:

$$\oint \mathrm{d}W \propto \oint \mathrm{d}Q$$

where $\mathrm{d}W$ is the incremental work done for an incremental change in heat $\mathrm{d}Q$, and $\oint$ is summation or integral of the changes around the cycle.

Removing the proportionality factor

$$\oint \mathrm{d}W = J \oint \mathrm{d}Q \tag{1.29}$$

where J is a constant, and its value depends on the units of conversion. Therefore, assuming that both sides of the equation have the same units, the equation becomes

$$\oint \mathrm{d}Q - \mathrm{d}W = 0 \tag{1.30}$$

The *first law of thermodynamics*: for any cycle of a closed system, the net heat transfer equals the net work. In summary, energy can be converted from one form to another but cannot be destroyed.

1.6.2 Internal energy as a property

The internal energy of a system is the sum of the internal potential and kinetic energies; this sum has a unique value for any particular state. It is considered one of the many properties of a specific system.

Although the first law of thermodynamics, as stated, applies to cyclic processes of closed systems, we can consider the changes within the system as it changes state. For example, let us consider a system undergoing a cyclic process from state 1 to state 2 and then back to 1, as shown in Figure 1.5.

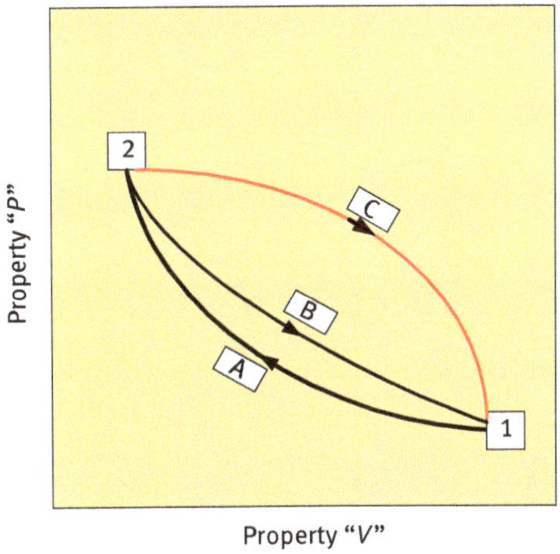

Figure 1.5: A cyclic system changing states with different paths.

Applying the first law of thermodynamics:

For path 1A–2B

$$\int_{1A}^{2} (dQ - dW) + \int_{2B}^{1} (dQ - dW) = 0$$

For path 1A–2C

$$\int_{1A}^{2} (dQ - dW) + \int_{2C}^{1} (dQ - dW) = 0$$

Therefore

$$\int_{2C}^{1} (dQ - dW) = \int_{2B}^{1} (dQ - dW)$$

The value of the integral $\int_1^2 (dQ - dW)$ is fixed by the end states and is therefore independent of the path or process. This represents a change in property and, by definition, is the *internal energy* of the system

$$dQ - dW = dU \tag{1.31}$$

Applying this to a *nonflow (NF) process*

$$dQ = dU + dW \tag{1.32}$$

Consider a constant pressure (NF) process where heat is transferred to the fluid.

Then $dW = Pdv$ (work done on the fluid)

Therefore

$$dQ = dU + Pdv \tag{1.33}$$

Integrating both sides

$$\begin{aligned} Q &= (U_2 - U_1) + P(V_2 - V_1) \\ &= (U_2 + P_2V_2) - (U_1 + P_1V_1) \end{aligned} \tag{1.34}$$

Since P, U, and V are all properties, it is acceptable to define another property in terms of these properties:

$$Q = H_2 - H_1 \tag{1.35}$$

where $U_2 + P_2V_2 = H_2$ and $U_1 + P_1V_1 = H_1$.

The property denoted by "H" is called enthalpy.

This concept can be extended to the interchange between heat and work transfer across a system boundary.

For the process shown in Figure 1.6, the equation can be written as

$$Q - W = H_1 - H_2 \tag{1.36}$$

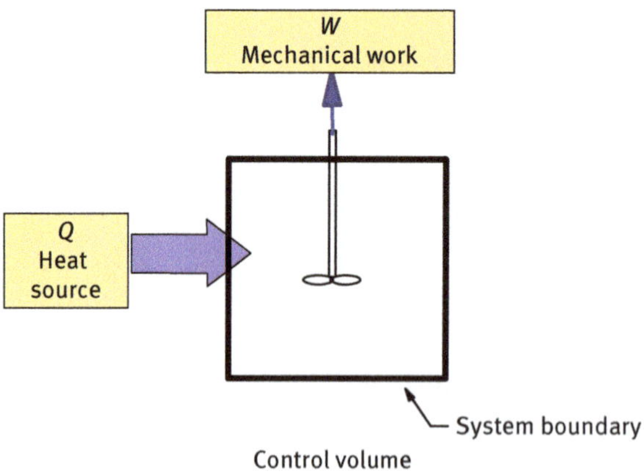

Figure 1.6: Heat and work transferred across a system boundary.

Equation (1.35) is used together with the Mollier diagrams to evaluate shaft power for several types of machinery.

1.6.3 Thermodynamic processes and systems

In Section 1.1, it was mentioned that by knowing any two properties of a substance, the state of the substance is known. Figure 1.1 highlighted this fact by showing a hypothetical fluid going from one state to the next. The change from one state to another is called a *process*.

A *system* is made up of one or more *processes* and depending on the type of system, it can either exchange energy and matter or not.

The three types of systems that will be mentioned are shown in Figure 1.7.

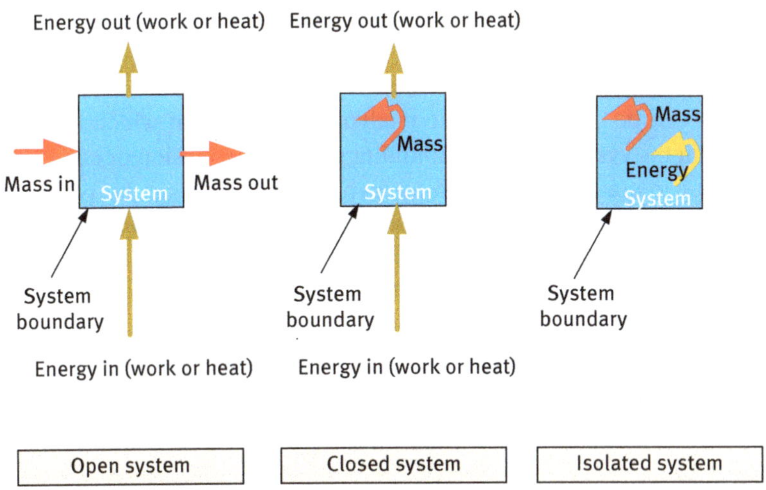

Figure 1.7: Three types of thermodynamic systems.

1.6.3.1 Open systems

An open system or control volume has mass as well as energy crossing the system boundary.

Examples are compressors, pumps, turbines, and heat exchangers.

1.6.3.2 Closed systems

In a closed system, the mass within the system boundary remains constant, and only the energy transfer takes place between the system boundaries (surroundings).

Examples are as follows:

- Air conditioning or refrigeration system. Heat exchanges take place in both the condenser and evaporator, and work is done on the refrigerant by the compressor. The fluid mass/refrigerant stays within the boundary of the system.
- The gas is trapped between a piston and the surrounding cylinder during the compression cycle of a reciprocating compressor. Work input is from shaft power, and heat output is from the temperature gained by the gas from the compression process.

1.6.3.3 Isolated systems

An isolated system is a closed system with no energy crossing the system boundaries. Among many possible examples are reactors within a chemical process unit.

1.6.4 Special thermodynamic processes

1.6.4.1 Adiabatic process

An adiabatic process can be defined as a thermodynamic process in which there is no heat exchange from the system to the surroundings during expansion or compression from, say, state 1 to state 2. For this to happen, the system must be perfectly insulated from the surroundings. An adiabatic process can either be *reversible* or *irreversible*. The *reversible* process is an ideal process that theoretically can be returned to its original state by inducing extremely small changes to some properties of the system. The return to its original state never occurs; however, this is a concept by which we can judge other processes. The *irreversible* process is the more natural process that is found in practical applications.

1.6.4.2 Isentropic process

A reversible adiabatic process is called an *isentropic* process. Isentropic processes are idealized thermodynamic processes in which work is transferred in a frictionless manner; there is no transfer of heat or matter, and the process is theoretically reversible. While not actually achievable, an idealized isentropic system is used in engineering as a model to compare with real processes.

1.6.4.3 Isothermal process

In an isothermal process, work is done due to the change in net heat content in the system, but the temperature of the system remains constant. For this to occur, the system will have to be in contact with an outside thermal reservoir. The change in the system will progress slowly enough to continually adjust to the temperature of the reservoir through heat exchange.

1.6.5 Second law of thermodynamics

The *first law of thermodynamics* expressed that energy can be converted from one form to another during a cyclic process in a closed system. Energy could be neither created nor destroyed. However, the first law did not address the fact that processes inherently have a measure of disorder. It left open the question whether it would be possible to convert all the energy going into the process.

The *second law of thermodynamics*, however, explains the types of heat and work transfers that are possible. The second law formulates clarifications in several ways:

1. Heat will not flow unaided from a low-temperature reservoir to a high-temperature reservoir, without work done to accomplish the flow.
2. It is impossible for an engine to convert all the heat supplied to it into work. The surplus energy is rejected as heat to the sink at a lower temperature. This formulation is credited to Sadi Carnot who in 1824 showed that there is an upper limit to energy conversion in heat engines.

The two statements are basically equivalent when viewed from the standpoint that for a process to be completely *reversible* even within a closed system, there can be no disorder within the system. We do know that practical systems are not 100% efficient and, therefore, will always have some disorder.

The measure of the disorder comes from the amount of change in the thermodynamic state property called *entropy*.

For a reversible cyclic process, *entropy* is expressed as

$$\oint \frac{dQ_R}{T} = 0 \tag{1.37}$$

or

$$\oint \frac{dS}{T} = 0 \tag{1.38}$$

where dQ_R is the incremental *reversible* heat exchange between states, T is the temperature of the surroundings in absolute units, and dS is the change in entropy from one state point to the next.

As with internal energy, it is the change in entropy that is most important to engineers. One way of looking at entropy is the more the entropy increases, the less energy is available to do useful work.

So far, both the first and second laws of thermodynamics have been derived based on an ideal *reversible* process; however, while some processes come close to being *reversible*, most (practically all) processes are *irreversible*.

To prove that the entropy for practical systems is almost always increasing, we will consider the two cycles in Figure 1.8. In both illustrations, a system changes from state 1 to state 2. The "change transmitter" is either a *reversible* or an *irreversible* process, which then returns to state 1 by the same reversible process.

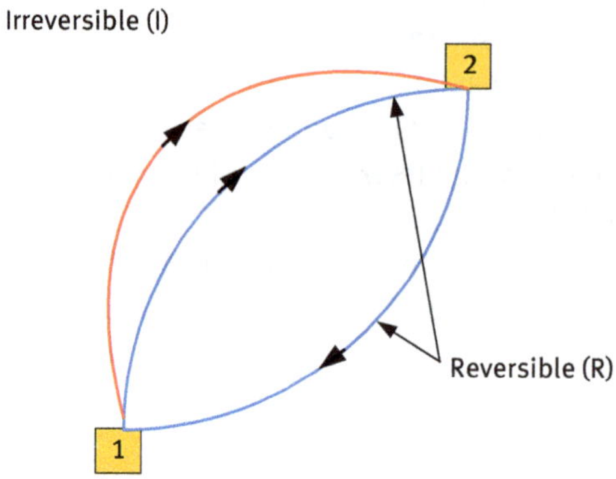

Figure 1.8: Cyclic process following a reversible path or a combination of irreversible and reversible paths.

From eq. (1.37):

Entropy for the *reversible* process $\oint(\mathrm{d}Q_\mathrm{R}/T) = 0$

$$\int_1^2 \frac{\mathrm{d}Q_\mathrm{R}}{T} + \int_2^1 \frac{\mathrm{d}Q_\mathrm{R}}{T} = 0 \qquad (1.39)$$

The equal sign is only possible when the process is considered reversible.

For the *irreversible* process $\oint(\mathrm{d}Q_\mathrm{I}/T) < 0$

$$\int_1^2 \frac{\mathrm{d}Q_\mathrm{I}}{T} + \int_2^1 \frac{\mathrm{d}Q_\mathrm{I}}{T} \leq 0 \qquad (1.40)$$

Subtracting eq. (1.40) from eq. (1.39)

$$\int_1^2 \frac{\mathrm{d}Q_R}{T} - \int_1^2 \frac{\mathrm{d}Q_I}{T} \geq 0 \qquad (1.41)$$

Therefore,

$$\int_1^2 dS \geq \int_1^2 \frac{dQ_I}{T} \tag{1.42}$$

We previously mentioned that the equal sign is only relevant for a reversible process; therefore, eq. (1.42) shows that entropy in an irreversible process is always increasing, or

$$\int_1^2 dS \geq 0 \tag{1.43}$$

1.6.6 Entropy relationships

Since

$$dS = \frac{Q_R}{T}$$

and from eq. (1.33), $dQ_R = dU + pdv$ for a reversible NF process; therefore

$$dS = \frac{dU}{T} + \frac{pdV}{T}$$

and rearranging the equation,

$$TdS = dU + pdV \tag{1.44}$$

also, since $h = U + pv$ then, by differentiating

$$dh = dU + pdv + vdp$$

$$= dQ_R + vdP$$

$$\frac{dQ_R}{T} = \frac{dh}{T} + \frac{vdp}{T}$$

Therefore,

$$Tds = dh - vdp \tag{1.45}$$

Equations (1.44) and (1.45) are the results of combining the two laws of thermodynamics. They relate to the six most important properties and are applicable to NF processes.

The entropy of an **adiabatic process** will remain constant if the process is considered reversible; however, the entropy will increase if the process is irreversible (Table 1.2).

Table 1.2: Comparing reversible adiabatic process with an irreversible adiabatic process.

Adiabatic reversible		Adiabatic irreversible
Change in entropy		Change in entropy
$dS = \frac{dQ}{T}$		$dS > \frac{dS}{T}$
	For any adiabatic process $dQ = 0$	
Therefore $dS = 0$		Therefore $dS > 0$
$S_1 = S_2$		$S_2 > S_1$

By definition, the change in entropy of an **isentropic process** is similar to a reversible adiabatic process, that is, $S_1 = S_2$.

1.7 Specific heats

1.7.1 Specific heat at constant volume

This is the amount of energy that crosses the boundary of a system containing unit mass so that its temperature changes by 1 °C while its volume remains constant.

For a constant volume NF process having unit mass $dW = 0$ and $dQ = du$ (eq. (1.32)).

From the definition of *specific heat at constant volume*:

$$dQ = C_v dT$$

Therefore $C_v dT = du$.

Rearranging the equation

$$C_v = \frac{du}{dt} \tag{1.46}$$

1.7.2 Specific heat at constant pressure

This is the amount of energy that crosses the boundary of a system containing unit mass so that its temperature changes by 1 °C while its pressure remains constant.

For unit mass in a constant pressure NF process, eq. (1.35) is rewritten as

$$Q = h_2 - h_1$$

where h_1 and h_2 are the enthalpies per unit mass at states 1 and 2.

From the definition of *specific heat at constant pressure*:

$$Q = C_p(T_2 - T_1) = h_2 - h_1 \tag{1.47}$$

$$C_p \mathrm{d}T = \mathrm{d}h$$

Rearranging the equation

$$C_p = \frac{\mathrm{d}h}{\mathrm{d}T} \tag{1.48}$$

Since $h = u + Pv$

$$= u + RT \text{ (substituting from } Pv = RT) \tag{1.49}$$

Differentiating eq. (1.49)

$$\mathrm{d}h = \mathrm{d}u + R\mathrm{d}T = C_v \mathrm{d}T + R\mathrm{d}T \tag{1.50}$$

Substituting $\mathrm{d}h = C_p \mathrm{d}T$ in eq. (1.50)

$$C_p \mathrm{d}T = C_v \mathrm{d}T + R\mathrm{d}T$$

$$C_p - C_v = R \tag{1.51}$$

1.7.3 Specific heat ratio

$$\frac{C_p}{C_v} = \frac{\mathrm{d}h}{\mathrm{d}T} \bigg/ \frac{\mathrm{d}u}{\mathrm{d}T} = \frac{\mathrm{d}h}{\mathrm{d}u} = k \tag{1.52}$$

or

$$k = \frac{MC_p}{MC_p - 1.99} \tag{1.53}$$

where k is the specific heat ratio (note that y is sometimes used instead of k) and MC_p is the molal specific heat at constant pressure.

For gas mixtures

$$MC_{pm} = Y_1 MC_{p1} + Y_2 MC_{p2} + Y_3 MC_{p3} + \cdots \tag{1.54}$$

where $Y_1, Y_2, Y_3, \ldots, Y_n$ are the mole fractions for the individual gases.

1.8 Simplification of the compression process

The aim of a compression process is to increase the pressure of a compressible fluid, usually a gas or a mixture of gases. Work is done on the gas to change its state from a low-pressure high-volume state to a high-pressure low-volume state. Theoretical compression processes function between *adiabatic* and *isothermal* compression.

Figure 1.9 shows a *P–V* diagram with entropy and isothermal lines superimposed. If the compression was an adiabatic reversible process, it would follow along a constant entropy line as the gas changed state, say from 1 to 2. Similarly, if the compression was isothermal, the process would follow along the path of a constant temperature line, say 1′ to 2. However, actual compression processes will fall somewhere in between adiabatic and isothermal. This type of compression process is called a **polytropic process**.

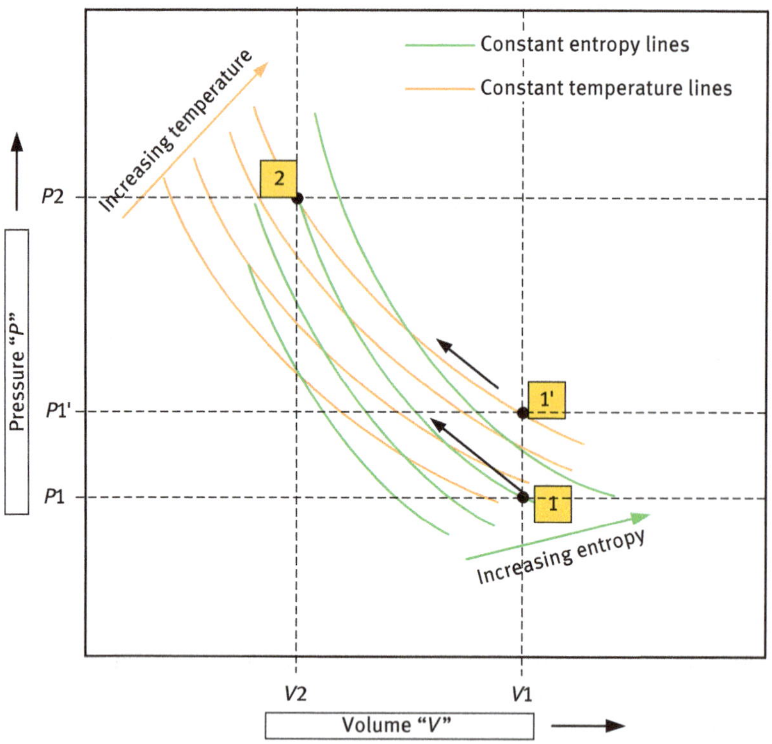

Figure 1.9: *P–V* diagram with constant entropy and constant temperature lines superimposed.

From eq. (1.45), $TdS = dh - vdp$

For a reversible adiabatic process, that is, an isentropic process, the equation reduces to $0 = dh - vdp$ that is,

$$0 = C_p dT - vdp \tag{1.55}$$

Restating the ideal gas equation

$$Pv = RT$$

Differentiating the ideal gas equation

$$RdT = pdv + vdp$$

Substituting for dT in eq. (1.55)

$$0 = \frac{C_p}{R}(pdv + vdp) - vdp$$

Rationalizing the equation

$$Vdp\left(\frac{R}{C_P} - 1\right) = pdv \tag{1.56}$$

Substituting R from eq. (1.51) into eq. (1.56)

$$Vdp\left(-\frac{C_V}{C_P}\right) = pdv$$

$$\frac{dP}{P} = \left(-\frac{C_P}{C_V}\right)\frac{dv}{V}$$

Integrating between states 1 and 2:

$$\log_e \frac{P_2}{P_1} = \log e\left(\frac{V_1}{V_2}\right)^{C_P/C_V}$$

but $C_p/C_v = k$

therefore $P_2 v_2{}^k = P_1 v_1{}^k = \text{constant}$.

In a similar manner, the relationship between P and v can be written for the polytropic process as $Pv^n = C$, where "n" denotes the polytropic index. Figure 1.10 shows how "n" varies for the different types of processes.

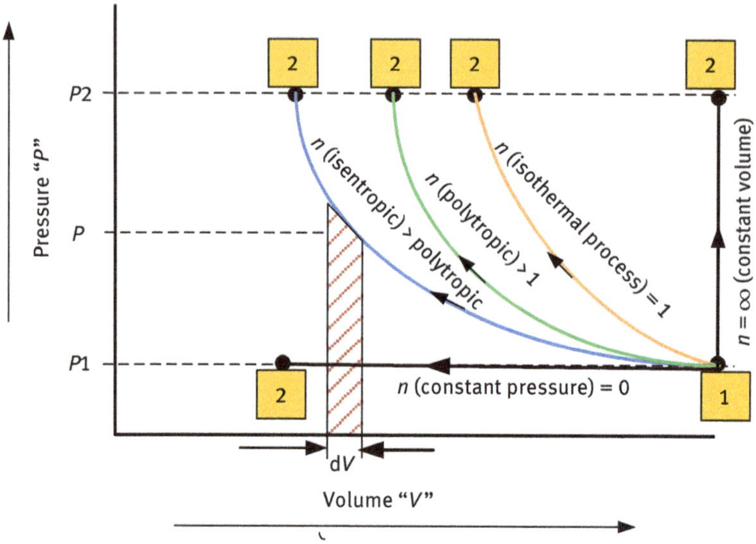

Figure 1.10: Different compression processes and how the exponent varies.

Therefore

$$P_1 v_1{}^n = P_2 v_2{}^n$$

or

$$\frac{P_2}{P_1} = \left(\frac{V_1}{v_2}\right)^n \tag{1.57}$$

We also know that for a perfect gas

$$\frac{Pv}{T} = C$$

Therefore,

$$\frac{P_1 v_1}{T_1} = \frac{P_2 v_2}{T_2} \tag{1.58}$$

By combining eqs. (1.57) and (1.58) and eliminating "v"

$$\frac{T_2}{T_1} = \left(\frac{P_2}{P_1}\right)^{n-1/n} \tag{1.59}$$

The compression ratio

$$r = \frac{P_2}{P_1}$$

For an NF process, the energy required to compress the gas from state 1 to state 2 is the integral under the P–V diagram:

$$W_{NF} = -\int_1^2 P\mathrm{d}v \text{ (the negative sign indicates work done on the gas)}$$

$$PV^n = C \text{ or } P = CV^{-n}$$

therefore

$$W_{NF} = C\int_1^2 V^{-n}\mathrm{d}v$$

$$W_{NF} = \frac{-C}{1-n}[V^{(1-n)}]_1^2$$

Since $C = Pv^n$

$$W_{NF} = -\left(\frac{PV^n}{1-n}\right)(V_2^{1-n} - V_1^{1-n})$$

$$= \frac{P_2V_2 - P_1V_1}{n-1}$$

Steady flow work $\quad W_{SF} = W_{NF} + \text{flow work}$

Flow work $\quad = P_2V_2 - P_1V_1$

therefore

$$W_{SF} = \frac{P_2V_2 - P_1V_1}{n-1} + (P_2V_2 - P_1V_1)$$

$$= \frac{n}{n-1}(P_2V_2 - P_1V_1) \tag{1.60}$$

Substituting $Pv = RT$ into eq. (1.60)

$$W_{SF} = \frac{nR}{n-1}(T_2 - T_1) \tag{1.61}$$

The steady flow work characterizes the amount of energy in the gas. This energy is called the "head" and the associated units are ft-lbf/lbm, N-m/kg, etc.

1.8.1 For an isentropic (reversible adiabatic) compression process

Where $n = K$ and using eq. (1.59) to eliminate T_2 from eq. (1.61)

$$W_{SF} = \frac{k}{k-1} RT_1 \left[r^{\frac{k-1}{k}} - 1 \right] \text{ [for an ideal gas]} \tag{1.62}$$

$$W_{SF} = \frac{k}{k-1} Z_{avg} RT_1 \left[r^{\frac{k-1}{k}} - 1 \right] \text{ [for real gases]} \tag{1.63}$$

where Z_{avg} is the average compressibility factor between states 1 and 2.

1.8.2 For a polytropic compression process

$$W_{SF} = = \frac{n}{n-1} RT_1 \left[r^{\frac{n-1}{n}} - 1 \right] \text{ [for an ideal gas]} \tag{1.64}$$

$$W_{SF} = = \frac{n}{n-1} Z_{avg} RT_1 \left[r^{\frac{n-1}{n}} - 1 \right] \text{ [for real gases]} \tag{1.65}$$

1.8.3 For an isothermal compression process

Where $n = 1$,

$$W_{SF} = RT \log_e r \text{ [for an ideal gas]} \tag{1.66}$$

$$W_{SF} = ZRT \log_e r \text{ [for real gases]} \tag{1.67}$$

Equations (1.62) through (1.67) represent the energy inside the gas. However, for practical purposes, input energy or shaft power entering into the control volume of the system is needed to calculate the efficiency of the process.

Considering the first law of thermodynamics, *energy in is equal to energy out, which* are quantified in Table 1.3.

Table 1.3: Energy balance of an open system assuming unit mass.

Energy in crossing at 1–1	Energy out crossing at 2–2
Internal energy: u_1	Internal energy: u_2
Flow work: $P_1 v_1$	Flow work: $P_2 v_2$
Kinetic energy: $C_1^2/2$	Kinetic energy: $C_2^2/2$
Energy transferred to system is work: W	Energy transferred out: $\boldsymbol{Q}$
Potential energy of incoming fluid: Z_1	Potential energy of outgoing fluid: Z_2

A conceptual open system is shown in Fig. 1.11

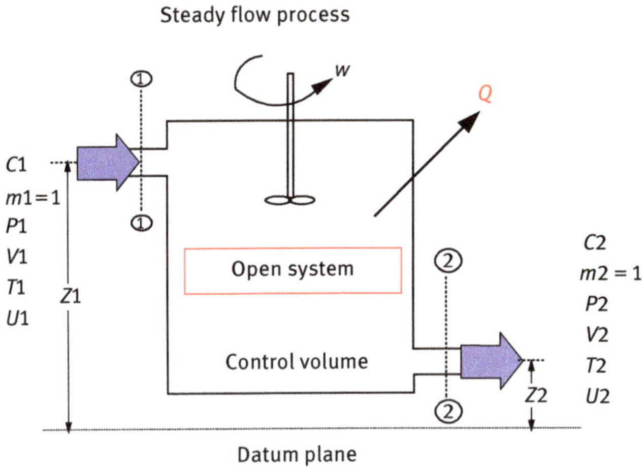

Figure 1.11: Conceptual open system with fluid flowing from state 1 to state 2.

Therefore

$$u_1 + P_1v_1 + \frac{C_1^{\ 2}}{2} + Z_1 + W = u_2 + P_2v_2 + \frac{C_2^{\ 2}}{2} + Z_2 + Q \tag{1.68}$$

$$h_1 + \frac{C_1^{\ 2}}{2} + Z_1 + W = h_2 + \frac{C_2^{\ 2}}{2} + Z_2 + Q$$

In *adiabatic* machine operations, such heat losses are negligible, and the values of inlet and exit velocities are of the same order of magnitude and therefore can be ignored:

$$Q - W \;\cong (h_1 - h_2)$$

For $Q = 0$,

Shaft work

$$W \cong \; - (h_2 - h_1) \tag{1.69}$$

The negative sign indicates the work done on the gas.

Rewriting eq. (1.69)

$$W \cong \; - \; C_p \; (T_2 - T_1) \tag{1.70}$$

Since $C_p - C_v = R$ and $C_p/C_v = k$

Substituting for C_p and rearranging eq. (1.70)

$$W \cong \frac{k}{k-1} RT_1 \left[\mathrm{r}^{\frac{\mathrm{n}-1}{\mathrm{n}}} - 1 \right] \tag{1.71}$$

1.8.4 Polytropic efficiency

$$n_p = \frac{\text{theoretical work to compress the gas from state 1 to state 2}}{\text{actual shaft work}}$$

$$= \frac{\frac{n}{n-1} RT_1 \left[r^{\frac{n-1}{n}} - 1 \right]}{\frac{k}{k-1} RT_1 \left[r^{\frac{n-1}{n}} - 1 \right]}$$

$$= \frac{\frac{n}{n-1}}{\frac{k}{k-1}}$$

Chapter 2
Brief overview of compression machinery

2.1 General overview

This section focuses on the different types of compressors used in industrial applications, their working principles, and range of applications. Appendix B follows up with the relevant aerothermodynamic fundamentals and derivations; Appendix B includes examples for selecting some of the common types of compressors used in industry.

Compressors are widely used in the process industries to move fluid within the boundary of a process and sometimes outside its boundary. The fluids can be any compressible fluid, either gas or vapor, and can have a wide range of molecular weights. Compression of gases and vapors is an important and prevalent operation in the chemical and petrochemical industries.

Compression processes deliver gas or vapor at pressures higher than the inlet pressures by progressively reducing the inlet volume. The mere movement of gases and vapors would not require compression; it could be accomplished with low-pressure blowers. However, in either of the two types of machines, inlet pressure levels can range from deep vacuum to high positive pressures. Discharge pressure can range from subatmospheric levels to numbers in the tens of thousands of pounds per square inch. The relationships of inlet to outlet pressures are based on thermodynamic laws and their interactions or derivatives; these were addressed earlier in Chapter 1.

Applications of compressed gas vary from consumer products, such as the home refrigerator, to large, highly complex petrochemical plant installations. Recorded molecular weights of compressed gases range from 2 (for hydrogen) to 352 (uranium hexafluoride).

Depending on the application, compressors are constructed and configured in various ways. They are broadly divided into two main groups, based on the mode of operation on how the gas is being compressed. The two basic operating modes are *intermittent* and *continuous*. The *intermittent* mode of compression is cyclic in nature. A specific quantity of gas is ingested by the compressor, acted upon, and discharged before the cycle is repeated. As implied by the adjective, the *continuous* compression mode is one in which the gas is continually being ingested and discharged without interruption of the flow at any point in the process.

Compressors operating in an intermittent compression mode are referred to as positive displacement compressors, of which there are two distinct types: reciprocating and rotary. In contrast, compressors operating in a continuous mode are referred to as dynamic compressors; the two distinct types within this category are centrifugal and axial.

Figure 2.1 shows the family of compressors used in industrial plants such as the petrochemical industry. Ejectors are shown under the dynamic category since they are characteristic of the continuous-mode type of compressor.

https://doi.org/10.1515/9783110678765-002

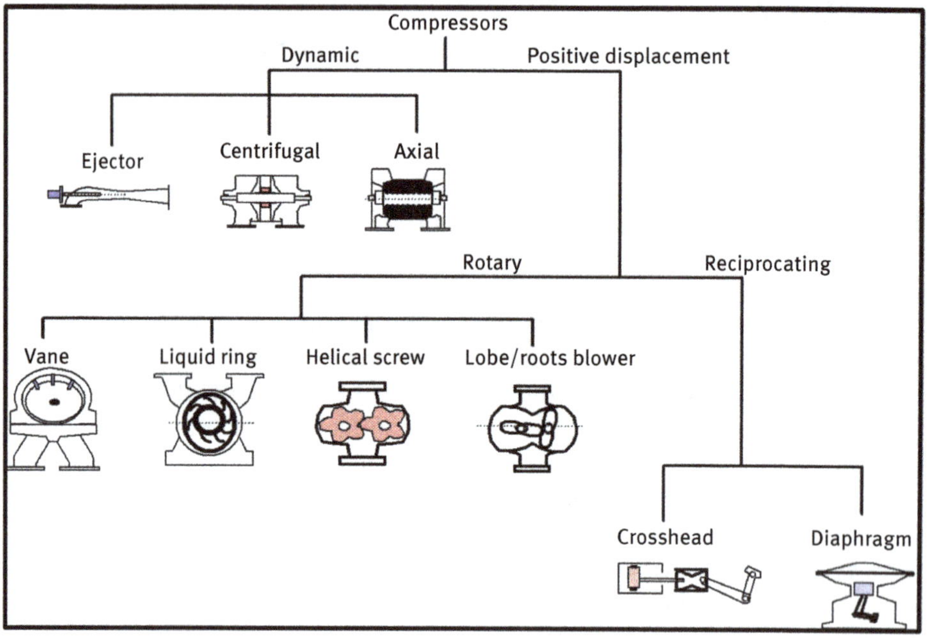

Figure 2.1: Family of compressors used in the process industries.

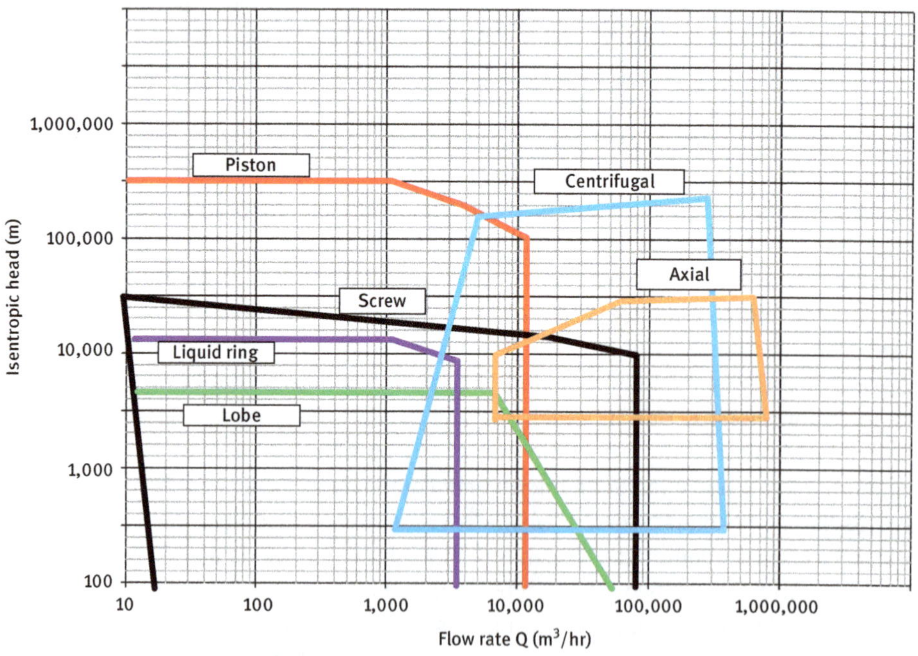

Figure 2.2: Selection charts for different types of industrial compressors.

Figure 2.2 shows the overlap between types of compressors. Therefore, during selection, the choice is largely influenced by safety considerations and life cycle costs. The scope of our discussion is limited to popular compressor types, ones that are commonly used in process industries.

2.2 Dynamic compressors

Dynamic compressors impart velocity to working fluids by the action of a rotor designed and manufactured with airfoil-contoured blades. The blades can be installed directly on either the rotor or the drum, as is the case with axial flow compressors, or on wheels or disks, as is done with centrifugal compressors. Figures 2.3a and b conveys the details better than words.

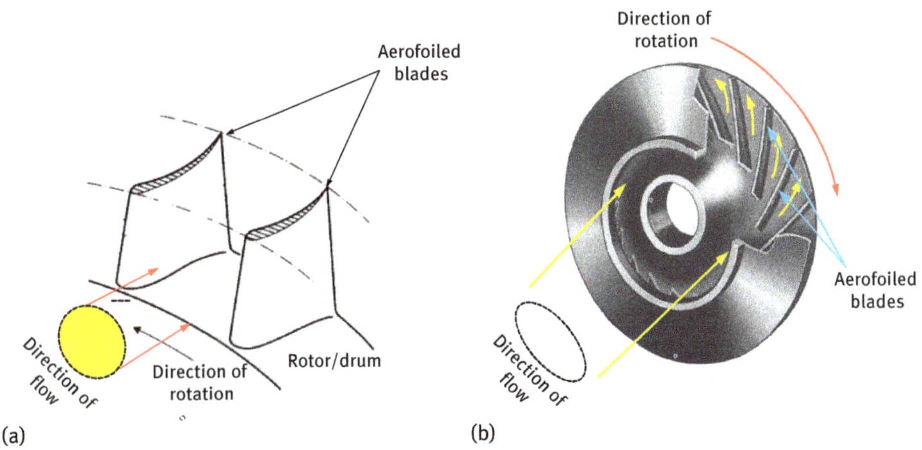

Figure 2.3: (a) axial compressor and (b) centrifugal compressor.

The kinetic energy developed in the rotating portion is converted into pressure rise in the stationary portion of the compressor. The two main types of dynamic compressors most widely applied in process industries are centrifugal and axial compressors.

2.2.1 Centrifugal compressors

Centrifugal compressors impart velocity to working fluids by moving the fluid radially outward in a controlled path.

The four main components in a centrifugal compressor are as follows:
- inlet channel
- centrifugal impeller
- diffuser
- discharge volute/scroll

2.2.1.1 Inlet channel

Modern inlet channels are generally configured with fixed inlet guide vanes (IGV); these direct the incoming flow by using optimum angles of twist. This flow is swirled in a specific manner to prevent nonlaminar flow entering the first impeller. The amount of swirl is designed to give an optimum incident angle to the first impeller. An optimum angle will allow the flow to transition smoothly as it enters the impeller eye.

The contoured flow path, including the fixed angle of the IGV, is generally designed for API specifications with the label "the guarantee flow," but this flow may not always be the preferred operating flow at the purchaser's facility; being interested in a range of actual flows, experienced machinery engineers often recommend that the equipment manufacturer performs a few computational flow dynamics (CFD) analyses. It makes much sense to have CFDs done at both high and low flows to understand their impact on the machine's performance. It is not unusual to require redesigned IGV angles to correct or alleviate performance problems.

2.2.1.2 Centrifugal impeller

The manner in which the impellers function is why we call the machines *centrifugal* compressors. Impellers can be either open or shrouded, as shown in Figures 2.4a and b.

The function of an impeller is to impart radial velocity to the gas immediately after it enters the eye of the impeller as shown in Figure 2.5. A rotating impeller imparts energy to the gas by increasing its angular momentum. As the gas gains energy from the rotation of an impeller, its static pressure and absolute velocity (relative to the stationary casing of the compressor) increase through the impeller passages as the gas enters the diffuser. The diffuser then converts the kinetic energy of the gas into static pressure by decelerating the gas. Gas decelerates when a volume flows into and through gradually increasing diffuser cross-sectional areas.

The energy imparted by the impeller to produce the required pressure ratio between suction and discharge is termed head. Units of head are ft-lbf/lbm (feet, for short), N-m/kg, and kg-m/kg (meters, for short).

Typically, open impellers can operate at higher tip speeds compared to the traditional closed impellers. Therefore, they can produce heads between 4,500 and 7,000 kg-m/kg (15,000–25,000 ft-lbf/lbm) compared to closed impellers that

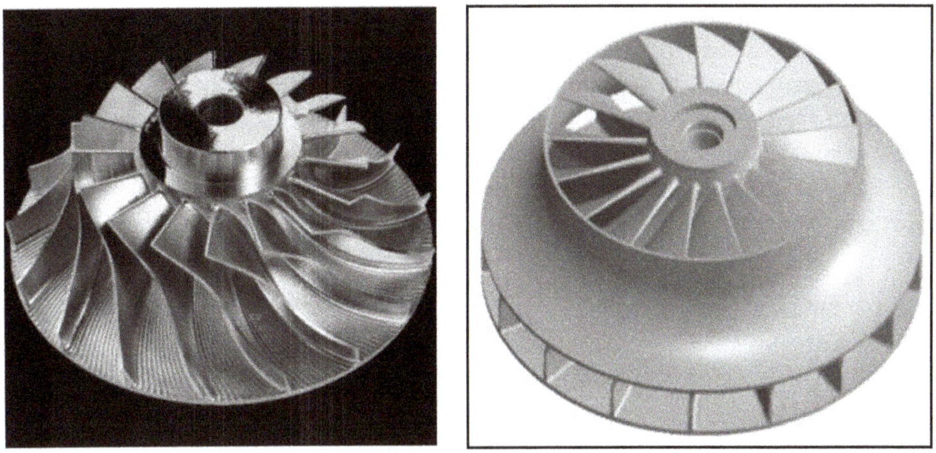

(a) (b)

Figure 2.4: (a) Open impeller and (b) closed impeller.

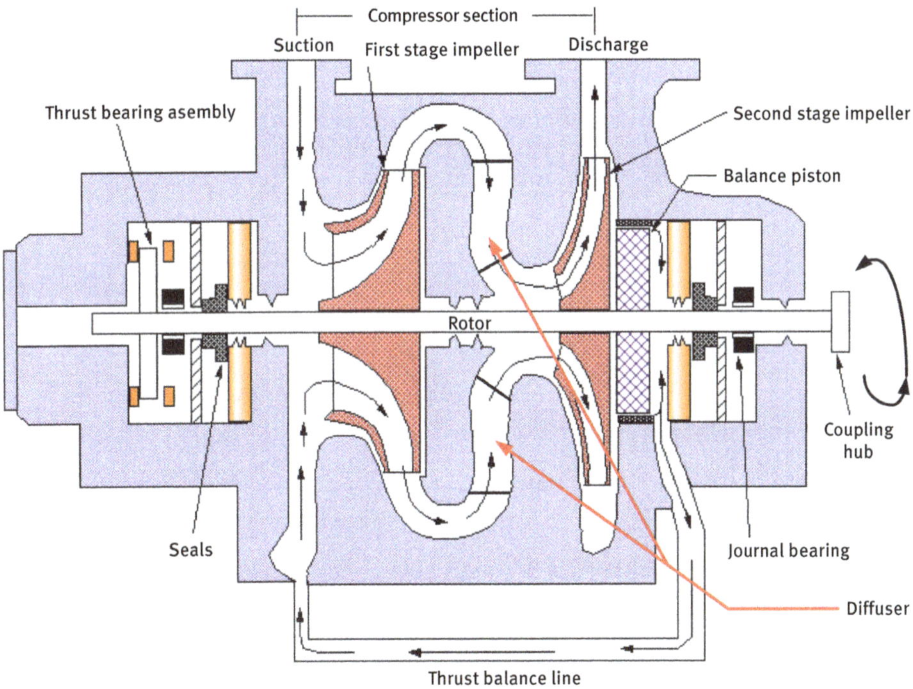

Figure 2.5: Fundamental components that make up a centrifugal compressor section.

traditionally limit heads to 3,048 kg-m/kg (10,000 ft-lbf/lbm). This will be discussed further in Appendix B.

The combination of an impeller with its corresponding diffuser is called a *stage*. One or more stages lined up in series to raise the pressure of the gas between the suction and the discharge is called a *section*.

Because of cost, physical dimension, and rotordynamic concerns, the number of stages for a centrifugal compressor is limited to no more than 10 per casing.

2.2.1.3 Diffuser

Diffusers are designed to convert the kinetic energy imparted to the gas or vapor into pressure. The conversion of kinetic energy into pressure is done by gradually slowing down the velocity of the gas. Gas velocity will decrease once it reaches the diffuser's smooth-contoured and incrementally enlarging cross-sectional flow areas.

Diffusers can be vaneless, vane-style ("vaned"), or an alternating combination. As denoted by their names, vaneless diffusers have no vanes, whereas vaned diffusers contain one or more rows of vanes, as depicted in Figure 2.6.

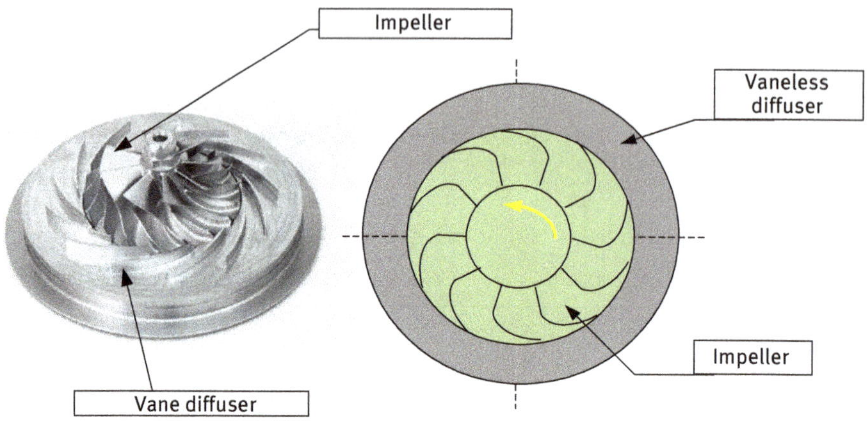

Figure 2.6: Vane-type and vaneless diffusers.

There are pros and cons for using either of these two types of diffusers. For example, because there are no vanes to interfere with the flow of gas as it passes through the diffuser, it offers a wider flow range than vaned diffusers. However, the recoverable pressure in a vaneless diffuser is less than the pressure that can be recovered in a vaned diffuser. Vaned diffusers, on the other hand, are more efficient but offer a smaller flow range, and the differences are shown in Figure 2.7.

The operating characteristics of vane-style (or vaned) diffusers are well understood. In addition, this style of diffuser can be subdivided into many different types, including wedge type, airfoil type, and plate type. Another notable term used

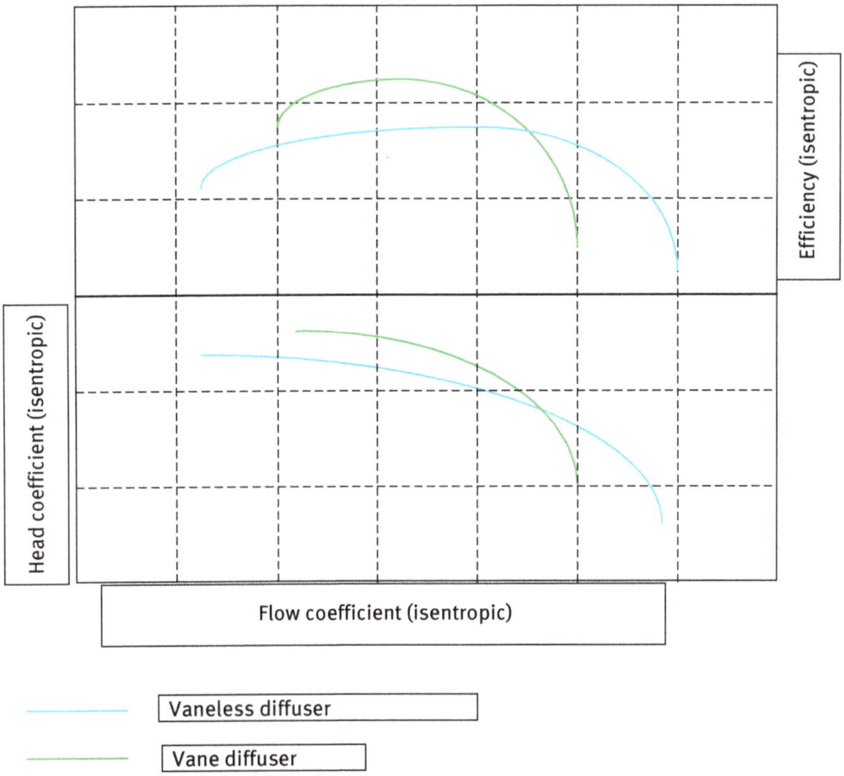

Figure 2.7: Graph illustrating difference between vaneless and vane-style diffusers.

in designing vaned diffuser is low-solidity vane diffuser (LSVD). Figure 2.8 further explains the term LSVD.

Diffusers have been studied for many years, and considerable literature on the subject has been published. Many papers and conference proceedings highlight the pros and cons of different designs. Accordingly, this book is limited to giving the reader information that is best described as foundational knowledge.

2.2.1.4 Collector (volute/scroll)

The collector in a centrifugal compressor is the last chamber into which the compressor finally discharges the gas. The gas is collected in the discharge volute annulus (the chamber or scroll), and from there it is sent to the compressor's discharge nozzle.

If a collector/volute is mismatched to a compressor, the compressor performance curves are shifted. Depending on the degree of mismatch and if the volute is undersized, the curve is shifted to the left. This produces lower flows. However, if

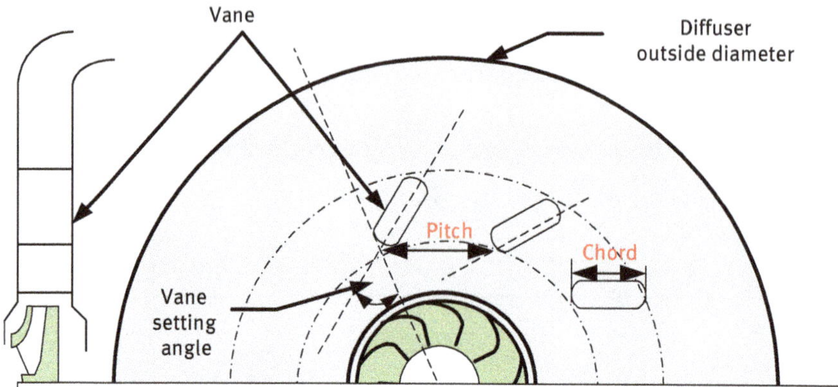

Low solidity vane diffuser (Senoo et al.)

- Chord/pitch ratio = 0.7 to 0.8
- No conventional diffuser throat
- Limited radial extent (10% to 30% of diffuser length)
- Little or no vane camber
- Full passage height
- Vane setting (stagger) angle fairly flat
- Can be multiple rows

Figure 2.8: Low-solidity vane diffuser (LSVD).

the volute is oversized, the performance curve is shifted to the right, and higher flows result.

The collectors/volutes can have various shapes. Depending on the shape, it can be called a *volute* or a *plenum*. Industrial process compressors can employ volute-shaped collectors as their discharge chamber. The volute shape is inspired by the shape of a snail's housing. Its flow area increases as it approaches the compressor's discharge nozzle.

Volutes can be installed with vanes or they can be vaneless, just like diffusers. The two basic volute configurations *symmetrical* and *overhung* predominate; they are shown in Figure 2.9. Both symmetrical and overhung types are complex to design and produce; moreover, they behave differently from a fluid dynamics point of view.

The symmetrical type delivers lower diffusion losses over a wide operating range; however, it requires larger casings compared to the overhung type. The overhung type volute can be either circular- or box-shaped.

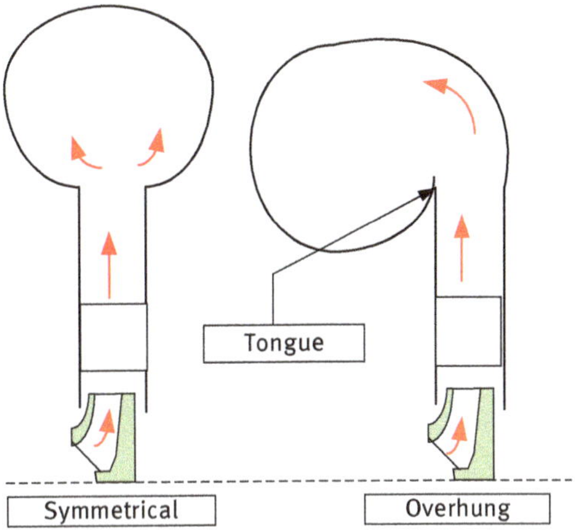

Figure 2.9: Symmetrical and overhung volutes.

Compressor manufacturers typically prefer overhung-type volutes because of installation geometric constraints. By carefully designing the tongue with the proper radius, a wider operating range can be achieved.

2.2.1.5 Horizontal and vertical split

Centrifugal compressors are constructed in two basic casing designs, namely horizontally split or vertically split casings. As both names imply, the casing split lines are horizontal or vertical, as illustrated in Figure 2.10. Generalized application ranges are shown in Figure 2.11.

2.2.1.5.1 Horizontally split casing

Horizontally split casings typically permit internal pressures of up to approximately 70 bar (1,015 psi) and volume flow rates of up to 700,000 m^3/h (410,000 ft^3/min) at lower pressures. The largest single-casing horizontally split compressors are presently (2020) rated at approximately 38 MW.

Both casing halves are sealed and bolted together by high-strength stud bolts. The stud bolts are passed through very thick flanges and tightened to a specific torque value. This type of construction offers great advantage in maintainability and access. Erection and dismantling are easily achieved by completely removing the upper-half casing and its associated stationary components. These may include large diameter pipes.

Because large diameter piping bolted to the upper casing half will have to be removed for access to the rotor, maintenance can become cumbersome. Avoiding such pipe removal is one of the reasons why horizontally split compressors are located on

Figure 2.10: Horizontal and vertical split casings.

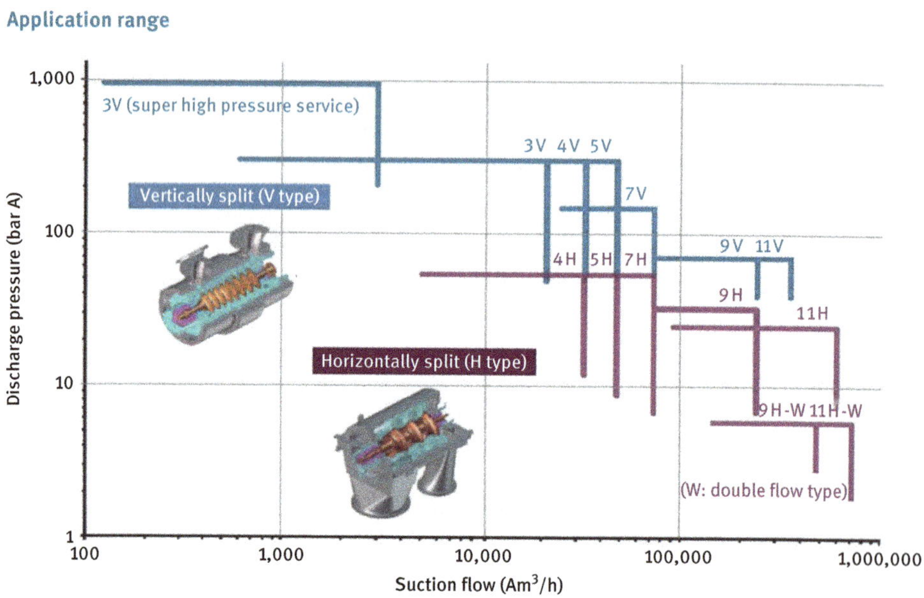

Figure 2.11: Application ranges for horizontally split and vertically split centrifugal compressor casings (*courtesy of Mitsubishi Heavy Industries – MHI*).

an elevated platform, a mezzanine. Mezzanine mounting also takes care of oil drainage by letting gravity cause oil to flow back into a supply reservoir located at grade.

Structurally, these compressors are generally centerline supported, thus greatly reducing vertical shifting of the compressor shaft due to thermal expansion. Available materials of construction cover a wide range of ferrous metals; selection is based on the specific application and may include chrome or nickel alloys.

The default seals in today's centrifugal compressors are dry gas seals (DGS). DGS systems can make use of a wide range of materials. Such systems make horizontally split compressors suitable for virtually all fields of application in modern industry. These fields include chemical plants, oil refineries, and petrochemical process facilities.

2.2.1.5.2 Vertically split compressors

In vertically split or barrel-type compressors, as they are often called, an inner barrel or bundle is surrounded by a pipe-like, thick-walled outer casing. However, unlike the outer casing, the inner barrel is split horizontally with both halves bolted tightly together. The outer casing is sealed radially by circular end flanges or cover plates at each end. A modern alternative to the massive but now outdated bolt-on cover plates (Figure 2.10) is depicted in Figure 2.12. This modern design uses end plates inserted into the outer casing past wide recesses (grooves) machined into each end of the thick-walled outer casing bore. So-called shear ring segments (Figure 2.13) are placed into each of the wide grooves and held in place with suitable locating screws. A very large, suitably dimensioned O-ring groove is machined into each of the slide-in end plates. To remove the inner barrel or bundle, the casing must be opened by removing either the massive cover plates shown in Figure 2.10 or the insertable end plates shown in Figure 2.13.

Figure 2.12: A modern vertically split ("barrel") compressor (source: *Mitsubishi Heavy Industries – MHI, Hiroshima, Japan*).

The barrel or bundle is pulled out axially until it clears the end of the outer casing. If placed in the middle position of a string of machine casings, the axial pull-out would necessitate removing the adjacent casing or compressor body. It is therefore desirable to place barrel compressors at the end of a string of machine casings.

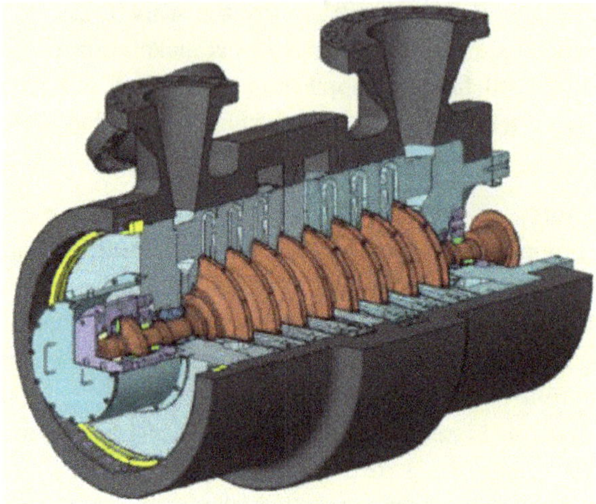

Figure 2.13: Vertically split centrifugal compressor casing, with shear ring segments shown in yellow color (source: *Mitsubishi Heavy Industries – MHI, Hiroshima, Japan*).

An often overlooked attribute of vertically split (barrel) compressors is that inlet, outlet, and side-stream nozzles can be welded into the casing at virtually any desirable angle. Since the casing remains undisturbed during overhauls or when servicing the compressor, purchasers occasionally prefer barrel machines. However, installation at grade level requires thoughtful engineering to obtain adequate oil drainage while only minimally assisted by gravity flow.

Barrel-type compressors are designed to operate at much higher pressures than horizontally split compressors. The upper images in Figures 2.11–2.13 show vertically split compressors designed to operate at pressures as high as 1,000 bar (14,500 psia). The cylindrical thick-walled pipe-like outer casing guarantees good stress distribution and, due to a suitably placed O-ring seal, extremely good gas tightness.

Among the main applications for barrel-type compressors are:
- gases rich in hydrogen. API 617 recommends this type of compressor for gases where the hydrogen partial pressure exceeds 13.8 bar (200 psi)
- hydrocracking
- synthesis of ammonia, urea, and methanol
- gas lift and reinjection
- transportation of gas in pipelines

2.2.2 Axial compressors

Axial flow compressors are dynamic compressors, and in some respects similar to, yet unlike centrifugal compressors. However, instead of moving the fluid in a radial manner, axial compressors move the fluid through the machine in a parallel path along the axis of the rotor. This is achieved by using successive rows of airfoil-type blading with – on a per row basis – rather low compression ratios. Each row consists of 20–50 airfoil-like rotating blades attached to the rotor/drum and a number of airfoil-like vanes installed in the stator. The combination of a single row of rotating blades and a row of stationary vanes makes up a stage.

The rotating blades accelerate the gas in both the axial and circumferential directions into the stationary blades. The stationary blades with their airfoil-type-shaped vanes convert the kinetic energy into static pressure. Stationary blades are designed to act as diffusers in a similar way to the diffusers in a centrifugal compressor. Diffusion is done by slowing down the gas and converting velocity into pressure. The gas is then redirected to the next stage by the adjacent stationary vanes. This process continues incrementally until the gas is finally discharged to the process.

The pressure ratios per stage for industrial-type axial compressors typically range from 1.15 to 1.25. By compounding the compression process, polytropic efficiencies can reach between 90% and 95%. Multiple stages allow industrial axial compressor to achieve overall pressures ratios up to 30:1. For axial compressors used in aerospace applications, overall compression ratios can reach as high as 40:1.

Recall that Fig. 2.2 depicted the overall operating envelope for axial compressors. Axial compressors are used in applications with flow rates between 100,000 m^3/h (58,700 acfm) and 1,500,000 m^3/h (878,500 acfm). Compressor manufacturers can usually offer six to eleven standard frame sizes that, in turn, incorporate two to twelve stages.

Figure 2.14 shows the variation of velocity, pressure, and temperature as an axial compressor compresses the gas from suction to discharge. The rise in pressure and temperature across the compressor also relates to the change in enthalpy.

For a multistage axial compressor (Figure 2.15), the density of the gas will increase as the pressure increases. To compensate for the increase in gas density, both blade length and annular area between rotor and shroud decrease gradually along the length of the machine. Constant axial velocity and optimal Mach number favorably affect efficiency. Inlet relative Mach numbers (defined in Section 2.4) range from 0.4 to 0.8 per stage for industrial axial compressors.

Axial compressors are used in many industrial applications wherever compression of large flows is required. These applications include the following:
- natural gas services
- large air separation plants
- fluid catalytic cracking
- propane dehydrogenation plants
- large refrigeration units
- large volumes of process gas

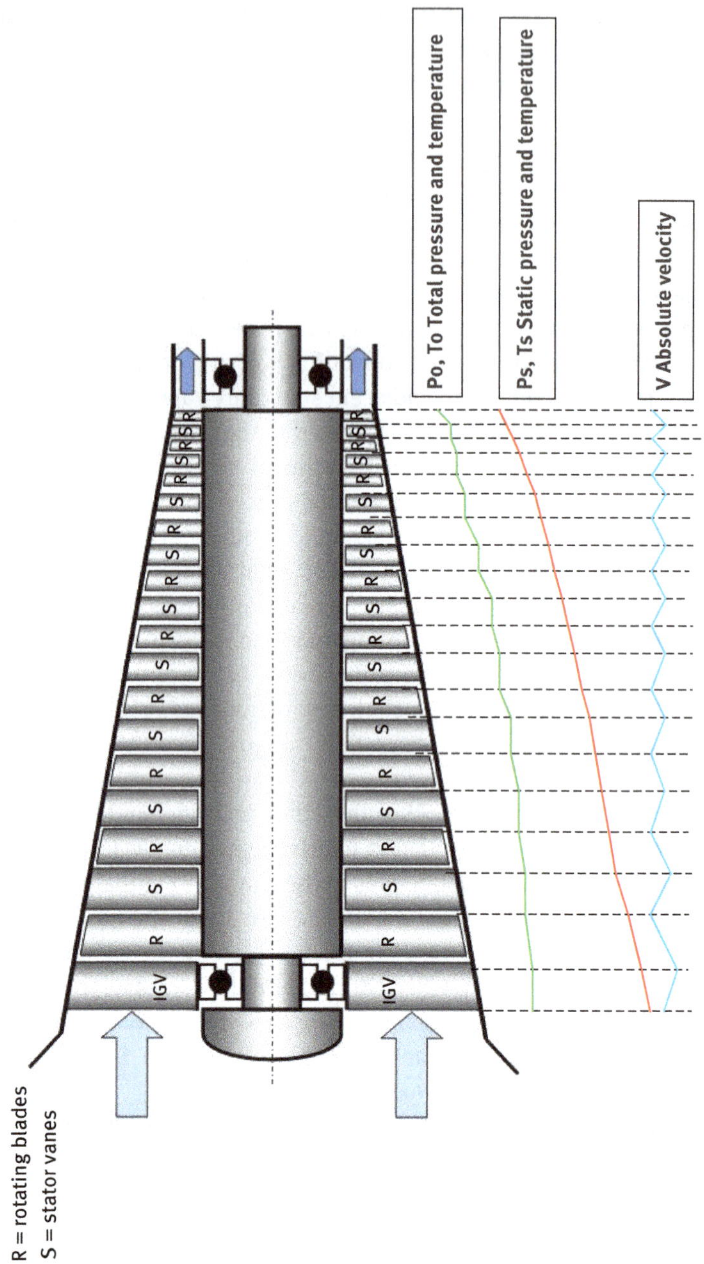

Figure 2.14: Velocity, temperature, and pressure changes through an axial compressor.

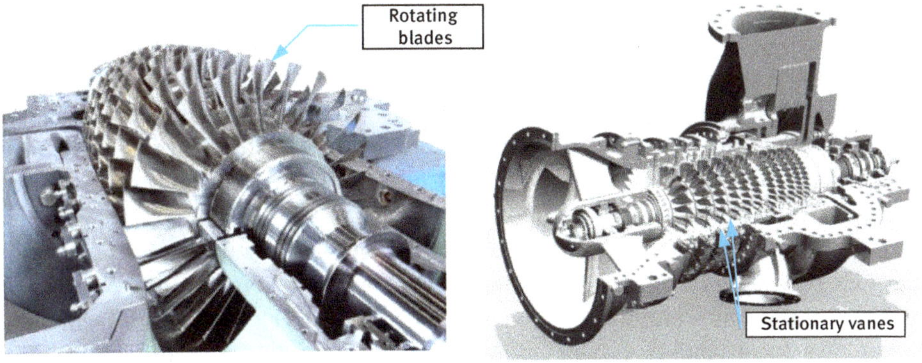

Figure 2.15: Cross section of the MAX-1 axial compressor (source: *MAN-ES, Germany*).

Axial compressors deliver large volume flows over a narrow range of pressure. To boost the discharge pressure, compressor manufacturers can install a centrifugal impeller at the inlet end of the rotor, as shown in Figure 2.16.

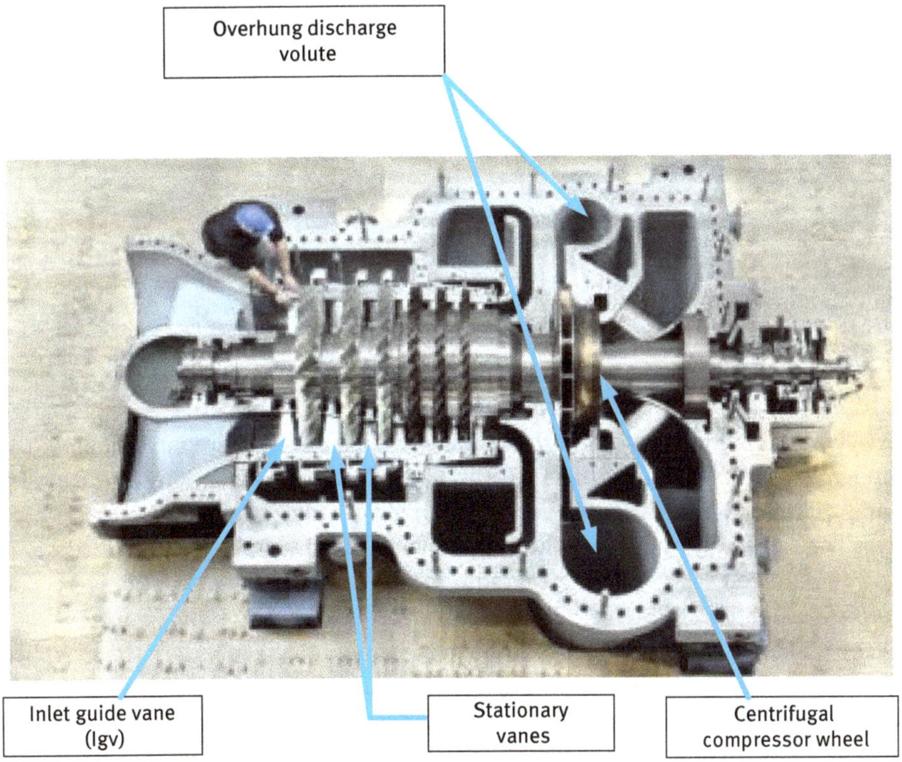

Figure 2.16: Axial compressor with first stage centrifugal impeller for boosting head (source: *MAN-ES, Germany*).

Traditionally, axial compressors were designed with fixed stator vanes and adjustable IGV. This basic design offered a fair degree of operating flexibility; it allowed varying the operating flow range, within limitations. However, modern designs are now frequently designed with manually and/or automatic-variable stator vanes. Fixed stator vanes are still available and are also suitable for variable speed applications. However, variable stator vanes offer better capacity control and wider operational flexibility.

This advanced capacity control strategy improves surge control and compressor start-up. For example, closing the vanes during start-up allows for little or no compression. If the vanes are then opened automatically to track speed during start-up, this allows for a smooth transition to normal operation. The reverse holds true while shutting down; surge prevention is greatly facilitated with automated stator vane adjustments.

2.3 Positive displacement compressors

This class of compressors increases the pressure of the working fluid by trapping it in a confined space, followed by progressively decreasing the space initially occupied by an inlet volume of gas. Many configurational arrangements are available that harmonize with this basic principle, and reciprocating compressors are the most prominent version.

2.3.1 Reciprocating compressors

Small air compressors are single acting; the air is being compressed during the forward stroke of the piston. The forward stroke is understood to be the stroke toward the head end of the cylinder. Reciprocating compressors for process gases use a double-acting cylinder–piston arrangement, which includes inlet and discharge valves on both the head end and the frame end (also called the "crank end") of the cylinder. Inlet valves act as check valves that allow the gas to enter the cylinder during volume expansion; the valves close off during compression. Discharge valves stay closed during the volume expansion and begin to open after the gas being compressed has reached a pressure in excess of the opposing (delivery side or process side) pressure.

The primary components of a double-acting reciprocating compressor are identified in Figure 2.17. The illustration helps visualize that during the compression cycle, the gas is trapped inside the cylinder with the valves closed. As the piston moves toward the clearance volume, the volume is decreased, and the pressure continues to increase. The clearance volume is the minimum volume of gas remaining in the cylinder when the piston has reached the end of its stroke. As the piston approaches the end of its stroke, the

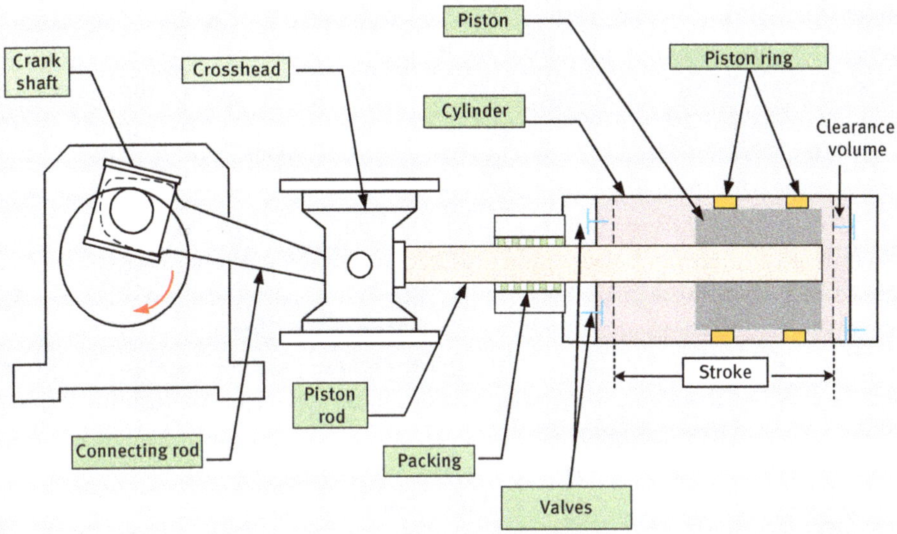

Figure 2.17: Primary components of a typical double-acting reciprocating compressor ("Not to scale").

high internal pressure overcomes the opposing pressure. This then causes the discharge valve(s) to open, and the gas gets expelled from the cylinder.

The back-and-forth movement of the piston inside the cylinder is done mechanically by means of a traditional slider-crank mechanism. With probably no exceptions, process industry services and gas transmission applications are double acting. Compression takes place on one of the two sides of the piston while gas is aspirated on the other side. Upon stroke reversal, the cycle is reversed, and the other side of the cylinder is now involved.

Reciprocating compressors offer several advantages, and four of these usually come to mind first:

- Design flexibility. These machines can be manufactured in vertical or horizontal configuration. The internal packing, including piston rings and rider bands, supporting and sealing the piston can be lubricated or nonlubricated. The piston can be configured with integral labyrinths for vertical reciprocating compressors, thus eliminating piston rings.
- Excellent efficiency even for small compressor sizes operating at high pressures and at partial loads.
- Operating flexibility over a wide range of gases and gas conditions for a given configuration.
- Wide range of throughput volumes is achievable by keeping suction valves open during all or part of a cycle.

The design and operating limits for reciprocating piston-type compressors were shown in Figure 2.2. Tables B1a and B1b in Appendix B also show the operating range of typical compressors found in the process industries. Reciprocating compressors have the widest pressure range in the compressor family; they range from vacuum to 60,000 psi (4,100 bar).

Design flexibility is one of the important features of reciprocating compressors. It is entirely feasible to connect several crank mechanisms to a single crankshaft. Many cylinders (or *crank throws*) can thus be installed on a single frame, all driven by one driver. This arrangement also allows multiple, and often entirely different, gas services to be incorporated in the same frame. The total number of throws is limited by torsional considerations and should be checked during design audits.

The main component parts of the typical process gas reciprocating compressor shown in either the elementary schematic of Figure 2.17 or in the installed (on-site) view of Figure 2.18 are as follows:
- cylinders
- piston (including piston rings and wear bands)
- piston rod
- packing rings
- crosshead
- connecting rod
- crankshaft

Figure 2.18: Arrangement showing six cylinders (*throws*) on a single frame (source: *Kobelco Mfg.*).

2.3.1.1 Cylinder

Compressor cylinders (Figure 2.19) are pressure vessels, which make them critical component parts of a reciprocating machine. Cylinder materials are selected based on application or service. Large size cast iron cylinders qualify for operation up to 1,200 psi (83 bar). Nodular iron and cast steel are used for operating pressures up to 2,500 psi (172 bar), usually with smaller size castings. The maximum allowable working pressure (MAWP) of a cylinder determines the relief valve setting. Applicable safety rules call for these valves to open when a value of 50 psi (3.5 bar) or 10% lower than MAWP is exceeded.

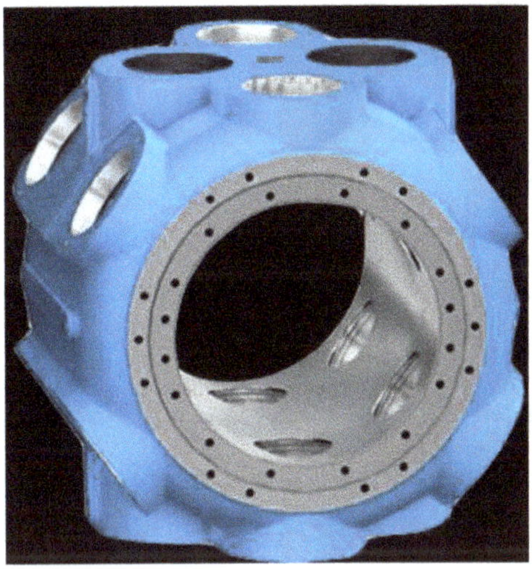

Figure 2.19: Reciprocating compressor cylinder showing valve ports and replaceable (insertable) cylinder liner.

Cylinders are detachable; they are separated from the frame by distance pieces. The component parts are piloted ("rabbeted" or recessed) for alignment and ease of fit-up purposes. The API 618 specification requires that replaceable liners be installed in the cylinder. Worn liners can be readily replaced without having to modify the compressor cylinder. Virtually all cylinder designs incorporate cast-in ("cored") cooling water jackets.

2.3.1.2 Piston, piston rings, and wear bands

Piston actuation is obtained through their respective crank mechanisms, and each piston imparts energy to the gas during the compression cycle. Pistons must have weight and strength, and all piston parts must be compatible with the gas being compressed. Materials of construction can be either aluminum or ferrous metals. aluminum pistons are often larger and favored in low-pressure applications.

Sometimes pistons are made hollow in order to gain weight savings. Trapped gases can make hollow pistons hazardous when undergoing maintenance. These gases must be vented before commencing piston maintenance.

The piston in Figure 2.20 incorporates piston rings to minimize leakage from the high-pressure side to the low-pressure side. Oil is injected to reduce friction between rings and cylinder; the oil film also serves as a sealing barrier.

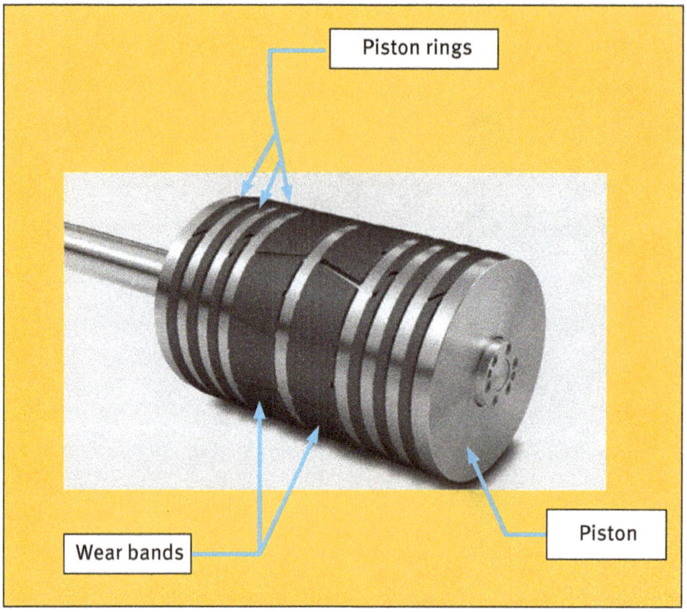

Figure 2.20: Piston with piston rings and wear bands/rider bands.

For lubricated cylinder applications, materials such as bronze or cast iron rings can be used. However, nonmetallic high-performance plastics are becoming the default ring material for both lubricated and nonlubricated cylinders. In nonlubricated cylinders, the material of choice should have low-friction properties. Glass-filled polytetrafluoroethylene is often selected because its Teflon® ingredient "rubs off" during initial run-in; it thereby infuses itself in the pores of opposing stationary parts.

Wrap-around wear bands are installed in grooves on the piston; they support pistons and rods in horizontal compressors. By distributing their weight over a large area, contact pressure is kept low, and wear is minimized.

2.3.1.3 Piston rod

There are different ways of fastening piston rods to crossheads and pistons. Rolled threads are much preferred, and engineered lock nuts secure piston rods to the crosshead at the frame end and to the piston at the head end.

The piston rod itself is highly stressed and designed to survive millions of stress reversals as the piston travels back and forth inside the cylinder. Rod loading must be kept within the limits set by the compressor manufacturer; relevant calculations are always based on the combined gas load and inertia loading. The favored rod material is AISI 4142 alloy steel.

Piston rod wear is greatly reduced by hardening its surface in the region where contact with cylinder packing is possible. Different metallurgies define the maximum hardness selected by the designer. Rods made from AISI 4142 alloy steel should be hardened to Rockwell C50 as a minimum. For piston rods made from widely available stainless steels, hardness should not exceed Rockwell C40.

2.3.1.4 Packing rings

Packing rings (Figure 2.21) are generally segmented and housed in a stuffing box. The rings can be cut radially or tangentially; they can have passages for lubrication or cooling, depending on the application at issue.

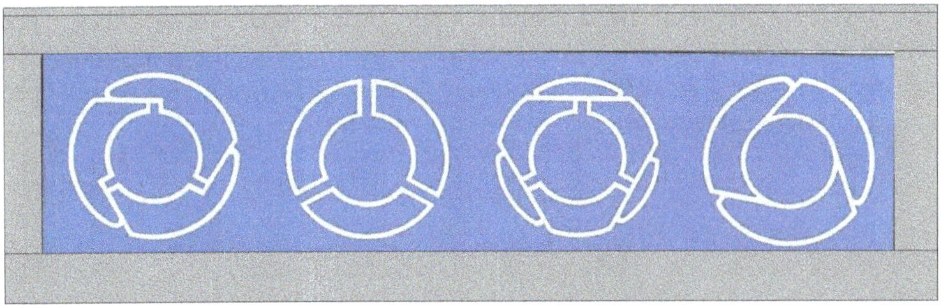

Figure 2.21: Different types of cuts for segmented packing rings.

Distance pieces are designed to prevent leakage from the cylinder into the crankcase. Pressure packing (Figure 2.22) is located where the cylinder is attached to the distance piece; an oil wiper box is located where the distance piece meets the compressor frame. The construction of the distance piece and its oil wiper box and pressure packing can be simple; it can also be more complex with leak-offs and purge compartments, depending on the hazardous nature of the gas being compressed.

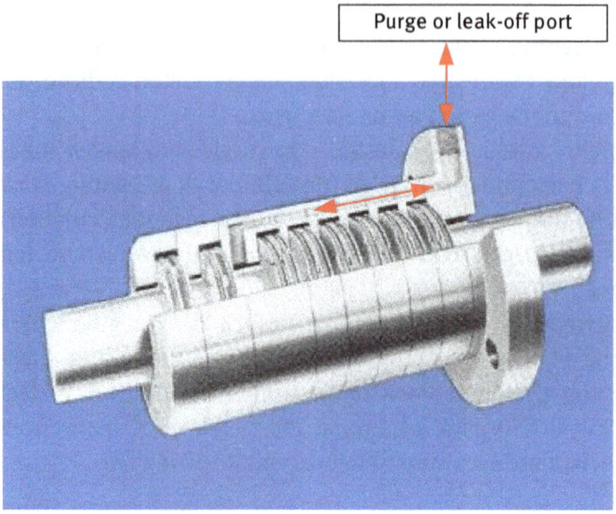

Figure 2.22: Simple stuffing box with port for leak-off or purge.

2.3.1.5 Crosshead

A crosshead (Figure 2.23) translates the rotating motion of the connecting rod to linear reciprocating motion of the piston rod, which, of course, is also called a stroke of the piston. Crossheads are moving back and forth; they are fitted with top and bottom guide shoes. The shoes are usually babbitted lubricated groove surfaces attached (Figure 2.23). Shoes can be shimmed to accomplish precise piston rod horizontality or alignment.

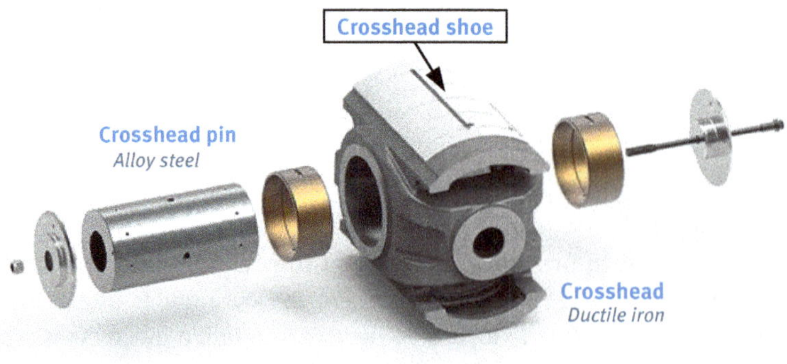

Figure 2.23: Crosshead assembly including shoe, bushings, and crosshead pin (source: *Ariel Compressors, Mount Vernon, Ohio, USA*).

The crosshead pin (Figure 2.23) is usually case hardened to about Rockwell C60 and surface finished to approximately 60 μin. Designs sometimes are straight fit and full floating; support by two bronze bushings is common.

Most industrial-type reciprocating compressor manufacturers today use nodular iron for crossheads. ASTM A536 60-40-18 is easily castable and rarely needs repairs. It has excellent fatigue strength and has proven quite reliable for seven or eight decades.

2.3.1.6 Crankshafts

Crankshafts (Figure 2.24) are generally made from carbon steel forgings or nodular iron castings. A carbon steel forging usually conforms to AISI 1020, and ASTM 668 for small shafts. For larger shaft sizes, manufacturers prefer AISI 1045 and ASTM 668 class F materials. Nodular iron crankshafts conform to ASTM A-536 grade 80-55-06.

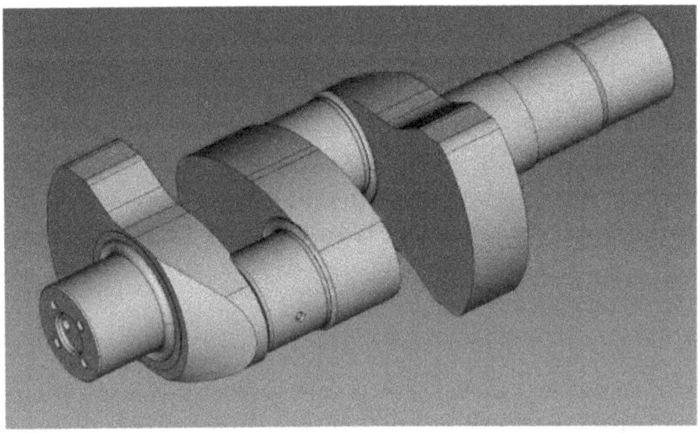

Figure 2.24: A two-throw crankshaft with rifle-drilled lubricating hole.

Crankshafts should not be surface hardened. Dynamic balancing is deemed necessary only if the rotative speed is equal to or exceeds 800 rpm.

2.3.1.7 Connecting rod and bearings

Connecting rods (Figure 2.25) convert the rotary motion of the crankshaft mechanism to reciprocating motion. As its name implies, a connecting rod connects the crankshaft to the crosshead assembly. Generally, connecting rods are fabricated from forged steel for machines >150 kW (200 HP) and ductile iron for machines <150 kW (200 HP).

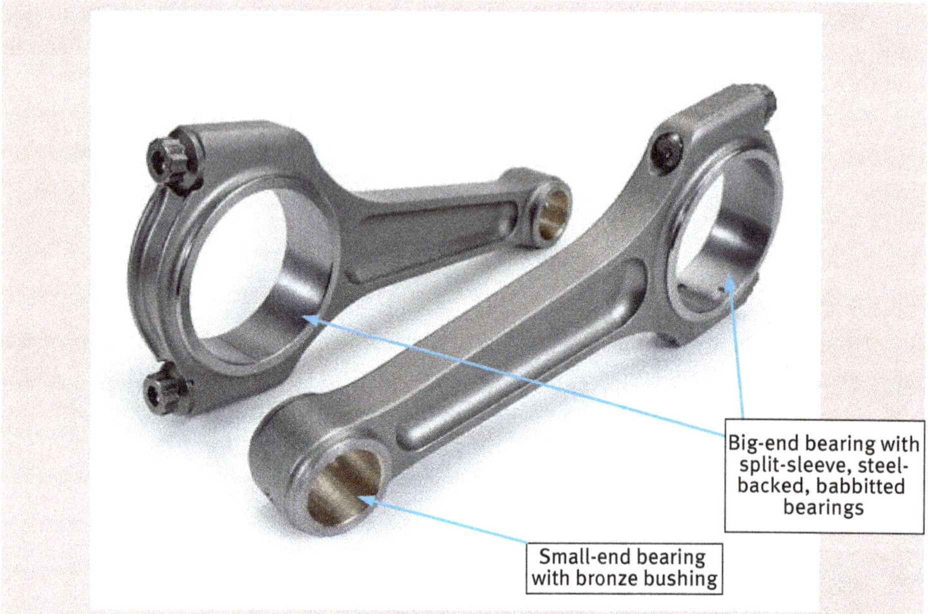

Figure 2.25: Connecting rod showing the types of bearings on each of the ends of the assembly.

Pressurized oil lubricates the bearings through rifle-drilled holes inside the crankshaft and the connecting rod. One such hole is visible in the two-throw crankshaft of Figure 2.24.

2.3.1.8 Valves

Reciprocating compressor valves are basically check valves. Most valves are spring loaded, but the most modern valves (Zahroof modular reed valves) do not incorporate springs. Not having to cope with springs can be decidedly advantageous.

Valves are installed in the cylinder's suction and discharge cavities, allowing gas to move in and out of the cylinder. Each valve is designed to open or close once per revolution of the crankshaft in less than one tenths of a second. Minimum resistance to flow is one of the designer's goals to avoid energy waste.

Valves will be discussed in much more detail in Section 4.2.2. Important new approaches to valve technology commenced in 2014.

2.3.2 Rotary compressors

Recall that Figure 2.1 showed the family of rotary compressors. The common mode of operation among this family of compressors is that the trapped gas is positively displaced by a rotating element to increase its pressure. The rotating element of each type of rotary compressor is uniquely designed to accomplish pressure increases to levels higher than the inlet pressures.

2.3.2.1 Screw compressors

Rotary screw compressors, sometimes called helical screw or spiral lobe compressors, belong to the rotary positive displacement machine category. For most users, they need little introduction or explanation; rotary screw compressors have been around for many decades and are very likely the "equipment of choice" for either oil-free or oil-wetted compression of air in mining, construction, industrial refrigeration, and a host of other applications where their relative simplicity, general reliability, and high availability are appreciated.

These compressors represent a subset of the machinery group making up rotating positive displacement compressors. Of the various machines available, twin rotary screw compressors are primarily used in higher pressure air and process gas services, whereas certain other rotary piston blowers and single-screw compressors are often used in lower pressure, high-volume applications. Large screw compressor models are frequently used in refrigeration services and process industry applications; one such twin-screw rotary machine is shown in Figure 2.26. The machine builder is working on one of the four carbon seal assemblies that, until relatively recently (2018), were typical for this compressor category. The carbon seal to which the builder is pointing is located on the compressor input rotor.

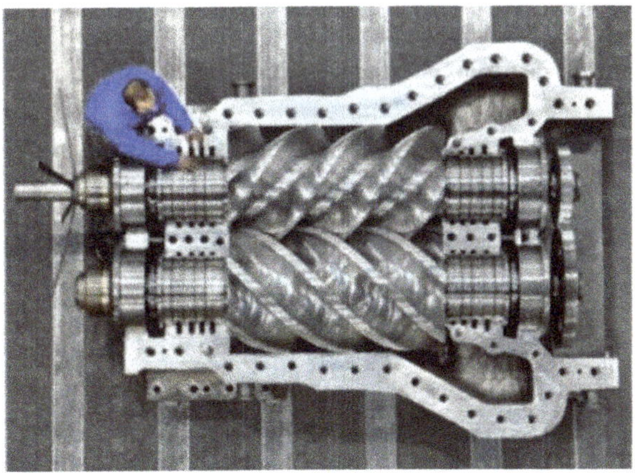

Figure 2.26: Twin-screw rotary compressor (*courtesy of MAN-ES, Germany*).

Rotary screw compressors are separated into two main categories: wet and dry screw machines. For both dry screw and wet screw compressors, typical rotor profiles are shown in Figure 2.27.

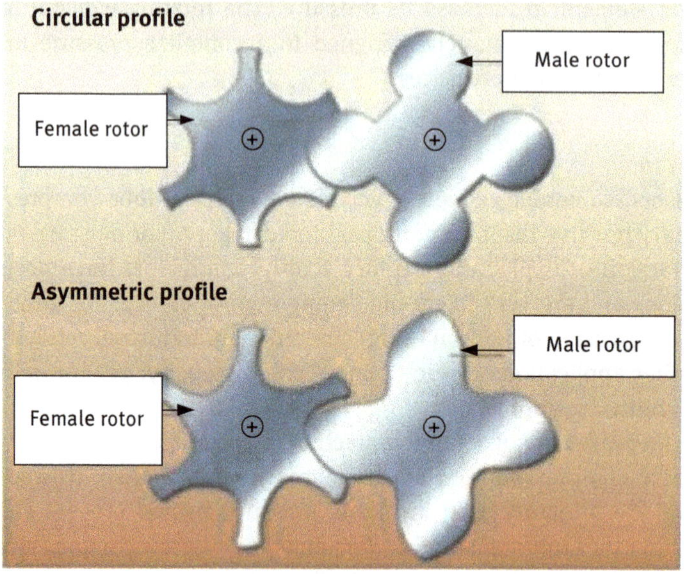

Figure 2.27: Typical rotor profiles for rotary screw compressors (source: *Kobelco Mfg.*).

2.3.2.2 Operating principles of oil-free (dry) and liquid-flooded (wet) compressors

Regardless of whether the screw compressor is designed and marketed for oil-free ("dry") compression or liquid injection (the "wet" method), the gas is compressed in chambers that progressively decrease in volume. While we will use the convenient description "oil injected" for wet screw machines, it is of great importance to note that the wet method is not confined to oil. In German coal gasification plants, a number of 7,000 HP water-injected twin-screw compressors have been operating with remarkable availability and reliability for decades. Other wet screw machines use glycol and yet others about a dozen other liquids.

Progressively advancing chambers in rotary screw machines are formed by the intermeshing action of two helical rotors and a surrounding close-clearance housing wall. However, dry machines incorporate timing gears that keep the two counterrotating screws in the exact relationship to each other. Oil-injected (sometimes called oil-flooded) compressors do not incorporate timing gears, and the driven male rotor interacts directly with the female rotor. Oil injected into the compressor cavity serves to provide massive lubrication while also allowing a large portion of the compression heat to be absorbed by the oil. Because the clearances between rotors and the cylinder (casing) wall are also filled with oil, reverse flow of compressed gas is prevented, and the overall compression efficiency is increased.

After leaving the compressor discharge flange, the gas–oil mixture exits through a check valve to the oil reservoir where most of the oil is separated from the gas. In oil-flooded compressors, the remaining oil is removed in a downstream separator, and only residual oil amounts of typically 5 ppm will remain in the gas stream. If needed, even this oil carryover can be further lowered by downstream cooling and final moisture separation. The oil separation unit must be properly maintained, and the pressure drop across the separator cartridges must be taken into account. This will determine the overall performance of the compressor package. It should also be recognized that the efficiency of oil separation changes as the separator elements become progressively more loaded with contaminants.

On small rotary screw compressors, the housing is vertically parted on the suction side. The compressor cylinder (generally called the "casing") and its discharge side plate are frequently combined in one housing. The housings of larger machines are usually parted horizontally for easy assembly. Rotors and shafts are milled from one piece of either forged or stainless steel. Some manufacturers provide rotors with nonmetallic coatings. In severe service where loss of coating occurs on the rotor edges, a rapid drop in compressor efficiency could be encountered.

Process gas machines are typically designed with the direction of gas flow from the top entry nozzle to the bottom discharge nozzle. This facilitates liquid removal from the compression space whenever liquid is injected into the rotor chamber for cooling or for cleaning during operation. Onstream cleaning is highly advantageous in services where gases are contaminated or tend to polymerize. The sealing area is equipped with connections for sealing medium supply and relief. In principle, it is possible to apply a cooling medium to the cylinder wall, but noncooled cylinder housings (or casings) can be used as well. Figure 2.27 illustrates the typical rotor combinations, including an asymmetrical rotor profile. The profile combination 4 + 6 means that the male rotor has four lobes and the female rotor has six. Due to this profile combination, the diameter of the rotor core is relatively thick. This allows for operation with large differential pressures.

2.3.2.3 Application ranges for rotary screw compressors

Application limits for rotary screw compressors are set by the pressure and temperature ranges and by the maximum allowable speeds of the machines. Oil-free rotary screw compressors can be mechanically loaded with pressure differences up to 12 bar (174 psi) and oil-flooded compressors up to 20 bar (290 psi). Higher pressure differentials are possible in special cases. Flow volumes up to 100,000 m^3/h (~59,000 acfm) have been accommodated in these compressors (Figure 2.28).

The maximum allowable compression ratio for one twin-screw compressor stage is the temperature rise that will not exceed the permitted value of 250 °C (482 °F). This compression ratio and the associated temperature will, to a very large extent, depend on the specific heat ratio (c_p/c_v) of the gas to be compressed. For example, where the specific heat ratio c_p/c_v equals 1.4, the maximum compression

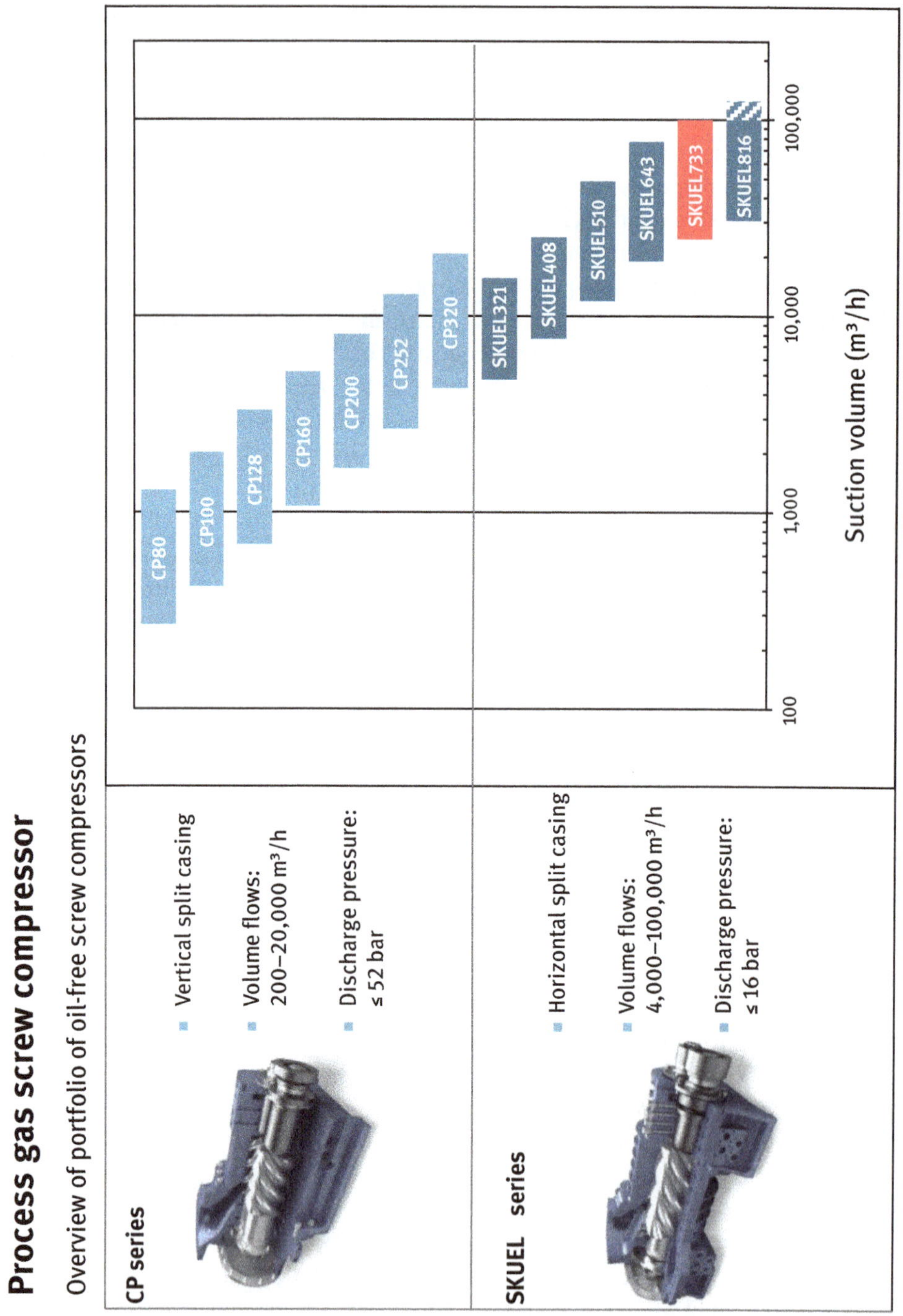

Figure 2.28: Application range for rotary screw compressors (source: *MAN-ES, Germany*).

ratio would be approximately 4.5, and where the specific heat ratio c_p/c_v equals 1.2, the maximum compression ratio would be approximately 10 for a single oil-free twin-screw compressor stage.

Multistage (multicasing) arrangements are not uncommon; they can achieve pressure ranges from approximately 0.1 bar absolute to 40 bar. Interstage cooling is used in many of these applications. Depending on the compressor size, speeds from 2,000 to 20,000 rpm can be encountered. The limiting factor is often the circumferential speed of the male rotor, which typically ranges from 40 to approximately 120 m/s and up to a maximum of 150 m/s for very light gases.

2.3.2.4 Oil-free versus liquid-flooded twin-screw compressors

Rotary screw compressors are available in oil-free or oil-flooded construction. Technically speaking, oil free refers to not having oil in the compression space, but bearings still require lubrication by a clean medium. This lubricant is typically a clean premium grade lubricating oil, although clean pressurized water can also be used [1]. In fact, while pressurized water represents less well-known advanced technology, it has nevertheless been applied for decades. In the hands of truly competent compressor manufacturers, water-flooded screw compressors have been remarkably effective and successful in some of the dirtiest services [1].

While oil-free twin-screw compressors are widely called “dry screw” machines, at least one prominent manufacturer defines and designates as “dry screw” any screw compressor equipped with timing gears. Therefore, whether or not the compression space is dry, oil-flooded, or water-injected makes no difference: With timing gears keeping the two screws synchronized, it should be labeled a dry screw machine. Without timing gears, it cannot function as a dry screw machine, because the resulting contact of mating rotors would destroy the machine. If there are no timing gears, a separating liquid must be used. Any separating liquid introduced to or circulating in the compression space will make it a wet screw machine [2].

Fields of application for oil-free machines include all processes that cannot tolerate contamination of the compressed gas or where the lubricating oil would be contaminated by the gas. Oil-flooded machines can achieve slightly higher efficiencies than “dry screw” machines and can utilize the oil for cooling as well [3]. The same statements could be made for water-flooded machines. In some instances, the bearing lube circuit must be totally separate from the fluid circuit used in the compression space. Whenever this requirement is disregarded, the purchaser–user may end up with either high maintenance costs or low equipment reliability.

Properly designed rotary screw compressors have no metallic contact whatsoever inside the compression chambers, either between the rotors themselves or between them and the walls of the housing. Although originally intended for air compression, rotary screw compressors are now found in numerous services in the petrochemical and related industries. These include air separation plants, industrial refrigeration plants, evaporation plants, mining, and metallurgical plants.

Practically all gases can be compressed; ammonia, argon, ethylene, acetylene, butadiene, chlorine gas, hydrochloric gas, natural gas, flare gas, blast furnace gas, swamp gas, helium, lime kiln gas, coking plant gas, and carbon monoxide gas have been compressed in screw machines. The same is true for all hydrocarbon combinations; coal gas/town gas, air/methane gas, propane, propylene, flue gas, crude gas, sulfur dioxide, oxide of nitrogen, nitrogen, styrene gas, vinyl chloride gas, and hydrogen gas can be found on the reference tabulations of experienced manufacturers. These reference tabulations become valuable if the purchasers do their due diligence and have serious conversations with the vendor's experts.

Many examples of the conversation agenda are questions on internal clearances between rotor tips and casing internals: Will the design clearances be different for different gases or are they strictly a function of the calculated temperature rise? Are different metallurgies used for certain services? Do their differing coefficients of thermal expansion necessitate different rotor clearances?

Some of the very best manufacturers of rotary screw compressors offer compressor packages similar to the ones shown in Figures 2.29 and 2.30. A two-stage oil-free compressor with intercooling is shown in Figure 2.29. An oil refinery in the United Kingdom uses the machine to compress 2,300 m^3/h of flare gas from 1.0 bar to a

Figure 2.29: Two-stage intercooled flare gas recovery package in use at a UK oil refinery (source: *Aerzener Maschinenfabrik GmbH, Aerzen, Germany*).

Figure 2.30: Single-stage PSA feed gas compressor package (source: *Aerzener Maschinenfabrik GmbH, Aerzen, Germany*).

discharge pressure of 10.1 bar. Each stage is equipped with double, self-acting DGS (Dry Gas Seals). Imagine what we could learn from discussing details with this manufacturer. The very fact that the flare gas is not usually very clean would be a subtopic in our conversations. DGS in this machine represent an innovation. What prompted this choice? Chances are that the owner–purchaser received bids or offers for less expensive machines but opted for the one with best available technology. We can learn much from exploring the forward thinking of this user-purchaser.

The single-stage pressure swing adsorption (PSA) feed gas compressor in Figure 2.30 takes 12,600 m^3 from 5.0 bar absolute to a discharge pressure of 14.5 bar absolute. It incorporates gas-purged restrictive carbon rings and double, oil cooled mechanical seals. PSA compressors take raw hydrogen from a great variety of sources, including reforming, partial oxidation, cryogenic purification, and methanol purge gases. It would be difficult to find a more suitable compressor than the twin-screw machine, and seal selection is a critical element of its design.

A large (rotor diameter 845 mm) water-injected compressor stage for about 50,000 m^3 of styrene off-gas (mainly hydrogen and water vapor) is depicted in Figure 2.31. It takes gas from 0.2 bar absolute to a discharge pressure of 1.9 bar absolute, a compression ratio of 9.5:1. The compressor incorporates steam-purged restrictive carbon ring seals on the process side.

2.3.2.5 Screw compressor volume control

In principle, it is necessary to consider the volume control options for dry-running and oil injection-type screw compressors separately.

2.3.2.5.1 Controlling dry screw compressors

Control by variable speed

Because screw compressors displace the medium positively, the most advantageous volume control strategy is to vary the speed. This may be done by using variable

Figure 2.31: Large compressor designed for styrene off-gas service (source: *Aerzener Maschinenfabrik GmbH, Aerzen, Germany*).

speed electric motors, steam turbine drives, hydraulic or hydromechanical torque converters, and other means.

Speed may be reduced to about 50% of the maximum permissible speed. Induced flow volume and power transmitted through the coupling are thus reduced in roughly the same proportion. The allowable turndown depends on the adequacy of bearing lubrication at low speeds and on the compressor discharge temperature. More than 50% reduction is possible in special cases. As mentioned earlier, there is no surge limit (a minimum flow below which the gas would alternate between forward and reverse flow) for these positive displacement machines.

Bypass

Using this method, the surplus gas volume is made to flow back to the intake side by using compressor discharge pressure controller. An intermediate cooler brings the surplus gas volume down to intake temperature.

Full-load/idling-speed governor

As soon as a predetermined final pressure is attained, a pressure controller operates a diaphragm valve that opens a bypass between the discharge and suction sides of the compressor. When this occurs, the compressor idles until pressure in the system drops to a predetermined minimum value. The valve will close once again on receiving an impulse from a pressure sensor. This brings the compressor back to full load.

Suction throttle control

Suction throttling is suitable for air compressors only. As in the case of the full-load/idling-speed control method, a predetermined maximum pressure in the system, for example in a compressed air receiver, causes pressure on the discharge side to be relieved down to atmospheric pressure. Simultaneously, the suction side of the system is throttled down to about 0.15 bar absolute pressure. When pressure in the entire system has dropped to the permissible minimum value, full load is once again restored.

2.3.2.5.2 Controlling liquid-injected screw compressors

Suction throttle control

Since the final compression temperature is governed by the injected oil, a greater range of compression ratios can be accommodated by liquid-injected machines. This permits the main flow volume to be varied within wide limits.

Built-in volume governor

Large compressors are frequently equipped with an internal volume-regulating device, and Figure 2.32 illustrates the principle. A slide valve that is shaped to match

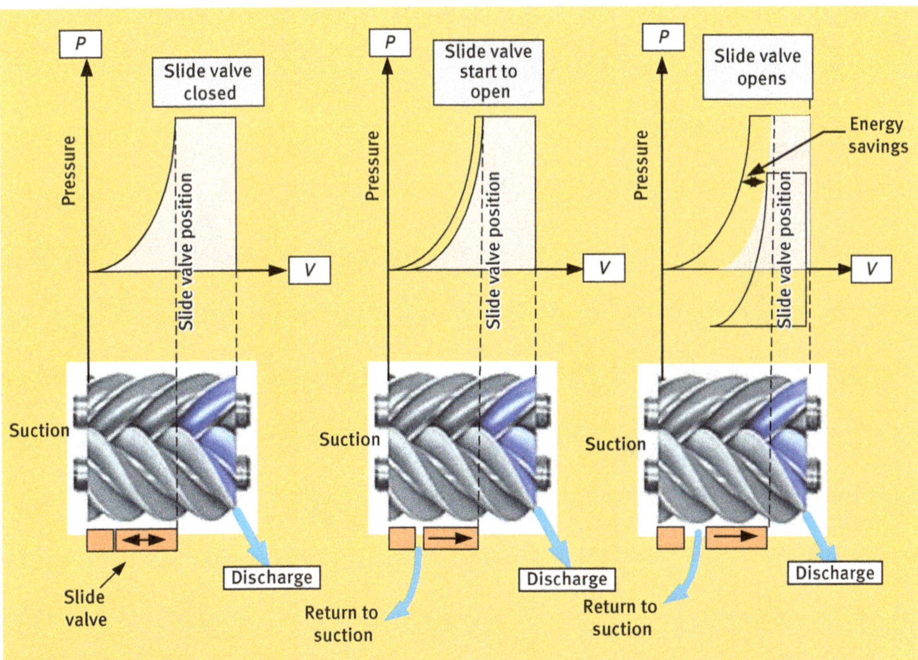

Figure 2.32: Slide valve operation.

the contours of the housing is built into the lower part of the housing. It is designed to move in a direction parallel to the rotors, whereby the effective length of the rotors can be shortened. The range of this control mode is typically between about 10% and 100%. Compared with suction throttling, this type of control offers more efficient operation.

2.3.2.6 Bearings

Although air machines are often equipped with rolling element bearings, most dry and wet process gas compressors are furnished with journal bearings and thrust bearings of the type commonly found in centrifugal process gas compressors. The service life of these bearings is practically unlimited, as long as proper lubricating and operating procedures are in effect. Rolling element bearings are acceptable for (relatively) light loads and where oil cleanliness is assured. Needless to say, selecting a separate closed-loop bearing oil circuit is a powerful step in the direction of maintaining a clean bearing environment.

If a vendor offers a wet screw machine with a single "same oil serves all" support system, the purchaser–owner may be taking a big risk – unless prepared to run up operating and maintenance costs by planning very frequent oil changes or major investment in oil purification equipment. If that is not in your plans, make sure you specify twin-screw compressors with separate oil circuits. In other words, do not buy wet screw machines where the oil introduced into the compression space also flows to the bearings. Separate circuits are used in the 5,200 kW coke gas compressors illustrated in Figures 2.33 and 2.34.

Figure 2.33: Rotary screw coke gas compressor train #1, around 1990 (source: *Aerzener Maschinenfabrik GmbH, Aerzen, Germany*).

Figure 2.34: Rotary screw coke gas compressor train #2, around 1990 (source: *Aerzener Maschinenfabrik GmbH, Aerzen, Germany*).

2.3.2.7 Seals

In many rotary screw compressor applications, it is best to provide a sealing barrier between the process gas and the bearings. Several different seal types are feasible (Figure 2.35): mechanical seals and carbon rings (Figure 2.35a), water as the sealing liquid (Figure 2.35b), and water in small quantities as the sealing liquid (Figure 2.35c).

At the compressor input shaft, manufacturers often opt for either labyrinth seals or double-acting mechanical seals with rotating springs. Carbon ring seals with connections for the injection and eduction of inert, clean gases are used in cases where leakage gas, even in connection with sealing gas, may enter into the bearing areas or into the atmosphere. The gas pressure is relieved across floating carbon rings at the edge of the seal chamber.

There are also barrier water floating ring seals and double-acting mechanical seals with stationary springs. On barrier water floating ring seals, barrier water enters the seal chamber and a small amount of water reaches the compression space. Most of the water is returned to the barrier water system for cooling, filtration, and reuse. Barrier water seals can fully prevent gas leakage and can provide valuable cooling and scrubbing duties. A double-acting stationary spring mechanical seal

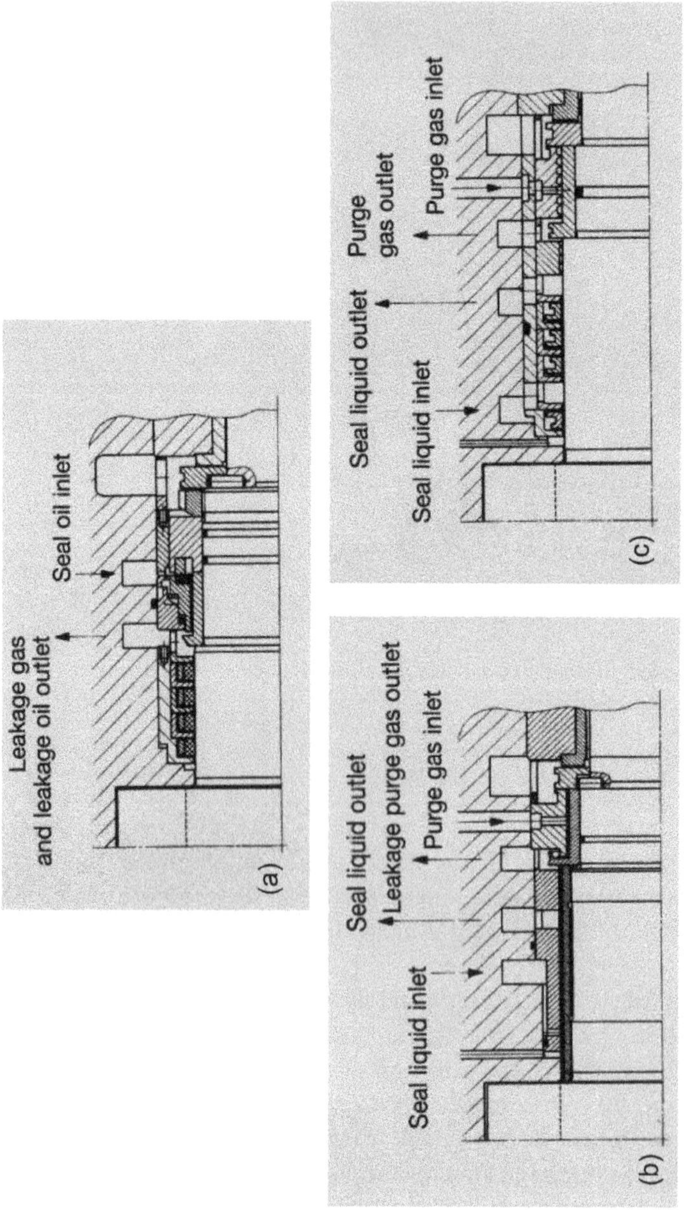

Figure 2.35: Three of many different sealing arrangements possible in screw compressors (source: *Aerzener Maschinenfabrik GmbH, Aerzen, Germany*).

and a combination mechanical and floating ring seal are primarily used for compression with high differential pressures.

Finally, the use of DGS technology that we alluded to in the earlier narrative opens up new frontiers. It is certainly worth examining the discussions between leading manufacturers of twin-screw rotary compressors and reliability-focused potential owner–purchasers.

2.3.2.8 Rotary sliding vane compressors

Rotary sliding vane compressors belong in the family of positive displacement rotary compressors. The rotating element is called the rotor; it is eccentrically located inside a cylinder (Figure 2.36).

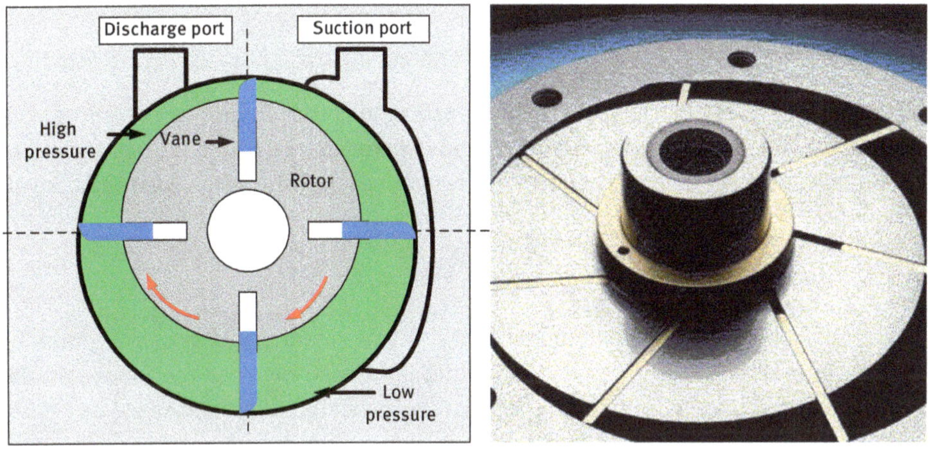

Figure 2.36: sliding vane compressor.

With vanes sliding in and out of a rotating element, the gas is trapped between successive vanes and the outer casing, and as the rotor turns, the space between vanes decreases from the inlet to the discharge sides of the compressor.

Rotary sliding vane compressors (Figures 2.36 and 2.37) can handle many different gas compositions, including sour gas, hydrogen, and heavy hydrocarbons. A key advantage of sliding vane compressors is that they can handle gases that are wet at their saturated vapor point. Single-stage machines can manage compressing ratios as high as 7:1 with proven operating discharge pressure up to 200 psi (13.6 bar). Inlet pressure can be as low as 28 in-Hg vacuum (5 kPa absolute). Flows range from 3 to 3,500 MSCFD (3.5 to 3,850 Nm^3/h).

Probably the key disadvantage of rotary sliding vane compressors is their relatively high maintenance frequency. Process plants accustomed to years of uninterrupted run

Figure 2.37: Sliding vane compressor (source: *Ro-Flo Compressors*).

lengths in their fluid machines may have expectations that sliding vane compressors may not be able to fully meet. Examine field experience to gain an understanding of where this is the "just-right machine" and where misapplying these compressors will disappoint.

2.3.2.9 Rotary lobed compressors

The rotary lobed compressor (Figure 2.38) has two rotating lobes with peripheries moving in opposing directions. Working fluid is trapped between the lobes and the outer casing as it moves from inlet to discharge. The lobes are synchronized by timing

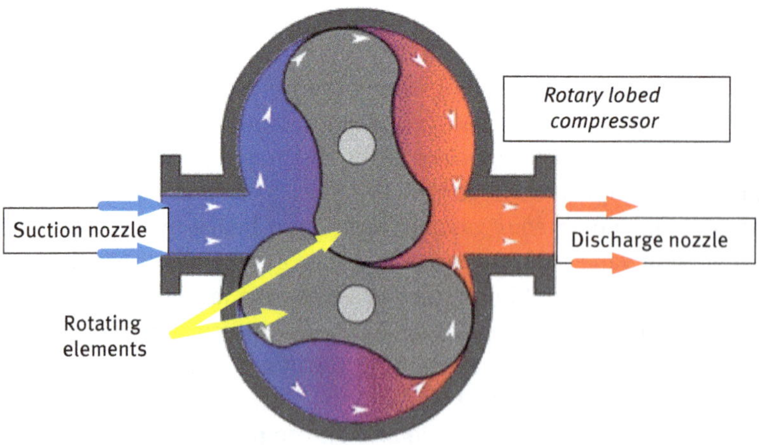

Figure 2.38: Rotary lobed compressor.

gears installed at the ends of the shaft, and therefore do not actually touch. There is some leakage from the high-pressure to low-pressure cavities, although the leakage rate is generally considered small.

In the process industry, these machines are typically known as "roots blowers" or "vacuum boosters." By widely accepted definition, blowers deliver discharge pressures above the atmospheric pressure. For vacuum boosters, both suction and discharge pressures are below the atmospheric pressure.

Rotary lobe compressor are oil-free machines, and most are built to comply with API standard 619. They can handle corrosive, flammable, and toxic gases for most process applications. These machines are designed to operate at pressures from high vacuum to 25 bar (360 psi), and flows ranging in capacity from 30 to 85,000 m^3/h (18 to 50,000 cfm).

2.3.2.10 Liquid ring/piston compressors

Liquid ring compressors (Figure 2.39) use the centrifugal whirling of a sealing liquid to create a series of piston-like water columns to trap gas or vapor between the impeller cells of a rotor installed eccentric to its casing. During operation, the trapped space occupied by the gas or vapor in each impeller cell progressively decreases around the rotor. By sealing the liquid being flung outward from the center of rotation, the volume of the gas is progressively reduced. The pressure thus increases and the gas finally leaves the discharge port at some elevated pressure.

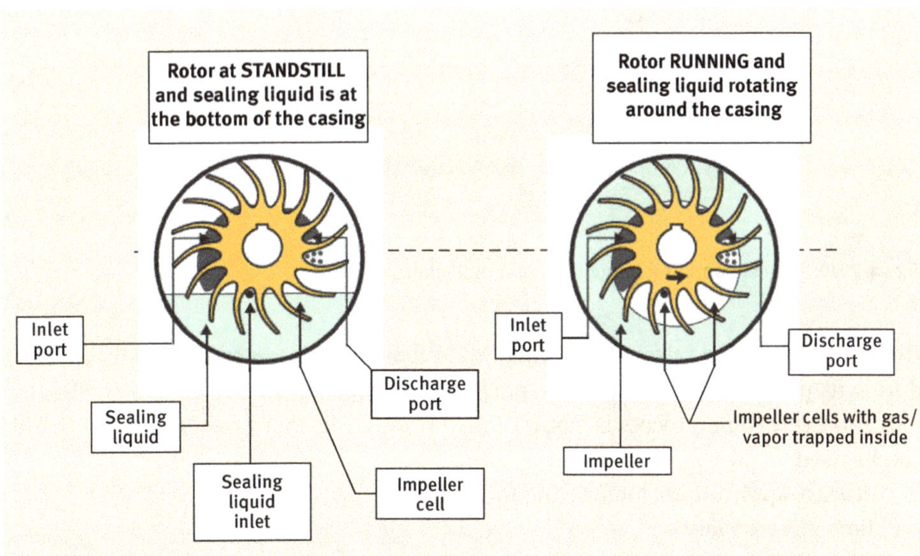

Figure 2.39: Schematic illustrating operating principle.

As the gas discharges, entrained vaporized liquid (originating from the sealing liquid) is discharged along with the gas. However, this loss of sealing liquid is continuously made up by a fresh flow of sealing liquid.

The overwhelming majority of these machines are in the range of 15–150 kW, although larger machines have been known to exist. Their traditional application range is shown in Figure 2.40.

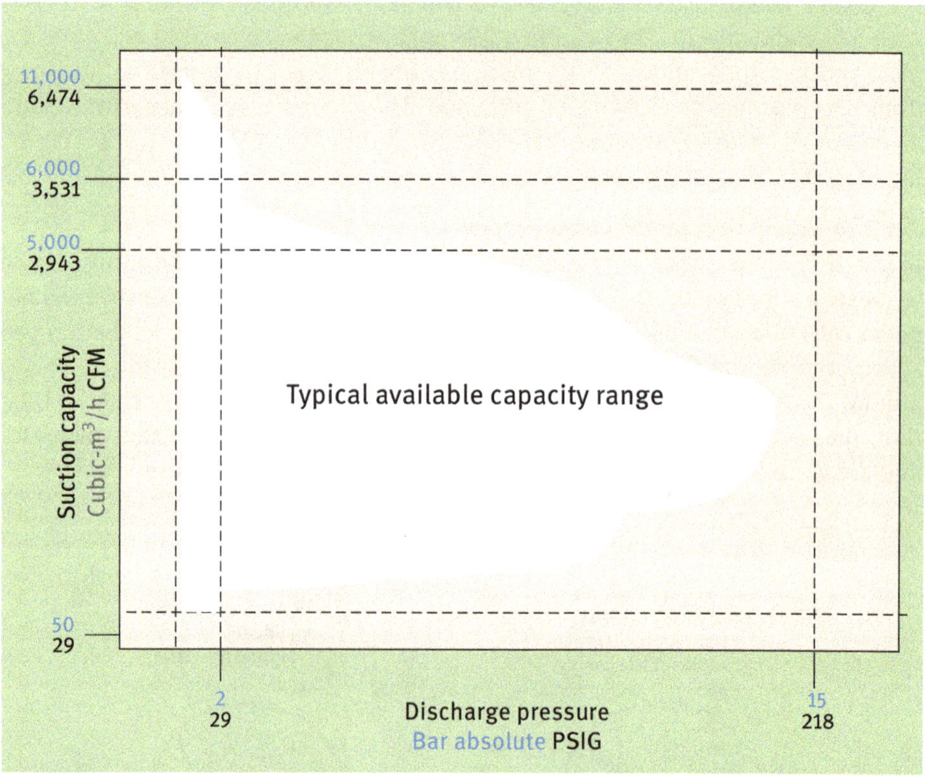

Figure 2.40: Typical operating range for liquid ring compressors.

Heat of compression raises the sealing liquid temperature to approximately 10–15 °F during its passage from entry to discharge. Water is generally used as the sealing liquid; however, some processes cannot tolerate water. In that case, substitute liquids can be used.

Process applications include the following:
- flare gas recovery
- flare gas compression
- vinyl chloride monomer recovery
- vent gas recovery

- vapor recovery
- hydrogen compression
- oxygen/ozone compression
- methane compression in wastewater treatment
- carbon dioxide compression
- waste gas compression, power plants

References

[1] Bloch, H.P. A Practical Guide to Compressor Technology, 2nd Edition, John Wiley & Sons, ISBN 0-471-727930-8, See also 1st Spanish Edition, (1998) McGraw-Hill, New York and Mexico City, ISBN 970-10-1825-7]. McGraw-Hill, New York, NY (Copyright 1996) [Also Spanish Ed., 2006].

[2] Bloch, H.P., Noack, P. Recent Experience with Large Liquid-Injected Rotary Screw Process Gas Compressors, Proceedings of 20th International Turbomachinery Symposium, Texas A & M University, Dallas, TX, 1991.

[3] Bloch, H.P., Soares, C. Process Plant Machinery, 2nd Edition, Elsevier Publishing, London-New York-Amsterdam, 1998, ISBN 0-7506-7081-9.

Chapter 3
Technical briefs on dynamic compressor technology

3.1 Axial compressors

3.1.1 Optimization of flow path; optimized blade profiles; comments on application requirements

Industrial axial compressors have been built since the early twentieth century. The main applications for industrial axial compressors are in large volume flow services, for example, blast furnace blowers in steel works, fluid catalytic cracking processes in refineries, nitric acid production processes, liquefied natural gas plants, and large air separation units (ASUs).

The different applications require a variety of customized solutions for the market; these include pure axial compressors without intercooling, and axial–radial compressors with intercooling for higher overall pressure ratios, and reduced specific power consumption. Intercooled compressors are mainly used in air separation units [1]. The air separation market for large train sizes is continuously growing, driven by future-oriented applications such as gas to liquid or coal to chemicals.

The growing market demands for large volume flow capacities and high-pressure ratios in the last 20 years led to the design and development of more compact and lighter industrial axial compressors. In principle, axials and other modern industrial compressors are the result of combining established aeroengine know-how with experiences derived from industrial compressors. In 2011, this compressor style was first introduced in a medium-sized ASU application in China. It was followed by the world's largest ASU train in 2014 with a flow capacity of approximately 1 Mio. m^3/h. This axial compressor features an inlet diameter of 3.2 m and a power consumption of approximately 65 MW. Figure 3.1 is a top view of this modern industrial compressor in an axial–radial configuration.

Industrial compressors are designed within a flexible modular system, providing individual solutions for a large range of flows and pressure ratios in many different applications. As of 2020, they cover flows ranging from 50,000 m^3/h to approximately 1.4 Mio. m^3/h. Available pressure ratios range from 2 to 10 for pure axial compressors and up to 21 for axial–radial-type compressors. The main driver for these high-volume rates and high pressures are the air separation, blast furnace, and FCC (Fluid Catalytic Cracking) industrial processes. Besides many other compressors, the world's largest blast furnace blower features a capacity of 730,000 m^3/h and a design pressure ratio of approximately 6 in a purely axial 10 stage compressor (see Figure 3.2).

https://doi.org/10.1515/9783110678765-003

Figure 3.1: Top view of a modern axial–radial compressor *(AR-MAX1 © of MAN Energy Solutions)*.

Figure 3.2: Top view of a modern pure axial compressor *(AG-MAX1 © of MAN Energy Solutions)*.

The maximum pressure ratio, and therefore maximum number of axial or axial–radial stages, is dependent on the overall shape and size of the rotor shaft. It is limited by rotordynamic effects. Generally, available compression ratios decline as compressor sizes increase. Nevertheless, the overall application ranges have been significantly increased with advances in blade design (Figure 3.3).

Both developmental progression and the design of individual elements leading to modern axial machines are of interest to the reader. The full concept is best described as a modular design with an emphasis on adaptability to the customer's process requirements. The implied flexibility of such designs requires forethought; clearly, the

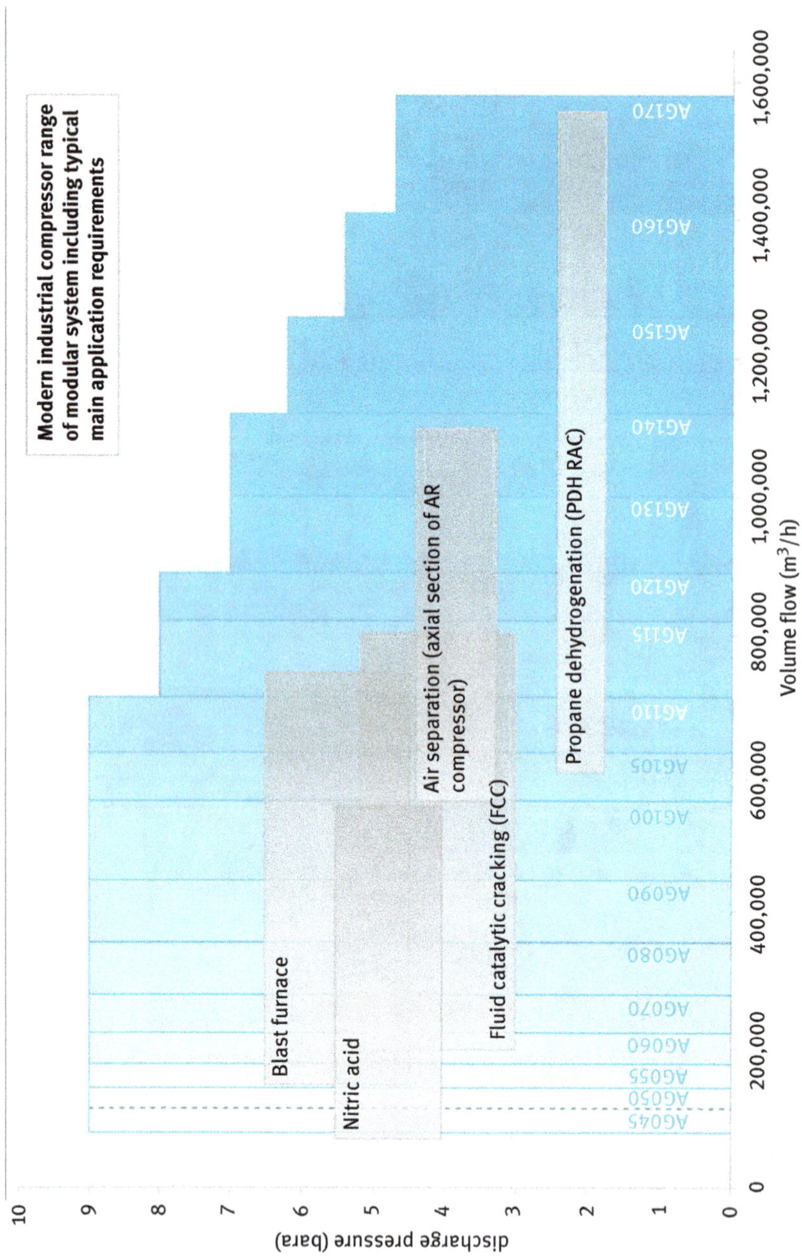

Figure 3.3: application range diagram of modern axial compressors (source: *MAN-Energy Solutions, Germany*).

importance of process and location-related factors such as altitude and ambient temperature is evident. A good design cannot overlook the fact that the user's processes being served will likely see a measure of optimization. Accordingly, an axial compressor's performance must be adaptable to reasonable shifts in pressure and/or flow requirements.

The compressor manufacturer needs to be flexible regarding these kinds of changes. In most cases, only relatively small adaptations are necessary. Experienced manufacturers are capable of transforming their standard designs into smooth-running, perfectly fitting machines. They can accommodate reasonably short delivery times without infringing on the safety and reliability criteria stipulated by world-scale user plants.

The first step requires defining frame sizes that cover the overall flow range required by the user–purchaser (see Figure 3.4). Operational range and flexibility will be provided by selecting blade height and the means selected to rotate the blades. Simply put, blade rotation changes the flow's angle of incidence. This angle affects flow range and power efficiency. In a second step, the design needs to accommodate the pressure requirements. Select (actually "preselecting" and later verifying) the overall stage count; stage count will also affect the total cost of the compressor. Further optimization can be obtained by varying the rotational speed and/or making tip diameter adjustments if radial stages are used in an axial–radial compressor.

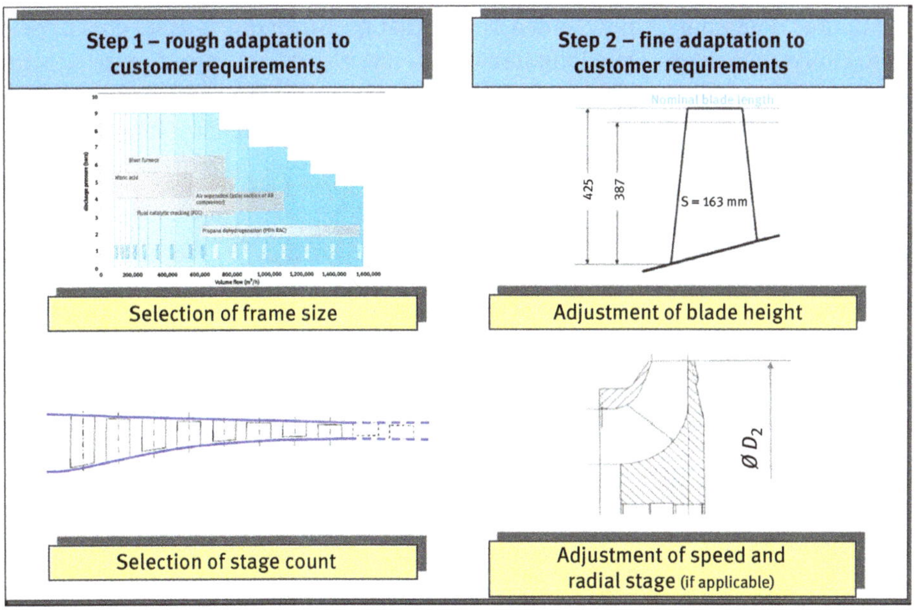

Figure 3.4: Modular system adaptation procedures (source: *MAN-Energy Solutions, Germany*).

Apart from these two main design steps, additional, considerably more detailed requirements comprise the basic performance-oriented goals. Like virtually all process compressors, axial machines require an overall operating range. The flow can vary 10% or more due to changing inlet conditions in a hot summer and/or a cold winter. Start-up conditions often require low flow, but low flow can be accompanied by unacceptable surge. Provision must be made to never allow operation in surge; a safe nonsurge zone must be established.

Maximum flow operation corresponds to maximum production capacity, so that the customer for business reasons wants to design this feasible operating range as large as possible. Considering all relevant effects might add up to a desired operating range of 30–50% flow at near-constant pressure. This is the main reason why a gas turbine axial compressor is not suitable for industrial applications. The Mach number level is so high that it leads to a narrow performance map and poor operating range. Aside from that, gas turbine axial compressors are designed only for operation on the working line. Close to choke and surge, flutter and high vibratory forces may occur, causing efficiency to drop sharply. The blades are thin with poor resistance to massive surge. In other words, gas turbine axial compressors are designed for small surge volumes since backflow volumes are defined by the relatively small combustor volume.

An industrial compressor is expected to operate continuously for a minimum of 5 years without interruption or repairs. Maintenance, if needed, must be feasible without shutting down. Therefore, redundant oil filters, speed governors, and on-stream lube oil purifications are among the built-in measures that contribute to long uptime. Reliability, safety, and a design that has ruggedness as its goal impart the so-called robustness. On a long-term basis, these goals are valued on a par with high overall efficiency (see Figure 3.5). This reasoning makes sense because once the compressor fails, the complete plant shuts down and production stops.

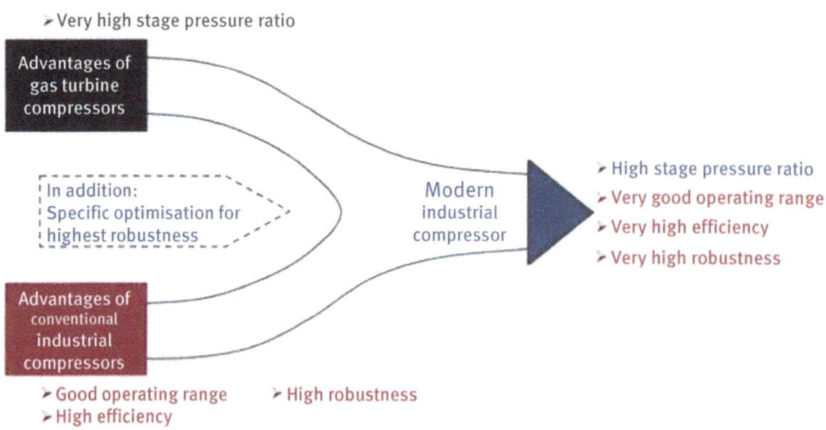

Figure 3.5: Hybrid concept of the modern axial compressor (source: *MAN-Energy Solutions, Germany*).

Depending on the size of the plant, forced shutdowns can be enormously costly for the end user and stakeholders. Unable to fill their contractual obligation to customers “A through G” with many thousand tons of products per day, the unfortunate user company will pay its opportunistic competition’s inflated price rather than to cough up the gazillion dollars in damages the jury awards to legions of profit-deprived customers.

All things considered, designing or optimizing industrial compressors today requires finding the best combination of several turbomachine designs. Gas turbine compressor technology incorporates high-stage pressure ratios leading to a reduced stage count and hence to a shorter compressor. The primary benefits of conventional industrial compressors are centered on wide operating ranges and highest efficiencies. This concept is referred to as a hybrid concept [2]. The compressor rotor needs to fulfill the rotordynamic API stability criteria and the blading has to provide a very high surge robustness [3].

A last design highlight of an axial compressor relates to the train concept. Instead of only a compressor, user–purchasers receive an entire turbomachinery train. Compressor and driver are positioned in-line and designed to fit each other. Some applications may require three or four in-line turbomachines, which increases the complexity of machine interactions. Figure 3.6 shows a four-body machinery train for a nitric acid process. Because the train incorporates a modern axial compressor, the customary gearbox was omitted; the entire assembly is now shorter and less complex.

Figure 3.6: Turbotrain for nitric acid process with “steam turbine, modern axial compressor, radial compressor, and axial expander” (source: *MAN-Energy Solutions*).

With few exceptions, the driver becomes the determining design criterion for an axial compressor. Electric motor drives favor compressors operating with fixed speed and variable guide vanes. Mechanical drive steam turbines, however, offer flexible speeds and thus control flow by varying turbine speed. A combination of vane and speed control will always increase the efficiency at part load and overload conditions. It deserves to be favored by sites with high power cost.

3.1.1.1 Design procedure

Designing a modern, fully optimized industrial compressor requires a multidisciplinary effort. Already during the conceptual phase and starting with a clean sheet, optimization iterations will be needed. In this context, optimization implies that rotordynamics, blade mechanics, aerodynamics, and power density all converge at the chosen design point on a performance map.

The flow path design results in a typical rotor shape. A conventional industrial axial compressor has low-stage pressure ratios resulting in a very long shaft. In comparison, a typical gas turbine compressor has very high-stage pressure ratios. Its rotor core (or hub) contour increases considerably from the front stages to the rear stages so as to obtain maximum circumferential velocity. High flow velocities are needed to achieve high-stage pressure ratios at the compressor outlet without sacrificing performance. Modern axial compressor designs represent a melding of certain elements of gas turbine and industrial compressor technologies. The resulting design has been termed as hybrid approach for a good reason.

From a practical view, the demanding operational range and off-design efficiency requirements of industrial compressor applications cannot be fulfilled with a high aerodynamic loading. Academic researchers deal with de-Haller numbers (low is good) and Lieblein diffusion factors (high is good). Both are well outside the scope of our book but are known as experts in flow path optimization. For optimization, the aerodynamic loading (diffusion) of the blading may only be slightly increased compared to a conventional industrial compressor (Figure 3.7). Thus, instead of increasing the head coefficient, the circumferential velocity can be increased to gain high-stage pressure ratios at nearly equal aerodynamic loading. This leads to a more compact, shorter compressor. The phi number must be lowered to compensate for the increase of Mach-number with higher speed. As a result, fewer stages are needed for a given pressure ratio in modern axials compared to conventional industrial compressors. To restate, maximizing circumferential velocity increases rotordynamic stability and allows higher overall pressure ratios.

Both highly loaded gas turbine compressors and the even more highly loaded aeroengine compressors make use of the earlier approaches. They thereby achieve, for more ruggedly designed industrial machines, highest pressure ratios, circumferential speeds, and high local diffusion factors.

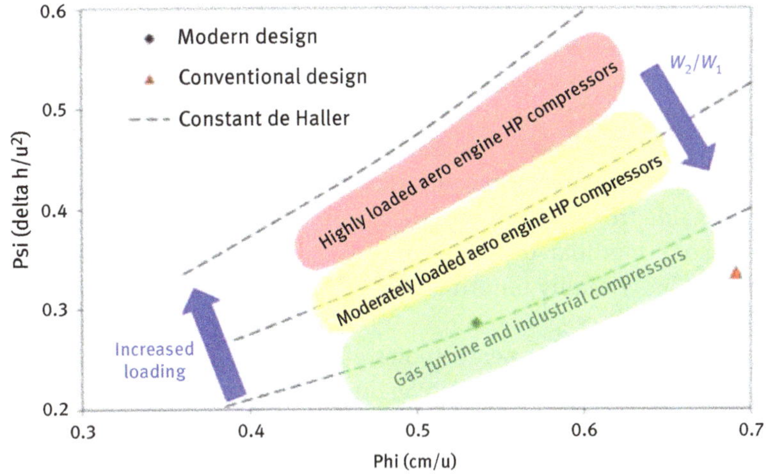

Figure 3.7: Comparison of different compressor design philosophies in the φ–ψ diagram [4] (source: *MAN-Energy Solutions, Germany*).

3.1.1.1.1 Flow path optimization

With owner–purchasers demanding a wide operating range, blade optimization is a first-order priority. Optimization must not be limited to a certain design point; it must extend or even cover the complete speed and vane control performance map. To define the overall parameters of the new blading, a sophisticated streamline curvature code is considered useful. The code should not be applied to calculating the design point but must include prediction of the various performance maps. The two modes speed control and guide vane control must be included on the map or maps. Individual stage pressure ratios are thus defined as the number of stages needed to achieve the overall pressure ratio. Overall pressure ratio and required operating range are specified by the user–purchaser as part of the initial inquiry.

Targeted operating range is also defined as any location between minimum and maximum flows at constant pressure ratio. The compressor manufacturer calculates the current requirements and anticipates (or estimates) the user's future demands. Best-in-class companies involve machinery quality assessment (MQA) engineers in specification, review, clarification, and assessing the imputed value of prepurchasing future upgrade possibilities. The concept is easily explained by the terms "blind stage" or "dummy stage." Suppose we received a bid for a machine priced at 1*X* dollars producing the full 100% of specified pressure and flow. But suppose it could not, in the future, be uprated to accommodate 110% of today's flow.

If for an amount of 1.01*X* dollars we could purchase a compressor with future uprate capability to 110%, we should do so. However, if such a machine were offered to us at 2*X* dollars, the incremental cost would be judged excessive and we would decline the offer. At best-in-class companies, an MQA engineer or similarly experienced subject matter expert advises managers on the merits of such options.

The operating range of an axial compressor should be optimum for both guide vane and speed-controlled applications. A measure of design flexibility is achieved by using identical stages in the discharge portion of the machine. This concept enables the adaptation of the requested pressure ratio by simply adding or omitting stages. Aside from frame sizes, which are simply scaled up or down, a modern compressor's modular design approach should allow a streamlined cut of the blading for individually tailoring blades to produce the specified flow volumes.

The concept phase starts with general studies of the overall flow-path configuration (Figure 3.8) that defines hub and tip contours of the blading. Compared to gas turbine or aeroengine compressors, more moderate stress loads and circumferential velocities are chosen; all must be safely below the limitations imposed by mechanical integrity, stage pressure ratio, and maximum inlet tip Mach numbers. The outlet Mach number of the blading is a compromise between a blade designed for high efficiency and losses incurred by its adjacent diffuser. The number of VGV-controlled stages and the VGV (variable Guide Vane)-sequencing schedule depend on stage loading; both can be defined based on experience with a large number of industrial compressor applications [5].

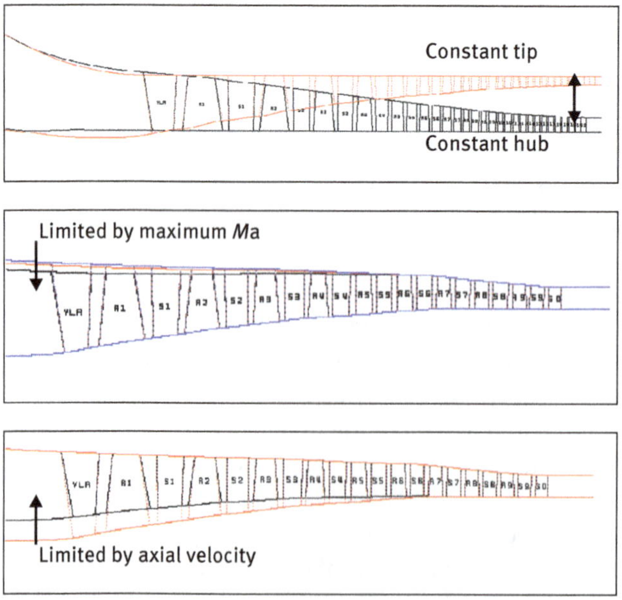

Figure 3.8: Flow-path variations and limiting parameters (source: *MAN-Energy Solutions, Germany*).

In addition to the prevailing aerodynamic and mechanical stresses, both hub contour and speed are subject to constraints imposed by rotordynamics. As is often the case, virtually all of the many different requirements are influenced by the specifics of a service or application. As before, the maximum possible pressure ratio for blading may not be an optimum. Therefore, conceptual studies of a number of different feasible compressor configurations and their effects on rotordynamics must be performed and studied by the designers. In most cases, compromises are needed between ideal aerodynamics, highest possible blade strength, unparalleled rotordynamics, and the ultimate theoretically achievable power density.

Blade stresses change throughout the speed range, and vibratory resonance conditions must be avoided. Mere blade mass would be an insufficient means of safeguarding the goals of efficiency and other engineering decisions. However, selecting a wide chord for all stages and combining it with generous blade thickness at the blade and vane roots are foremost among the steps taken for enhanced ruggedness.

3.1.1.1.2 Detailed stage optimization

Following the general design of the overall flow path, the individual stage data distributions must be optimized. During this design phase, stage and span-related swirl distributions are determined (Figure 3.9). These parameters mainly influence the deflection angles and Mach numbers of different blade sections. Staging and swirl distributions strongly influence loading; they therefore determine operating range and resulting efficiency at design and off-design conditions. Several different distributions for the degree of reaction in the front and rear sections should be evaluated in order to determine the specific stage and span-related swirl distributions. Blade numbers and chord distributions along the span can be chosen on the basis of the diffusion factors stipulated by Lieblein to provide an optimized operating range [6].

Extraordinary efforts should be devoted to detailed stage parameter variations. Seemingly small variations are helpful in designing blading with the limited Mach numbers often found in typical industrial compressors. Finding the optimal flow path design and stage data requires many iterations and parameter variations. In short, the design engineer needs time and patience, perhaps even a desire to at least *aim* for perfection, although perfection will elude even the best of us.

3.1.1.1.3 Multiobjective aerodynamic design stage behavior

Once we have found a close enough solution to near-optimum flow path and stage pressure distributions, the final three-dimensional design of the compressor blading will be initiated. At this point in time, blade geometry is defined by blade sections. To confirm that the desired pressure rise and work distributions for the design point are reached, a succession of computational fluid dynamics (CFD)

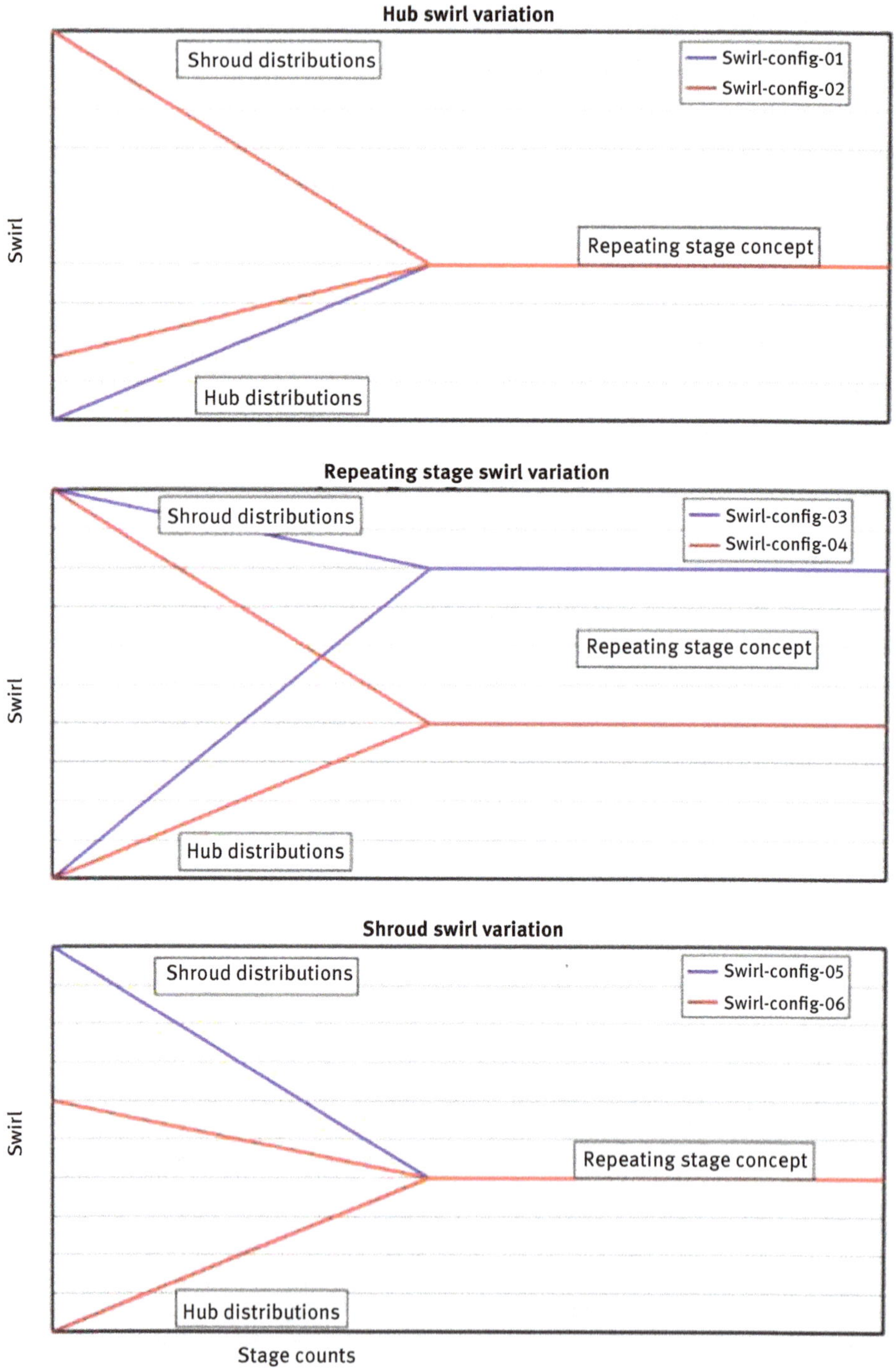

Figure 3.9: Examples of swirl distribution investigations (source: *MAN-Energy Solutions, Germany*).

calculations is used to verify the profiling process. Concurrently, the engineers involved conduct and reconfirm mechanical strength calculations. Static- and vibration-induced blade stresses are being evaluated to ascertain structural integrity. Here, special attention must be given to avoiding critical blade resonances, including, but not limited to, torsional and lateral critical modes. In the event that a non-sensitive design cannot be found, harmonic response analyses must be conducted for resonance node locations [3].

Up to this point neither the design process nor overall approach varies much from common design techniques. However, a multiobjective optimization approach makes the process and its progression unique. Using the initial results obtained for various blade configurations, their off-design behavior for minimum and maximum flow positions in the speed and guide vane-controlled compressor maps need to be analyzed. The analyses are based on CFD calculation (Figure 3.10). Knowledge gained about the different responses of profile sections as well as stage matching at extreme flow conditions can be used to optimize the operating range and off-design efficiency of the compressor.

The diagram on the left of Figure 3.10 shows an example of the stage ψ distribution (PSI; Greek letter Psi) and diffusion factors reported by Lieblein for rotor blade rows. The plots identify four operating points in the speed control map; they vary from minimum to maximum flow at constant pressure ratio. The right side of the plot shows the same diagram for guide vane control operation. At speed control, all stages undergo a near-linear shift to higher ψ values with reduced speed. The same response can be observed for the diffusion factors. As expected, the first blade row (rotor 1) will define the surge margin.

However, at guide vane control operation a completely different stage behavior can be observed. At maximum open VGV position, the highest ψ numbers are found in the front stages and reduced values are now in the rear stages. At minimum VGV position, the first stage is completely unloaded due to the closing of the vanes, whereas the last stages experience the greatest pressure rise. In terms of diffusion factors, the rear stages limit the operating range.

In addition to investigating the overall stage characteristics, variations along the blade span need to be checked. An example for such an analysis for "rotor 1" blade is presented in Figure 3.11. At speed control operation, only slight variations of rotor pressure ratio occur at midspan; stronger and inverse variations exist in the hub and tip regions. Diffusion factors generally rise with lower speed and reach their maximum values in the tip region while operating close to the surge line. The pressure ratio along the span at guide vane control operation shows its greatest deviations in the hub region, whereas lower deviations are noted near the tips. As a result, the tip region for guide vanes (particularly while in speed control mode) suffers under an increase in diffusion factors. This region must receive thorough consideration; the effort is aimed at a design that incorporates a sufficient margin of safety.

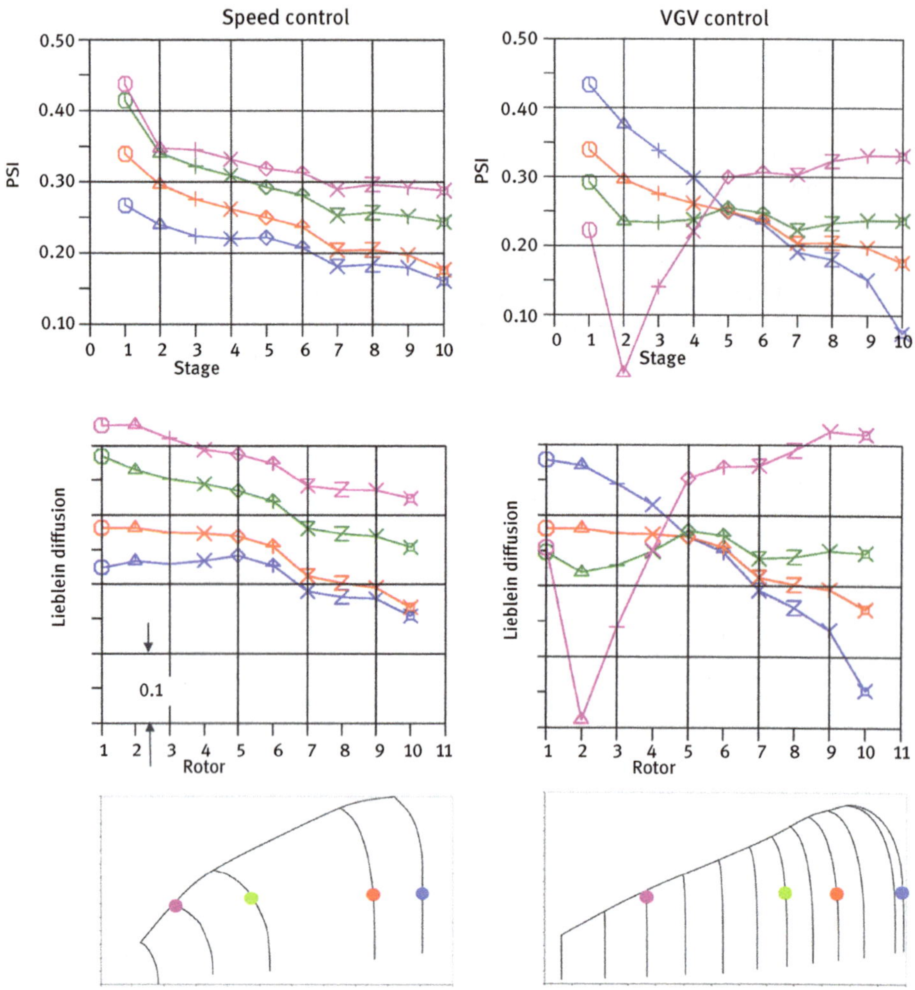

Figure 3.10: Variation of stage pressure ratios between minimum and maximum flow for speed and VGV control (source: *MAN-Energy Solutions, Germany*).

With this knowledge, reasonable compromises in stage performance can be entertained. The load distributions of the various stages can be investigated and "tweaked" to optimize the overall loading for both control approaches. However, minor modifications of the flow path geometry may still be necessary at this phase of the design process.

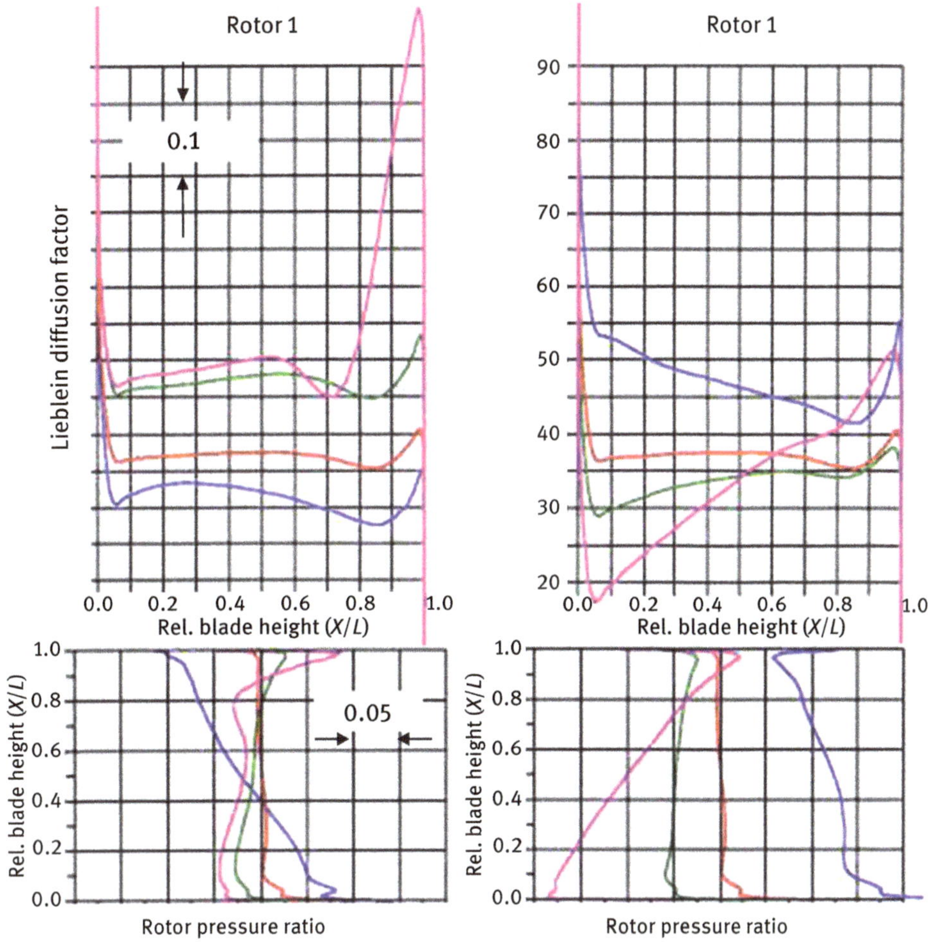

Figure 3.11: Spanwise distributions of a rotor 1 pressure ratio and diffusion factor (same control points as in Figure 3.10) (source: *MAN-Energy Solutions, Germany*).

3.1.1.1.4 Profile section behavior

The next major step consists of an individualized shaping of the various profile sections performed in an iterative manner. During this profiling process the characteristic behavior of a profile section is individually tailored to the specified operating range. Optimal behavior concerning the reduction of losses and incidence range of each profile section is thus achieved. Individual shaping is constrained by the necessity of robustness to improve structure mechanics. During this phase of the design process, numerous CFD analyses are conducted for several control points between minimum and maximum flow. The outcome is a tailored shape of profile surface Mach number distributions for the individual profile sections.

An example for the individual tailoring of the profile sections for a rotor blade is given in Figure 3.12. The shape of the profile sections shows a substantial increase in thickness in the hub region, which results from efforts to increase mechanical strength (ruggedness or "robustness") of the blading. Inlet and outlet radii were mostly designed with ruggedness (or strength) in mind. Thus, in contrast to conventional designs, the maximum profile Mach number in the hub region exceeds the Mach number in the tip section. Tapering of the blades can be used to widen the operating range and thus reduce diffusion factors in the tip region. Compared to the tip region, the Mach number distribution in the hub section will differ greatly from the distribution in the tip section. This demonstrates the individual tailoring of the various profile sections. Such individual tailoring adapts their incidence range to the requirements needed to accommodate the wide spread between the compressor's minimum and maximum flows.

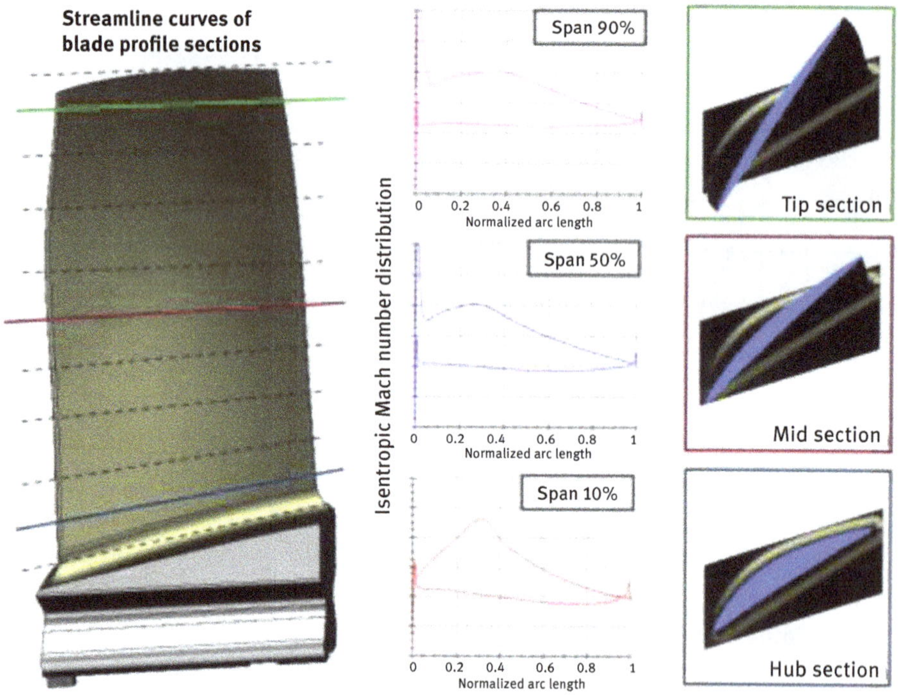

Figure 3.12: Example for the individual tailoring of the different profile sections (source: *MAN-Energy Solutions, Germany*).

Special care is needed to acknowledge issues concerning the tip Mach number of a rotor 1 blade. That number must be low enough to achieve a sufficient margin in the event of exposure to regions where shocks and subsequent higher profile losses occur. Thus, efficiency can be kept high over a wide flow range. For the actual example in Figure 3.13, the Mach number variation for the tip region at constant pressure shows a

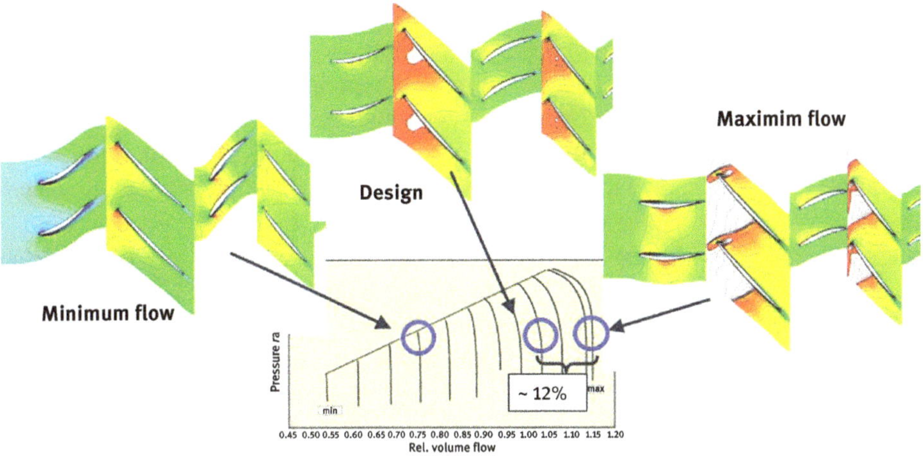

Figure 3.13: Mach number variation of a rotor 1 blade tip region for minimum design and maximum flow (source: *MAN-Energy Solutions, Germany*).

margin of close to 12% of flow until shocks impinge at the tip region of this rotor 1 blade. The shock configuration shown on the right at maximum flow is similar to what one would expect for the design point in a standard gas turbine compressor service.

3.1.1.1.5 Overall performance and testing

In a final step, the overall expected performance is to be verified. Numerical uncertainties can be reduced by using different CFD codes for the final performance verification. Inlet and exhaust configurations can be investigated in a 360° model. Inlet losses strongly influence overall pressure rise. Depending on the overall Mach number level in the rear stages, the pressure rise (and losses in the diffuser and exhaust casing) requires detailed investigation as well. Always keep in mind that the industrial compressor performance is measured flange to flange over the entire performance map. This includes operation close to surge and in choke conditions.

After all the design calculations and analyses have been completed, an experimental verification is still recommended. There are many effects that even today cannot be perfectly modeled by CFD codes. For instance, there are VGV cavities, internal leakages, radial clearances of variable guide vanes, or other real geometric effects. The overall numerical diffusion or the numerical accounting for flow past the rotor–stator interface differ from code to code. These are just two examples that will influence the overall performance and stage matching of an axial compressor. The efficiency will therefore depend on the numerical setup and needs to be calibrated by a true-to-scale and real geometry flange-to-flange test.

The most prominent uncertainty regarding the operating range of dynamic compressors is their surge line. Surge lines are determined by every other known uncertainty and can still not be predicted correctly today. The measured surge line can shift by up to 10% compared to CFD predictions. In some instances, start-up points are simply too close to actual surge, and experimental in-place operation to detect surge is not advisable. Compressors with high compression ratios will require start-up bleed (bypassing) to safely start in a pure speed-controlled operation.

Ideally, the exact shift of the surge line and the interaction with upstream and downstream blades need to be verified. Is the compressor only controlled by variable guide vanes the minimum position for start-up and the corresponding power consumption needs to be tested as well? In many of these cases, unstable operating conditions, such as rotating stall, can occur close to the surge line. They can seriously degrade and even totally wreck the mechanical integrity of the blading. Depending on the row, the end result may be full or partial "Schaufelsalat" – a neatly descriptive tongue-in-cheek German word. The closest translation "mixed blade salad" is not even close to fully doing it justice.

The topic of blade ruggedness can ultimately be answered only by a final test, and most compressor manufacturers echo these sentiments. Nevertheless, there are good approximations that we have found of great value. We call them simplified approaches.

3.1.2 Simplified approaches to evaluating mechanical integrity of axial blading

3.1.2.1 Methods to validate blade resonant frequencies

The fundamental work by Armstrong and Stevenson [7] provides a good overview of the different types of rotor blade vibration phenomena and their specific sources and influence parameters. Traupel [8], who received his doctorate from the acclaimed Eidgenoessische Technische Hochschule in Zürich/Switzerland in 1942, published the standard methods of calculating resonant stresses for the fundamental vibration modes of rotor blades.

Various types of vibration modes must be taken into account when designing modern axial compressor blading as in the present case. These types of vibration modes include the fundamental bending and torsional modes (e.g., modes 1F and 1T) and also higher modes such as chord-related modes with and without radial nodes (e.g., modes 1C and 1C1R). For the prediction of resonant stresses manifesting themselves in the fundamental vibration modes, the well-established stimulus approach can be used. It is based on one of Traupel's equation [8]:

$$\sigma_{\mathrm{a}}(\mathrm{res.}) = \kappa \cdot \sigma_{\mathrm{Gb}}\,(\mathrm{n}_{\mathrm{res.}})$$

where κ is a function of excitation source, excitation order, damping (e.g. due to oil film thickness or related properties), material, and Mach number.

The proportionality factor κ relates the resonant stresses σ_{a} to the gas bending stresses σ_{Gb} at the speed $n_{\mathrm{res.}}$ of the specific resonance (crossings in the Campbell diagram). Traupel's method distinguishes between excitations from struts or vane wakes, and flow disturbances from – for instance – radial intakes. Furthermore,

Traupel's expressions take into account the excitation order, blade geometry, blade material, and flow conditions, respectively.

Today, the finite element analysis (FEA) method provides a better means to predict blade vibration modes. A modal analysis is a *linear* dynamic analysis [9]; it determines the vibration characteristic (natural frequencies and mode shapes) of a structure or machine component. Natural frequencies and mode shapes are important parameters in the design of a structure for dynamic loading conditions. One can also perform a modal analysis on a prestressed structure, such as a spinning compressor blade (see Figure 3.14).

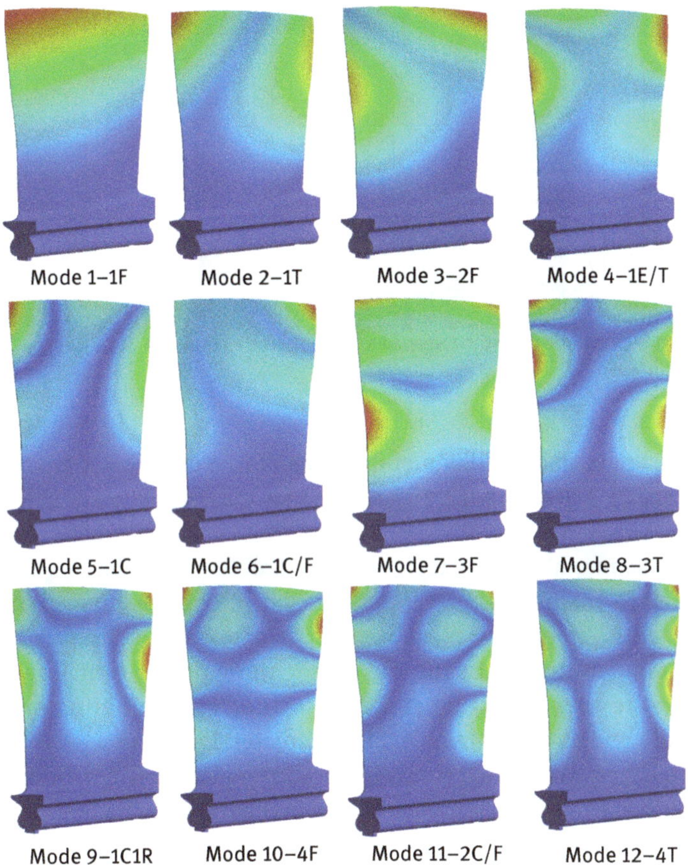

Figure 3.14: Overview of different vibration modes of a compressor front stage rotor blade (source: *MAN-Energy Solutions, Germany*).

A prestressed modal analysis requires performing a static structural analysis first (see Figure 3.15). For the static design of a compressor rotor blade, a *nonlinear* FEA [9] must be performed for the blade airfoil and root sections and also the corresponding

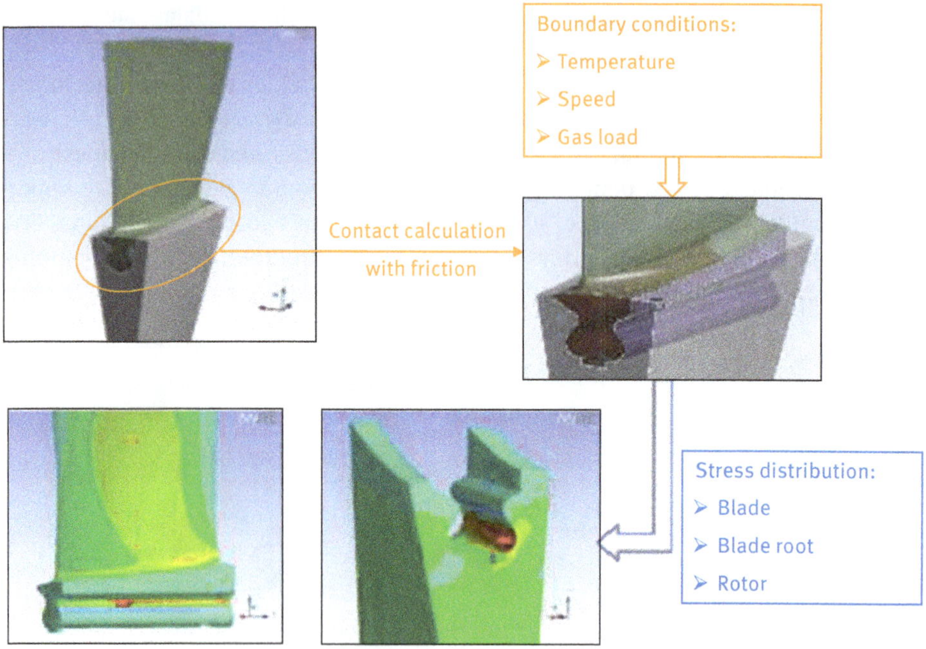

Figure 3.15: FEA scheme for the calculation of static stresses in rotor blades (source: *MAN-Energy Solutions, Germany*).

groove in the rotor. Here, the contact condition of the dovetail root in the rotor groove is of great importance. Figure 3.15 shows the FEA scheme for the calculation of static stresses in rotor blades.

Due to the unavoidable manufacturing tolerances of compressor blades, a quality verification is needed to ensure that blade-to-blade variations are minimized. Conventionally, only "cold" frequency measurements, for example, frequency measurement at room temperature and without static loading due to centrifugal forces, are made. Shaker testing is another standard method for evaluating prevailing compressor blade and vane frequencies. It is depicted in Figure 3.16.

As shown from the previously calculated mode shapes and frequencies (from the calculation procedure described earlier), these match well with the real vane frequencies.

3.1.2.2 Optimizing blade design to prevent excitation

Blade vibrations may occur for a variety of reasons when operating over the entire compressor map. Figure 3.17 shows a typical compressor map for either vane or speed-controlled operation. While resonances may occur at any point in the map, their origins deserve to be understood. Resonance is encountered whenever the natural frequency of a blade is being excited by pressure wakes from the up- or downstream

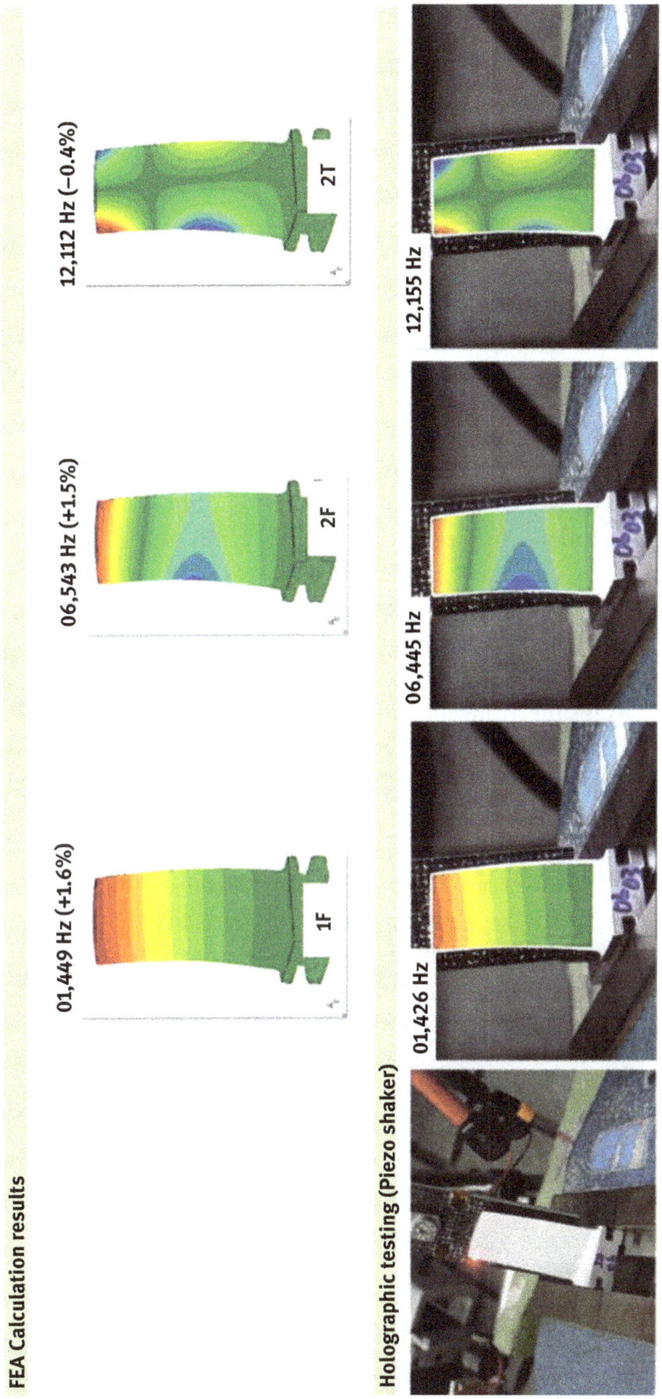

Figure 3.16: Holographic testing of a cantilever vane with a Piezo shaker (source: *MAN-Energy Solutions, Germany*).

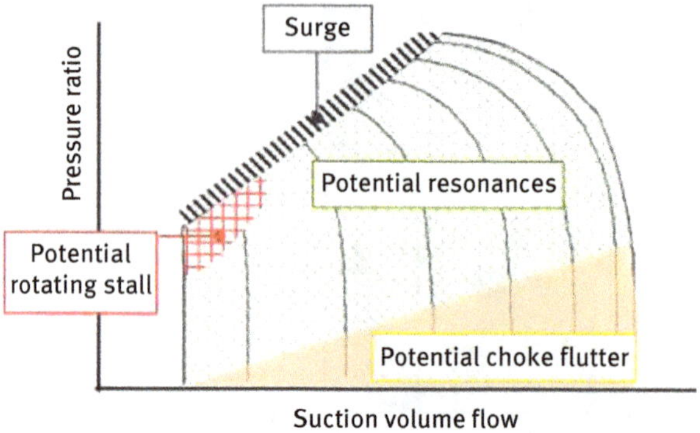

Figure 3.17: Limitations of compressor performance maps (speed or guide vane control) (source: *MAN-Energy Solutions, Germany*).

stator vanes. Blades can usually be designed to avoid concurrent frequencies. In the rare cases where complete freedom from resonance cannot be achieved, resonant stresses in the blades must be evaluated numerically and/or by stress measurements. While sometimes elaborate, stress measurements are feasible and accurate.

Significantly more difficult to predict are blade vibrations due to unsteady flow phenomena such as flutter and surge. Flutter may occur when operating a compressor close to the surge line (rotating stall) or near the choke line (choke flutter). In these regions, stochastic (randomly occurring but statistically captured) flow disturbances may lead to self-excited blade vibrations and subsequent high blade stresses. Even higher stresses may occur during compressor surge, when the sudden change in flow direction will impose extreme loads on axial blading. Due to the complexity of these flow phenomena, blade vibration from flutter and/or surge is extremely difficult to predict as of 2020. This explains why rugged blade design and sufficient safety margins are common to reliability-focused thinking. Only the implementation of a reliability-focused design can avoid blade failure.

In the past decades, standardized aero profile series like NACA or Göttinger profiles were used for the blading of axial industrial compressors. These blading types or styles are referred to as conventional compressor blading in many references in various machines and applications. In the 1990s, Eisenberg et al. started to develop a new blading type with increased performance and better robustness, referred to as wide chord blades [5]. This blading type was designed as controlled diffusion airfoils (CDA). These airfoils made use of controlled deceleration of the flow on the airfoil suction side and low risk of separation of the boundary layer. These wide chord blades have successfully proven their performance and rugged designs in the field. Survival of surge conditions was quite significantly increased by the rugged design inherent in wide chord front stages.

Based on the positive experience of the wide chord blade as front stage, modern developments transfer those features into all stages of the axial compressor [10]. Next to such aerodynamic design targets as high efficiency, broad operating range,

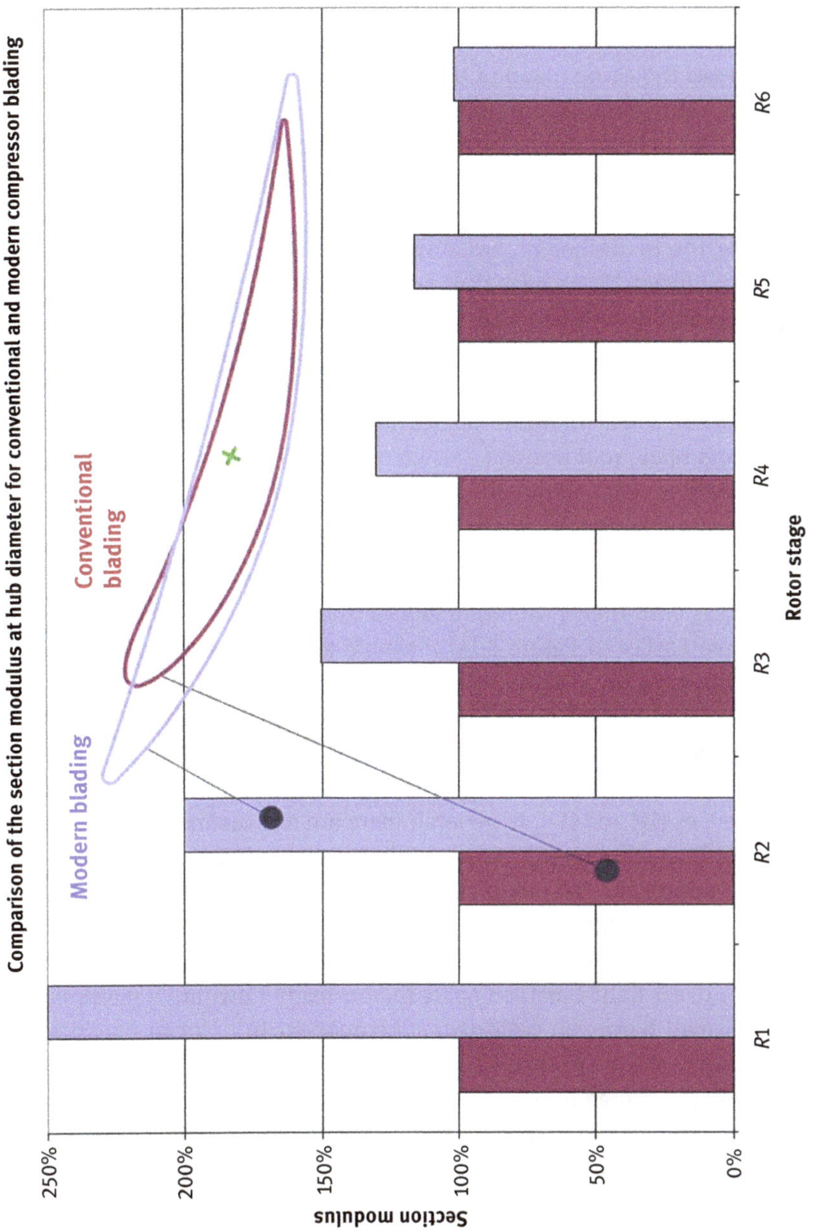

Figure 3.18: Comparison of the section modulus at hub diameter for conventional and modern compressor blading (source: *MAN-Energy Solutions, Germany*).

and high power density, the main design target is an even further strengthening of the entire compressor.

Figure 3.18 shows the comparison of the section modulus of the rotor blades at the hub section for a conventional and modern type of axial compressor.

The section modulus in all stages is significantly increased leading to increased ruggedness and lower stresses in all modern rotor blades. Furthermore, other dimensional features, such as trailing edge thickness, distribution of chord length and thickness over the blade height, twisting of the airfoil, and maximizing fillet radii, have been optimized. There have thus been significant improvements in overall compressor availability (uptime) and reliability (freedom from random failure). Compared to predecessor designs, the speeds, corresponding circumferential velocities, and centrifugal loads on the rotor blades are higher with modern design strategies. Still, the stress levels in rotor blades are kept lower with the resulting optimized airfoil and blade root designs. Design optimization steps were taken, and the goals are achieved for each of the different rows of rotor blades.

3.1.2.3 Different modes to be examined to validate excitation avoidance

In case of resonances with higher vibration modes (mixed modes or modes with radial nodes as shown earlier in Figure 3.14) a simple approach with proportionality assumptions will not give the desired results. Overly simplified approaches do not accurately represent the physical processes that accompany blade resonance. In these cases, fluid structure interaction (FSI) methods are more suitable to predict the level of resonant stresses. Today, various FSI methods exist to solve these problems, for example, references [11] and [12]. In general, there are two distinct approaches for solving unsteady turbomachinery flows: The nonlinear time-advancing approach and the time-linearized approach. The time-linearized methods are generally much more efficient than the nonlinear time-advancing approach. The latter has benefits when investigating more complex flow phenomena [13]. Furthermore, a method described by Kahl [14] can be used; Kahl's method splits the unsteady flow into a mean steady component and a small harmonic perturbation component in order to linearize the equations leading up to the mean steady solution.

The broad operating range imposed on industrial axial compressors adds complexity to the prediction of resonant stresses. While blade resonances occur at a particular speed, the flow conditions at that speed can vary drastically; they depend on the operating point in the performance map. One important influencing parameter for blade resonant stress excited by a vane wake is – in case of a variable guide vane as the excitation source – the position of the VGV. It makes a difference whether the vane is fully opened or fully closed. The influence of the VGV position on the blade resonant stress is shown in Figure 3.19. The illustration depicts corresponding FSI results for a modern front stage rotor blade with mode 3 T.

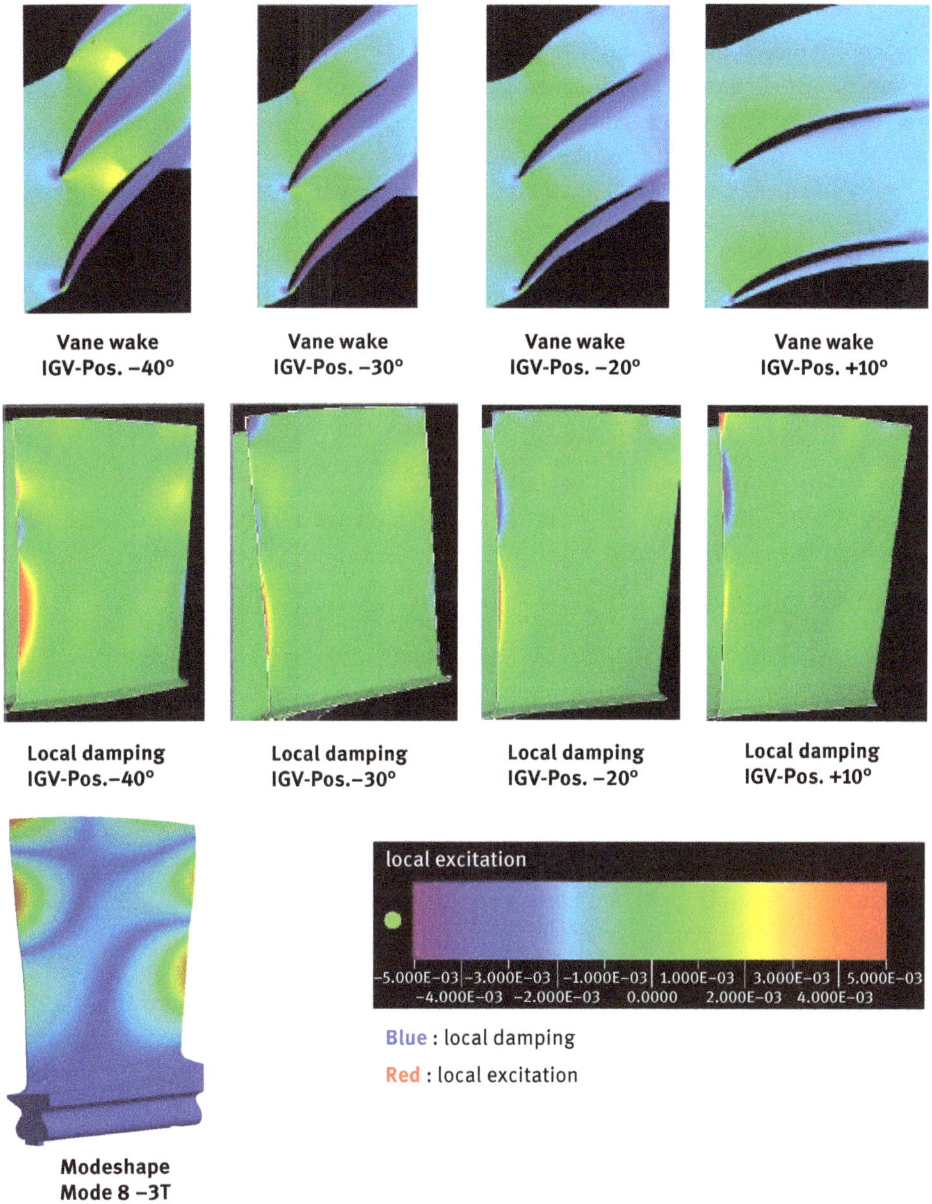

Figure 3.19: FSI calculation of local damping/excitation on a rotor blade depending on the variable guide vane position (source: *MAN-Energy Solutions, Germany*).

It can be observed that the local excitation on the rotor blades changes significantly with guide vane position. The more the vane is closed, the larger are the vane wakes. Then there will be expanded zones of excitation on the blade's surface. As a result, the overall damping of the specific resonance mode – in this case mode

3 T – is reduced with the closing of the guide vanes. Moreover, Figure 3.19 shows the aerodynamic excitation/damping that must be superimposed on mechanical damping. Such damping results from friction in the contact zones (blade root, see Figure 3.15) and material damping. As a general rule, resonant stresses increase with further closing of the guide vanes.

For the design of modern axial blading, all resonances found in the compressor map need to be validated by means of the FSI method described earlier. Validations must include different guide vane positions and varying pressure ratios. This is done by moving the operating point on the specific performance line. Again, the main objective of any optimization process is defining a mechanically strong blade airfoil and a root design with low nominal and peak stresses. Achieving these goals in all blade rows will lead to significantly extended blade service life.

3.1.2.4 Stress margin verification by test

The structural integrity of all stator vanes and rotor blades in newly designed axial compressors must be verified by test. Although this testing can be extensive, it cannot be skipped or overlooked. Verification tests are done in the manufacturer's shop with the entire compressor being operated at conditions that yield dynamic similitude. Similitude involves matching of scaled blade vibration frequencies, flow conditions, and Mach numbers with full-scale operational conditions. The test rig itself should contain all geometric features of the large-scale machine. Axial intake, rotor blades and stator vanes, diffuser, and volute must be manufactured to scale. The test rig must be heavily instrumented; it will resemble the test rig of an aeroengine compressor. Instrumentation duplicates aero instrumentation for flow field and performance measurements, pressure transducers to measure unsteady flow effects, and instrumentation for vibratory stress measurements. Strain gages are placed on blades and vanes in each stage to measure vibratory stresses during operation. Telemetry is used quite extensively.

Modern testing methods strive to employ state-of-the-art technology known as "tip timing" [15, 16], which allows measuring of vibratory motion of each blade in each blade row. The exact time when a blade tip passes an excitation originator is measured and compared to its undeflected position. This modern method is also capable of measuring higher orders of blade vibration. It will require placing enough probes in the axial direction over the blade tip to physically observe angles and distances of tip deflection. Sensitive instrumentation can spot tip deflection and determine various vibratory modes.

Strain gage telemetry together with tip timing represent the most comprehensive monitoring. Because two independent methods are used, compressor manufacturer and end user can be confident that the results are accurate. All vibratory stresses are captured, and field experience – once established – adds icing to the cake, so to speak. While industrial axial compressor blades are always subject to resonances or complex flow phenomena, modern technology defuses old assumptions. Blade vibration can be understood if we make the effort to understand it. Strengthening blades by optimized design is a quantum step in the right direction. Thorough numerical analysis is another valuable step, although broadening the operating range of axial compressors has added to the complexity of calculations that cover the full and enlarged performance map. Verification tests are needed, and in-depth experience checks are helpful to close the investigative, computational, and test-related loop.

3.2 Centrifugal compressors

3.2.1 Novel approaches to mitigate fouling spearheaded by Elliott-turbo

3.2.1.1 Surface treatment technologies

Corrosion and fouling of centrifugal compressors during operation are major concerns for process operators within the petrochemical industry. Corrosion and its effects are readily visible and well understood, and in most circumstances, can be minimized by selecting a base material or welding overlay that is suitable for the process gas application. Fouling is more process dependent as it is normally the result of a polymerization reaction intrinsic to the compression process or a carryover of a contaminant into the compressor. Over time, fouling leads to the buildup of solids on the compressor's internal surfaces, which limits the flow through the compressor and can alter the compressor's aerodynamics. It can also reduce efficiency over time, increase vibration, and result in an unplanned shutdown.

Centrifugal compressor operators try to minimize fouling within the compressor through process control and injections. Polymerization in many hydrocarbon services has been reported starting at temperatures above 90 °C [1, 17]. While this does not occur in all applications, the compressor should be designed to keep the process temperature low throughout. Liquid injections of oil or water assist in maintaining a lower process temperature, and anti-polymerization additives can be added to the process as well. Process systems also attempt to keep contaminants from being carried over into the compressor.

Despite all of these preventive efforts, fouling still occurs in many hydrocarbon compressors. To help minimize the effects of foulant buildup, antifouling coatings can be applied to the components that make up the aerodynamic flow path of the compressor.

3.2.1.1.1 Organic topcoat systems

The first successful antifouling coatings applied to centrifugal compressors used the compound polytetrafluoroethylene, more commonly known as PTFE or Teflon®. PTFE was widely used in numerous industrial and commercial applications because of its excellent "nonstick" properties. These nonstick properties also provided an effective antifouling coating, but required the Teflon to be mixed with a resin binder to allow it to adhere to the complicated geometries of a centrifugal compressor flow path [18]. An aluminum-filled chromate–phosphate ceramic coating is applied directly to the base materials of the compressor aerodynamic flow path to provide corrosion protection in case the topcoat is compromised. A prerequisite of any effective antifouling coating is that it must be a corrosion-resistant coating.

This layered antifouling coating system was developed in the late 1980s and is still used today in specific applications. Elliott's layered coating system "Pos-E-Coat®" (Figure 3.20) has a base layer that is an aluminum-filled chromate–phosphate coating, an intermediate coating that is a protective primer, and the topcoat is a PTFE-filled resin (Figure 3.21). Each layer is applied by a spray process and requires a separate heat cure afterward. Coating application equipment that can reach inside the narrow internal passageways of an impeller or diaphragm is used, and specially designed furnaces are necessary to avoid distortion of the compressor rotors during the heat curing cycle. The Pos-E-Coat system has a proven history of use in hydrocarbon compressors.

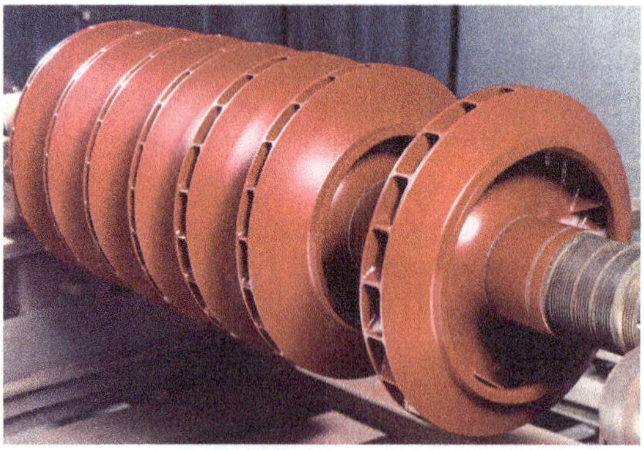

Figure 3.20: Centrifugal compressor coating with Pos-E-Coat system (source: *Elliott-Turbo*).

To improve Pos-E-Coat's antifouling characteristics, Elliott evaluated the components that make up the topcoat. To improve the coating's nonstick properties, fluorinated ethylene propylene was used in place of PTFE. The resin was changed to a harder and more durable epoxy, which is less permeable and allowed for application of the epoxy topcoat directly to the sacrificial basecoat, eliminating the need for an intermediate primer. Elliott called this improved coating system Pos-E-Coat Plus (Figure 3.22). The cross-sectional photomicrograph in Figure 3.23 displays the layers that make up the Pos-E-Coat Plus system. Each layer of the Pos-E-Coat Plus system is applied through a spray and bake system similar to the Pos-E-Coat coating.

The durability of the *Pos-E-Coat* and *Pos-E-Coat Plus* coating systems was compared using a falling sand abrasion test or solid particle erosion test as per ASTM D968. The Pos-E-Coat Plus system is over three times more durable than the first-generation

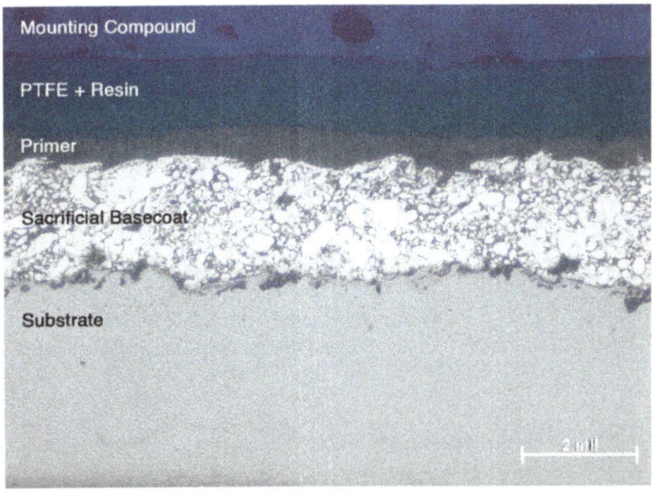

Figure 3.21: Cross-sectional photomicrograph of Pos-E-Coat system (source: *Elliott-Turbo*).

Figure 3.22: Centrifugal compressor coating with Pos-E-Coat Plus system (source: *Elliott-Turbo*).

Pos-E-Coat system, as Table 3.1 illustrates. While the ASTM D968 falling sand abrasion testing is not a "fit-for-service" test that can be directly applied to centrifugal compressor operation, it provides a relative comparison between the coating systems.

While the Pos-E-Coat and Pos-E-Coat Plus coating systems effectively minimized fouling in hydrocarbon compressors, these systems did not work in all applications.

Compressors are expected to operate from 5 to 7 years between turnarounds (occasionally called inspection and repair downtime, or IRD). Cracked gas and charge gas processes typically involve liquid injection for cooling and cleaning

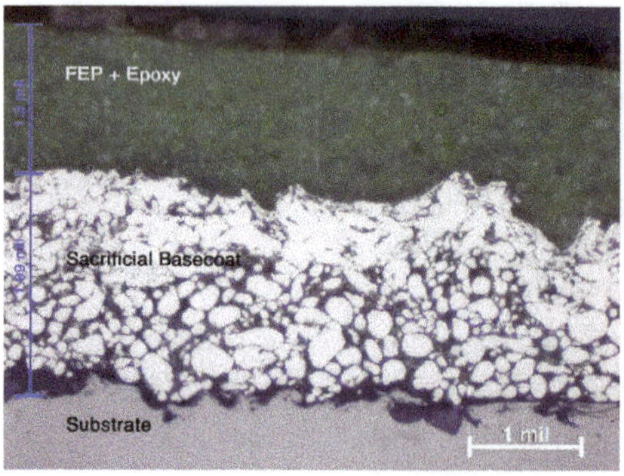

Figure 3.23: Cross-sectional photomicrograph of Pos-E-Coat Plus system (source: *Elliott-Turbo*).

Table 3.1: ASTM D968 falling sand test results.

ASTM D968 Abrasion resistance (L/µin)				
	Spot 1	**Spot 2**	**Spot 3**	**Average**
Pos-E-Coat	1.6	1.2	2.0	1.6
Pos-E-Coat Plus	5.2	5.2	6.0	5.5

purposes. It has to be done under controlled conditions because high-velocity liquid droplet impacts on the aerodynamic surfaces will eventually wear away the coating system's organic topcoat. Chemical additions to the processes either react with the epoxy or resin topcoats or permeate the topcoats and result in blistering. The push for an antifouling coating that could maintain compressor efficiency for a 5-to-7-year operational period caused Elliott to look beyond standard coating systems that rely on organic topcoats.

3.2.1.1.2 Electroless nickel coating

Elliott's Pos-E-Coat 523 coating is different from the industry standard spray and bake type of coating (Figure 3.24). Pos-E-Coat 523 applies a nickel-phosphorus glass to the surface of the components being treated. Besides its antifouling benefits, this coating system has numerous advantages in terms of durability. The metallic glass coating has a hardness of over 40 HRC, which makes it more durable than the compressor impeller base metal, allowing the coating to withstand liquid injections. The amorphous nickel coating is relatively chemical resistant to many process gas environments and chemical applications.

Although the coating is metallic, there are no grain boundaries; the coating is therefore not susceptible to intergranular corrosion. Electroless nickel performs very well in the presence of hydrogen sulfide. Electroless nickel can become even more corrosion resistant since a thin but tenacious layer of nickel sulfide forms at the surface.

Figure 3.24: Rotor coated with Pos-E-Coat 523 after 6 years of operation in a crack-gas compressor with the coating still intact (source: *Elliott-Turbo*).

Nickel sulfide is extremely effective at blocking additional corrosion reactions. By blocking corrosion and providing an antifouling surface, this coating makes liquid injections in the compressor extremely effective in removing foulant buildup in the compressor.

The coating is applied by a submersion process whereby the nickel-phosphorus reacts directly with the base material. This application process forms an extremely strong chemical bond and allows for application to complicated geometries with narrow tip openings (Figure 3.25).

Figure 3.25: Cross-sectional photomicrograph of Pos-E-Coat 523 system (source: *Elliott-Turbo*).

3.2.1.1.3 Measuring antifouling benefits of centrifugal compressor coatings in hydrocarbon service

Centrifugal compressor operators in hydrocarbon services need to know the benefits of applying an antifouling coating, which is information they can use to justify the cost of applying the coating. Measuring the antifouling benefits of a coating in a laboratory will be a challenge. Laboratories cannot simulate the exact operating conditions inside a hydrocarbon gas compressor. Every centrifugal compressor has a different process gas composition and temperature; the amount of liquid injected can vary significantly, and chemical injections likely differ from site to site.

However, a relative ranking of the foulant release abilities of different coatings against a standard hydrocarbon foulant can be obtained. Elliott tested and compared the Pos-E-Coat, Pos-E-Coat Plus, and Pos-E-Coat 523 coating systems for foulant release.

The first step in the testing process is to apply a substance that can form a foulant similar to that found in hydrocarbon compressors. Elliott developed an imitation foulant and applied it to a bare steel sample (Figure 3.26) as well as to coated samples, and baked the samples in a curing oven until a thick tar-like foulant formed on the surfaces. Each sample was then placed in a test rig (Figure 3.27) that scrubbed the sample against a soft pad, and the amount of foulant removed was measured by weight at set intervals. The bare steel samples did not release the foulant after 2,000 cycles, which demonstrates the typical problem encountered in hydrocarbon compressors; their efficiency cannot be increased regardless of the amount of liquid injection used to clean the compressor. The Pos-E-Coat and Pos-E-Coat 523 coated samples released the "gunk" after 1,250 cycles, while the Pos-E-Coat Plus system released the foulant after a mere 150 cycles (Figure 3.28). This testing helped to demonstrate the significant antifouling benefits these coatings can provide.

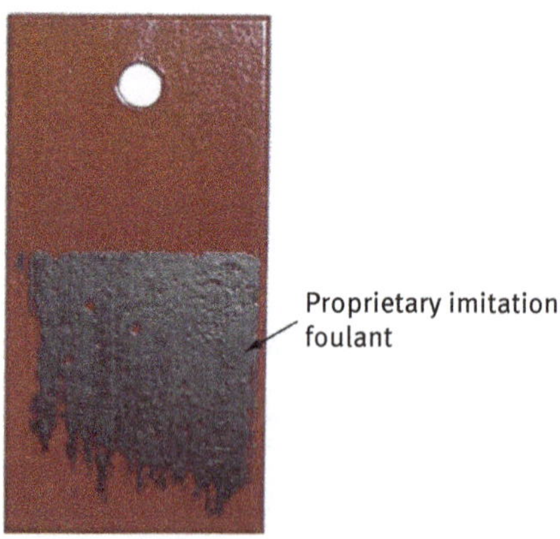

Figure 3.26: Coated steel sample with imitation foulant formula used to test coating foulant release (source: *Elliott-Turbo*).

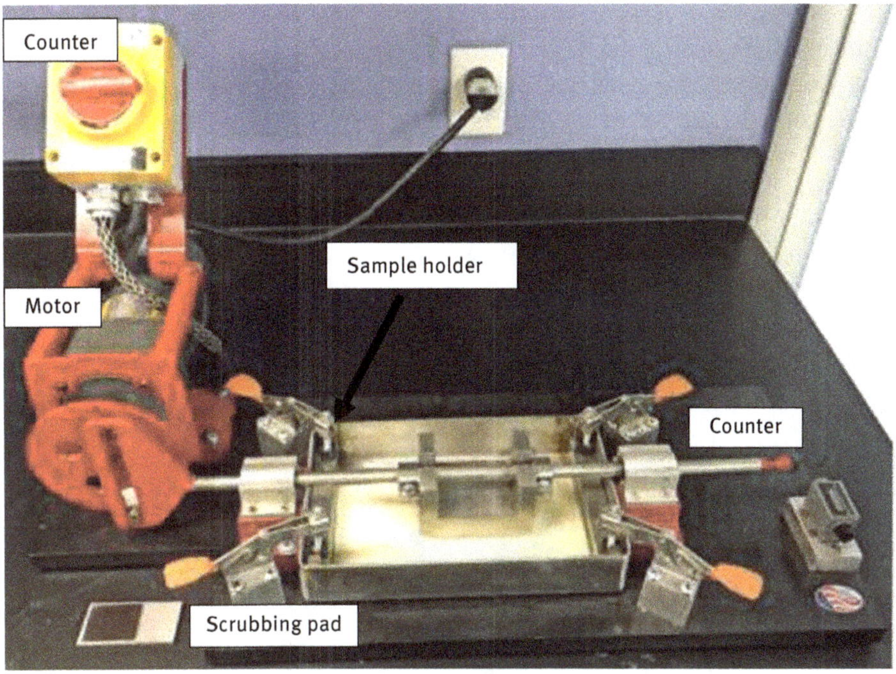

Figure 3.27: Foulant release testing apparatus (source: *Elliott-Turbo*).

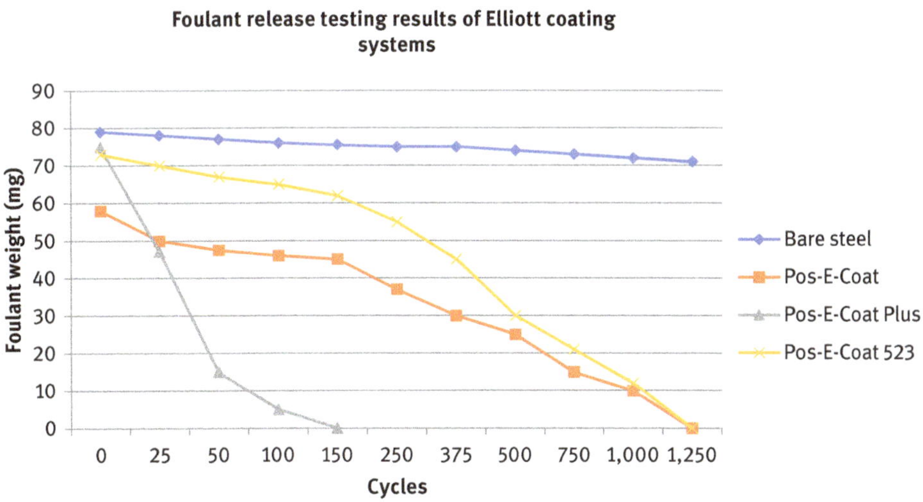

Figure 3.28: Fouling release testing results for Elliott coating systems (source: *Elliott-Turbo*).

3.2.1.1.4 Hydrogen recycle compressors

Centrifugal compressors in hydrogen recycle service operate in environments that present unique corrosion and fouling challenges. Salts, such as ammonium chloride, carried over from upstream processes adhere to metallic surfaces and reduce compressor efficiency over time (Figure 3.29). Underneath the ammonium chloride deposits, pitting corrosion can occur rapidly on carbon and stainless steel components, creating the perfect initiation sites for fatigue cracks. If not addressed, this can eventually result in operating issues or even equipment failure.

Figure 3.29: Ammonium chloride deposits inside the impeller flow path of a hydrogen recycle compressor (source: *Elliott-Turbo*).

Although compressor coatings traditionally focused on minimizing fouling, there were no coatings specifically designed and tested for hydrogen recycle compressors. Liquid injections are not typically used in hydrogen recycle compressors because ammonium chloride is not soluble in many solvents. The chlorides are soluble in water; however, liquid water can accelerate the corrosion potential of the deposits.

Elliott developed a foulant adherence test to simulate ammonium chloride deposits in order to test the effectiveness of various coatings (Figure 3.30) for hydrogen recycle compressors. The results revealed a very clear distinction between coatings that were effective in this environment and standard industry coatings that did not perform well against ammonium chloride deposits (Figure 3.31).

As previously mentioned, a prerequisite of any effective antifouling coating is that it must provide corrosion protection. The corrosion resistance against ammonium chloride is readily tested through electrochemical impedance spectroscopy with an ammonium chloride electrolyte. The coating degradation is tracked over time and provides a numerical method for ranking the coatings (Figure 3.32).

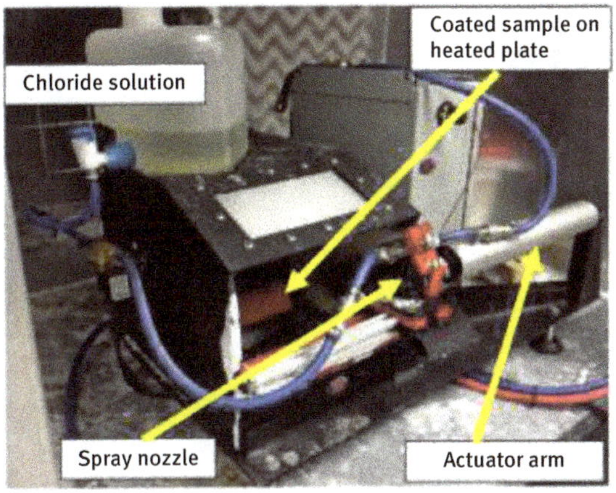

Foulant adherence test

- Plate sample is mounted to holder with a heater behind it
- Gravity feed spray nozzle is mounted to an actuator
- The actuator moves and a solenoid valve is opened every 30s to intermittently spray foulant. Creates a repeated wet/dry cycle in a chloride environment.
- Every 5 cycles, an air blade blasts air at a low angle to simulate the gas flow in the compressor
- The test is run for a set time
- Used as a screening test since this is the attribute of primary importance for an antifoulant coating

Figure 3.30: Foulant adherence testing of ammonium chloride deposits (source: *Elliott-Turbo*).

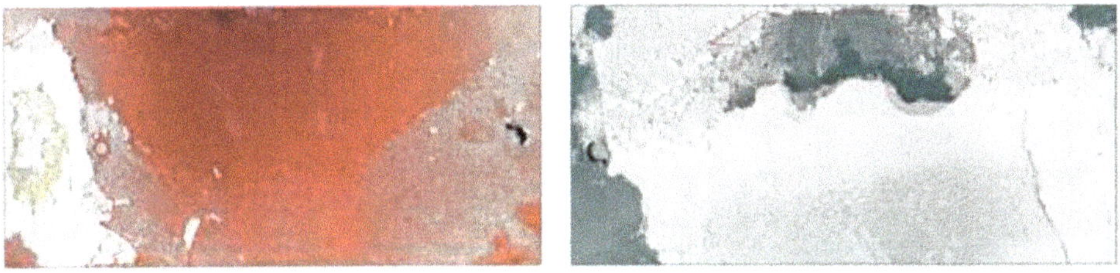

Pos-E-Coat Protect — Ineffective coating against ammonium chloride

Figure 3.31: Results of coating effectiveness from foulant adhesion testing (source: *Elliott-Turbo*).

Finally, the durability of any coating is of great importance to centrifugal compressor applications. A falling sand test as per ASTM D968 allows for silicon carbide particles to hit a coated sample. This testing provides a standard method comparing the durability of various coatings; the impingement of solid particles against the surface is an indication of the type of erosion that a coating will experience in a centrifugal compressor. The results are reflected in Figure 3.33.

These testing methods were applied to the development of Pos-E-Coat Protect, which is specifically intended for hydrogen recycle compressors. This is a multilayer coating applied through a spray and cure method to assembled compressor rotors and diaphragms in the same manner as the Pos-E-Coat and Pos-E-Coat Plus systems (Figure 3.34). The antifouling properties of Pos-E-Coat Protect minimizes the amount of ammonium chloride that adheres to the flow path of a centrifugal compressor and protects the base material from pitting corrosion (Figure 3.35).

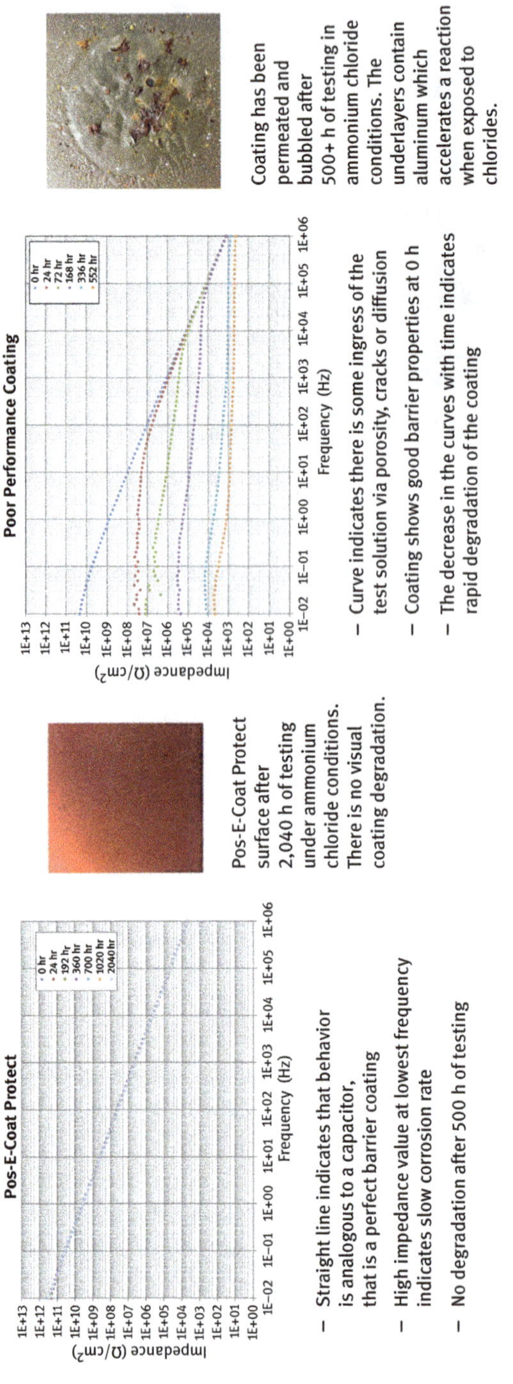

Figure 3.32: EIS testing results under an ammonium chloride electrolyte (source: *Elliott-Turbo*).

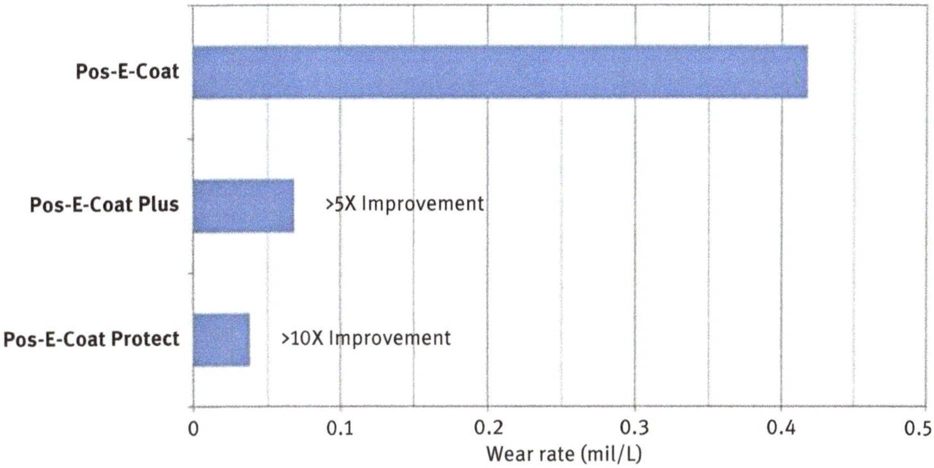

Figure 3.33: Results of ATM D968 method B falling sand tests (source: *Elliott-Turbo*).

Figure 3.34: Centrifugal compressor rotor with Pos-E-Coat Protect system (source: *Elliott-Turbo*).

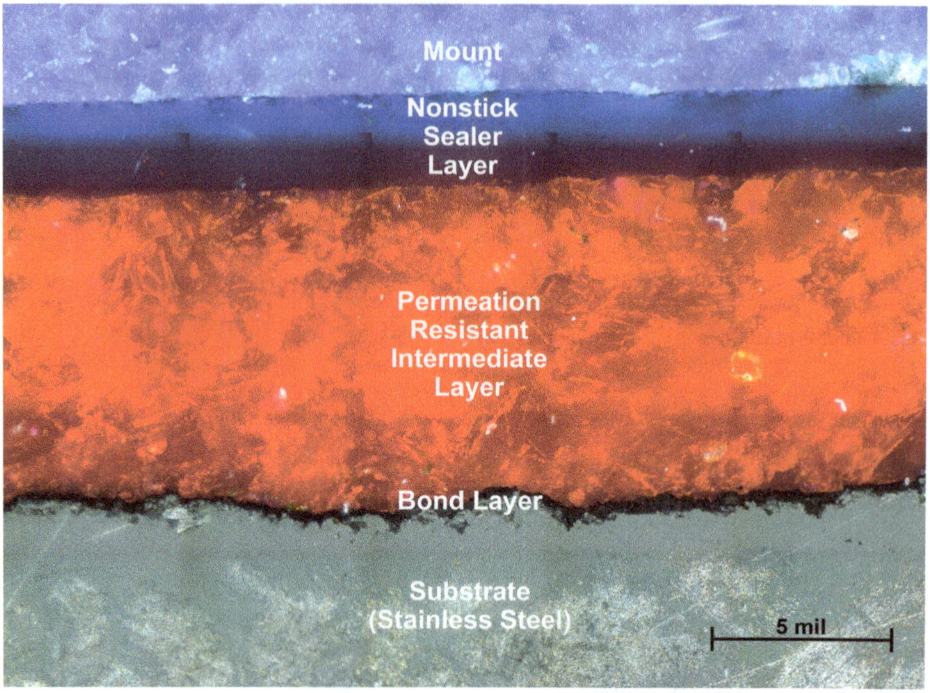

Figure 3.35: Cross-sectional of Pos-E-Coat Protect system (source: *Elliott-Turbo*).

3.2.1.1.5 Conclusion

The petrochemical industry faces hurdles that must be overcome regarding compressor fouling; moreover, each operating condition is unique. Compressor coatings that best suit a customer's specific operating conditions and help extend compressor service life are available. In addition to extending service life, well-selected coatings will maintain optimum operational efficiency. The selection of the correct coating should be based on the operating conditions of the compressor and the type of fouling that is known to occur.

3.2.1.2 Discussion on effective wash nozzle arrangements

3.2.1.2.1 Liquid injection

Traditionally, metered liquid injection through spray nozzles has been used with moderate success to mitigate fouling in process gas compressors. Injection is typically done at suction piping upstream of the compressor suction nozzle and diffuser return bends; however, it is difficult to uniformly disperse the fluid around the diaphragms. Because of this, fouling still takes place locally, even with the best efforts of continually applying detergent-type wash fluids. With commercial considerations aiming for extended run lengths of these compressors, rethinking the design and dispensing the wash fluid more uniformly led to a new approach.

3.2.1.2.2 Fouling mechanism

As is well known, the temperature of a gas increases during the compression process. Depending on the temperature reached and the types of hydrocarbon constituents present in the gas, polymerization can occur. In time, these polymers become tightly packed and difficult to dislodge. The mass or "foulant" sticks to surfaces and thereby restricts compressor flow passages. A frequent result is reduced plant output and, possibly, reduced component life.

The chemical mechanism that generates polymerization is complex; however, Figure 3.36 depicts – in a simplified form – what typically occurs during polymerization and fouling. Monomers making up part of the process gas react as shown in Figure 3.36; they become polynuclear aromatic compounds (PNA). Over time and with dehydration, these PNA compounds will polymerize into coke-like or tar-like substances. The amount of polymer formation depends on the concentration, pressure, and temperature of monomers in the process gas. The onset of polymerization is generally noted at process gas temperatures above 90 °C (194 °F), and the rate of polymerization proceeds exponentially as temperatures are increased.

Deposition of polymers on compressor components is mitigated by making the surfaces in the gas flow path relatively smooth; surface smoothness and "slipperiness" can be imparted by applying modern diffusion-conversion techniques. As an example, some of the surface treatment technologies mentioned in Section 3.2.1.1 hold promise as of 2020 and are worth exploring. Some diffusion-conversion treatments are

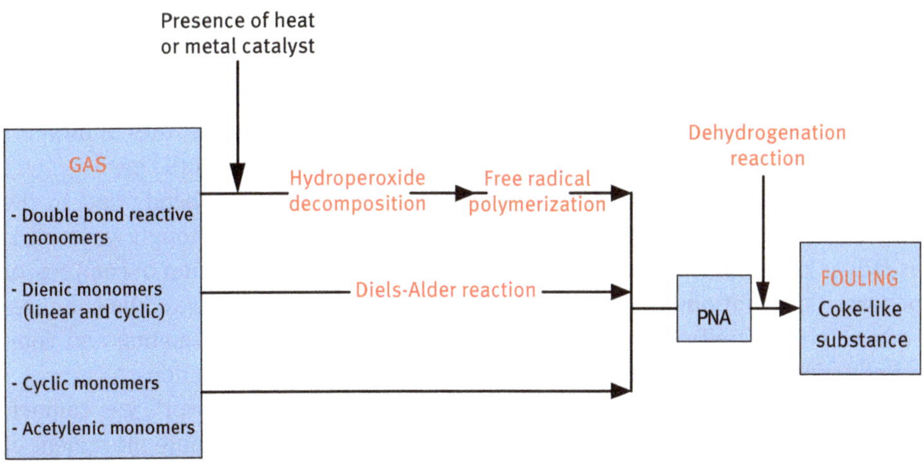

Figure 3.36: Simplified illustration of compressor fouling mechanism *(Mitsubishi Heavy Industries Compressor Corporation [MHI], Hiroshima, Japan).*

known to impart high hardness (wear resistance) to low-alloy steels while potentially reducing the ability of foulants to adhere or "stick" to impellers and other compressor components.

3.2.1.2.3 Liquids used to mitigate fouling

The most popular liquid media injected into gas streams are demineralized water or process-compatible oils. Wash oils ("washing oils") are selected based on their solvency or detergent capabilities. Wash oils must be compatible with the process and serve to dissolve and flush away polymers formed in the compressor gas flow path. For greatest effectiveness, wash oil injection should commence before the layer of polymer becomes too thick or gets to a hard-to-dissolve consistency.

The main purpose of continuous flush liquid injection is to reduce the discharge temperature of the gas. Ambient temperature condensate injected at or near the compressor suction flange can prevent polymerization of gas constituents. The rate of injection varies with compressor size; however, the targeted gas discharge temperature should be lower than 90 °C (194 °F). Whenever possible, water injection is used for cost reasons. Still, hydrocarbon "oil" injection may have to be used for its detergent-type properties. In many instances, "intermittent" injection of liquids may be ineffective, and continuous flush injection will be needed.

Excessive amounts of flush fluids are not beneficial. There would be erosion due to high-velocity impingement of liquid droplets and the metallic surfaces in the flow path. Therefore, injecting water only into the compressor suction piping but not into the return bends may be preferred so as to minimize direct impingement on compressor internals.

Water (or oil) is usually injected into the compressor suction (50%) and sometimes (partially, about 50%) into the diffuser return bends. The amounts injected are largely governed by owner experience and gas composition. The liquid injection hardware items (pipes, valves, nozzles, etc.) are typically dimensioned for 5% of gas weight flow. End users, compressor manufacturers, and engineering and procurement contractors (EPCs) often use generalized guidelines and "rules of thumb." These usually assume injection rates of 2–3% of the weight flow of the gas.

Traditionally, wash nozzles are installed at diffuser return bends and in the suction piping just upstream of the compressor inlet flange. Although wash oil is usually injected on a batch or intermittent basis to offset degradation of compressor performance, continuous washing is often more effective. Either way, the compressor's polytrophic efficiency trend is monitored on suitable instruments, and computer projections can readily determine optimized wash-cycle duration.

Continuous liquid injection is employed to keep the compressor's gas contacting wetted surfaces. Wetted surfaces are sufficiently slick to promote movement of the polymers through the compressor, at the same time the wash oil will provide solvent action. The economics of intermittent versus continuous wash injection are examined before choosing one or the other. Figure 3.37 shows the typical fouling condition for a cracked gas (steam cracker process gas) compressor without oil injection; minimal fouling with oil injection is depicted in Figure 3.38.

Figure 3.37: Significant fouling condition in a cracked gas compressor operating without oil injection *(Mitsubishi Heavy Industries Compressor Corporation [MHI], Hiroshima, Japan).*

Although Figure 3.38 shows the compressor internals to be much cleaner with wash oil injection than without, it should be noted that the residual polymers are still present. It is this residual buildup of polymers that recently drew the attention of compressor manufacturers.

New fouling prevention techniques were studied back in 2015 because even relatively small efficiency gains can be worth considerable money. Again, diffusion-conversion-based technologies may be of interest in compressor applications. Although previously used as turbocharger wheel surface enhancers in the hot exhaust gases of truck diesel engines, diffusion-conversion-based technology soon migrated from vehicle applications to the process gas compressor field. In the past and as of now, the technology serves to successfully impart extreme hardness and "slipperiness" to hundreds of thousands of turbocharger wheels. This experience record explains why the technology is rapidly being adopted by other branches of manufacturing.

Figure 3.38: Insignificant fouling is experienced in a cracked gas compressor with oil injection using effectively placed nozzles *(Mitsubishi Heavy Industries Compressor Corporation [MHI], Hiroshima, Japan).*

As mentioned previously, wash oil nozzles are traditionally installed in the suction pipe and diffuser return bend of each compression stage to mitigate fouling during a wash cycle. However, the nozzle locations must be chosen with considerable forethought. Fully optimizing the location of wash oil nozzles in compressor flow paths is physically difficult, especially since flow distributions within these paths are unknown during wash cycle events. Compressor manufacturers are occasionally challenged in this regard, and third parties have been asked to predict flow conditions during wash oil injection.

Meanwhile, compressor manufacturers use analytical tools such as computational flow dynamics (CFD) to study liquid injection at various flow conditions. Figure 3.39 shows a typical CFD analytical model for simulating wash liquid injection. In this particular case, the model examined flow distribution from the compressor inlet nozzle to the inlet of the second impeller of a particular process section by varying the operating conditions. Figure 3.40 shows the study results for three different simulations; pressure and velocity distribution through flow passages were among the parameters studied. Injection nozzles can then be optimally placed to accommodate a number of different operating conditions (Figure 3.41).

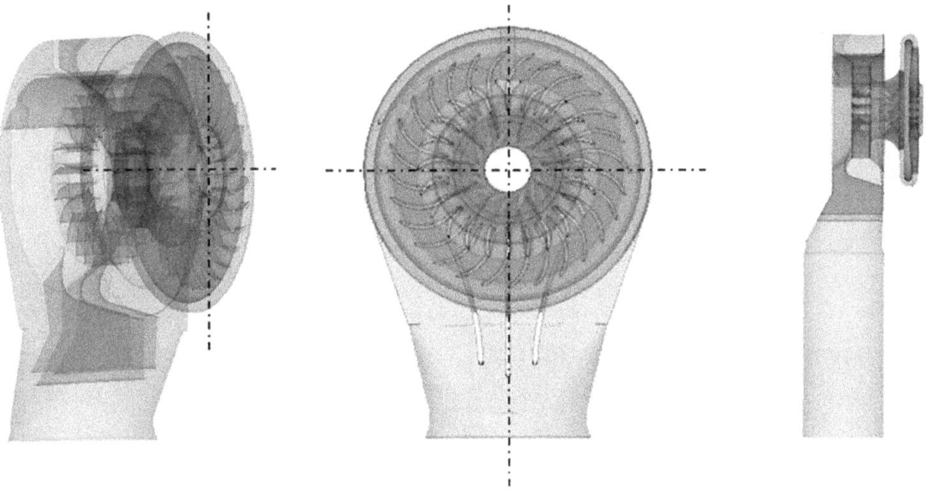

Figure 3.39: CFD analytical model used in studying wash liquid injection *(Mitsubishi Heavy Industries Compressor Corporation [MHI], Hiroshima, Japan).*

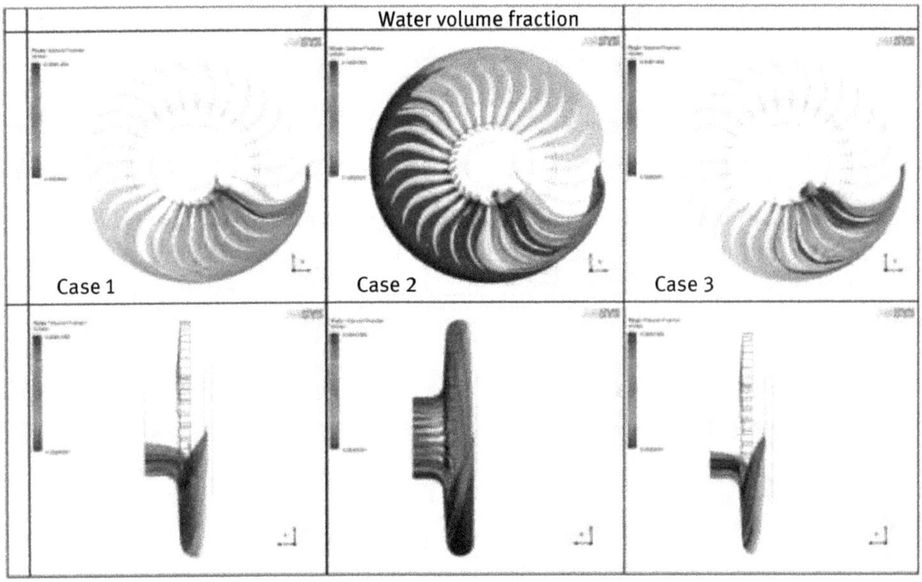

Figure 3.40: Study results for optimizing nozzle effectiveness and locational arrangement *(Mitsubishi Heavy Industries Compressor Corporation [MHI], Hiroshima, Japan).*

Figure 3.41: Some compressor internals after successful continuous liquid wash fluid injection (*Mitsubishi Heavy Industries Compressor Corporation [MHI], Hiroshima, Japan*).

A given compressor manufacturer uses their own unique design guidelines as they apply study results to their compressors. End users and equipment manufacturers are encouraged to collaborate during data gathering, data analysis, and field validations.

Ideally, optimized wash oil introduction would be through nozzles installed 360° around each compressor stage or to have other effective means of uniformly distributing wash oil at each stage.

3.2.2 Modern design and manufacturing considerations

Demand for increased plant capacity is usually reflected in larger compression equipment. As components become larger, compressor manufacturers are progressively moving toward best available design and fabrication techniques. These techniques will accommodate higher flow coefficients. By definition, a compression stage consists of an impeller, a diffuser, a return bend, and a return channel. A major concern for these large compression stages is flow distortions within the flow path; such distortions would lead to poor aerodynamic performance. The resulting performance deficiencies are narrow operating range and low polytrophic efficiency.

Figure 3.42 shows the configuration for typical high flow coefficient and low flow coefficient compression stages. The concerns and effects of these coefficients regarding larger compression stages are highlighted and compared in Table 3.2.

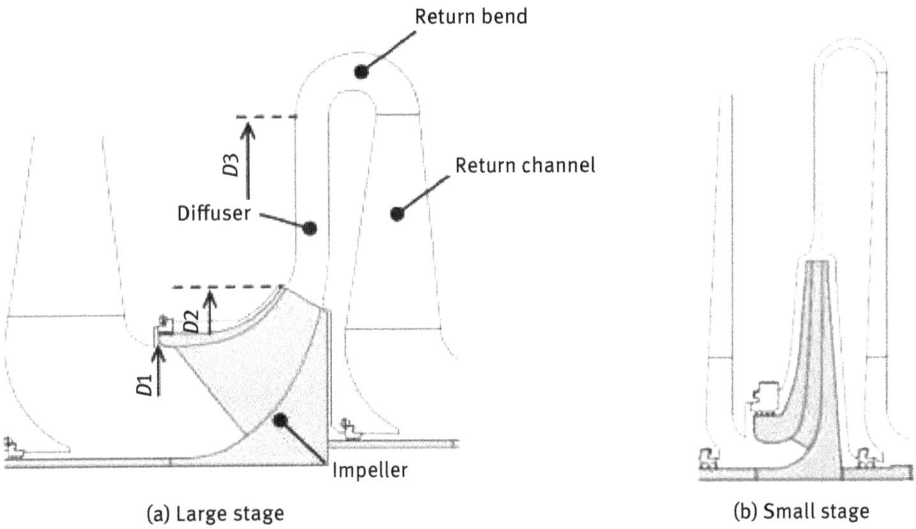

Figure 3.42: Compression stage components *(Mitsubishi Heavy Industries Compressor Corporation (MHI), Hiroshima, Japan).*

Table 3.2: Features and concerns of large compression stage *(Mitsubishi Heavy Industries Compressor Corporation (MHI), Hiroshima, Japan).*

Part	Feature	Concern	Effects
Rotational part (impeller)	Large ratio of $D1/D2$	High Mach number at shroud	Development of flow separation
		Low degree of reaction	Low pressure rise within impeller
	Wide flow paths	Sensitive to flow pattern at inlet	Development of flow distortion
Stationary part(diffuser, return bend, and return channel)	Wide flow paths	Flow distortion	Influence on downstream stage performance

As flow coefficients become greater, the impeller shroud diameter at inlet "D1" becomes larger in comparison to the tip diameter "D2." Therefore, the relative Mach number at the inlet shroud is somewhat high. High Mach numbers can cause flow separation at the inlet shroud, which in turn can significantly influence stage aerodynamic performance. Moreover, because of the now larger $D1/D2$ ratio, the centrifugal flow path from inlet to outlet along the impeller shroud is shortened. Therefore, for this type of configuration, pressure rise within the impeller is relatively small, and static pressure recovery in the diffuser becomes critically important for proper stage performance (Figure 3.43).

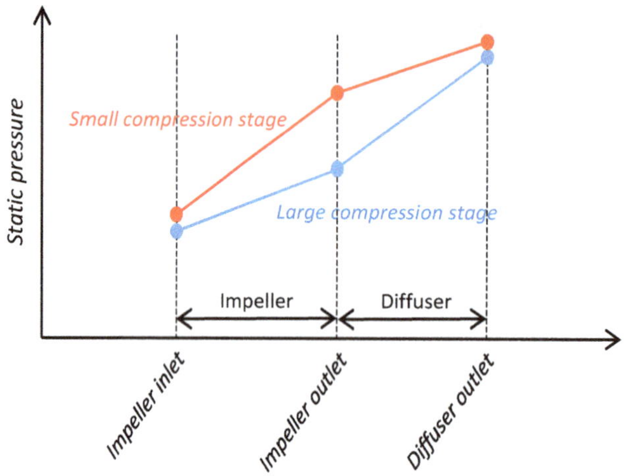

Figure 3.43: Pressure development in a compression stage *(Mitsubishi Heavy Industries Compressor Corporation (MHI), Hiroshima, Japan).*

Another point on impeller design is that impellers with higher flow coefficients have wider flow paths, which also increases its tendency to promote uneven flow patterns. Uneven flow patterns at the impeller inlet lead to the development of flow distortion; secondary flow streams will appear within the impeller. This flow distortion cascades through the stationary parts (diffuser, return bend, and return channel); it can cause further flow separation at the return bend, as shown in Figure 3.44. These undesirable flow patterns will cause poor aerodynamic performance not only in the stage primarily causing it, but also in other downstream stages. The overall effect is poor aerodynamic performance for the entire section. It follows that compression stages with high flow coefficients require very special attention.

Consideration must be given to impeller design as well as the design of stationary components. CFD is employed as a prominent tool in avoiding the various negative effects mentioned earlier. Particular attention is paid to inlet guide vanes or return channel – ensuring uniform flow pattern at impeller inlet.

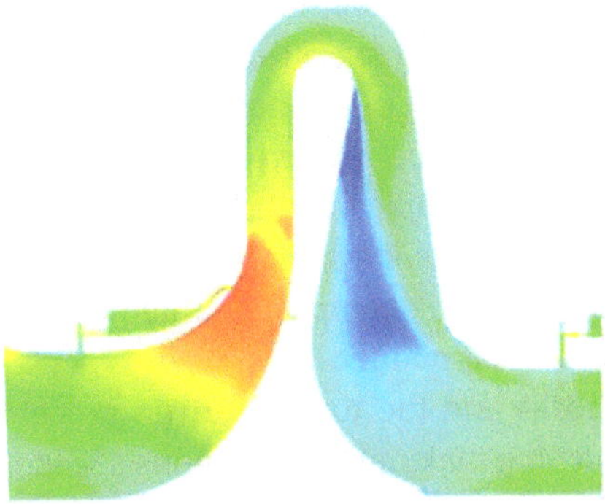

Figure 3.44: Typical CFD result of separation flow in return channel *(Mitsubishi Heavy Industries Compressor Corporation (MHI), Hiroshima, Japan).*

Proper diffuser contouring will promote optimum static pressure recovery at the return bend. Likewise, suitable inlet flow angles to return channels and optimized flow width assist in achieving smooth flow patterns.

Compressor design must harmonize with applicable industry standards. On physically large compression stages, the rotor dynamics criteria imposed by various standards demand attention. As compressors are scaled up, the axial span of higher flow coefficient impellers increases, and the longer impellers will be heavier than previous designs. It follows that rotor dynamic parameters such as critical speed, rotor stability, and bearing loads tend to be affected; these parameters must be reassessed whenever rotors become longer and heavier.

These are just a few items to be considered very early during the design of large flow coefficient compression stages. They must be formulated into questions by MQA engineers.

3.2.3 A new approach to uprateability of compressor trains

Properly assessing the uprateability of new centrifugal compressor trains can be problematic for machinery engineers. For example, when building grassroots plants, project managers will often assert that it is not their intention to spend money on any form of preinvestment on compressor trains. At the same time, it is not unusual for these managers to insist that the new plant should have sufficient "creep capacity" to readily achieve at least 15% and sometimes even 30% additional or uprate throughput. These conflicting requirements ask machinery engineers to draft equipment purchase specifications that require considerable skill, to say the very least. Follow-up must be arranged in the form of MQA visits by the owner's representative (see index for "MQA").

To compound the issue, many compressor manufacturers convey the flawed understanding that the uprateability of each compressor frame matches the throughput capacity limits shown in their sales literature. By making this claim, some manufacturers show as much as 50% uprateability relative to the stipulated or specified rated capacity.

Experience shows that the capacity normally shown in vendors' literature is very rarely achievable. Therefore, uprate capacity should be reviewed based on factors other than mere throughput capability. As an example, in the 1980s a well-known operating company rerated its process gas compressor train in the expectation of achieving at least 120% of initial capacity. The operating company wanted to take advantage of favorable market projections but was disappointed when only 15% was added. Greater throughput required speed increases and the turbine-driven process gas compressor train became unstable before reaching maximum continuous speed.

For illustration purposes, we can examine Figure 3.45 and the associated Table 3.3. Assume that we are dealing with compressor frame size 8HS and wish to determine its uprateability.

We assume a rated flow of 29,600 acfm. Anticipate that the compressor vendor will try to show that by changing out rotor and diaphragms; the uprateability of this particular frame is 148.6%. The compressor manufacturer obtained this number by dividing 44,000 by 29,600 × 100%. The larger of the two numbers is the maximum flow shown in the catalog. However, the true uprateability of the compressor frame depends on several other physical and/or dimensional constraints that cannot be changed during the life of the machine.

3.2.3.1 Rotor flexural stiffness limitation

There are several bearing spans and shaft sizes that can be installed in a particular compressor frame. However, bearing span and shaft size determine the rotor flexural stiffness, which becomes a physical constraint for that particular compressor. Rotor stiffness is proportional to $[D^4/L^3]$, where D is the shaft diameter, and L is the bearing span. Longer bearing spans result in less stiff rotors, and larger shaft diameters increase rotor stiffness. Oil film behavior also plays a role here. Refer to Figure 3.46 for an experience-based plot of rotor flexural stiffness vs. maximum continuous speed.

Rotor flexural stiffness is somewhat analogous to a guitar string: Shorten it and the vibratory frequency goes up. A longer string vibrates at a lower frequency. Rotor flexural stiffness and oil film govern the rotordynamics for each compressor frame; these three parameters often limit the operating speed range for an uprate. Rotor flexural stiffness is one of the key parameters that determine rotor critical speeds and stability, and hence a threshold speed for compressor rerates. This threshold speed is determined through lateral and stability analyses carried out as per API 684 guidelines.

Incremental speed studies determine how much incremental speed can be added for improved capacity before impinging on applicable API guidelines. Uprateability is then viewed as a percentage speed increase to get to the threshold speed divided by the specified rated speed. The threshold speed should also be used to check to see if bearing journal peripheral speeds exceed the values shown in Table 3.4.

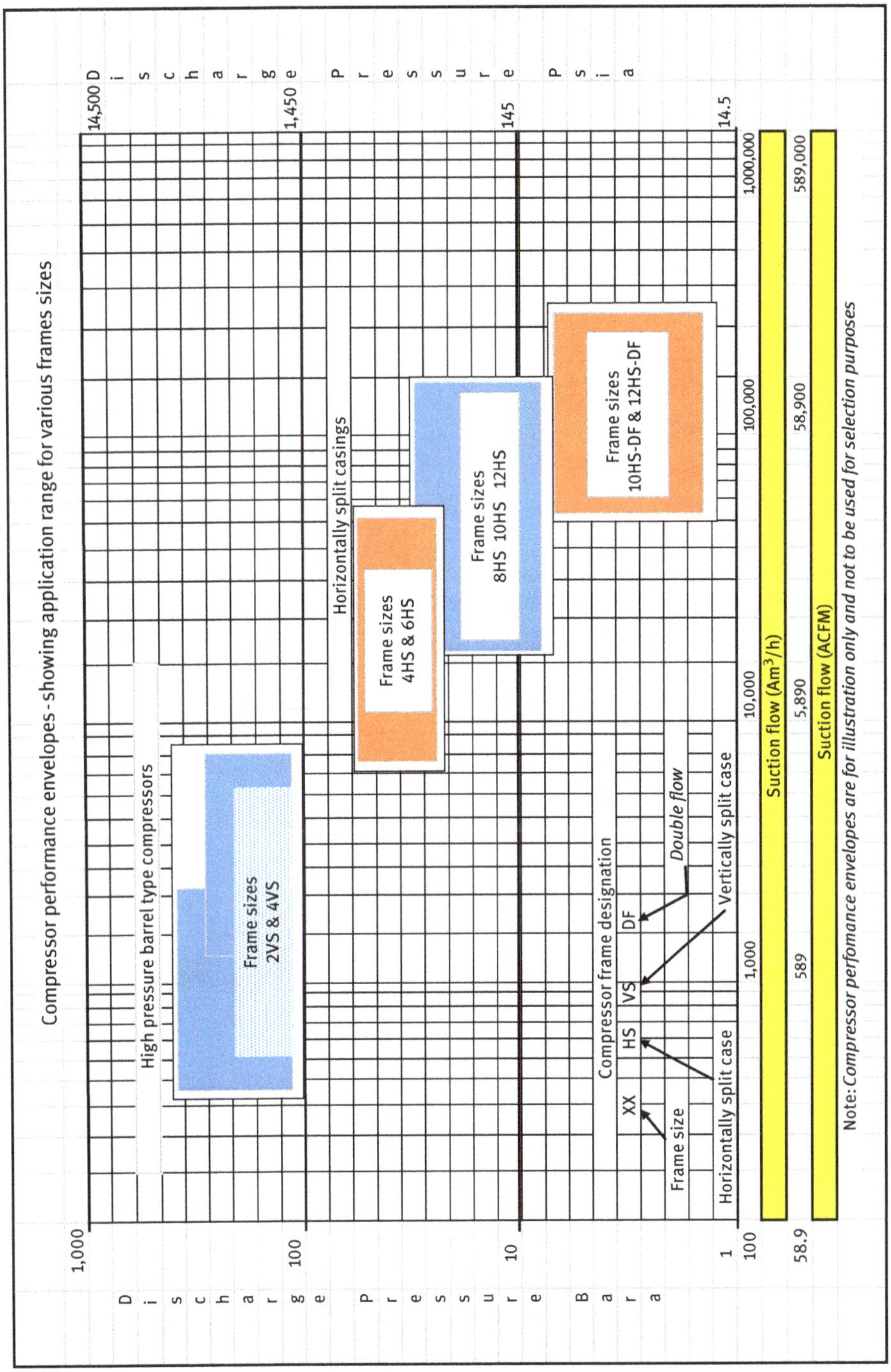

Figure 3.45: Typical performance envelopes from compressor vendors.

Table 3.3: Capacities for horizontally split casing shown in Figure 3.45.

Frame	Flow range (single flow) (acfm)	Stages per casing	Max casing pressure (psig)	Max speed (rpm)	Nominal polytrophic efficiency	Bearing span min–max (ins)	Diameter bearings min–max (ins)
4HS	3,000–7,000	6	1,000	20,000	0.80	36–70	3–4
6HS	6,000–14,000	8	1,000	15,000	0.80	40–80	3–5
8HS	18,000–44,000	8	782	8,000	0.84	80–155	3–5
10HS	25,000–88,300	8	600	4,500	0.84	100–250	6–8
12HS	30,000–118,000	8	426	4,100	0.84	120–280	8–10
10HS-DF	59,000–190,000	4 + 4	128	3,700	0.84	110–260	6–8
12HS-DF	118,000–294,000	4 + 4	128	3,200	0.84	140–310	8–10

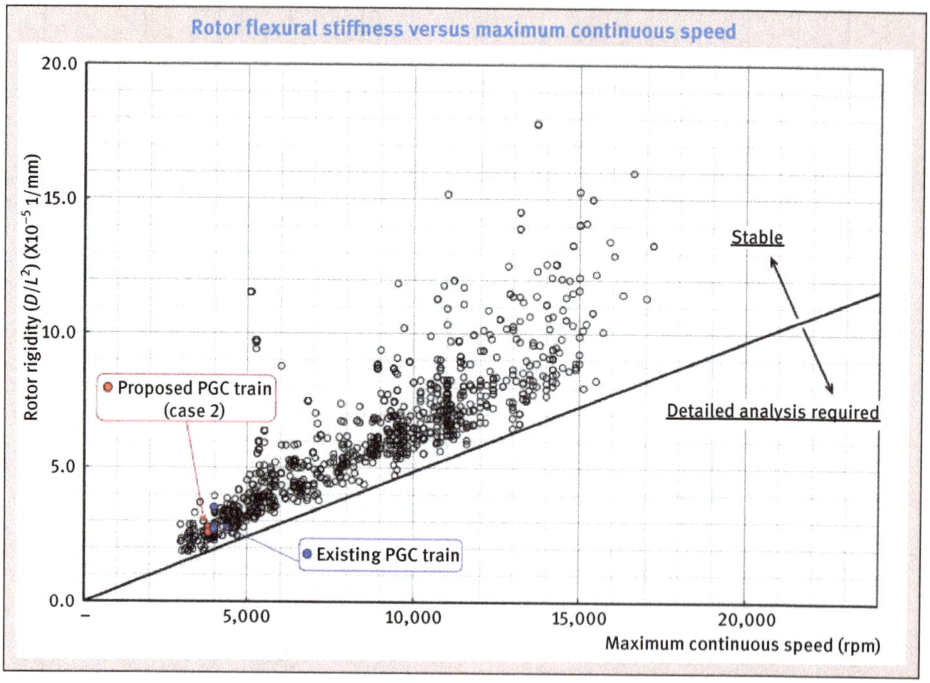

Figure 3.46: Example of one compressor vendor's experience plot of rotor flexural stiffness versus maximum continuous speed.

Table 3.4: Critical design parameters used to set limits for calculating uprateability parameters.

Critical parameters	Units	Limiting criteria	Comments
Maximum steam velocity at pipe flange	ft/s	<200	Calculate the velocity at rated conditions, compare it with the limiting criteria as a percentage of rated.
Speed	rpm		The limiting operating speed is determined by performing incremental rotor dynamics studies by increasing compressor speed to increase compressor capacity until a threshold speed is reached. The threshold speed is determined from lateral and stability analyses done as per API 684 guidelines during the incremental study.
Gas velocity at inlet nozzle flange	ft/s	<140	This can be used as an initial check to appreciate the magnitude of the velocity entering the nozzle; we still recommend using nozzle Mach number as the ruling criterion
Mach number at inlet nozzles		0.11 to 0.16	
Peripheral speed of bearing journal	ft/s	<197	
Combined shaft end stresses		<80% yield stress	
Coupling hub slip ratio		<1.5	

3.2.3.2 Nozzle limitation

Since nozzles are part of a compressor casing, they are installed for the life of the compressor. A proper analysis needs to be done to determine nozzle limitations.

Higher velocities result in higher pressure drops through the nozzles; they cascade through the compressor. Increased pressure drops result in lower overall compressor efficiency and performance. Low compression efficiency manifests itself in higher power requirements. To address the issue, most compressor vendors use an upper velocity limit of 140 ft/s through suction nozzles. This velocity is based on air at 80 °F but process gases will inevitably have different molecular weights and *k* values. Therefore, a better approach is to use the Mach number of the particular gas passing through the nozzle to determine nozzle limitation. For example, the acoustic velocity of air at 80 °F is 1,139 ft/s (347 m/s). For an inlet nozzle velocity of 140 ft/s, the inlet nozzle Mach number is 140/1,139 or 0.123. This is used as the basis for comparison to other gases at other conditions. The first step is to calculate the actual inlet nozzle Mach number at design/operating condition and then ratio this with respect to 0.123.

As an example, we calculate the inlet nozzle velocity for a compressor having a 40 in nozzle and operating at the following conditions:

Speed	7,150 rpm
Nozzle diameter	20 in
Flow through inlet nozzle	12,565 acfm
Mole weight of gas "Mw"	42
Specific heat ratio "k"	1.209
Inlet temperature "T"	−45.4 °F
Compressibility	0.91615
Universal gas constant "R_u"	1,545 ft lbf/lb mol R

Calculate

Nozzle area "A"	$\pi\,(20/12)^2$ x1/4	= 2.18 ft^2
Nozzle velocity	(12,565/2.18) × 1/60	= 95.98 ft/s
Static temperature "Ts"	T + 460	= 414.6 °R
Specific gas constant "R"	R_u/Mw	= 36.79
Sonic velocity (from eq. (18.1))	$(kgRTs)^{0.5}$	= 755.66 ft/s
Nozzle Mach number	95.98/755.66	= 0.127

Some compressor manufacturers recommend that the ratio of the calculated Mach numbers to 0.123 should never exceed 1.35. Another recommended guideline for nozzle inlet Mach number is ≤0.11 for new machines and 0.11–0.16 for machines undergoing rerates.

3.2.3.3 Shaft end limitation

The stresses transmitted from the coupling hub to the shaft end limit compressor loading; they therefore have to be calculated for any rerate application or operating condition. The principal parameter that limits the loading of the shaft end is the maximum combined stress, which should not exceed 80% of the material yield stress. Also, a parametric or "what if" study should be done to determine maximum uprateability based on power transmitted at the uprated conditions. The uprated power is incremented until the coupling hub overload slip factor reaches approximately 1.50 (overload slip factor = coupling hub slip torque/max continuous torque at the uprated condition) or the combined stress reaches approximately 80% of yield stress. Whichever limiting factor is reached first during the incremental study will set the uprateability. It is usually expressed as a percentage of the specified rated load.

Table 3.4 highlights some of the critical design parameters discussed earlier that limit compressor uprateability. These should be used during the uprateability assessment. To perform an uprateability analysis a parametric study is done on each of the compressor's physical constraining components to determine their individual uprateability. The critical parameter for each constraining component is calculated at rated condition; the calculated value is then compared (as a percentage) to the limiting criteria mentioned in Table 3.4. Final uprateability of the compressor is governed by the lowest number calculated for each limiting constraint.

Some large compressor trains are driven by steam turbines, and so we have included a recommended limiting criterion for maximum steam velocity at the pipe flange. By including this parameter, we determine the train uprateability factor for steam turbine-driven trains as a whole and not just the uprateability of the compressor by itself.

Getting back to the example mentioned earlier and assuming that the 8HS compressor in Table 3.5 was driven by a steam turbine, a proper uprateability analysis of the train was done; the data are typically tabulated as shown in Table 3.5.

Table 3.5: Compressor train uprateability parameters.

Compressor frame	(1) Steam velocity at inlet flange	(2) Casing capacity	(3) Gas velocity at nozzle flange	(4) Speed limitation of bearing span (% speed increase)	(5) Shaft end strength	(6) Percent frame loading = min (lowest number from columns 1 through 5)
8HS	122%	149%	117%	128%	149%	117%

The train uprateability parameters are referenced with respect to the specified rated case and show that the train is only capable of 117% uprate, and not 149% as the vendor would like you to believe.

For the multibody compressor trains shown in Figure 3.47, uprateability is first assessed one body at a time, and then looking at the train as a whole (see Table 3.6).

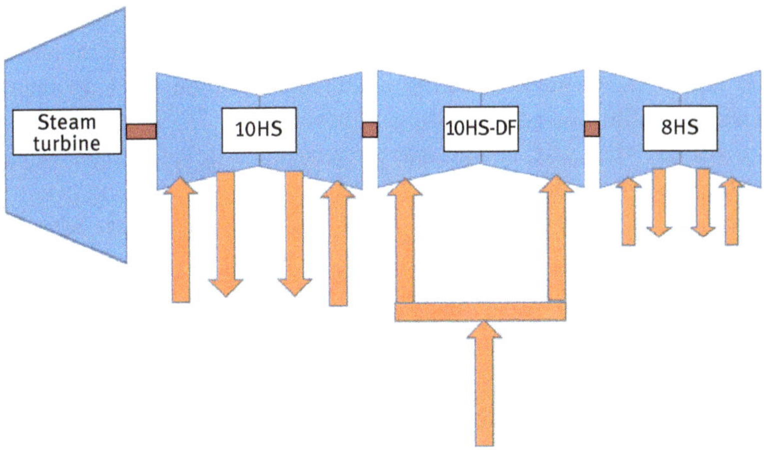

Figure 3.47: Multibody compressor train.

Table 3.6: Uprateability for compressor trains with three compressor bodies.

Compressor frame	(1) Steam velocity at inlet flange	(2) Casing capacity	(3) Gas velocity at nozzle flange	(4) Limitation of bearing span (available speed increase)	(5) Shaft end strength	(6) Percent frame loading = min (1,2,3,4,5)
Steam Turbine	122%					
10HS	______	161%	139%	123%	157%	123%
10HS-DF	______	188%	169%	154%	154%	154%
8HS	______	149%	117%	128%	149%	117%

Uprateability of the train is therefore 117%. assessment of train uprateability should be done during contract negotiations with the compressor vendor and not after final procurement. Recall the importance of MQA in the assessment process.

3.2.4 Using available database to assist design audits and selection

Over the last 40 years, centrifugal compressor manufacturers have compiled a vast array of impellers and associated stationary components allowing for higher efficiency staging. Within this array, there are different families of impellers and stationary parts that allow application engineers to make selections to match various process conditions. They have achieved this through newer and more sophisticated machining technologies and by consistently using more accurate CFD analyses. However, even with these advanced technologies, the matching of internals such as diffusers and return channels with new and improved impellers can be difficult. In instances where trial and error must be used, the end results will be accurate but it will take time to get there. Geometric scaling is still the common scale-up technique; it makes use of similitude or proportionality principles.

To reconfirm: For the most part, compressor manufacturers do get it right when matching stages to process conditions. However, problems arise when scaling is used to extrapolate stages outside the vendor's normal experience or comfort zone without proper CFD validation techniques. Therefore, user–purchasers must check a manufacturer's proposed staging when purchasing centrifugal compressors. The purchaser should ensure that compressor manufacturers submit experience lists for these stages. Alternatively, or if the user has data from past purchases, in-house experience could be utilized. Early delivery of the compressor train, field erection with attention to the proper placement of temporary, but fully approved measuring instruments, and full scale field testing is often possible and should be on the purchaser's list of options.

Figure 3.48 shows two trend lines of head per stage versus molecular weight obtained during a 40 years period of data mining. The database covers molecular weight applications ranging from 4 to 104, using compressors from different manufacturers. These trend curves have been used in the past as a rapid checking tool for proposals from various compressor manufacturers. Both new and revamped (re-rated and upgraded) compressors are represented in Figure 3.48.

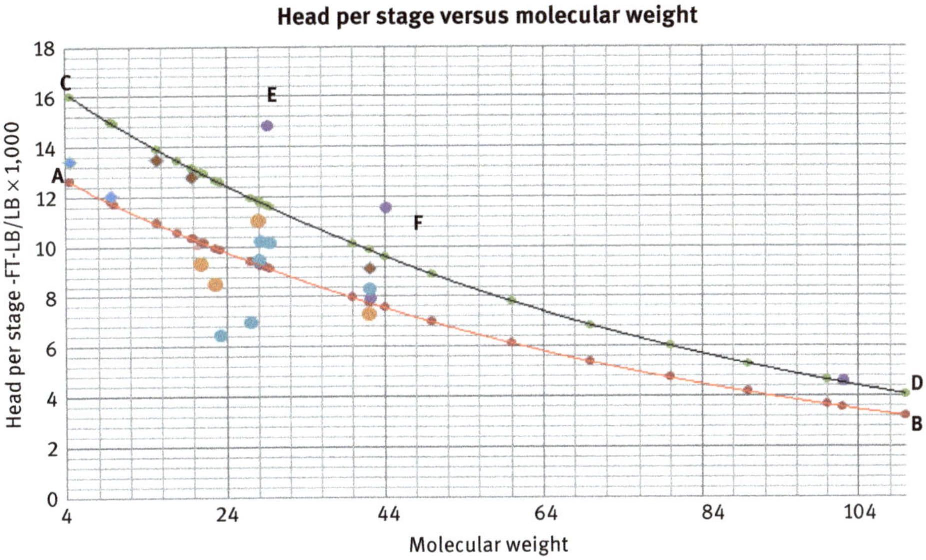

Figure 3.48: Head per stage versus molecular weight curves obtained during 40 years of data mining.

Trend line A–B indicates compressors manufactured pre-1990, and trend line C–D is for compressors manufactured post-1990 to about 2019. Points E and F are two case histories where the compressor manufacturer initially did not perform proper CFD to match the high coefficient impellers. Several iterations had to be done using high-fidelity CFD techniques to be able to match the stationary internal components with the proposed impeller.

Over the years, trend lines A–B and C–D have shown that if a compressor manufacturer's compression stage design fits on or below these lines, the compressor will very likely perform as designed. These lines are considered suitable audit tools to either have confidence in a proposed offer or to raise questions on the compressor manufacturer's proposed staging. However, these lines are not considered foolproof and should only be used to raise questions when designs deviate from the lines. Experienced third parties are sometimes brought in to assess certain risks or to assist in resolving different opinions before they become more serious disputes. Engaging in structured MQA will be of great benefit to the user–purchaser.

3.2.5 Rethinking driver selections for ethylene refrigeration trains

For most ethylene plants, the major compressor trains are a process gas compressor (PGC), a propylene refrigeration compressor (PRC), and an ethylene refrigeration compressor (ERC). Some facilities also have a methane refrigeration compressor train, depending on the licensor's technology.

Historically, the ERC train is physically smaller and requires 30–50% of the power and/or footprint of the PGC and PRC. The trend from the late 1970s until 2020 has been for existing plants to go through successive capacity increases. Many of these increases have shifted an occasional plant from originally 300,000 metric tons per annum (MTA) to as much as 1,500,000 MTA. In the 2015 time frame, new grassroots plants often started with screening capacities in the range from 1.5 to 2.0 million MTA.

The ERC trains at the lower plant capacities were small enough to use standard fixed speed induction motors in combination with a gear speed increaser as the default drive train. However, as plant capacities grew larger all trains were sized up proportionately, and shaft power for these ERC trains went from approximately 5,000 HP to now in excess of 70,000 HP – again depending on the licensor's technology. As a result, the approach to driver selection for larger ERC trains has become more complex and deserving of attention. Engineers selecting drivers for these larger ERC trains look at the power threshold and a plant's energy balance to determine where a particular driver type can be employed most effectively.

3.2.5.1 Driver criteria

Engineering options for driver selection must take into account the pros and cons of each option. A simple overview might include the following:

Induction motors with gear speed increaser:

Pros

- Use of a standard induction motor allows for train simplicity
- Not much associated auxiliary equipment
- Simple to maintain
- Simple start-up procedure and process control logic

Cons

- Process control is usually done by suction throttling; this results in lower than average thermal efficiency
- Lack of soft start capability results in high inrush currents. After a defined number of start-up attempts, the motor must be allowed to cool. Restarting could be delayed.
- Additional instrumentation is needed for monitoring the gearbox
- This type of drive train can be a suitable choice whenever the required power is relatively low

Variable speed motor using variable frequency drives:
Pros
- Soft start capability
- Variable speed allows better process optimization with wider operating range

Cons
- Complex circuitry
- Additional space requirement in motor control center (MCC) for drive system control panel
- Risk of possible torsional interactions from subsynchronous electrical frequencies
- Higher than normal cost

Large fixed speed induction motor with variable speed gearbox:
Pros
- Soft start capability
- Variable speed allows better process optimization with wider operating range

Cons
- Large footprint
- Complexity of maintenance
- Higher cost consideration

Steam turbines:
Pros
- Soft start capability
- Variable speed allows better process optimization with wider operating range
- Higher power availability compared to induction or synchronous motor as per Figure 3.49
- Effective choice for plant steam system efficiency optimization. Steam balance can be optimized with the extraction nozzle and/or admission nozzle over a wide operating range
- High reliability

Cons
- Cost consideration of incidental facilities such as boilers, condensers, and piping
- Over the last 20 plus years, steam turbines continue to be the driver of choice for these large ERC trains. On the other hand, variable speed motors or fixed speed motors with variable speed gears, although occasionally chosen as drivers for these larger ERC trains are yet not common enough as a driver for the larger ERC trains.
- Although steam turbines require a lot more incidental facilities (boiler, condensers, piping, etc.), ethylene plants usually have these facilities available as

integral parts of their utility systems and, therefore, would have enough steam. To date, there are many large ethylene plants that have installed steam turbine drivers for their ERC train; as of 2020, it appears to be a continuing trend.

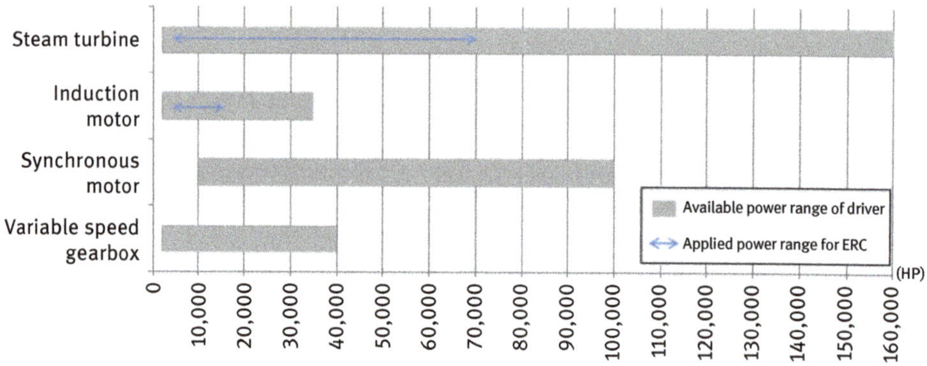

Figure 3.49: Applied power range of ERC (source: *Mitsubishi Heavy Industries Compressor Corporation (MHI), Hiroshima, Japan*).

In any case, the type of driver that is used for the ERC should be determined by considering the following factors:
- utility/energy costs and availability
- construction cost and time needed
- reliability of operation
- power requirement
- operating speed and range

Table 3.7 is a typical comparison between drivers used in ethylene plants. It is shown here for the purpose of illustration and is greatly simplified. Nevertheless, the table and our narrative are demonstrations of the concerns that must go into compressor scale-up and driver selection.

The process licensor will make known their experience-based recommendations or, in some instances, stipulate rather definitive requirements. While EPC firms play a major role, the owner–purchaser's project executives must depend on the firm support of equipment reliability professionals whose understanding of numerous interacting factors will be of great importance.

Table 3.7: Typical comparison of drivers *(Mitsubishi Heavy Industries Compressor Corporation (MHI), Hiroshima, Japan).*

Driver type	Advantages	Disadvantages
Steam turbine	– High reliability – Wide operating range – Easier process Optimization – Small footprint	– Additional auxiliary facilities required
Electric motor	– Easy start-up	– Complexity of electric power facilities (a) Consideration for inrush current (b) Consideration for controlling inrush current (c) Consideration for dealing with short circuits (d) Overall stability of plant electrical system during start-up
Fixed speed with gearbox	– Simple design – Low cycle maintenance – Small footprint	– Low overall mechanical efficiency (gearbox plus motor) – Low process efficiency using suction throttling – Small operating range
Variable speed motor	– Wide operating range	– Space and utility requirement of drive system control panel – Consideration to subsynchronous torsional interaction
Fixed speed with variable speed	– Wide operating range	– Complexity of maintenance – Large footprint

References

[1] Eisenberg, B., Voss, H. Realization of Large Scale ASU Turbo Compressor Machine Trains in Mega Process Plants. IMechE, C633/25/2006, 2006.

[2] Ziegler, K., Eisenberg, B., Hörmeyer, K., Emmrich, R., Ressing, H. Development and Verification of a New Axial Compressor Generation for Industrial Application – Part 1: Compressor Design and Performance. ASME-Paper No. GT2014-26294, 2014.

[3] Anding, D., Ressing, H., Hörmeyer, K., Pisch, R., Ziegler, K. Development and Verification of a New Axial Compressor Generation for Industrial Application – Part 2: Blade Mechanics. ASME-Paper GT2014-26246, 2014.

[4] Broichhausen, K., Ziegler, K. Supersonic and Transonic Compressors: Past, Status and Technology Trends. Proceedings of the ASME, GT2005-69067, 2005.
[5] Eisenberg, B. Development of a New Front Stage for an Industrial Axial Flow Compressor, Transactions of the ASME, 604, 1994, 116(4), 597–604.
[6] Cumpsty, N. Compressor Aerodynamics, Krieger publishing company, Malabar, Florida, 2004.
[7] Armstrong, E., Stevenson, R. Some Practical Aspects of Compressor Blade Vibration, The Journal of the Royal Aeronautical Society, 1960, 64(591), 117–130.
[8] Traupel, W. Thermische Turbomaschinen, Springer-Verlag, Berlin, Heidelberg, New York, 2001.
[9] Bathe, K.-J. Finite Elemente Procedures, Prentice Hall, Pearson Education, Inc, 1995.
[10] Chellini, R. MAX1 – A New Generation of Air Compressors, CompressorTechTwo, 2010, XV(VII), 28–32.
[11] Gottfried, D., Fleeter, S. Fluid-Structure Interaction Analysis of Blade Row Unsteady Aerodynamics and High Cycle Fatigue. Proceedings of the 9th International Symposium on Unsteady Aerodynamics, Aeroacoustics and Aeroelasticity of Turbomachines and First Legendre Lecture Series, Lyon, France, September 4–8, 2000.
[12] Mayorca, M. Numerical Methods for Turbomachinery Aeromechanical Predictions, Doctoral Thesis, KTH Stockholm, Sweden, 2011.
[13] Ning, W., He, L. Computation of Unsteady Flows Around Oscillating Blades Using Linear And Nonlinear Harmonic Euler Methods, ASME Journal of Turbomachinery, 1998, 120(3), 508–514.
[14] Kahl, G. Application of the Time Linearized Euler Method to Flutter and Forced Response Calculations. ASME-Paper 95-GT-123, Proceedings of the International Gas Turbine and Aeroengine Congress and Exposition, Houston, Texas, June 5–8, 1995.
[15] Knappett, D., Garcia, J. Blade Tip Timing and Strain Gauge Correlation on Compressor Blades. Proceedings of the Institution of Mechanical Engineers, Part G: Journal of Aerospace Engineering, 2008, 222, 4, 497–506.
[16] Lawson, C., Ivey, P. Compressor Blade Tip Timing Using Capacitance Tip Clearance Probes. Proceedings of ASME Turbo Expo 2003, Atlanta, Georgia, 2003, June 16–19.
[17] Chow, R., McMordie, B., Wiegand, R. New Coatings Extend Compressor Service Life, Hydrocarbon Processing, (October 1995), 71–74.
[18] Wang, W., Dowson, P., Baha, A. Development of Antifouling and Corrosion Resistant Coatings for Petrochemical Compressors. 32nd Turbomachinery Symposium. Houston, TX, 2003.
[19] Engber, M., Rüd, K., Ardey, S., Gier, J., Waschka, W. Advanced Technologies for Next Generation Regional JETS – SURVEY of Research Activities at MTU Aero Engines. ISABE-2007-1282, 2007.
[20] Wiedermann, A., Frank, D., Orth, U., Beukenberg, M. Computational and Experimental Analysis of an Industrial Gas Turbine Compressor. ASME-Paper No. GT2011-4633, 2011.
[21] Kügeler, E., Nürnberger, D., Weber, A., Engel, K.. Influence of Blade Fillets on the Performance of a 15 Stage Gas Turbine Compressor. ASME-Paper No. GT2008-50748, 2008.
[22] Zielinski, M., Ziller, G. Noncontact Blade Vibration and Tip Clearance Measurement System for Aero Engine Application. ISABE-2007-1109, 2007.

Chapter 4
Technical briefs on positive displacement compressors

4.1 Screw compressors

4.1.1 Using innovative new rotor profile to maximize performance of screw compressor rotor in general

Rotary screw compressors were not widely used in the 1960s; they were perceived to have low efficiency in relation to competing types of compressors. The inefficiencies of these compressors stemmed from the complex helical shape of the rotor. Nonoptimized rotors were tried by some vendors; it was said that they attempted to cut costs. Likely some were trying to use the same clearance regardless of casing and rotor materials. Of course, thermal growths will vary when different alloys are involved. Moreover, in decades past, there were leakage paths when the rotor-to-stator fits were too open; there also were occasional rotor seizures when clearances were too tight, and real problems arose when a manufacturer was reputedly attempting to use a "standard" clearance for all conceivable applications.

However, by thoroughly studying the sources of leakage, competent engineers have improved screw compressor efficiencies to be no different from those of other types of compressors. Optimizing screw compressor profiles to fully accommodate specific process applications was responsible for much of the improvement.

Singh and Patel [1] and Singh and Bowman [2] identified and then quantified the influence of the various leakage paths on screw compressor performance. The three researchers showed the deleterious effects of previously neglected leakage paths. It goes without saying that leakage affects the performance of all fluid machines.

4.1.1.1 Identifying the factors influencing screw compressor efficiency

Twin-screw compressors use a pair of male and female helical screw rotors (Figure 4.1). The centerlines of the rotors are parallel. Their operating principle is very straightforward and best explained by assuming a rotor length of, say, 1,000 mm. Assume that as the pair of rotors turn, a volume of 2,000 mL of gas is trapped and that the volume weighs 10 g. The next turn forces the 2,000 mL into a 1,950 mL space, the next one into a 1,900 mL space, the next one into a space of 1,850 mL, and so forth. Finally, after advancing the full 1,000 mm of available rotor length, a trapped volume of now only 400 mL is forced into a receiver at what we call the discharge pressure. All the while, the mass that is being advanced per trapped volume stays the same, for example, 10 g per turn. However, the pressure in the final 400 mL volume then equals discharge pressure; also, its volume reduction is 2,000/400 = 5:1. The compression

https://doi.org/10.1515/9783110678765-004

Figure 4.1: Male and female rotors in mesh showing sealing points (source: *KOBELCO, Japan*).

ratio and temperature will have increased in harmony with the gas laws explained in an earlier chapter.

Because the *gas working volume* is enclosed by a common housing (or casing), the clearances between rotors and the surrounding housing bores take on critical importance. Considering that each trapped volume of gas advances and produces progressively increasing pressures during operation, it is easy to visualize how backflow (or leakage to the adjacent lower pressure regions) can occur.

If we look at a single trapped volume of gas, the leakage paths where the *gas working volume* leaks in and out can be identified in Figure 4.2 and as follows [3]:

- **Path #1.** This leakage path is where the gas leaks from a trapped volume (also called an "enclosed cavity") back to suction across a *contact line* between male and female rotors. While such leakage has very little influence on machine capacity, continuous recompression of the leaked flow causes power loss and increasing gas discharge temperature.
- **Path #2.** This leakage path allows gas to leak across a *sealing line* into the adjacent ever-moving trapped volume (or cavity) or region of lower pressure. The leakage takes place along the bore of the compressor housing.
- **Path #3.** Leakage occurs through the triangular-shaped area called a cusp. The cusp (or *blow hole)* is formed by the male–female rotor lobes and the housing on the high-pressure side. Cusps (*blow holes)* are the inevitable consequence of rotor profile geometries that engineers try to minimize. By definition, a cusp is the smallest triangular area bounded by the housing and the two rotors. The cusp exists in a plane perpendicular to the primary leakage flow path between two adjacent cavities. The leakage triangle does not lie in a Cartesian plane. It can only properly be defined in curvilinear coordinates.

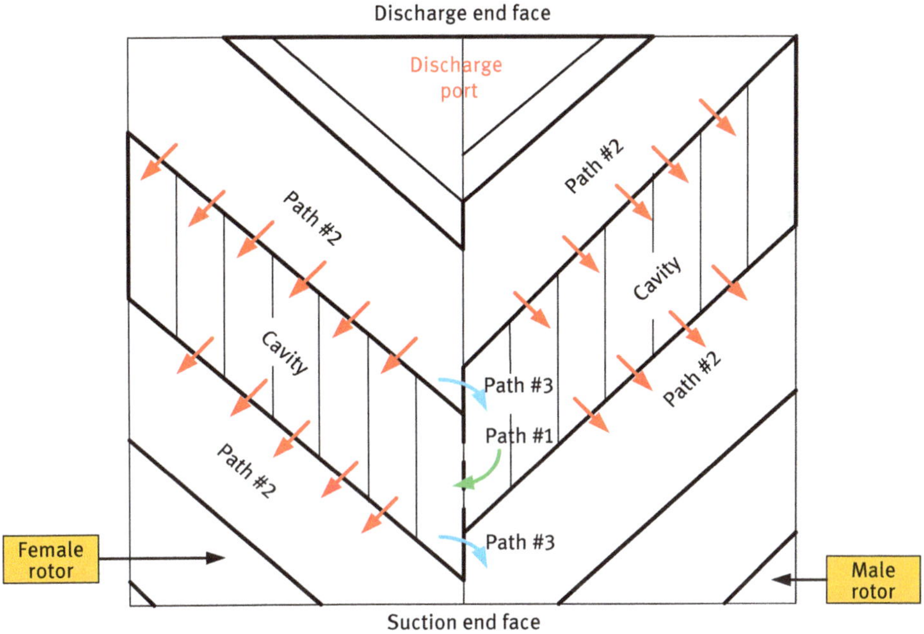

Figure 4.2: Leakage paths of an enclosed cavity for twin-screw compressor [3].

- **Path #4.** At the start of compression, a cusp is created and runs along the housing cup on the low-pressure side for about 30–40° of the male rotor rotations. This is called the *compression-start cusp (blow hole)*, and the gas leaks back to suction.
- **Path #5.** Leakage to suction through the clearance between the suction end wall and the rotor end face.
- **Path #6.** Leakage through the clearance between the discharge end wall and the rotor end face back to the preceding (or adjacent) trapped volume.

4.1.1.2 Considerations for minimizing leakage paths

Knowing how these leakage paths are formed, screw compressor profiles can now be designed with better mathematical modelling and numerical simulations using computational fluid dynamics (CFD). These CFD models run performance simulations by examining ways to optimize the screw compressor's profile by adopting well understood basic design practices; these include the following:

1. Shortening the rotor sealing-line length per advancing unit volume. A unit volume is represented by a "per turn" volume. Leakage across the rotor sealing line generally has the largest influence on adiabatic efficiency.
2. Minimizing the passage way called the cusp or "*blow hole*" between the male/female rotors and casing. Any improvement in profile should concentrate on

reducing the cusp (blow hole) to improve adiabatic efficiency, even though its influence on volumetric efficiency (VE) is small.

Changing the wrap angle of male rotor will affect both the *sealing length* and cusp (*blow hole*). However, changing wrap angle will always be a compromise effort based on the following:

- The greater the *wrap angle,* the smaller the cusp (*blow hole)* and the smaller the *sealing line* length [4]. However, the *contact-line* length will increase. The cusp's contribution to overall losses, although relatively high compared with other losses, will decrease significantly with increasing speed.
- The adiabatic efficiency decreases as the wrap angle decreases.

The effects of rotor profile geometry cannot be overstated, although other geometric factors such as rotor *L*/*D* ratio and number of lobes must also be considered; they, too, will influence performance. Screw compressors have advanced from symmetric profiles with large cusp (*blow hole*) regions to asymmetric profiles with smaller cusps. Studies show performance gains of 8–10% as industry moved from symmetric to asymmetric profiles despite an increase in rotor-to-rotor contact, thus increasing the potential leakage region.

Although optimizing screw compressor performance by designing smaller cusps or, having no cusps at all, would be ideal but is not a practical option. The various interacting or influencing factors cannot be optimized independently. The application and its operating conditions should be considered whenever a new rotor profile is designed. It is, therefore, not easy to meet all the conditions since many parameters are interdependent.

For example, one would think that by reducing both the cusp region and the contact-line length, one might improve the compressor's performance. However, for a given rotor profile reducing the cusp region results in an increase in the length of the contact line. The designer must, therefore, compromise between size and thermodynamic efficiency.

Mathematical models on how to minimize the cusp (blow hole) region have been addressed by P. J. Sing, et al. [5]

4.1.1.3 *L*/*D* considerations

The *L*/*D* ratio is the length-to-diameter ratio of the rotor. The merits of producing compressors with equal diameter rotors are not based on design considerations but mostly likely on cost and manufacturing efficiencies. In practice, the manufacturer will choose equal diameters, and rotor profile pairs will have four male lobes and six female lobes (Figure 4.1). The design will not be ideal throughout; to accommodate the 4–6 combination, relatively large cusp regions are required.

At higher tip speeds, leakage has less of an influence on performance for all L/D ratios.

L/D ratios between 1.65 and 1.8 typically give the best overall performance.

4.1.1.4 Lobe combination

The most common industrial lobe combination is 4 + 6 as shown in Figure 4.3(a)–(c). This combination covers a wide range of applications and includes both oil-free and liquid-flooded screw compressors.

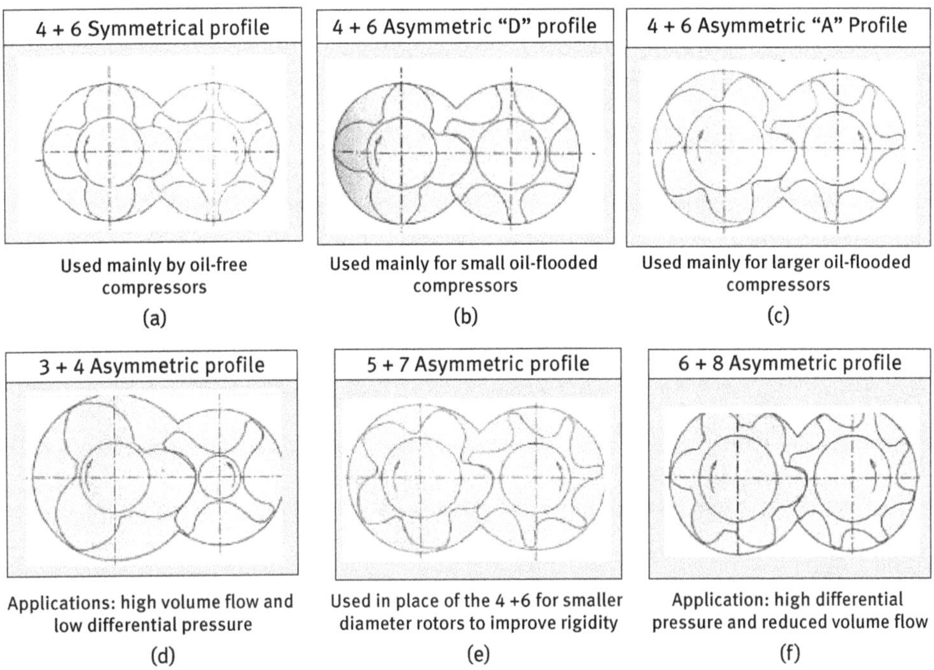

Figure 4.3: (a)–(f) Recommended profiles for different applications.

Increasing the number of lobes generally decreases the velocity at the discharge port. On the other hand, screw compressors with profiles having a smaller number of lobes experience higher discharge port losses. Furthermore, discharge port losses become significant at higher tip speed that, combined with a lower number of lobes, will cause a decrease in the machine's performance.

A parametric study was performed by You et al. [6] on twin-screw refrigeration compressors. The objective was to determine how various geometric parameters

affect performance; the researchers used four different lobe combinations and five different L/D ratios and wrap angles. The study was done with the following:
- Lobe combinations (4 + 5), (4 + 6), (5 + 6), and (5 + 7)
- L/D ratios from 1.0 to 2.2
- Male rotor wrap angles ranging from 250° to 300°

Based on the results of the study, they concluded the following (Figures 4.4 and 4.5):
- Lobe combination (4 + 6) and (5 + 7) profiles are better choices for high-pressure applications.
- (5 + 6) lobe profile had relatively higher isentropic efficiencies for L/D ratios ranging from 1.0 up to 1.7.
- Performance of the (5 + 7) and (5 + 6) pairs was nearly identical for L/D ratios in excess of 1.7.

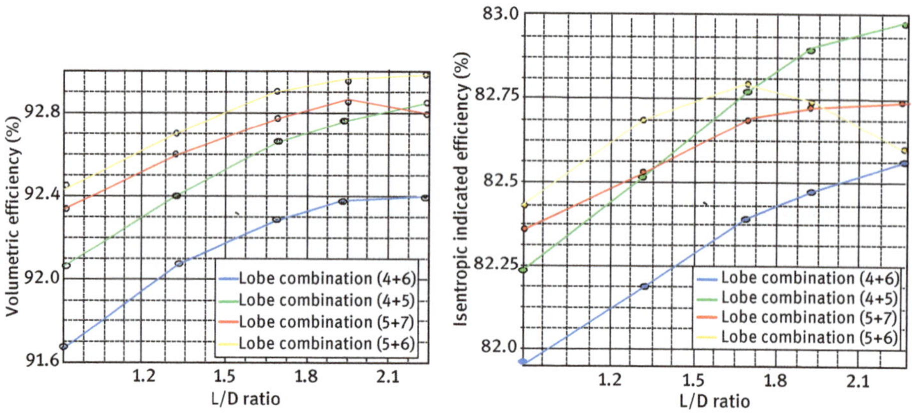

Figure 4.4: Effects of L/D on volumetric and isentropic indicated efficiencies for different profile combinations [6].

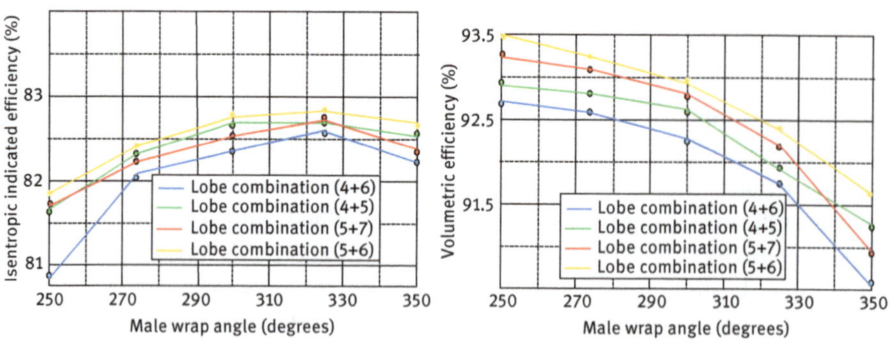

Figure 4.5: Effects of male rotor wrap angle on volumetric and isentropic indicated efficiencies for different profile combinations [6].

Other profile combinations not mentioned in the study, such as a (3 + 5) combination, are suitable for larger capacity and low-pressure service.

4.1.1.5 Rotor profile generation and manufacturing complexities

Producing these complex profiles required better manufacturing techniques [7–9]. Using newly developed CAD-connected machine tools, the desired optimized profiles can be machined within 1% of their narrowly specified tolerances. Advancements in numerical-controlled machining centers have greatly reduced the time previously invested in prototyping efforts.

The rotors of *oil-free* screw compressor are manufactured from forgings to obtain consistent and fully predictable mechanical properties. There are additional savings in machining time because rotor blanks are generally forged with dimensions and contours that closely approximate the overall shape of the rotor.

One of two materials, carbon steel S35C-N (API/NACE standard) or stainless steel, 13Cr–3.4Ni (API/NACE corrosive gas/water injection standard) are generally selected for dry (oil-free) twin-screw compressors. Although oil-free screw compressors have no oil to act as a corrosion inhibitor against aggressive gases, their rotors can be manufactured from carbon steels or stainless steels. The designer's choice depends on the gas being compressed.

Some oil-free rotors have an internal hollow chamber along their length. In rare cases, hollow or "cored" passages are used to pass oil along for "cooling" purposes. While this seems like a paradox, such designs do exist.

Depending upon the discharge temperature generated in an *oil-free* screw compressor, the rotor can be machined with a taper along its entire length to allow for expansion at the elevated temperatures near the discharge end. Oil-free rotors also require precision timing gears to ensure that the rotor clearances are maintained during operation.

As is the case for oil-free compressors, oil-flooded screw compressor rotors are also manufactured from carbon steel; moreover, certain rotor sizes and types are manufactured from forgings to ensure consistency of mechanical properties.

Typical rotor materials selected for oil-flooded screw compressors are again carbon steel S35C-N (API/NACE) for low-pressure machines. Carbon steel S45C-N (API/NACE) is customarily used for high-pressure machines or machines that demand high-strength material for other reason.

In summary, and to optimize screw compressor performance, rotor profiles are designed with large cross-section areas, short sealing lines, and the smallest practical cusp (blow hole) regions. Today's screw compressor rotor profiles are optimized using advanced computer modelling.

4.1.2 Optimizing slide valve design to improve capacity control

4.1.2.1 Screw compressor slide valves in general

A slide valve (Figure 4.6) is used to efficiently control compressor capacity while the screw compressor is running. Costly variable speed systems are not needed, and control is stepless. Slide valves are very common for *oil-flooded* screw compressors. A single such valve is installed at the bottom of the compressor casing, parallel to the rotor bores, and located within (or close to) the high-pressure cusp region. It can be installed facing one or both rotors.

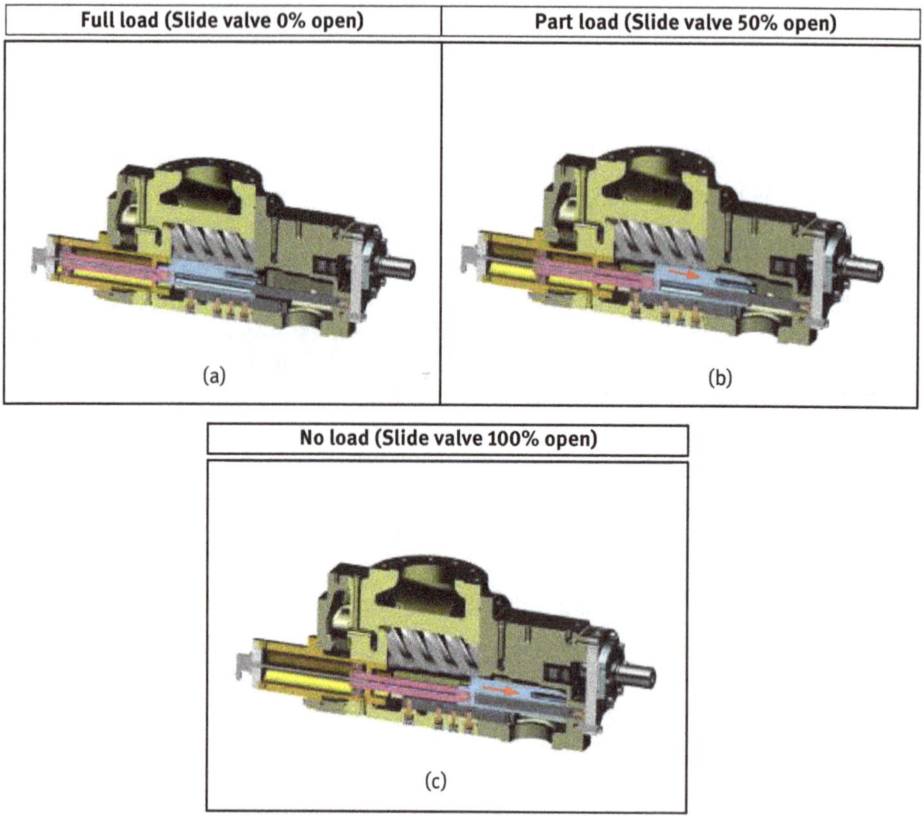

Figure 4.6: (a)–(c) Sequential loading and unloading with slide valve (source: *MAN-Energy Solutions, Germany*)..

4.1.2.2 Different slide valve controlling methods

Slide valves can be used in different ways to control capacity. Some common methods are controlling discharge port length, and/or controlling discharge port and volume ratio. The effects are depicted in the *P–V* diagram of Figure 4.7.

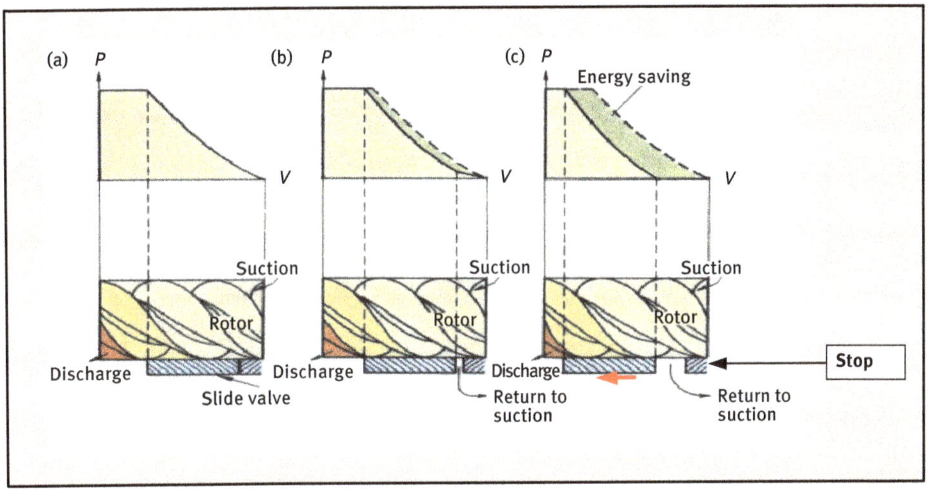

Figure 4.7: Capacity control using slide valve. Note the influence of slide valve position on resulting *P–V* curves (source: *KOBELCO, Japan*).

4.1.2.2.1 Controlling the discharge port

This is the most common type of capacity control used for screw compressors. It is possible to adjust capacity (throughput) between 10% and 100%.

When the slide valve is moved to the discharge side, clearance is created between the slide valve and the casing; this opens a port that returns gas to the suction. Since the return port pressure is the same as suction pressure, gas entering the compressor is not compressed. With the slide valve open, gas entering the compressor is transported toward the interlobe space of the rotor until it reaches the slide valve.

Depending on where the slide valve is positioned, some of the gas will return to suction, and the remainder will be compressed. The slide valve position reduces the effective length of the rotor; fewer turns of the rotor or lobe pair are needed to move the trapped volume from end to end. Both capacity and required shaft power input are decreased as depicted by the areas under the *P–V* curves in Figure 4.7. This method is very efficient since it avoids having to first compress and then bypass (decompress) a portion of the gas.

4.1.2.2.2 Controlling discharge port and volume ratio

In the configuration shown in Figures 4.6 and 4.7, the slide valve is designed with a movable stop. Having a movable stop accommodates stepless slide valve positions. Different positioning provides different volume ratios and capacity adjustments. Once again, this arrangement offers improved energy efficiency for loads down to approximately 40% of full load.

4.1.2.2.3 Power saving combination by increasing suction pressure

In operating conditions where inlet pressures fluctuate significantly, as might be the case in the gas pipeline and fuel gas compression duty shown in Figure 4.8, the actual inlet pressure sometimes exceeds the design point. This then forces the compressor to operate at a low compression ratio with higher inlet pressure and higher than normal discharge pressure. During this type of operation, an oil-flooded screw compressor equipped with slide valve can reduce power consumption.

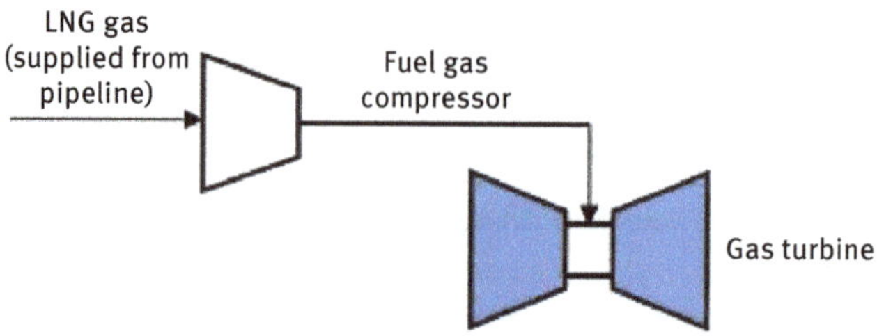

Figure 4.8: Typical arrangement for supplying fuel gas. Oil-flooded screw compressors serve well in fuel gas compression duty (source: *KOBELCO, Japan*).

The relationship between capacity and power load characteristics when the suction pressure changes at constant discharge pressures is shown in Figure 4.9.

Oil-flooded compressors operate with reduced power and increased standard flow volume when the inlet pressure becomes high. This effect is more pronounced in machines with low compression ratios. Compressor frames are usually selected and designed with consideration given to minimum inlet pressure. In actual operation, suction pressures are usually higher than a machine's as-designed capability. Capacity control using a slide valve can save significant energy during actual operating conditions.

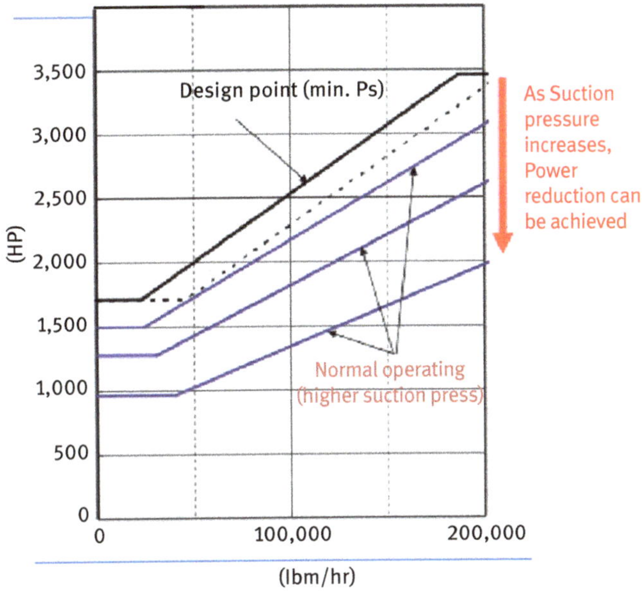

Figure 4.9: Typical performance characteristics of oil-flooded screw compressors (source: *KOBELCO, Japan*).

4.2 Reciprocating compressors

4.2.1 Understanding various industrial interpretations of rod loading

4.2.1.1 Frame loads

4.2.1.1.1 A discussion of reciprocating compressor "rod load" [10]

Reciprocating compressors are usually rated in terms of horsepower, speed, and rod load. Horsepower and speed are easily understood; however, the term "rod load" is interpreted differently by various users, analysts, original equipment manufacturer (OEMs), and so on. "Rod load" is one of the most widely used but least understood reciprocating compressor descriptors in industry. Typical end users know that rod load is a factor used to "rate" a compressor, but they do not generally have a good understanding of how this rating is developed and how to utilize it for machinery protection.

This discussion includes the various definitions of rod load, including historical and current API-618 definitions, manufacturers' ratings, and various user interpretations. It also explains that there are really load limits based on the *running gear* (moving parts such as pistons, rods, crosshead, and crank throw) as well as load limits based on the *stationary components* (frame, crosshead guide, etc.).

The basic kinematics and forces acting on a slider-crank mechanism will be reviewed to provide a better understanding of the various definitions that are used. Analytical results and field rod load measurements will be compared to illustrate the various factors that influence rod load in typical compressor installations.

4.2.1.1.2 Basic theory

Consider the typical double-acting compressor cylinder geometry illustrated in Figure 4.10. The loads (forces) that are generally of concern include the piston rod loads, the connecting rod loads, the crosshead pin loads, the crankpin loads, and the frame loads.

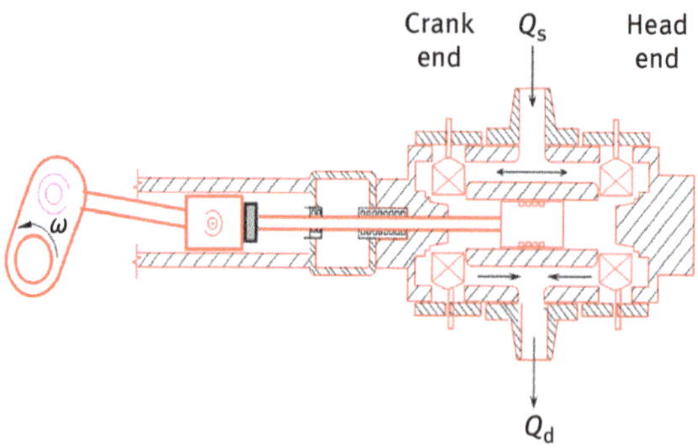

Figure 4.10: Double-acting compressor cylinder *(Engineering Dynamics Inc., San Antonio, Texas).*

As the crankshaft undergoes one revolution, all of these loads vary from minimum to maximum values. These loads are generated both by gas and inertia forces, as discussed in the following paragraphs.

Gas loads

As the compressor piston moves to compress gas, the differential pressures acting on the piston and stationary components result in gas forces as illustrated in Figure 4.11. An ideal pressure versus time diagram for a typical double-acting compressor cylinder is shown in Figure 4.12. The pressures acting on the piston faces (head end and crank end) result in forces on the piston rod. The force acting on the piston rod due to the cylinder pressures alone alternates from tension to compression during the course of each crankshaft revolution. It is straightforward to compute the net force on the piston rod due to pressure. A plot of this force versus crank angle for the ideal *P–T* diagram is shown in Figure 4.13. The magnitude of the tensile (+) gas force is less than the compressive (–) gas force because of the area of the piston rod. The forces due to pressure also act (equal and opposite) on the stationary components.

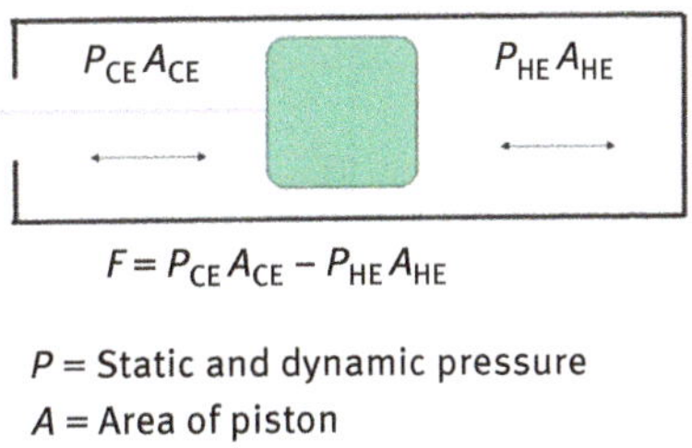

Figure 4.11: Gas loads *(Engineering Dynamics Inc., San Antonio, Texas).*

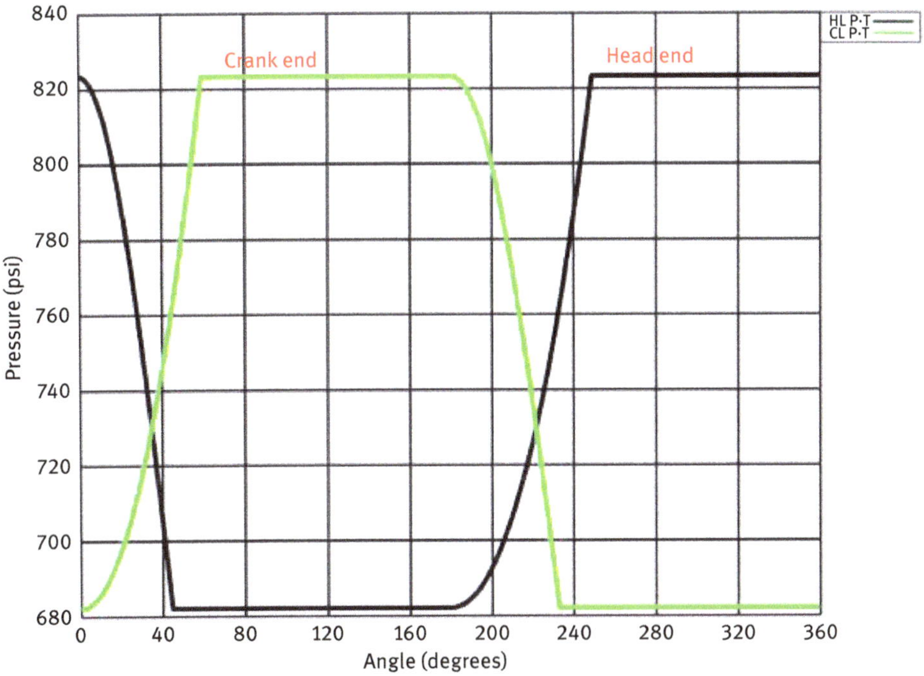

Figure 4.12: Ideal *P–T* diagram *(Engineering Dynamics Inc., San Antonio, Texas).*

The maximum compressive force due to pressure occurs when the head end is at discharge pressure and the maximum tensile force due to pressure occurs when the crank end is at discharge pressure. Therefore, the equation shown in Figure 4.11 is often evaluated at the extremes as follows:

$$F_{Tension} = (P_{Discharge} \times A_{CE}) - (P_{Suction} \times A_{HE}) \tag{4.1}$$

$$F_{Compression} = (P_{Discharge} \times A_{HE}) - (P_{Suction} \times A_{CE}) \tag{4.2}$$

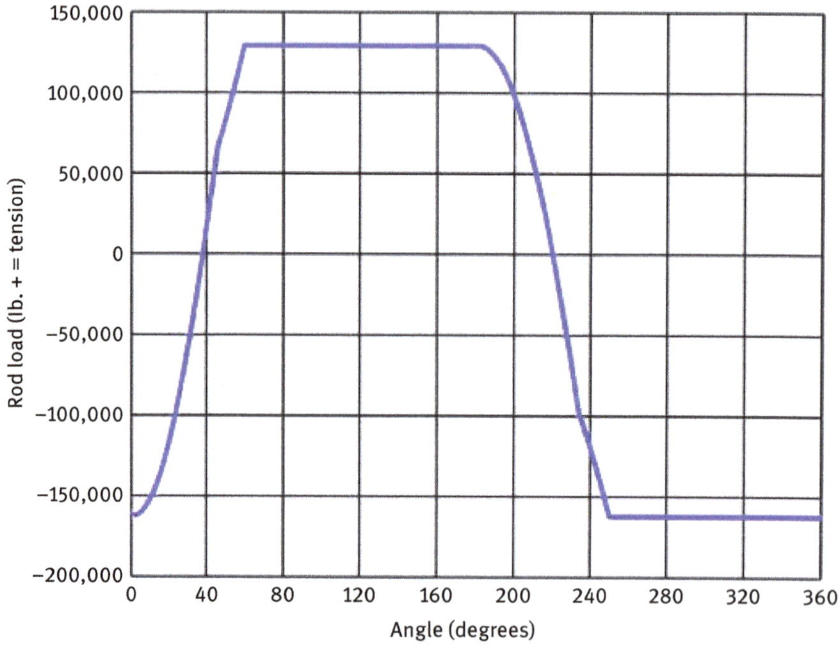

Figure 4.13: Gas loads based on ideal *P–T* diagram *(Engineering Dynamics Inc., San Antonio, Texas).*

Now consider a more realistic pressure versus time diagram as shown in Figure 4.14. "Line pressure" refers to the pressure at the line side of the pulsation bottle (suction or discharge). "Flange pressure" refers to the pressure at the cylinder flange. As shown, the in-cylinder discharge pressure exceeds the nominal discharge line pressure, and the in-cylinder suction pressure is less than the nominal suction line pressure due to several effects:

1. Pressure drop due to valve and cylinder passage losses (typically 2–10%)
2. Pressure drop due to pulsation control devices (typically <1%)
3. Pulsation at cylinder valves (typically <7%)
4. Valve dynamics (inertia, stiction, flutter, etc.)

API-618 specifies that the internal pressures must be computed, but does not define any procedure for the calculations. Different methods can be used to account for the non-ideal effects. A common method is to model the valve as an orifice and then the pressure drop across the valve (valve loss) is proportional to the square of the piston velocity (flow). This is illustrated in Figure 4.14. Theoretically, it would be more accurate to use the results of the valve dynamics analysis coupled with the digital pulsation simulation to model the instantaneous pressure at the valves. This is not practical to do until all of the piping and valve details are known. In any case, the difference should be small provided the losses are within the typical values listed earlier.

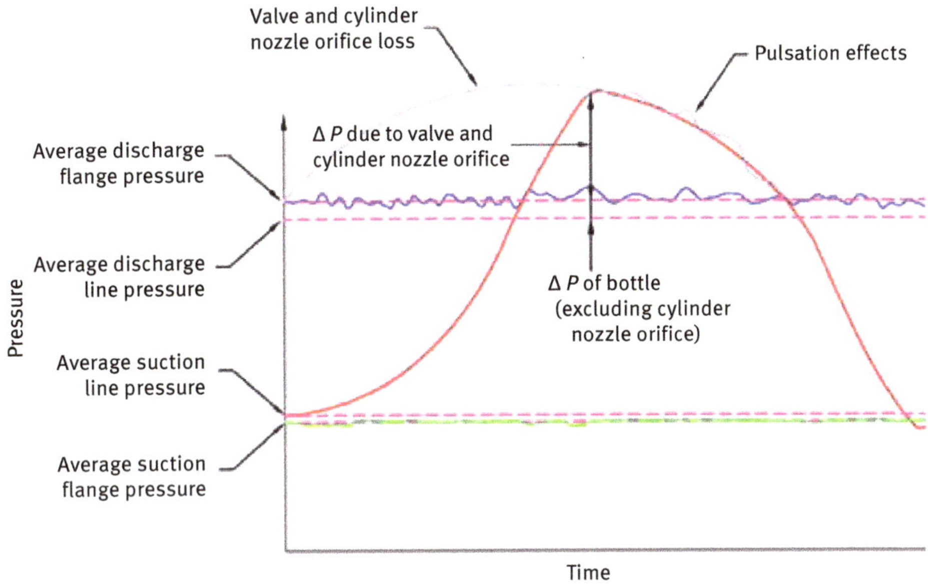

Figure 4.14: Nonideal *P–T* diagram *(Engineering Dynamics Inc., San Antonio, Texas).*

Because of these effects, the forces due to differential pressures are higher on both the running gear and the stationary components than those calculated based on nominal line pressures. However, eqs. (4.1) and (4.2) are still applicable as long as the appropriate pressures (discharge pressure higher than nominal discharge pressure; suction pressure lower than nominal line pressure) are used. If the *nominal* pressures at the suction and discharge cylinder flanges are used for P_{Suction} and $P_{\text{Discharge}}$, then these tension and compression forces represent the term "flange loads" as interpreted by some users. Equations (4.1) and (4.2) are easy to evaluate and for many years were the basis for rating "rod loads" of reciprocating compressors.

Of course, for the general nonideal compressor cylinder, the maximum discharge pressure on the head end will not necessarily occur at the same instant that the minimum suction pressure occurs on the crank end and vice versa. Therefore, it is common to evaluate the gas forces versus crank angle at discrete steps (e.g., every 5° or 10°). The history of these types of calculations is discussed later, but computing the instantaneous force due to differential gas pressures is easily accomplished with computer-based software. If the actual in-cylinder pressures are used and the extremes are evaluated, these forces are then the "gas loads" referred to in the API specifications.

Piston rod loads

The basic slider-crank mechanism is illustrated in Figure 4.15. The exact equation for the position of the crosshead with respect to the x-direction shown is

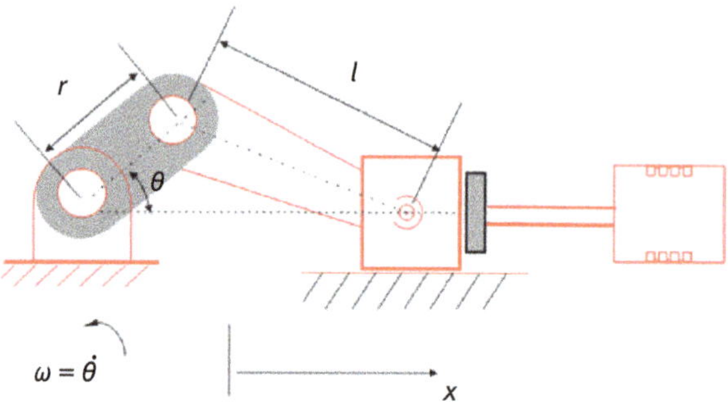

Figure 4.15: Slider-crank geometry *(Engineering Dynamics Inc., San Antonio, Texas).*

$$x = r\cos\theta + l\sqrt{1 - \frac{r^2\sin^2\theta}{l^2}} \tag{4.3}$$

The piston (crosshead) motion is usually approximated using the first two harmonics of the Taylor series as follows:

$$x = r\cos\theta + l\left(1 - \frac{r^2}{2l^2}\sin^2\theta\right) \tag{4.4}$$

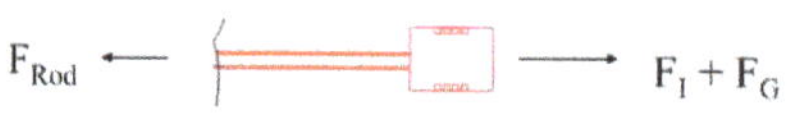

Figure 4.16: Piston rod load *(Engineering Dynamics Inc., San Antonio, Texas).*

The piston rod loads can be evaluated by considering the free body diagram in Figure 4.16. The forces acting on the piston rod are the gas forces due to differential pressures acting on head-end and crank-end piston areas *plus* the inertia forces due to the reciprocating mass. If the reference point is chosen as the crosshead end of the piston rod, then the reciprocating weight will include the piston rod and the piston assembly (piston, rings, rider bands, etc.). The reciprocating inertial force ($F{=}ma$) can be computed using the following equation:

$$F_I = m_{\text{recip}} r\omega^2\left[\cos(\omega t) + \frac{r}{l}\cos(2\omega t)\right] \tag{4.5}$$

where m_{recip} is the mass of reciprocating components, r is the crank radius, ω is the angular velocity, l is the connecting rod length.

The combined piston rod load is the sum of the gas force and the inertial force. In accordance with API-618, this value is routinely calculated in the design stage and used along with the rod area at the minimum cross section to compute tensile and compressive stresses in the piston rods. The stress in the piston rod is one factor to consider in the design, and in some cases, it may be the limiting factor or the "weakest link in the chain." However, this load is not the *rod load* to which API-618 refers to.

Crosshead pin loads

The free body diagram for the system including the crosshead pin is shown in Figure 4.17. Here the mass of the crosshead assembly (crosshead, balance weights, crosshead shoes, etc.) must be considered, but the same equations apply. The combination of the gas loads and inertia loads evaluated at the crosshead pin *in the direction of piston motion* are the "combined rod loads" to which API-618 refers. This load does not consider side forces on the crosshead or the 1/3 of the connecting rod weight that is usually considered to be reciprocating. Thus, "rod load" by API definition is not really a rod load but actually a pin load.

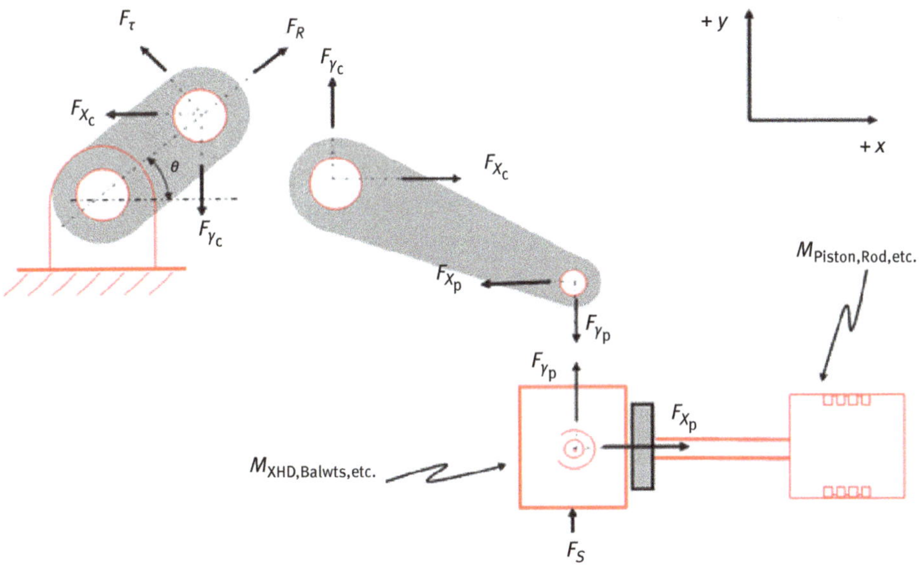

Figure 4.17: Forces acting on crosshead pin (*Engineering Dynamics Inc., San Antonio, Texas*).

Crankpin loads

If the loads and torques throughout the system are evaluated, then the rotating and reciprocating inertias as well as the side forces are included. Equations are applied for computing x and y components of crankpin and wrist pin loads, crank throw torques, main bearing loads, and so on. The typical output of the computer program used to evaluate these loads is shown in Figure 4.18. All of these loads are typically considered at the design stage. Different OEMs evaluate the loads as per their own experience. API guidelines are discussed in the following section.

4.2.1.1.3 History of "rod loads"

The first edition of API-618 was published in 1964 (34 pages). It included no definition of what was meant by the term "rod load." However, the data sheets did call for the compressor manufacturer to specify the "max allowable rod loading" and "rated rod loading." So, the definition of what that meant was left up to the compressor OEM.

In the 1963 edition of a prominent compressor manufacturer frame ratings guide, the "piston loads" were defined. This document stated that "piston load" is frequently referred to as "rod load," which is a misnomer as it implies that the piston rod is the only limit in the establishment of a compressor load rating. It defined the piston load as the nominal pressure at the cylinder flange times the area of the piston. These loads were easily calculated from simple equations presented earlier (eqs. (4.1) and (4.2)). The guide stated that the actual rod load would include the effect of inertia and valve losses, but these effects were considered in the piston rod load ratings, that is, the piston rod load ratings were necessarily conservative. This approach served the industry well, but perhaps resulted in "overdesigned" machinery.

This was before the advent of electronic calculators and digital computers, so combined rod load was tedious to calculate. The practice at the time was to look at and report simply the nominal gas load only without considering valve losses or inertia loads. On rare occasions if it were judged necessary due to a combination of high gas loads, high inertia forces and high VE (which can cause the gas load and inertia load to be additive), a manual calculation of combined rod load (gas + inertia + valve losses) would be done. This would consist of drawing a $P–V$ diagram including valve losses, using a planometer and slide rule to determine area (horsepower) and gas pressures at discrete degrees of rotation increments. Then inertia forces were calculated at each point and added to determine the combined rod load at the crosshead pin, forces in the connecting rod and crankshaft, and torque on the crankshaft. For a six-throw compressor, it would typically require six engineers (one cylinder each) and 1 week to perform this task.

The second edition of API-618 was published in 1974 (39 pages). The committee pushed the compressor manufacturers to advise how rod loads were calculated and to ensure that everyone would calculate rod loads the same way. This established

Crank Ang. (theta)	Local Crank Ang. (alpha)	X-M.Pin* force (lb)	Y-M.Pin* force (lb)	Comp. Piston rod force (lb)	Master side force (lb)	Torque* (In-lb)	Press. Torque* (In-lb)	Press. H.E. (psi)	Press. C.E. (psi)
0	0	−47,279.7	0	−47,597.9	0	0	0	1,698.2	822.5
1	1	−47,156.7	139.3	−47,474.9	139.3	7,698.6	18,334.4	1,697	822.6
2	2	−46,788.7	276.5	−47,106.9	276.5	15,273.7	36,526.8	1,693.4	822.8
3	3	−46,178.8	409.2	−46,497	409.2	22,603.7	54,437.1	1,687.4	823.2
4	4	−45,332	535.4	−45,650.2	535.4	29,570.7	71,928.8	1,679.2	823.8
5	5	−44,255.2	653.1	−44,573.5	653.1	36,061.9	88,871.1	1,668.6	824.5
6	6	−42,957.2	760.4	−43,275.4	760.4	41,971.6	105,139.8	1,655.9	825.4
7	7	−41,448	855.4	−41,766.2	855.4	47,202.2	120,619.5	1,641.2	826.5
8	8	−39,739.3	936.7	−40,057.6	936.7	51,665.6	135,204.3	1,624.4	827.7
9	9	−37,844.1	1,002.7	−38,162.3	1,002.7	55,283.8	148,798.9	1,605.8	829.1
10	10	−35,776.1	1,052.3	−36,094.3	1,052.3	57,990.2	161,319.3	1,585.5	830.6
11	11	−33,550	1,084.4	−33,868.2	1,084.4	59,729.2	172,693.2	1,563.7	832.3
12	12	−31,180.8	1,098.3	−31,499	1,098.3	60,457.3	182,860.5	1,540.3	834.2
13	13	−28,684.1	1,093.3	−29,002.3	1,093.3	60,142.2	191,773	1,545.7	836.2
14	14	−26,075.4	1,069	−26,393.6	1,069	58,763.3	199,394.4	1,490	838.5
15	15	−23,370.3	1,025.1	−23,688.5	1,025.1	56,310.8	205,699.8	1,463.2	840.8

Figure 4.18: Design calculation results *(Engineering Dynamics Inc., San Antonio, Texas).*

Note: Asterisked numbers are calculated values.

the term "allowable rod load" and "actual rod loading." The actual rod load was defined as the force due to the differential pressure across the piston plus the inertia of the reciprocating parts transmitted through the piston rod. It also stated that the actual rod load calculated on the basis of cylinder relieving pressure (RV setting) shall not exceed the vendor's maximum allowable rod load.

By this time mainframe computers and programmable calculators were in widespread use. This allowed for more precise engineering calculations and the elimination of some of the conservatism in the design process. Practice was to calculate compressor performance and "gas load" using a programmable calculator, since the computations were relatively simple. Basic compressor sizing and feasibility studies used these methods.

The final performance including actual rod load (combined rod load) was obtained using mainframe computers (punch cards, overnight batch processing, etc.). Gas loads were still calculated and reported based on nominal cylinder flange gas pressures, but actual rod load included the effect of valve pressure drop and inertia loads. There was variability between various users and OEMs on the reference points used for the calculation of combined rod load. At IR and Worthington, the reference point was the crosshead pin; so all inertia outboard of the pin bearing was included in the combined rod load calculation. There was also a lack of consistency over the relief valve pressure. Some users and OEMs (including IR) used the final relief valve pressure rather than each stage RV (Relief Valve) setting.

In the third edition (1986), API-618 grew to 111 pages. The term maximum allowable combined rod load (MACRL) was defined. The combined rod load was defined as the algebraic sum of the differential gas pressure on the differential piston area plus the inertia force. The reference point for inertia loads was defined as being at the crosshead pin. Additionally, API established a minimum rod load reversal criterion (to ensure proper lubrication, 3% and 15°), but that issue is outside the scope of this book. Gas load still was reported based on nominal cylinder flange pressures, and the relief valve pressure calculations were still inconsistent.

In the fourth edition (1995), API-618 was at 166 pages. The calculation of rod load was defined much more precisely. The terms maximum allowable continuous combined rod load (MACCRL) and maximum allowable continuous gas load (MACGL) were established. This was the first time that load limits based on running gear and load limits based on the stationary components were explicitly separated in the specification. Combined rod load was defined the same way as the third edition but with the clarification that it was to be at the crosshead pin, and only the component in the direction of piston motion was included. Note that the load in the connecting rod is higher due to geometry. Gas load was defined as being the gas pressure inside the cylinder (cylinder flange pressure minus valve and passageway losses). Combined rod load and gas load had to be calculated every 10° of rotation. These loads had to be calculated and must be less than the manufacturer's MACCRL/MACGL limit at the RV setting of each stage and the minimum pressure for each stage.

Computer capabilities had increased to the point that the combined rod load calculations were readily available using PC-based software, and most machinery analyzers had the capability of computing the combined rod loads in real time as long as the measured in-cylinder pressures and the weights of the various components were properly utilized and interpreted.

In API-618 fifth edition, the rod load definitions had only one minor change, they are to be calculated every 5° instead of every 10°.

Glossary of terms

A brief glossary of terms will be helpful.

Rated rod load. This is the term used in the first edition of API-618 but without definition. At IR interpretation was gas only load not including valve losses. This should not exceed on any normal operating load step (specified operating pressures, not relief valve setting).

Maximum allowable rod load (MARL). This is the term used in the first edition of API-618 but without definition. At IR interpretation was gas only load not including valve losses. This should not exceed at any upset including final discharge relief valve pressure. Operation at rod loads exceeding the MARL voided the warranty.

Actual rod load. This is the term used in the second edition of API-618 with definition. This includes gas + valve losses + inertia loads in the calculation but did not define the reference point for the calculation.

Maximum allowable combined rod load. This is the term used in the third edition of API-618. This is the same calculation as actual rod load (includes gas + valve losses + inertia loads). Load was defined at the crosshead pin. This is the maximum load that can be applied at any load step including final RV pressure.

Maximum allowable continuous combined rod load. This is the term used in fourth and fifth editions of API-618. Similar definition as MACRL except load to be calculated every 10° (5° in fifth edition), and load at the pin is the component in the direction of the piston motion. This is the current definition of the rated load that applies to the running gear.

Maximum allowable continuous gas load. This is the term used in fourth and fifth editions of API-618, which includes gas pressures inside the cylinder (flange gas + valve passageway losses). This is the current definition of the rated load that applies to the stationary components.

Internal gas load. This term is commonly used with no API definition, which typically means the gas load based on pressure inside the cylinder, that is, includes the valve and passageway losses. It is the same as the current API-618 definition of **MACGL.**

Crosshead pin load. This term is commonly used with no API definition, which typically means the same thing as the current API-618 definition of **MACCRL.** It is referred to as the crosshead pin load to avoid the frequent misconception that the combined rod load is at the crosshead end of the piston rod rather than the API-defined crosshead pin reference point for inertia calculations.

Total rod load. This term was used by Ariel with no API definition. It is used to define the maximum allowable load in tension plus compression that can be applied. Note that the *total rod load* limit is typically less than the sum of the internal gas load limit in compression and in tension. Ariel also lists a *rod load* limit that is defined as an internal gas load. A *combined rod load* limit is not published; however, *combined rod load* is calculated on the performance sheet and is used in determining acceptable rod load reversals at the crosshead pin.

Combined load exceeds gas load

Many users mistakenly assume that the combined rod loads (gas plus inertia) will always be lower than the gas loads. This is not true for low-ratio (high VE) applications. As an approximate rule, if the discharge VE exceeds 50%, the gas load will reach a maximum prior to 90°, while both inertia load and gas load have same sign; thus they are additive. If the discharge VE is less than 50% then the gas load does not reach a maximum until after 90° and so it is opposite in sign to the inertia load, and the combined rod load will be less than the gas load. This is illustrated in Figure 4.19, which shows measured rod load data for a 250 rpm single-stage compressor in natural gas transmission service.

Distorted pressure measurements= distorted rod loads

Another common mistake is to report distorted rod loads based on distorted pressure measurements. This is most often due to the "channel resonance" effect present in nearly all in-cylinder pressure measurements. This effect is illustrated in Figures 4.20 and 4.21. The rod load plot without channel resonance correction is shown in Figure 4.20, while the corrected plot is shown in Figure 4.21. The reported rod load is higher when the channel resonance is present.

4.2.1.1.4 Conclusions

Compressor rod load ratings are often misunderstood and misapplied. It is important to understand that the API definitions of **MACCRL** and **MACGL** are not actually rod loads but refer to crosshead pin loads and gas loads, respectively. The API definitions help to avoid confusion, but these ratings are not always available for pre-1995 vintage machines.

"Measured" rod loads are actually computed rod loads based on measured pressures. The forces based on the measured pressures are combined with inertia forces based on the input weights of reciprocating components in the analysis software. If the pressure measurements are distorted and/or the reciprocating weights are not accurately known, then the combined rod loads reported will be erroneous.

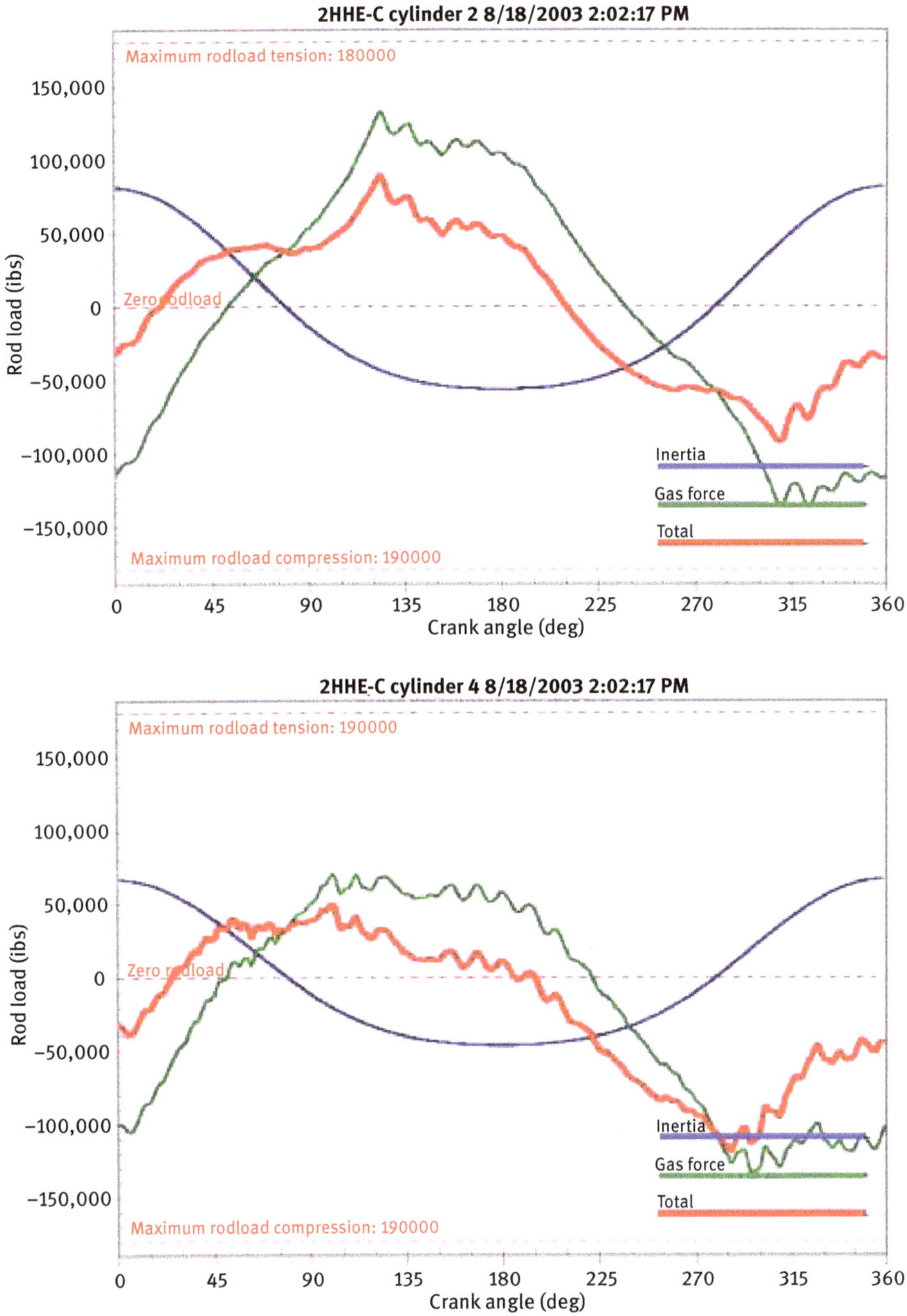

Figure 4.19: Measured rod loads *(Engineering Dynamics Inc., San Antonio, Texas).*

Compressor rod load report

Unit name: 2HHE-C   Model: HHE-VL   Date: 8/18/2003 2:02:17 PM
Location: Bravo Dome CO2 Plant   Unit MT: Ingersoll Rand   Serial no.:

Load Step: 2												
Cyl	Rod motion from baseline at 240 deg. ATDC		Allowable rod load (ibs)		Rod load based on line pressures				Rod load based on cylinder pressures			
					Absolute load (ibs)		percentage of actual to max. Allowable rod load		Absolute load (ibs)		Percentage of actual to max. Allowable rod load	
	probe loc.	(mil)	Comp.	Tension	Comp.	Tension	Comp.	Tension	Comp.	Tension	Comp.	Tension
Stage: 1	Stage 1											
1	Bottom	0.000	180,000	180,000	98,872	65,585	55%	36%	115,654	85,048	64%	47%
2	Bottom	0.000	180,000	180,000	79,510	65,670	44%	36%	87,212	84,191	48%	47%
Stage: 2	Stage 2											
3	Bottom	0.000	180,000	180,000	106,764	72,631	59%	40%	119,122	88,680	66%	49%
Stage: 3	Stage 3											
4	Bottom	0.000	180,000	180,000	105,973	38,551	59%	21%	115,058	46,245	64%	26%

Notes:
1. Line pressure rod loads are based on the apparent suction and discharge pressures as determined from the *P–V* and/or *P–T* curves. Many compressor manufacturers determine their allowable for load limits using this technique.
2. Cylinder pressure rod loads are determined from the maximum differential pressure across the rings as determined from the *P–V* and/or *P–T* curves. Cooper.Bessemer determines their allowable rod load limits using this technique.
3. Marker Type: Once Per Turn (OPT) and Trap Type: 9240.
4. Channel resonance correction (CRC) applied: 1H 1C 2H 2C 3H 3C 4H 4C
5. Corrected VE applied, PS and PD values Corrected: none
6. Forces due to Inertia are accounted for in both line pressure and cylinder pressure calculations.
7. Tandem configuration: none

Analyst signature: 3/3/2005 2:30:56 PM

Figure 4.20: Rod load report *(Engineering Dynamics Inc., San Antonio, Texas).*

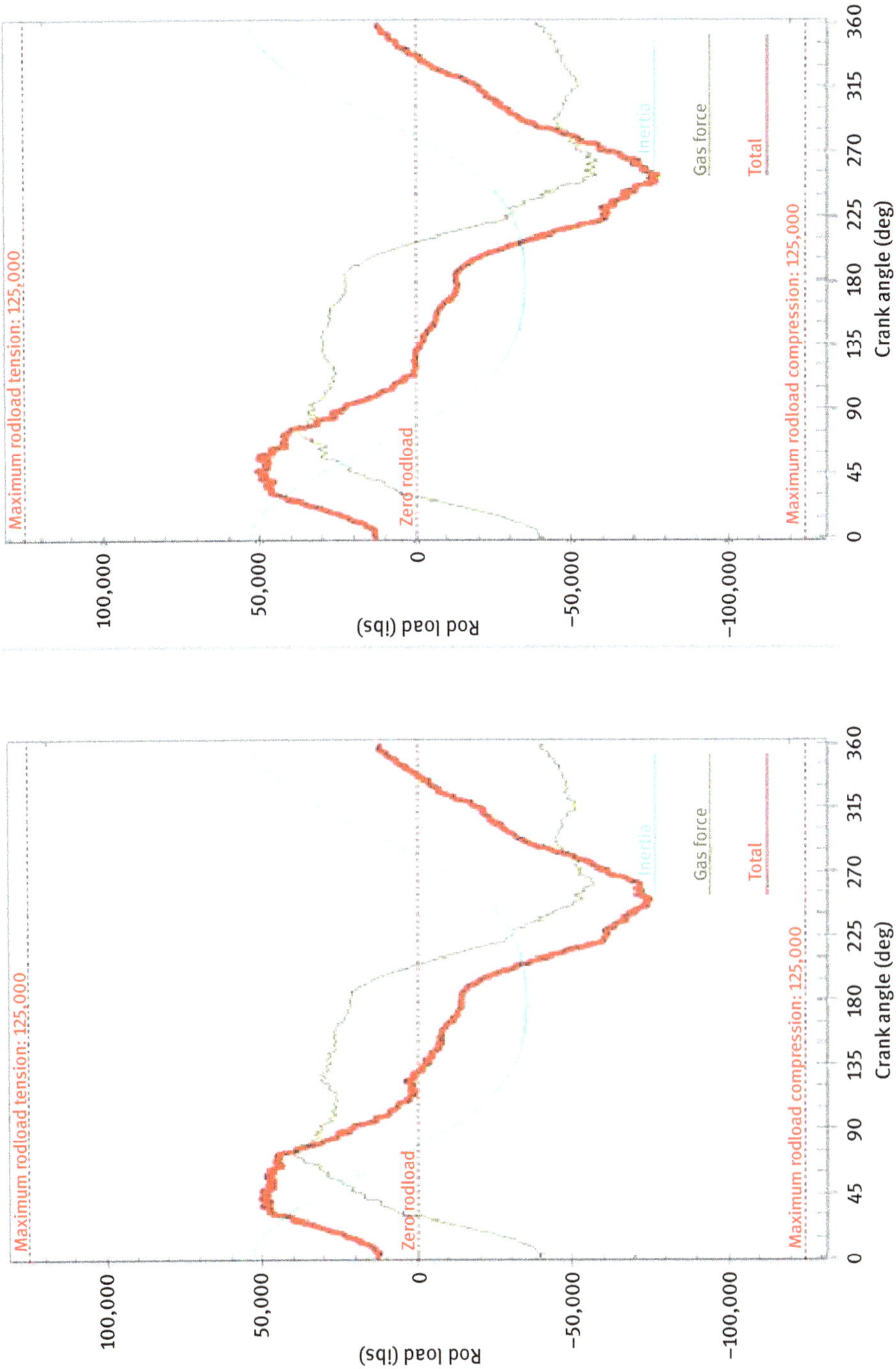

Figure 4.21: Combined pin load exceeds gas load *(Engineering Dynamics Inc., San Antonio, Texas).*

There is some logic in using the simplified gas rod load calculations presented in eqs. (4.1) and (4.2). The trends will be correct (i.e., higher differential pressure results in higher rod load). However, if nominal flange pressures are used to rate a compressor, care must be taken to include enough margin to account for the maximum possible in-cylinder pressures due to pressure drop, valve losses, pulsation, and valve dynamics. These effects vary for each application.

The user must make a decision when a compressor is revamped on whether to use current API definitions and ratings or the ratings in effect when the machine was first installed. Again, it is imperative that the OEM be included when evaluating design options because as an end user, we are not privy to all design limitations on a machine.

Finally we should be aware that pressure measurements can be distorted (Figure 4.22) and may have to be corrected (Figure 4.23).

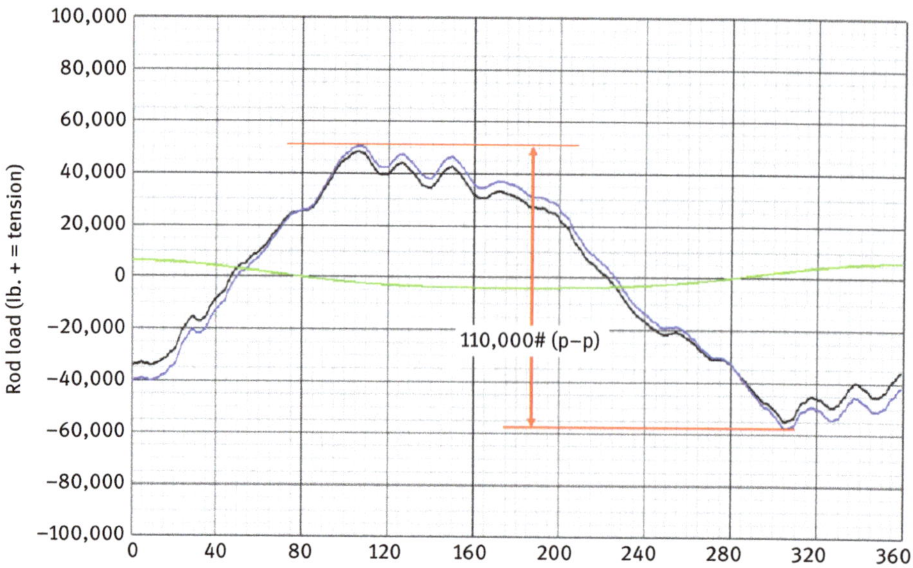

Figure 4.22: Distorted pressure measurements *(Engineering Dynamics Inc., San Antonio, Texas).*

4.2.2 Recent developments, valve selection, and technical advances

4.2.2.1 Overview

The main function of reciprocating compressor valves is to control flow in and out of the compressor cylinders. Valves are critically important components for reciprocating compressors; their functional integrity usually governs the availability and efficiency of these displacement compressors. Table 4.1 lists them at the top of failing components in reciprocating compressors.

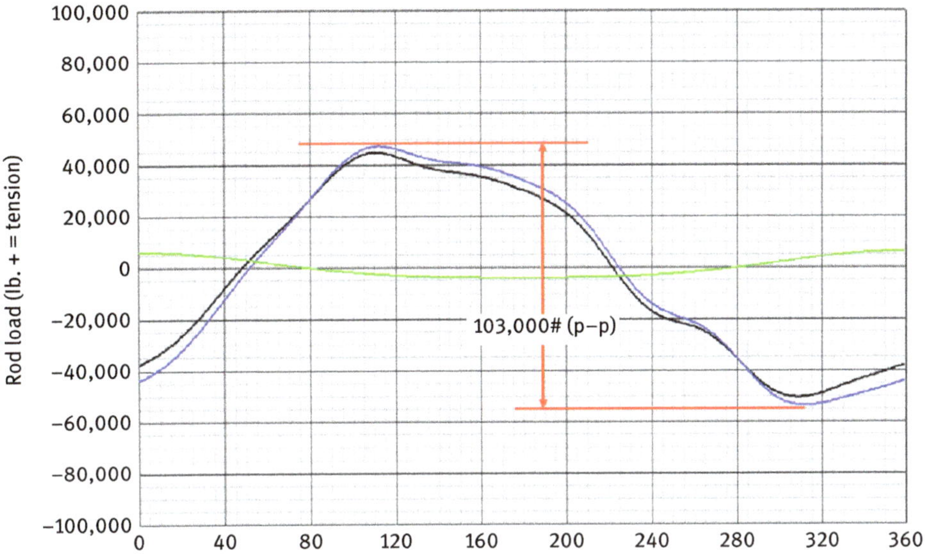

Figure 4.23: Corrected pressure measurements *(Engineering Dynamics Inc., San Antonio, Texas).*

Table 4.1: Top 10 failing components in reciprocating compressors.

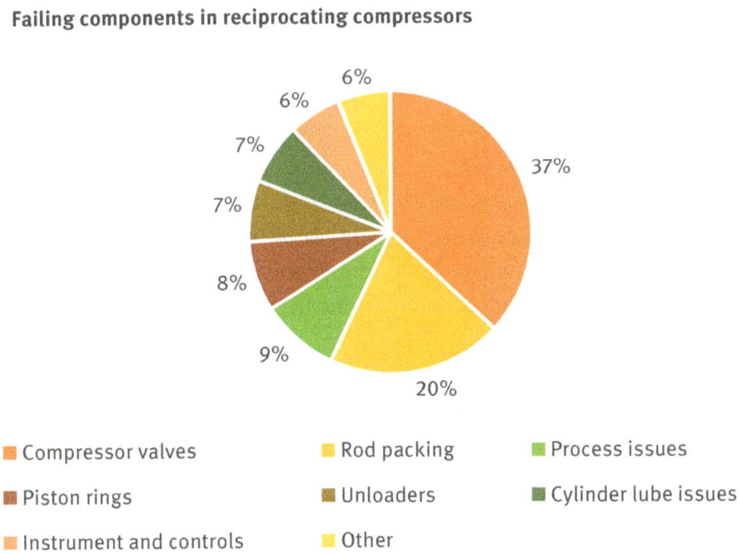

Four or five types or styles of valve configurations are normally used in reciprocating compressors. Except for suction valve *unloaders,* which use mechanisms to physically keep a valve open during part or the entire stroke length (displacement) of a piston, valves are actuated by differential gas pressure. Valves open when the pressure on one side of the cylinder is greater than the pressure on the opposing side. Most living creatures breathe air simply based on the same principle.

The compressor designer's and knowledgeable user's choice of valve depends largely on the type of process application. While valve type and size should be specified by the compressor manufacturer, the user–purchaser should likewise be involved. Although new or novel approaches are not entering the market place every year, these products nevertheless exist and are well worth considering.

Reliability-focused users will be interested in understanding why certain radically new modular reed compressor valves (Figure 4.24) have been making inroads since about 2014 when they were used on offshore platforms. As of 2017 and after initially concentrating on oil exploration and production facilities, these modular design straight flow-through valves have made excellent progress in oil refineries and other downstream facilities. Many modular reed valves are presently installed in hydrogen compressors with some individual machines approaching 20,000 bhp.

Being aware of a wide spectrum of readers and their different levels of experience, we opted to start this chapter segment with a brief compressor valve refresher. The most prevalent "traditional" valve types are described here. Note that these would often be labeled legacy valves and that some of these have probably changed little, if any, in the five decades from 1970 to 2020.

4.2.2.2 Types of valves

Channel and reed valves: Channel and reed valves (Figure 4.25) are now considered obsolete; they are used only in small air compressors. While they can be used at the highest practical speeds, they will require higher than average maintenance. Moreover, they exhibit below-average efficiency, meaning above-average power losses.

Ring valves: Ring valves (Figure 4.26) are generally used in slow-speed, high-pressure process applications. These valves use concentric rings, which open and close the valve ports. Ring valves share with plate valves the distinction of best suitability for high speeds – 1,000–1,800 rpm.

Plate valves: Plate valves (Figure 4.27) are typically used for high-speed separable compressors. (In this context, "separable" means compressors not integral to a gas engine driver.) Plate valves reassemble ring valves, but with the rings connected by ribs. Instead of individual elements opening and closing, all valve ports open and close in unison. They share the (above) high-speed suitability comment with ring valves.

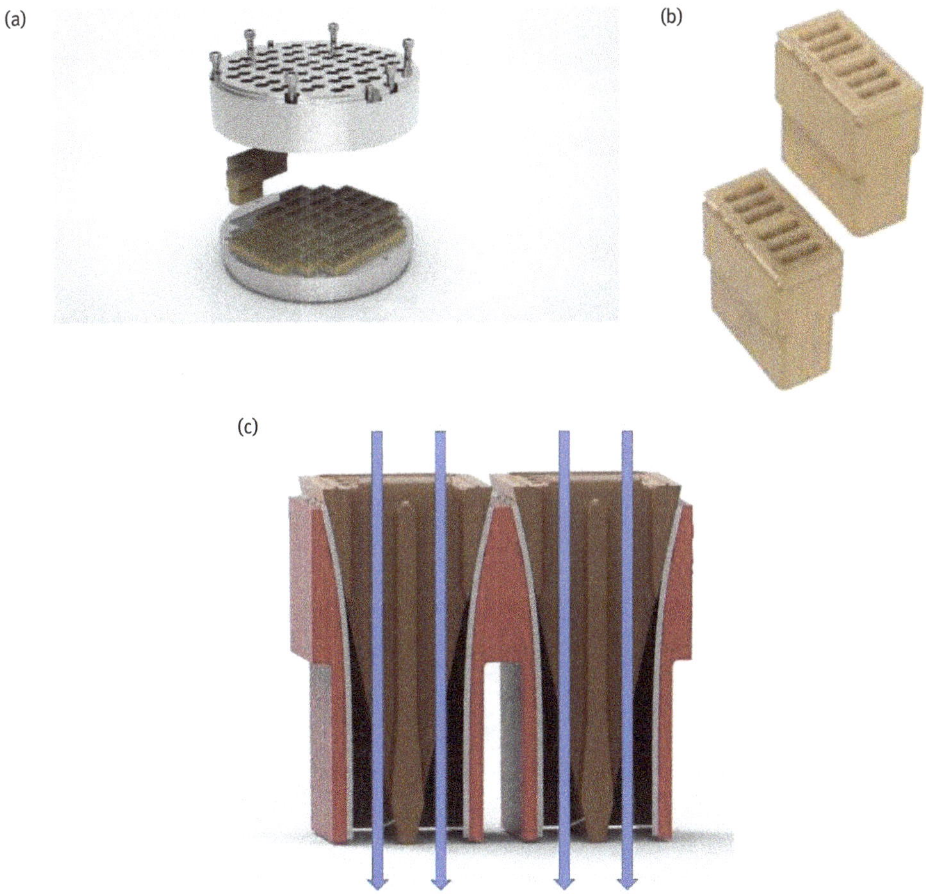

Figure 4.24: (a) StraightFlo® valve, (b) close-up of high-performance polymer straight-through flow elements, and (c) two through-flow ("StraightFlo®) elements illustrate the design principle (source: *Zahroof Valve Inc.*).

Poppet valves: Poppet valves find frequent use in low-compression ratio applications, such as pipeline booster compressors. They are preferred at low speeds, 250–600 rpm. These valves (see Figure 4.28) function by each individual poppet lifting with very low differential pressure action; they then allow gas to pass through the flow openings in the stop plate. Standard poppet valves are generally used in the speed range from 250 to 600 rpm; modified poppet valves are in use up to 1,200 rpm.

Design considerations: Because many factors will affect reliability, straightforward valve comparisons are difficult. We advocate that reliability-focused owner–purchasers look for demonstrated experience in virtually identical services elsewhere. Arguing that such comparisons infringe on someone's rights are as specious as either party viewing a free marketplace as restraint of trade. The point is an educated user

Figure 4.25: Channel or reed valve, legacy style.

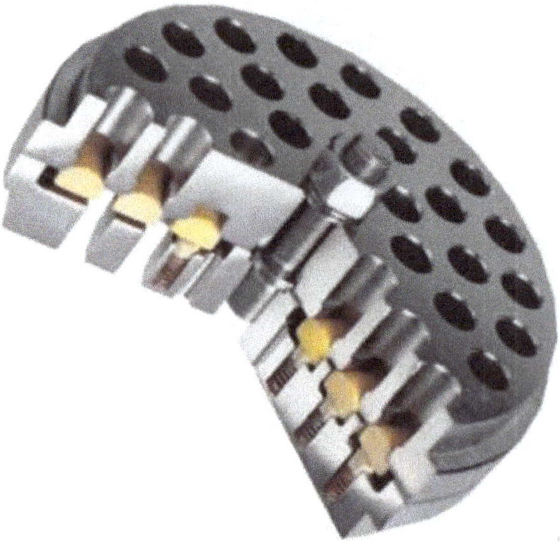

Figure 4.26: Ring valve.

should insist on viewing comparison data. A cooperative vendor or manufacturer will not withhold this information from reliability professionals engaged in due diligence.

In the pursuit of comparison, a well-managed user–purchaser applies the principles of up-front machinery quality assessment (MQA) (see index). Among many other tasks, the MQA engineer verifies that the vendor's material selection is appropriate for the purchaser's gas conditions, operating temperature profile, and the end user's expectations of low power losses. Corrections and/or knowledge updates

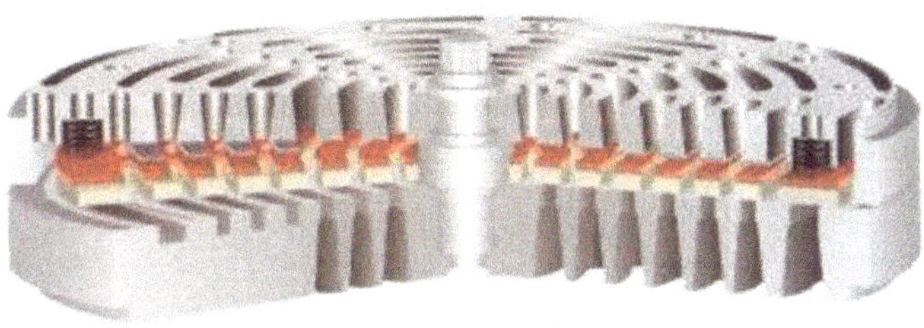

Figure 4.27: Plate valve.

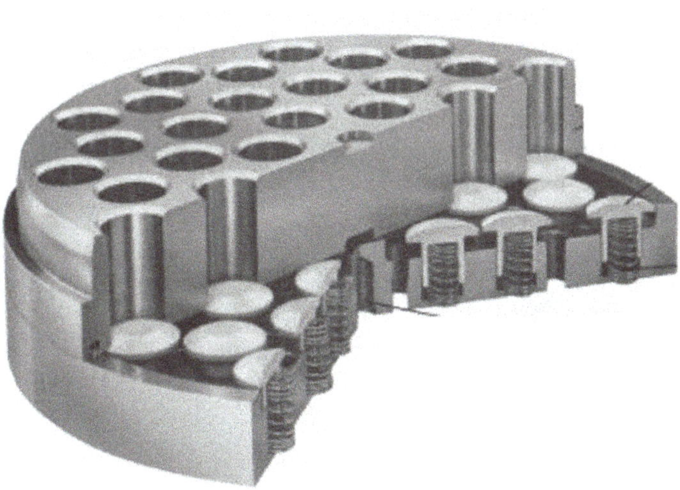

Figure 4.28: Poppet valve.

may be required by either party; reaching early agreement during MQA will save time and settle potential issues long before the compressors leave the factory.

A gain of 0.5% efficiency in a 6,000 kW reciprocating compressor where power costs $0.06/kwh and where machines are expected to accrue 8,700 operating h/year is worth more than one might think. Downtime avoidance also has value and should be part of the payback calculation. Again, clarification obtained during MQA should not be viewed as burdensome; it will prove valuable to both owner–purchaser and manufacturer–vendor.

Sharing knowledge: During scheduled pre-award meetings, a competent valve vendor or manufacturer will explain all about design compromises, gas velocity limitations, the meaning of valve lift (distance from seated to fully open and absence of valve flutter), tolerance for liquids and solid impurities carried in the gas, and overall valve operation in clean versus fouling services. Recall that there are bone-dry gases in some, and oil carryover in quasi-moist services. Explore if and how compressor valves tolerate these extremes. Find out before a sequence of premature failures tells you that experimentation with operating machinery is rarely acceptable when important assets are involved.

4.2.2.3 Examining modular straight-through flow valves

A new approach to compressor valve technology was introduced to refineries and petrochemical plants in 2017. The old legacy valves, Figures 4.25 through 4.28, can now be replaced by modular reed valves developed by Houston-based manufacturer Zahroof. Modular reed valves can serve as drop-in replacements for virtually any predecessor valve. The novel technology is shown in Figures 4.24 and 4.29 through 4.33; the modules (or modular elements) are an enhanced reed design incorporating high-performance polymers under compressive stresses. The inventor chose the generic term modular straight-through flow valve and selected "StraightFlo®" as the trade name for these modern products.

Figure 4.29: Double deck HiFlo® Valve, one of many possible drop-in replacement valves (source: *Zahroof Valve Inc.*).

The authors found considerable evidence that the manufacturer's advertised service ease improved compressor flow rates, and other claims were correct. Servicing by lower skilled workforce members, operating performance over an unusually wide speed range, and generally well above-average availability/reliability were factually proven. While the first applications were in upstream facilities (oil exploration and gathering), modular straight-through valves have since made considerable inroads in downstream facilities such as oil refineries and petrochemical plants.

The success of these valves is largely attributable to their straight-through design. Other advantages are intuitively evident; these include self-cleaning, an estimated 80% reduction in inventory, and validation done through rigorous analyses. Test reports are among the various items of backup documentation; they reaffirm that the inventor's understandable enthusiasm was (and is) fully justified. A high-flow version is shown in Figure 4.29; a stacked valve is depicted in Figure 4.30.

(a)

(b)

Figure 4.30: (a) Stacked straightFlo® valve and (b) staggered element/insert arrangement in a stacked straightFlo® valve (source: *Zahroof Valve Inc.*).

Taken together, the seemingly small advantages in this or that measurement add up. Both standard and integral cage modular flow-through valves are shown in Figures 4.31 and 4.32. Either style adds value that cannot be ignored.

The flow area ratios of two widely used valves are compared with three modular straight-through flow (StraightFlo®) valves in Table 4.2; the percentage area

Figure 4.31: Standard straightFlo® drop-in replacement valves (source: *Zahroof Valve Inc.*).

improvements are significant. Elementary physics explains that, for a given volume of gas, larger flow areas have less pressure drop than smaller flow areas. Lower maximum pressures in cylinders reduce pressure ratios by a small amount. These ratio reductions and larger areas of gas passage save energy and increase compressor throughput.

We were able to verify that the success of modular reed technology in "StraightFlo® valves" is linked to well-instrumented tests. It could be ascertained that the reports did not refer to comparisons of brand-new modular straight-through flow valves with old and/or worn valves. Instead, the comparisons were for unused old-style valves versus unused new-style valves.

The tests also showed sizeable improvements in energy efficiency, valve flutter, vibration, noise, pressure drop needed to activate valves, trouble-free operating time, and necessary maintenance outlays. The self-cleaning attributes of StraightFlo® valves were convincingly demonstrated in Figure 4.33.

(a)

(b)

Figure 4.32: (a) A typical integral cage modular StraightFlo® valve and (b) one of many less typical integral cage modular StraightFlo® valves (source: *Zahroof Valve Inc.*).

Table 4.2: Area ratio comparisons.

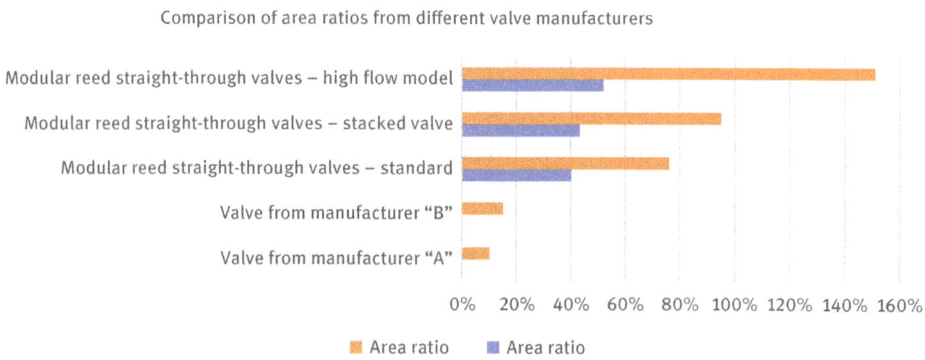

Figure 4.33: (a) Badly fouled legacy type valve, (b) StraightFlo® valves with self-cleaning attributes, (c) badly foulded legacy-type valve with no self-cleaning, and (d) StraightFlo® valves with considerably fewer fouling deposits (source: *Zahroof Valve Inc.*).

References

[1] Singh, P.J., Patel, G.C. A Generalized Performance Program for Oil Flooded Twin Screw Compressors, Proceedings of the 1984 International Compressor Engineering Conference, Purdue University, School of Engineering, West Lafayette, Indiana, USA.

[2] Bergadano, E., Giuffrida, A., Gorini, U., Tornatora, C. A Study on the Main Geometric Characteristics of Twin-Screw Compressors, AIP Conference Proceedings 2191 (2019), Published online, 17 December 2019

[3] Huang, C.-Y. New Twin Screw Compressor Design by Deviation Function Method, University of California, Los Angeles, CA, 2015.

[4] Hsu, R.-H., Yu-Ren, W., Chen, Z.-A. A Numerical Method for Reversely Generating a Pair of Conjugated Rotor of Twin-screw compressors from the measured Profile data, The 50th CIRP Conference on Manufacturing Systems, Procedia CIRP 63, 2017, 483–487.

[5] You, C.X., Tang, Y., Fleming, J.S. Optimum Rotor Geometrical Parameters in Refrigeration Helical Twin-Screw Compressors, International Compressor Engineering Conference, 1996, I, 1–6, Purdue University, School of Engineering, West Lafayette, Indiana, USA.

[6] Singh, P.J., Bowman, J.L. Effects of Design Parameters on Oil-Flooded Screw Compressors Performance, Proceedings of the 1986 International Compressor Engineering Conference, Purdue University, School of Engineering, West Lafayette, Indiana, USA.

[7] Fleming, J.S., Tang, Y. The Analysis of leakage in Twin Screw Compressors and its Application to performance improvement, Division of Dynamics and Control, Department of Mechanical Engineering, University of Strathclyde, Glasgow, Scotland.

[8] Singh, P.J., Bowman, J.L., Company, I.-R. Calculations of Blow-Holes Area for Screw Compressors, International Compressor Engineering Conference, 1988, Purdue University, School of Engineering, West Lafayette, Indiana, USA.

[9] Yu-ren, W., Fong, Z.-H. Rotor Profile Design for the Twin-Screw Compressor based on the Normal-Rack Generation Method, Journal of Mechanical Design, ASME, April 2008.

[10] Atkins, K.E., McCain, B. A Discussion of Various Loads Used to Rate Reciprocating Compressors, Presented at the 2005 Gas Machinery Conference in Covington, KY, October 2–5, 2005.

Chapter 5
Factory testing of centrifugal compressors

5.1 Brief overview of ASME PTC-10 thermodynamic test procedure

5.1.1 Introduction

As centrifugal compressors become larger due to demands from end users, equipment manufacturers must comply with more stringent specifications in order to stay competitive.

To be competitive, these equipment manufacturers have pushed the limits of several design parameters that were regarded as design limiting in 1995 or even in 2000. As an example, in 2018 or 2020, a few equipment manufacturers were manufacturing centrifugal compressor wheels well beyond the 0.14 flow coefficient normally accepted as a design limit two decades earlier. As of 2020, these equipment manufacturers are reaching flow coefficients beyond 0.2, while in many cases achieving polytropic efficiencies in excess of 85%.

In order to push the limits on these machines, equipment manufacturers are, as of 2020, employing highly sophisticated software tools. Although these tools do help with the design complexities of modern centrifugal compressor wheel designs, occasionally some of the compressors designed by these tools still fail factory performance test. Because of this, it is now a common practice for end users to specify both a factory thermodynamic performance test as per ASME PTC-10 [1] and mechanical run tests as per API-617. End users no longer restrict themselves to the choice between doing or not doing a thermodynamics performance test based on budget constraints. In other words, the cost of testing is in the budget and will be part of the allocated project cost.

As of 2020, many end users specify a mechanical running test of the spare rotor first, followed by a mechanical running test and performance test of the job rotor. By sequencing the tests in this manner, the job rotor will be performance tested as well as mechanically tested and will be left in the compressor casing.

Four main purposes can be cited in support of requiring an in-factory thermodynamic performance test as per ASME-PTC-10:

- To verify that the centrifugal compressor will perform as per the manufacturers design while also fully satisfying the end users' specified conditions.
- Such testing offers the end user a way of confirming the actual shape of the compressor head-flow curve.
- It validates equipment the manufacturer's predicted efficiency and flow range at various speeds.
- It can serve to expertly bring performance anomalies back into compliance and, if found to exist, the needed corrections can be made without undue delay.

https://doi.org/10.1515/9783110678765-005

5.1.2 Class of performance test

The ASME PTC-10 Power Test Code establishes a set of procedures and conditions based on which performance accuracy can be determined. The code gives equipment manufacturer and end user choices based on the category or type of tests and the accuracy of each test. The ASME PTC-10 code was revised in 1997; the revision consolidated the previous three classes of tests to only two classes. Class I and II tests are now called type 1 and 2 tests, respectively. The former class III test designation has been removed from the 1997 code.

The difference between Classes II and III testing in the 1965 version of the ASME PTC-10 test code is that Class II allows the option of using either real or ideal gas laws to compute gas properties. In contrast, Class III tests had to use "real gas" laws and equations of state (EOS) to compute gas properties.

Type 1: This test is performed with the specified gas at or near the specified operating conditions. Acceptable deviations between test and operating conditions are shown in Table 5.1. The gas can also be treated as an ideal gas or a real gas. Limits of departure from ideal gas laws are shown in Table 5.2.

Table 5.2 is based on a classic technical paper written by Schultz, circa 1962 [2]. The paper improves on Nelson–Obert's generalized compressibility factor charts for compressibility by introducing two additional functions X and Y. These two functions are plotted as generalized functions in charts A-4 and A-5 of Appendix A. Schultz then relates the functions X and Y to the main compressibility function Z and continues by relating isentropic and polytropic analyses.

Schultz's work was presented to ASME in 1960 and adopted later by the Power Test Codes committee. However, in 1990 the Power Test Code committee recognized certain inaccuracies in using the Schultz's generalized charts for real gas compression. So far, the ASME has developed no other method to determine compressibility factors. Therefore, knowing there are inaccuracies using the generalized values for Schultz compressibility factors, many equipment manufacturers now use other well-known EOS to determine gas properties, including polytropic exponents and polytropic work.

Type 2: This test is performed with either the specified gas or a substitute gas; it may use real or ideal gas laws in the various calculations of gas properties. This test is used whenever the test conditions cannot meet the specified conditions with the specified gas or a substitute gas. Test conditions are often limited by the compressor's aerodynamic design; therefore, permissible deviations from normal operating conditions are outlined in Table 5.1.

In general, the code is followed in its entirety by most equipment manufacturers; however, the manufacturers' practices may deviate in the specific method of determining gas properties. Many equipment manufacturers have become more comfortable using some of the well-known EOS to gain more accuracy.

Table 5.1: Permissible deviation from specified operating conditions for type 1 test (extracted from ASME PTC-10).

Variable	Symbol	Units	Permissible deviation
Inlet pressure	pi	psia (kPa)	5%
Inlet temperature	Ti	°R (°K)	8%
Speed	N	rpm	2%
Molecular weight	Mw	lb/lbmol (kg/kgmol)	2%
Cooling temperature difference		°R (°K)	5%
Coolant flow rate		gal/min (cu-m/h)	3%
Capacity	Qi	cu-ft/min (cu-m/h)	4%

General notes

(a) Type 1 tests are to be conducted with the specified gas. Deviations are based on the specified values where pressures and temperatures are expressed in absolute values.

(b) The combined effect of inlet pressure, temperature, and molecular weight shall not produce more than 8% deviation in the inlet gas density.

(c) The combined effect of the deviations shall not exceed the limits of Table 5.2. Cooling temperature difference is defined as inlet gas temperature minus the inlet cooling water temperature.

Table 5.2: Limits of departure from ideal gas laws of specified and test gases (extracted from ASME PTC-10).

Maximum ratio		Allowed range for function X		Allowed range for function Y	
Pressure ratio	kmax/kmin	Min	Max	Min	Max
1.4	1.12	−0.344	0.279	0.925	1.071
2.0	1.10	−0.175	0.167	0.964	1.034
4.0	1.09	−0.073	0.071	0.982	1.017
8.0	1.08	−0.041	0.050	0.988	1.011
16.0	1.07	0.031	0.033	0.991	1.008
32.0	1.06	−0.025	0.028	0.993	1.006

General notes

(a) $X = T/v\,(\partial v/\partial T)_p - 1$ and $Y = P/v\,(\partial v/\partial P)_T$, see Charts A-4 and A-5 in Appendix A.

(b) Maximum and minimum values of k shall apply to both the specified and test gas over the complete range of conditions.

(c) When these limits are exceeded by either the specified gas or the test gas at any point along the compression path, real gas calculation methods shall be used for that gas. Ideal or real gas method may be used if these limits are not exceeded.

5.1.3 Selecting the test gas

The test gas for an ASME PTC-10 performance test must be carefully chosen in order to comply with the limits shown in Table 5.3. Table 5.3 compares the performance of the compressor when running on a test gas with the performance running on the specified gas. The four basic parameters are as follows:
- Specific volume ratio q_i/q_d (same as the design density ratio).
- Capacity–speed ratio q_i/N (this is proportional to flow coefficient). The key to this parameter is determining the equivalent test speed for the test gas.
- Machine Mach number M_m is equal to U/a_i.
- Machine Reynolds number.

Table 5.3: Permissible deviation from specified operating parameters for types 1 and 2 tests (extracted from ASME PTC-10).

Variable	Symbol	Limit of test values as percent of design values	
		Min	Max
Specific volume ratio	q_i/q_d	95	105
Flow coefficient	$\emptyset$	96	104
Machine Mach number	M_m		See Figure 5.1
Machine Reynolds number [Note 1]			See Figure 5.2
Note 1: Minimum allowable test machine Reynolds number is 90,000.			

The most desirable substitute gas for test purposes is air. Air tests work quite well for many centrifugal compressors running on light hydrocarbons. This fortuitous acceptability can be attributed to the design density ratio (same as specific volume ratio), and proximity to design Mach numbers. These can be achieved when compressing air at equivalent speeds below the design speed. However, for gases heavier than air, unsafe stresses can be reached because the equivalent test speed on air will be above the design speed on many hydrocarbon gases.

The properties of heavier-than-air gases make it difficult to arrive at an equivalent test speed that would match both the design density ratio and design Mach number. Therefore, a suitable substitute gas is used for these cases.

Table 5.4 shows some of the typical substitute gases used for performance testing of centrifugal compressor sections for several specified gases and conditions. Note that whenever the molecular weight of the gas is less than air, using air as a substitute gas gives good correlation.

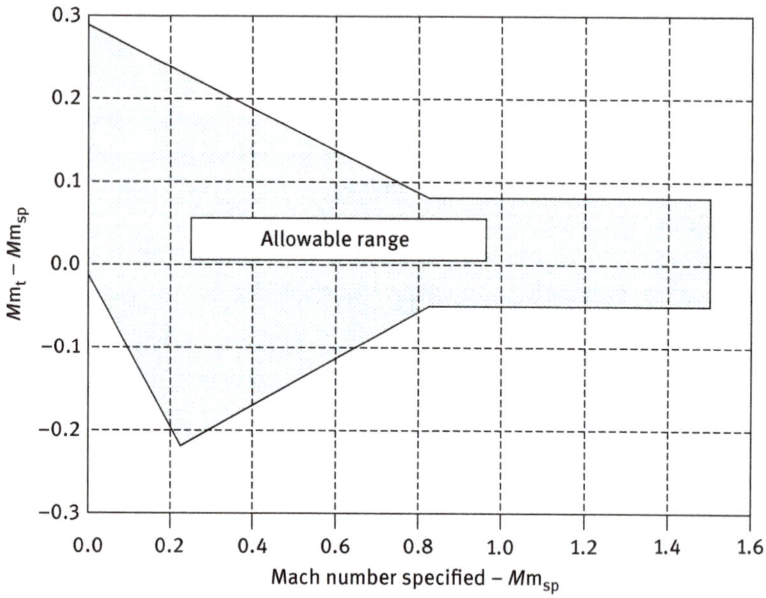

Figure 5.1: Allowable machine Mach number departures – centrifugal compressors [1].

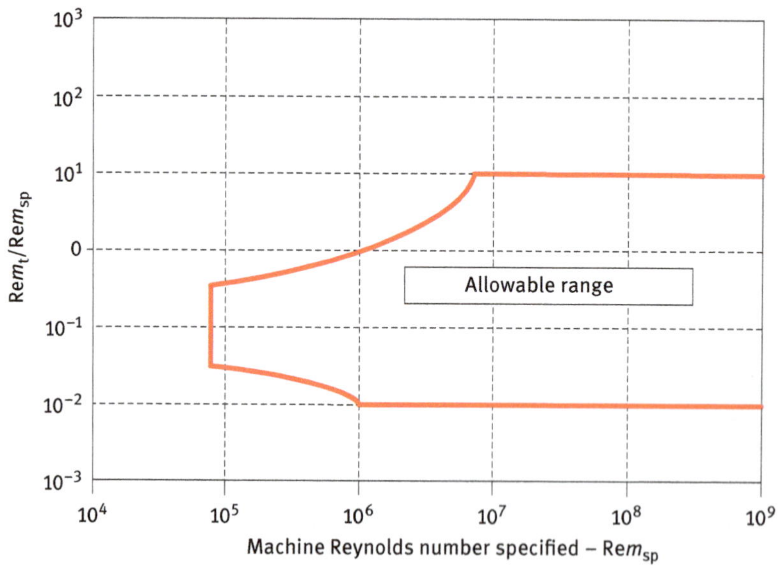

Figure 5.2: Allowable machine Reynolds number departures – centrifugal compressors [1].

5.1.4 Instrument and calibration

The ASME PTC-10 test code gives detailed requirements for the instrumentation to be used in performance testing of centrifugal compressors. Any test should use appropriate instrumentation to determine the inlet and discharge end states of the gas, flow rate, and speed. Depending on the type of test selected, section 4 of the code outlines the location, type, and quantity of each instrument.

Prior to the start of the test, the end user's engineer or inspector should do a walk-through to examine the test loop. Check that the piping matches the ASME PTC-10 recommended arrangement, including spacing of instrumentation and location of flow straighteners. Review the calibration records of each instrument that will be used during the test. The time since calibration records were last validated should not exceed 3 months.

5.1.5 Loop testing

Open-loop-type performance testing can be done with air if the properties of the available air closely match those of the specified gas. However, for other substitute gases, a closed-loop test is considered necessary for basic safety. The closed-loop test circuit has proven to give greater accuracy in determining the test properties.

Figure 5.3 represents a typical arrangement for closed-loop shop performance testing of centrifugal compressors.

Before starting the test, the loop must be checked for leaks; contaminants must be purged before commencing the test. The test loop piping, including compressor and cooler, is pressurized with air to approximately 3.9 bar (57 psi) and checked for leaks. After leak checking, the system is evacuated to approximately −0.98 bar (−14.2 psi) by a vacuum pump to purge the unit.

After purging the loop, the loop is again pressurized, only this time it is done with the substitute gas, and the speed of the compressor is gradually increased to the test speed.

The actual test begins when the suction pressure, temperature, and speed are at the values anticipated for the test. The suction flow rate is then adjusted to match the calculated value for the test.

Table 5.4: Examples of substitute gases used for different specified gases and conditions.

Test #	**1**	**2**	**3**	**4**	**5**	**6**	**7**
Specified gas	**Hydrogen and HC mix**	**Coker gas**	**Natural gas**	**HC mix**	**Propylene**	**Ethylene**	**HC mix**
Molecular weight	8.51	12.00	17.37	28.98	42.09	28.06	29.72
C_p/C_v	1.300	1.400	1.300	1.165	1.212	1.344	1.187
Inlet temperature (°F/°C)	150/66	95/35	60/16	98/37	−35.14/−37.3	−101.4/−/−74.1	106.7/42
Pressure ratio	1.43	2.02	2.63	3.50	1.83	2.47	1.895
Specific volume ratio	1.278	1.574	1.970	2.824	1.664	1.820	1.693
Substitute gas	Air	Air	Air	**Carbon dioxide**	**Refrigerant R-134a**	**Refrigerant R-134a**	**Carbon dioxide**
Molecular weight	28.97	29.97	29.97	44.01	102.3	102.3	44.01
C_p/C_v	1.3950	1.4000	1.4000	1.2628	1.1019	1.1015	1.2734
Inlet temperature (°F/°C)	70/21.1	60/16	60/16	104/40	113/45	43/6.1	113/45
Pressure ratio	1.498	2.018	2.905	3.820	1.829	1.997	2.013
Specific volume ratio	1.278	1.574	1.970	2.719	1.718	1.825	1.692
Equivalent speed as percentage of design speed	54%	62%	84%	87%	76%	61%	88%
Error in Mach number ($M_{m_t} - M_{m_{sp}}$)	0.028	0.000	0.035	0.020	0.020	0.151	0.018

Note: Performance testing is done for each compressors section. Therefore, the conditions shown in the table are for compressor sections.

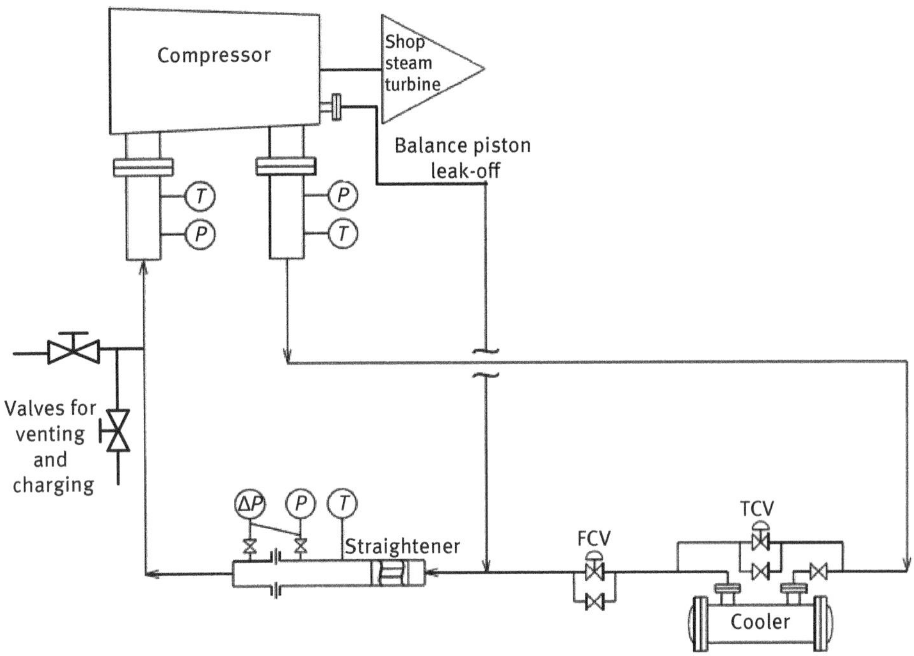

Figure 5.3: Typical arrangement of a test loop for performance shop testing.

5.1.6 Calculation procedures

5.1.6.1 Determination of the test speed

The API 617 standard stipulates that the test values and specified values should correlate closely; the variation should not exceed 5%. To ascertain that the specific volume ratios of tested and specified gases correlate closely, the following steps should be used to calculate the equivalent speed for the test gas:

Step 1

The polytropic exponent n_{sp} is determined from the operating conditions of the specified gas:

$$n_{sp} = \frac{\ln P_d / P_i}{\ln V_i / V_d}$$

Step 2

Suction condition properties of the test gas such as C_p, C_v, q_t, X, Y, and Z are determined from the suction temperature, pressure, and molecular weight. Once these properties are determined, then n_t and polytropic temperature exponent m_t of the test gas are obtained from the following equations:

$$m_t = \frac{ZR}{JC_p}\left(\frac{1}{\eta_p} + X\right)$$

where η_p the polytropic efficiency is the assumed value for the design condition, and J is the mechanical heat equivalent

$$n_t = \frac{1}{Y - m_t(1+X)}$$

Step 3

After obtaining n_t and n_{sp}, the compression ratio of the test gas at which the test volume ratio is equal to that of the design condition is determined as follows:

$$\text{Let "}b\text{"} = \frac{n_t}{n_{sp}}$$

$$(r_p)_t = (r_p)_{sp}{}^{b}$$

The test discharge pressure is also calculated

$$(P_d)_t = (P_i)_t \times (r_p)_t$$

Step 4

The temperature ratio and discharge temperature of the substitute gas are calculated from the following two equations:

$$\left(\frac{T_d}{T_i}\right)_t = (r_p)_t^{mt}$$

$$(T_d)t = (T_i)t \times (r_p)_t^{mt}$$

Step 5

Knowing the discharge pressure and temperature of the test gas, other discharge properties such as C_p, C_v, q, X, Y, and Z are obtained, and their mean values are calculated.

The temperature exponent m_t is calculated at the test gas discharge condition similar to what was done previously in Step 2. The value obtained should be similar to that at the suction condition of the test gas, since this exponent is constant along the P–T line. If, however, the values are not equal, the value of m_t is calculated at the discharge conditions to determine n_t using the equations in Steps 3 and 5.

Step 6

Once the appropriate values of m_t and n_t are obtained, the polytropic work W_p of the test conditions is calculated:

$$W_p = \frac{n_t}{n_t - 1} f P_i V_i \left[\left(\frac{P_d}{P_i} \right)^{n_t - 1/n_t} - 1 \right]$$

$$f = \frac{h_d' - h_i}{\frac{n_s}{n_s - 1} \left(P_d V_d' - P_i V_i \right)} J \times 10^{-3}$$

Step 7

After rationalization and substitutions, the test speed N_t is calculated by

$$N_t = N_{sp} \sqrt{\frac{(W_p)_t}{(W_p)_{sp}}} = N_{sp} \sqrt{\frac{Z_t Mw_{sp} T_t}{Z_{sp} Mw_t T_{sp}}} \times \sqrt{\frac{\left[\frac{n}{n-1}\right]_t \left[(P_2/P_1)^{\frac{n-1}{n}} - 1\right]_t}{\left[\frac{n}{n-1}\right]_{sp} \left[(P_2/P_1)^{\frac{n-1}{n}} - 1\right]_{sp}}}$$

5.1.6.2 Determination of the test gas volume flow

Inlet volume flow $(q_i)_t = (q_i)_{sp} \times (N_t/N_{sp})$

Inlet weight flow $W_t = (q_i)_t/(v_i)_t$

Discharge volume flow $(q_d)_t = W_t \times (v_d)_t$

5.1.6.3 Compression power required for test gas

$$\Delta h_t = \frac{(C_p)_t (T_d - T_i)_t}{1 + \eta_p X_t}$$

$$(P_p)_t = \frac{W_t \times \Delta h_t}{3,600} J \times g \times 10^{-3}$$

The total shaft power P_t is determined by adding the mechanical losses.

5.1.6.4 Determining the machine Mach number

The sonic velocity of the test gas "a" is determined by

$$a = \sqrt{\frac{k_i g Z_i R T_i}{Y_i}}$$

Machine Mach number $M_m = U/a_i$

where $U = (\pi DN/60 \times 10^2)$ and D = first stage impeller diameter (mm).

5.1.7 Test points

After charging the system with the test gas, the speed of the compressor is gradually increased to the test speed. Data should be taken at 5 points along the performance curve at 100% speed, including the guarantee point. The test starts from the overload side of the curve all the way to the surge point, as shown in Figure 5.4.

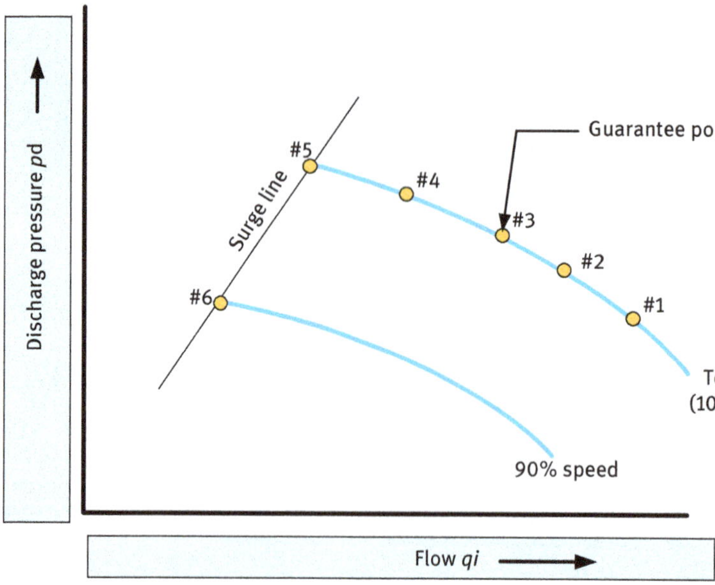

Figure 5.4: Acquisition sequence of test data points.

Data acquisition for each test point is done when pressure and temperature are stabilized at the acquisition point. The unit is generally operated for whatever it takes to obtain and record three or more consecutive readings at each test point. These readings must satisfy the fluctuation limits outlined in ASME PTC-10 (Table 3.4).

Confirmation of surge line

The surge point at 100% test speed must first be confirmed by reducing the flow rate until the flow becomes unstable (it is sometimes accompanied by the surge sound). If the surge does not occur at the predicted point, flow should be reduced further, but to not more than 10% beyond the predicted surge point. Data should be taken at this point and recorded as the surge point.

However, if at 100% speed, surge occurs prematurely (meaning before the predicted surge point), flow should be increased and then reduced gradually confirming and designating the unpredicted surge point as the actual surge point.

After completion of the test at 100% speed, the surge point no. 6 at 90% speed is obtained in a similar manner as earlier, and the surge line is drawn accordingly.

5.1.7.1 Side stream compressors

Large refrigeration process compressors are invariably designed as side-load centrifugal machines. These types of compressors are arranged with each compressor section oriented in the same direction in one single casing. One or more side streams are mixed internally with the discharge flow of the upstream section. Cooling can occur through the mixing of the predominantly lower temperature side stream. Side streams from these sections can also be extracted from the compressor casing.

Because of the construction and nozzle location of these side-load compressors, the test loop arrangement is very different from the test loop arrangement for straight-through-type compressors. There are several variations of the test loop setup; however, the basic loop arrangement or concept is shown in Figure 5.5.

Each section is tested separately while making sure that the other sections are all stable during the test. During the test, side streams and main inlet flows should be maintained as close as possible to predicted flows. This will prevent cascading flow instability of the sections.

Just as important as flow stabilization of the individual sections are their relevant instrumentation parameters. In addition to the external instrumentation shown in Figure 5.5, the compressors need to be instrumented to allow monitoring of internal flow mixing and potential temperature stratification in the mixing zones.

The PTC-10 code does not give definitive requirements for the number and types of instruments (or sensors) to be installed internally during the test. Instead, the code recommends a zone between location “A” and location “B” along the gas path (“aeropath”) of each stage where mainstream instruments should be located during the test. Figure 5.6 shows the zone that starts at the return bend and that moves along the diffuser to the mixing zone. The instrumentation includes several different types of pressure and temperature transducers.

However, since about 2010 and as equipment manufacturers have pushed the design envelope, the recommended zone has been increased. The zone expansion accommodates more instrumentation for the detection of stall cells and premature surging, among other undesirable flow disturbance events.

The authors are aware of a few side-load compressors that failed their performance tests either due to premature surging or from stall cells created while attempting to determine the surge point. All had common designs with high flow coefficients (i.e., beyond 0.15), without having been subjected to analyses involving high-fidelity computational flow dynamics.

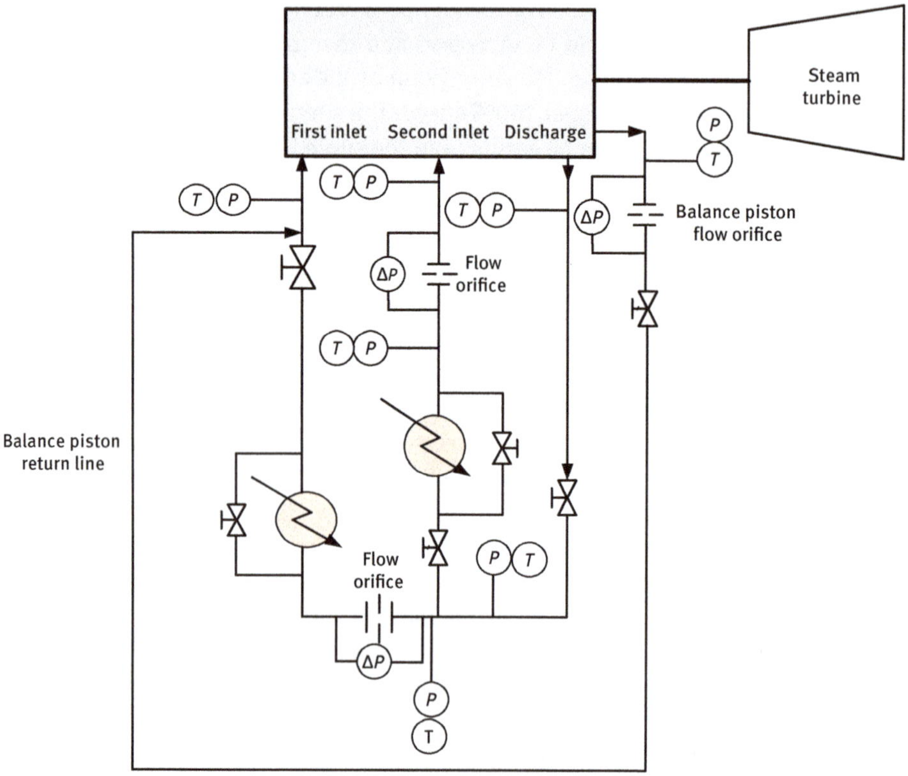

Figure 5.5: Test loop for testing side-load centrifugal compressors.

5.1.8 Performance evaluation

During each test, data are sent to an interactive computer system. This system is programmed for one of the popular EOS needed to evaluate the test data.

The computer system is involved to perform the following functions:

- Acquire the raw pressure and temperature data.
- Correct the raw data to reflect actual values. This is done by conditioning the effects of instrument fluctuations while testing is in progress.
- Use the uncertainty analysis mentioned in the code to interpret and evaluate the test results. This analysis is a measure of quality of the test. It is recommended that an uncertainty analysis be done prior to the test to ensure that the test objectives can be met. The uncertainty analysis is repeated for the actual test.

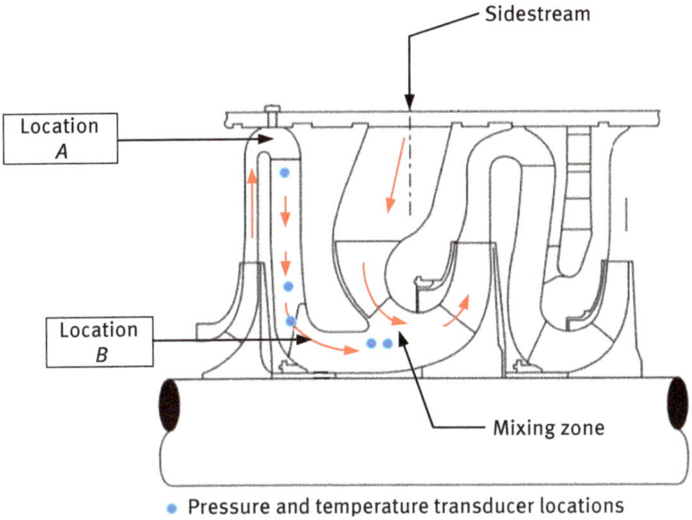

Figure 5.6: Cross section of a side-load centrifugal compressor showing the ASME PTC-10 recommended zone for locating test instrument during performance test.

5.1.9 Test report

Owner–purchasers receive the raw data; included are instrument fluctuation and performance calculation printouts from each specified point. Owner–purchasers will thus be able to confirm performance before moving on to the next specified point.

Final conversion to actual gas performance will be reported to the owner–purchaser after the performance test is completed.

5.2 Discussion of executing a mechanical run test as per API 617

5.2.1 Guidelines and test setup for no-load and full-load mechanical tests

Performance predictions are sooner or later followed by field performance verification. If a user wishes to verify performance in actual operation on-site, it will be necessary to place requisite and approved instrumentation at predefined locations at either the compressor or the piping or both. In this context, "approval" means contractual agreement between vendor/manufacturer and owner/purchaser.

Because warranty-supported and contractually binding on-site testing can be costly and/or time consuming, shop testing is far more common than field testing. Moreover, all vendors or manufacturers are equipped to do shop testing. Two different testing methods are typically used for mechanically testing large process gas compressors. It should be noted that mechanical tests differ from performance tests.

If a machine design is proven, a mechanical running test under no-load condition is chosen. The minimum setup of Figure 5.7 is typical for no-load tests. If, on the other hand,

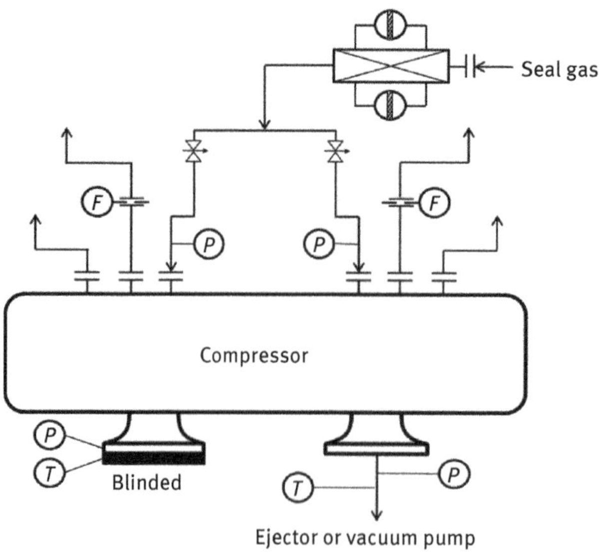

Figure 5.7: Typical setup for no-load mechanical running test (source: *Mitsubishi Heavy Industries Compressor Corporation (MHI), Hiroshima, Japan*).

the machine represents a step-out or if certain critical mechanical parameters push the boundaries of its normal operating envelope in an effort to comply with customers' requirements, full-load testing is recommended. This testing is then done in a closed-loop arrangement with an inert gas (Figure 5.8) so that the mechanical integrity under loaded conditions can be verified.

For either test arrangement, a contract driver and/or all compressors constituting the full compressor train may be arranged as shown in the test setup of Figure 5.9. Use of the contract coupling is recommended but is not mandated by API 617. However, if a shop coupling is used, its overhung moment should be within 10% of that of the contract coupling. Moment simulators are sometimes used to achieve the required overhung moment, especially on the uncoupled ends of drive-through shafts. With respect to the machine itself, API 617 requires, as a minimum, that contract seals, contract bearings, and contract instrumentation be used. Such contract instrumentation will usually include temperature sensors embedded in bearings; it also requires probes and transducers for shaft vibration in two directions and for monitoring the axial displacement of a rotor. Again, these requirements pertain to no-load as well as full-load mechanical run tests.

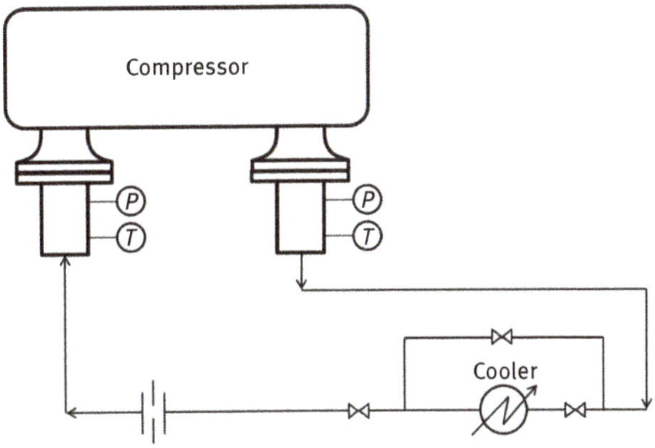

Figure 5.8: Typical setup for loaded mechanical running test (source: *Mitsubishi Heavy Industries Compressor Corporation (MHI), Hiroshima, Japan*).

Figure 5.9: Shop test arrangement (source: *Mitsubishi Heavy Industries Compressor Corporation (MHI), Hiroshima, Japan*).

Testing is preferred with a lubricating oil of the same ISO viscosity grade as the oil ultimately used during normal operation. However, this is not always practical for the vendor's shop tests. Acceptable industry practice would be to adjust the oil temperature for shop tests so that its viscosity is a close match to that at design conditions. Note also that the difference between oil brands does not affect the results of mechanical running tests; accordingly, the brand of shop oil will not necessarily have to be identical with the oil ultimately used by the customer.

5.2.1.1 Test sequence

Mechanical running tests are performed typically in accordance with the time–speed plot illustrated in Figure 5.10. In the initial run-up sequence, the machine speed is increased in increments of 10% from zero speed up to maximum continuous speed (MCS). Pauses or hold points are interspersed at minimum and rated speeds, allowing stabilized vibration and bearing temperature data to be obtained. While going from zero to MCS, it is important to rapidly traverse through shaft critical speeds; this avoids or prevents excessive shaft vibratory amplitudes as the machine passes through "quasi-resonant" frequencies. Once the machine's mechanical condition is stabilized at MCS, mechanical data are again recorded, and the machine is accelerated to trip speed. Operating a machine at its trip speed for approximately 15 min will ensure the robustness of the machine at the projected (lower) operating speed.

The run-up test phase concludes by rapidly decreasing the speed to approximately 500 rpm while recording vibration in preparation for the 4 h running test scheduled next. Both the overall amplitude and synchronous vibration amplitude with phase angle versus speed are plotted as the speed is decreased (Figure 5.11). The data of interest are usually presented in polar form as well (Figure 5.12).

5.2.1.2 The 4 h running test

Speed is again increased up to MCS, and the machine is operated at this speed for 4 h without interruption. During the 4 h running test, the oil supply pressure and temperature may vary within the range of applicable design conditions to ensure the mechanical integrity of the rotor-bearing system under varying conditions.

Operating data such as running speed, lubricant oil conditions, shaft vibrations, and bearing temperatures are recorded generally at 30 min intervals (Figure 5.13). Additionally, frequency analysis on shaft vibration is performed (Figure 5.14), seal oil (or gas) flow is measured, and sound-level measurements are obtained. In the speed-down period following the completion of the 4 h running test, data are again taken for the speed decrement similar to that acquired during the run-up test.

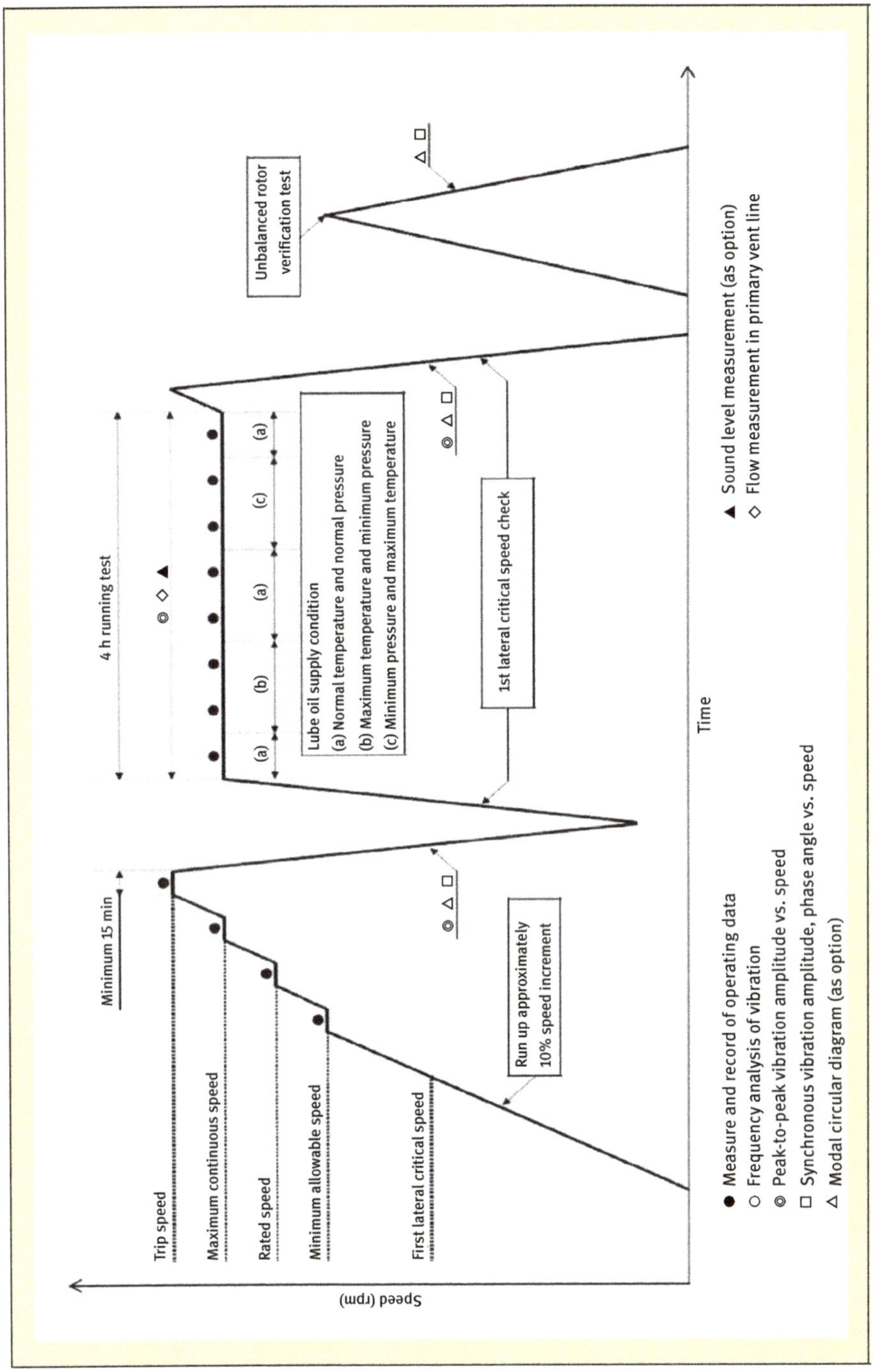

Figure 5.10: Typical sequence for mechanical running test (source: *Mitsubishi Heavy Industries Compressor Corporation (MHI), Hiroshima, Japan*).

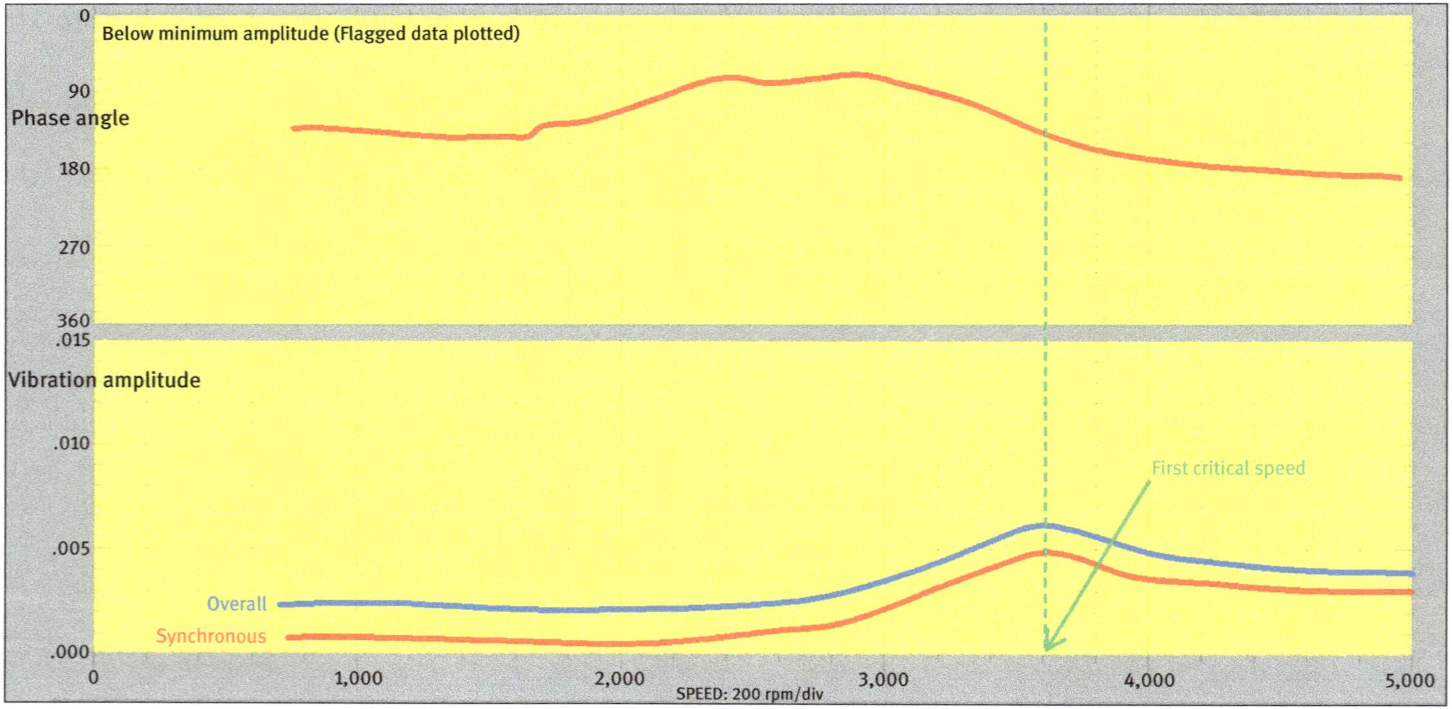

Figure 5.11: Overall/synchronous vibration amplitude with phase angle versus speed (source: *Mitsubishi Heavy Industries Compressor Corporation (MHI), Hiroshima, Japan*).

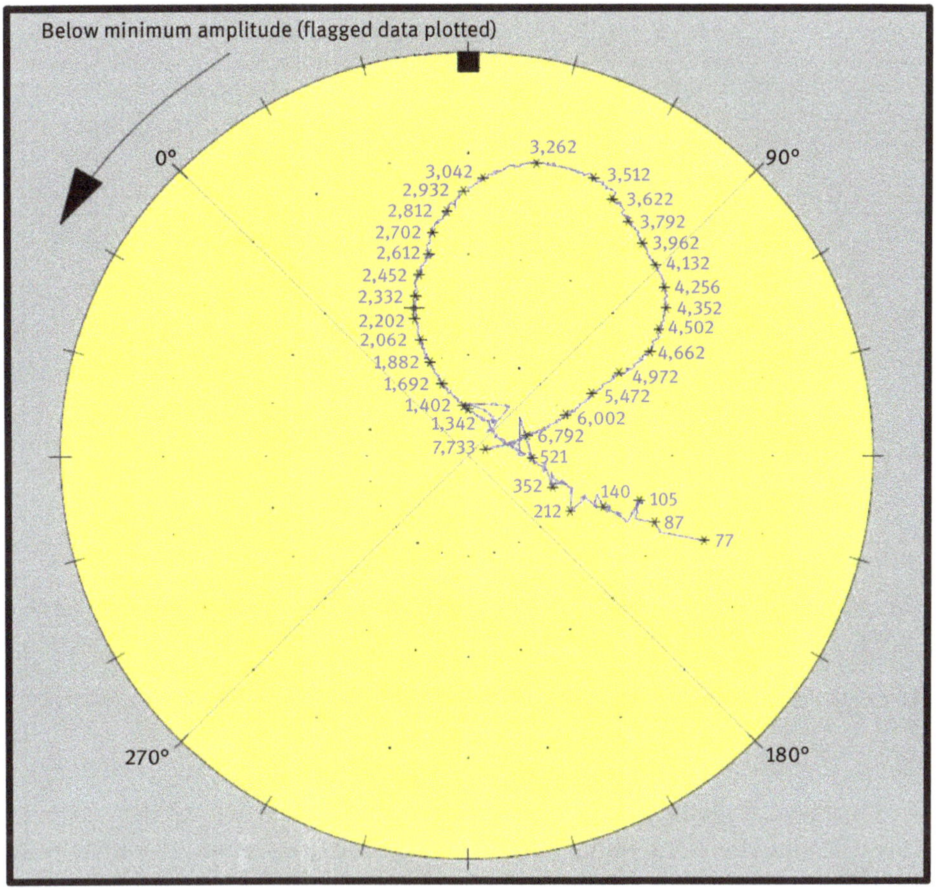

Figure 5.12: Polar plot (source: *Mitsubishi Heavy Industries Compressor Corporation (MHI), Hiroshima, Japan*).

5.2.1.3 Unbalanced rotor verification test

An unbalanced rotor verification test is performed to validate the accuracy of the analytical model; an analytical model will have been used in the damped unbalance response analysis for the mechanical design. As much as four times the API-recommended residual unbalance is usually added to the rotor analytical model. That physical unbalance should later be added to the coupling by removing or attaching bolts and/or washers. The machine must then run up again; it probably passes through the critical speed to approximately minimum governor speed and is then tripped. Coast-down data are recorded, among the data should be synchronous vibration amplitude and phase angle versus speed. The latter is to be recorded and displayed in polar format.

Item	Tag. No.	unit	Time				
			** : **	** : **	** : **	** : **	** : **
Speed	*****	rpm					
Lube oil supply pressure	*****	barG					
Lube oil supply temperature	*****	deg-C					
Lube oil drain temperature	*****	deg-C					
Lube oil flow	*****	ltr/min					
Shaft vibration	*****	um					
	*****						
	*****						
	*****						
Axial position	*****	mm					
	*****						
	*****						
Journal bearing temperature	*****	deg-C					
	*****						
	*****						
	*****						
Thrust bearing temperature	*****	deg-C					
	*****						
	*****						
	*****						

Figure 5.13: Typical record format (source: *Mitsubishi Heavy Industries Compressor Corporation (MHI), Hiroshima, Japan*).

Unbalance response is determined by vectorially subtracting the data taken on the rundown of the 4 h running test from that of this coast-down. Using the polar plot, the critical speed is determined from the speed at which the vibration shows maximum peak along with an associated phase change. Both calculated vibration response and critical speed must be compared with actual test results.

The results of the unbalance test and the predicted results of the analytical model should satisfy API acceptance criteria. In case the test results are not in conformance with the API acceptance criteria, the analytical model will have to be corrected.

5.2.1.4 Acceptance criteria

The acceptance criteria for the mechanical running test are referred to API-617. However, from a practical standpoint, the acceptance criteria for loaded mechanical running test can often be relaxed from those of the API-617 requirements. Knowledgeable users often allow more liberal acceptable limits, ones that are more representative of true site conditions.

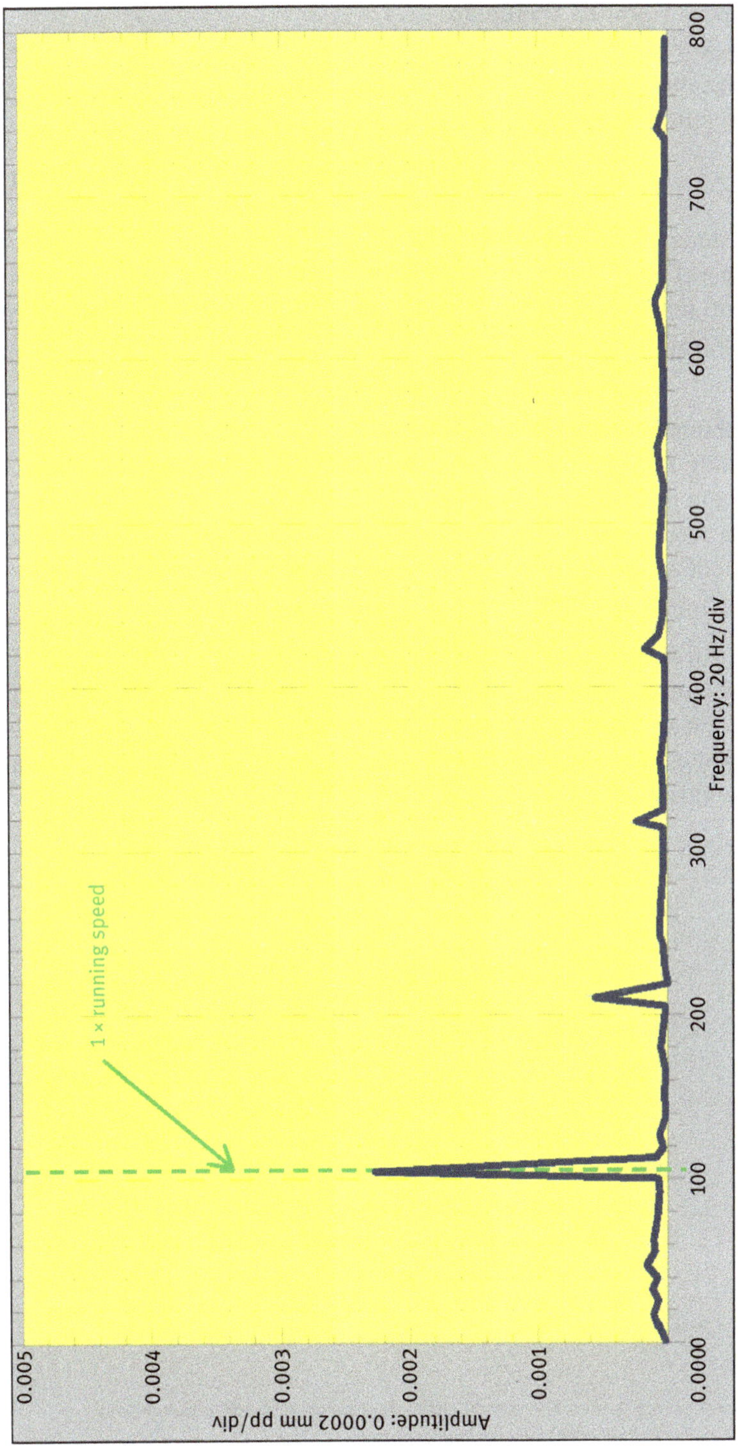

Figure 5.14: Frequency analysis (source: *Mitsubishi Heavy Industries Compressor Corporation (MHI), Hiroshima, Japan*).

5.2.1.4.1 Peak-to-peak vibration amplitude

The unfiltered peak-to-peak vibration amplitudes "*A*" on any vibration probes within the operating speed range shall not exceed the following value calculated by eq. (5.1) or 25 μm (1 mil), whichever is less:

$$A = 25.4\sqrt{12,000/N}(\mu\text{m}) = \sqrt{12,000/N}(\text{mil}), \tag{5.1}$$

where *N* is the maximum continuous speed (MCS).

At speeds greater than the MCS, up to and including the trip speed, the vibration amplitude shall not increase more than 12.7 μm (0.5 mil) above the maximum value recorded at the MCS.

5.2.1.4.2 Nonsynchronous vibration amplitude

The general guideline from API 617 for the amplitude of any discrete nonsynchronous vibration should not exceed 20% of the allowable overall vibration defined by eq. (5.1). If the test results do not meet this guideline, discussions should be held for the requirement of additional testing or acceptability of the machines as it is.

5.2.1.4.3 Lateral critical speed and separation margin

The location of any lateral critical speeds below the trip speed shall be verified, and the separation margin shall conform to the requirements of API-617. A polar plot is usually made available for evaluating lateral critical speeds, amplification factors, and separation margins (see Figures 5.15 and 5.16).

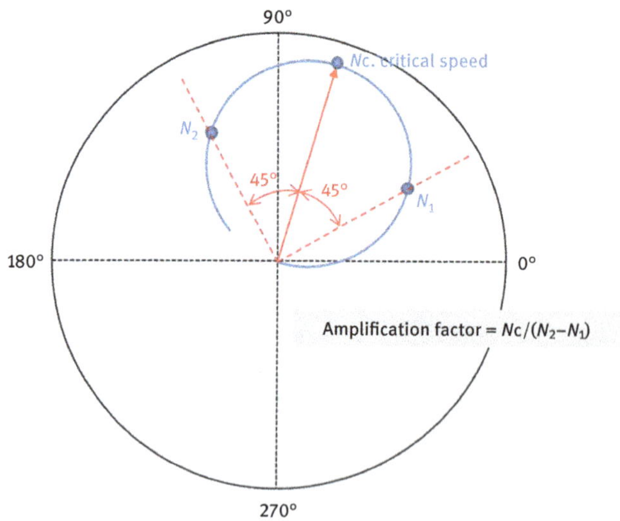

Figure 5.15: Evaluation of critical speed and amplification factor (source: *Mitsubishi Heavy Industries Compressor Corporation (MHI), Hiroshima, Japan*).

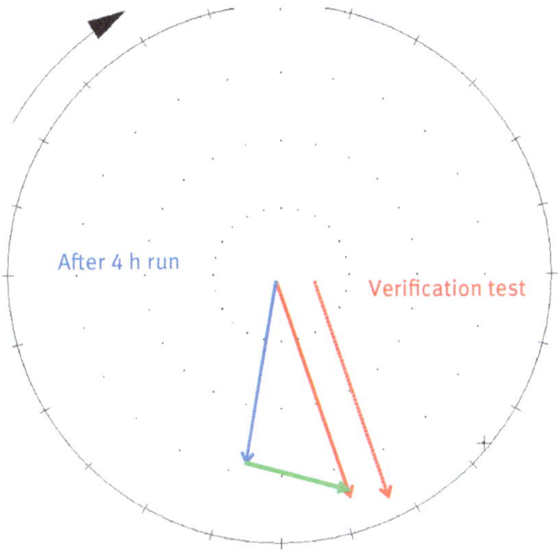

Figure 5.16: Typical unbalanced rotor response test result (source: *Mitsubishi Heavy Industries Compressor Corporation (MHI), Hiroshima, Japan*).

5.2.1.4.4 Unbalanced rotor response

(a) The actual critical speeds determined by the test shall not deviate from the corresponding predicted critical speeds by more than 5%.
(b) The actual peak response shall not exceed the predicted values.

5.2.1.4.5 Additional

In addition to the criteria from API-617, bearing temperatures within the operating speed range are generally subject to acceptable criteria for mechanical running test. The criterion is usually set between 110 and 120 °C.

5.2.2 Post test inspection

After the running test and to ensure the successful operation, visual inspections on compressor internals, bearings, seals, and shaft ends are performed.

5.2.2.1 Bearings and seals

Bearings and seals are removed and visually inspected. If minor scuffs and scratches are observed on bearings or seals, minor cosmetic repairs are allowed, and the repairs are not a cause for rerunning the test.

Shaft-end dry gas seals are sometimes overhauled, inspected, reassembled, and retested by dry gas seal vendor in accordance with customer's requirement.

5.2.2.2 Compressor internals

Compressor is dismantled and internals are visually inspected to ensure that there are no abnormalities on both rotor and internal static parts.

5.2.2.3 Hydraulic coupling fit

The shrink fits of hydraulically mounted couplings are inspected by comparing hub/shaft match marks to ensure that the coupling hub has not moved on the shaft during the running tests.

5.2.3 Rotor stability test

For high-pressure applications, vibration perturbations can result from large aerodynamic forces causing rotor instability. In this regard, a rotor stability test is sometimes requested by customers to be performed during mechanical running test under loaded condition before the machine is shipped.

In this test, the rotor vibrations are excited by external forces under loaded operating condition, and the transient vibration responses are measured to ensure that the rotor-bearing system has sufficient damping characteristics against the excitation forces (Figure 5.17).

To accomplish the stability test, two different methods are usually done.

5.2.3.1 Casing excitation

The hydraulic exciters are mounted on the compressor casing, and the rotor vibration is excited though the casing (Figure 5.18). Stability tests using this method require no modification of the rotor and, therefore, since the rotor-bearing system will be identical to that of the job rotor, results can be more easily validated.

5.2.3.2 Rotor excitation

In this stability testing method, modification of the rotor is necessary. An extension piece is installed on the rotor to accommodate a magnetic bearing assembly shown in Figure 5.19. The rotor-bearing system is directly excited by a function generator through the magnetic bearings, resulting in shaft vibration. The transient behavior of the rotor-bearing system can then be evaluated. The advantage of using this method is that the rotor can be easily excited, and the rotor response is clearly observed.

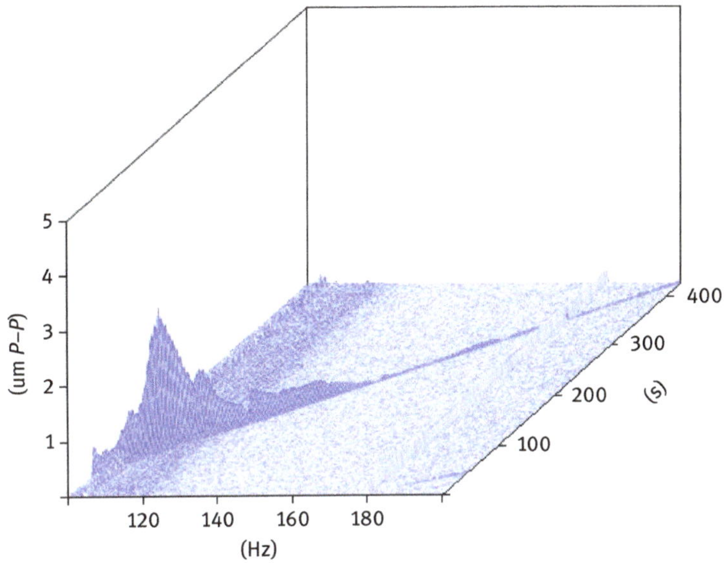

Figure 5.17: Typical vibration response on rotor stability test (sweep excitation test) (source: *Mitsubishi Heavy Industries Compressor Corporation (MHI), Hiroshima, Japan*).

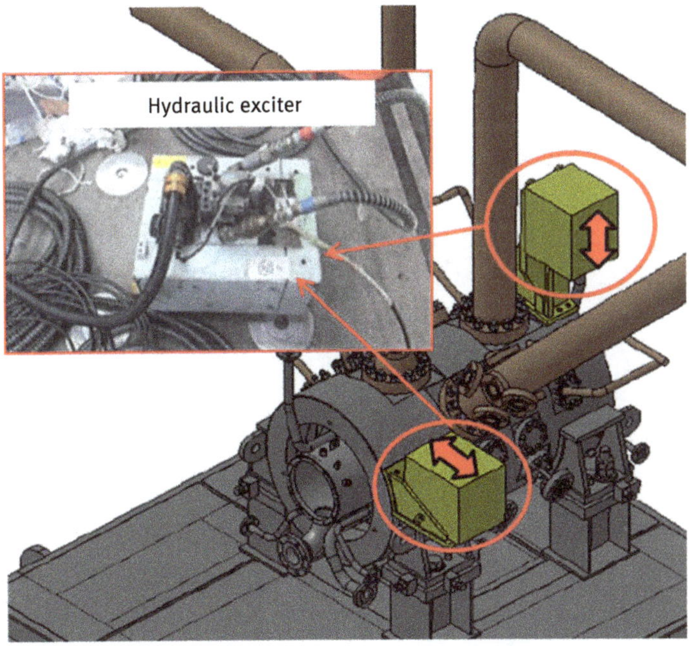

Figure 5.18: Casing excitation (source: *Mitsubishi Heavy Industries Compressor Corporation (MHI), Hiroshima, Japan*).

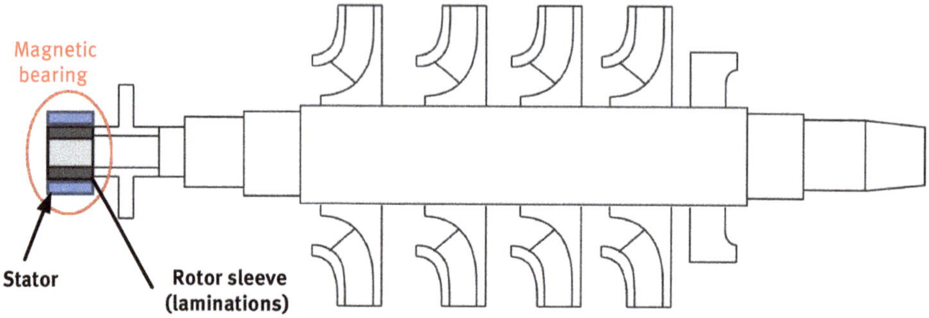

Figure 5.19: Shaft excitation (source: *Mitsubishi Heavy Industries Compressor Corporation (MHI), Hiroshima, Japan*).

References

[1] ASME PTC-10 -1997 Performance Test Code on Compressors and Exhausters.

[2] Schultz, J.M. The Polytropic Analysis of Centrifugal Compressors, ASME Journal of Engineering for Power, 1962.

Chapter 6
Measuring of train performance using torque meters

6.1 Understanding the basic principles of different measuring techniques

To understand the various techniques used for measuring torque and shaft power, we will first examine the mechanical principle of applying torque to a shaft.

If, for example, a circular shaft (hollow or solid) is subjected to a turning moment or torque "T," an angle of twist "ϕ" will exist across the section as shown in Figure 6.1. The angle of twist is proportional to the applied torque and the length "L" of the shaft. The twist will also result in a shear stress across the section proportional to the radius. A maximum shear stress "τ_{max}" will exist at radius "r" with a corresponding maximum shear strain "γ_{max}."

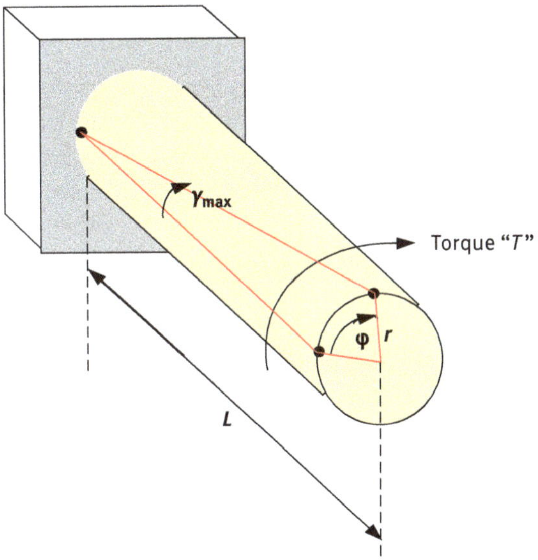

Figure 6.1: Shaft showing deformation when subjected to torque.

The maximum shear strain and angle of twist can be calculated from the following equations:

$$\gamma = \frac{\tau_{max}}{G} = \frac{Tr}{JG} \tag{6.1}$$

https://doi.org/10.1515/9783110678765-006

$$\varphi = \frac{TL}{JG} \tag{6.2}$$

where G is the material modulus of rigidity, J is the polar moment of area of the section, and L is the length of the section.

Manufacturers of torque meters will measure torque either by directly observing the shear strain using strain gages mounted along the axis of the shaft, or they will measure the angle of twist by monitoring the change in angularity/phase relationship between two-toothed wheels mounted at each end of the shaft. Either technique will determine the applied torque, which is used to calculate shaft power as follows:

$$\text{Shaft power} = 2\pi NT \tag{6.3}$$

where N is the rotational speed and T is the torque.

Both methods for measuring torque are discussed in Sections 6.2 and 6.3.

6.2 Measurement using phase change between fixed points on a rotor

Torque meters of the type generally known as the "Torquetronic™ system" measure torque as the shaft twists between a pair of toothed flanges made integral with a coupling spacer. Pairs of phase-displaced sinusoidal signals are captured by multiple pickups – actually internally toothed rings. These rings surround fully encapsulated circumferential coils that are energized to form a low-level toroidal flux path. The torque required to twist the shaft precisely through one tooth pitch will correspond exactly to a 100% phase displacement.

The toothed rings are permanently fixed to each end of a rigid stator tube that is supported clear of the rotating shaft from the coupled machines. The stator tube and associated split casings often replace an existing coupling guard or will make it unnecessary to purchase such guards for new compressor installations (Figures 6.2 and 6.3).

The readout unit shown in Figure 6.4 is a specialized digital phase meter that measures both speed and phase displacement of the two signals. The unit then converts the signals to torque, speed, and power in engineering units.

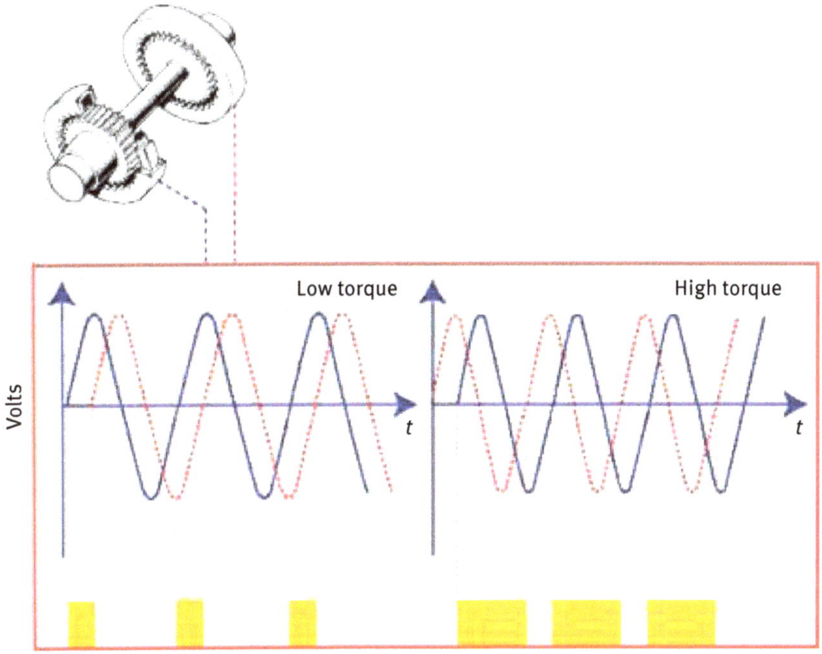

Figure 6.2: Torquetronics torque-metering principle showing a pair of phase-displaced sinusoidal signals *(courtesy of Torquemeters Ltd., Northampton, UK)*.

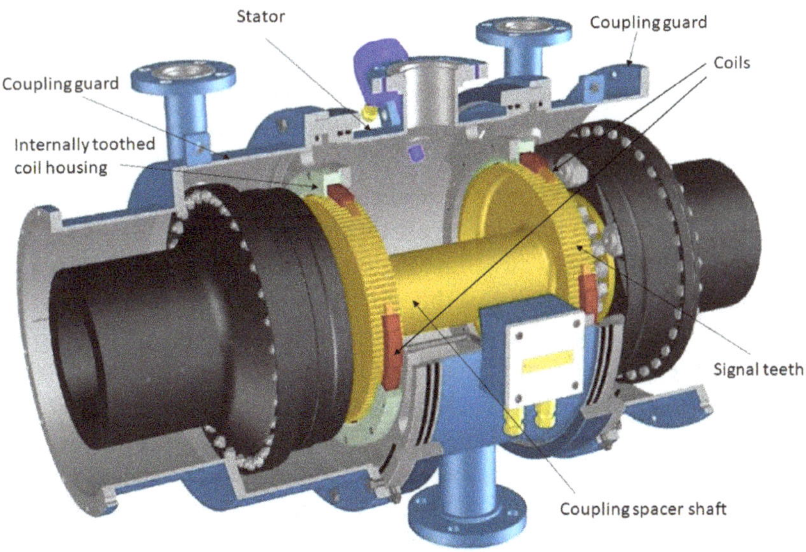

Figure 6.3: Cutaway view of torquetronics torque-sensing components (source: *Torquemeters Ltd., Northampton, UK*).

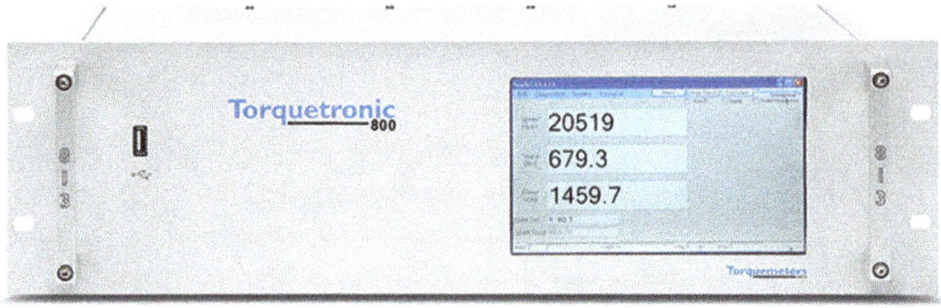

Figure 6.4: Torquetronics readout unit converts phase signal to speed, torque, and power (source: *Torquemeters Ltd., Northampton, UK*).

6.3 Measurement using strain gages

There are many torque-monitoring devices that utilize strain gage technology; they range from basic low-cost and fairly accurate systems to very precise and highly reliable systems.

The US-based Indikon/Riverhawk company has combined Indikon's torque metering know-how with Riverhawk's strong background in coupling design. Their approach takes advantage of strain gage technology to measure torque, utilizing a rotary transformer to power in-shaft electronics and gages as well as transmit data (see Figure 6.5). This requires minimal modifications to the coupling, less added weight, and reportedly lower cost compared to deflection-based torque-metering systems (TMS).

Basic operation involves a coupling or shaft fitted with strain gages, associated electronics, and a rotary transformer (Figure 6.6). The rotary transformer operates on the same principle as an ordinary transformer, but in this case the secondary set of windings is attached to the rotating shaft while the primary windings are fixed relative to the machine frame. The air gap between primary and secondary windings is large enough to accommodate worst-case misalignment.

The rotary transformer serves two functions: to excite the strain gages and electronics inside the shaft and to receive the signals representing torque from the shaft. From here the signals make their way to local electronics for proper conditioning before being sent to the monitor (or display) through the associated cable arrangement (Figure 6.7). Exposure to oil spray or oil mist does not affect the torque meter's operation.

The monitor processes and displays data on a digital readout as speed, torque, and power. Standard analog outputs (4–20 mA, 1–5 V, etc.) are available for each of these real-time values. An Indikon/Riverhawk system provides two independent channels (strain gages, electronics, transformers, and cabling) for 100% redundancy at the coupling for reliable, long-term operation.

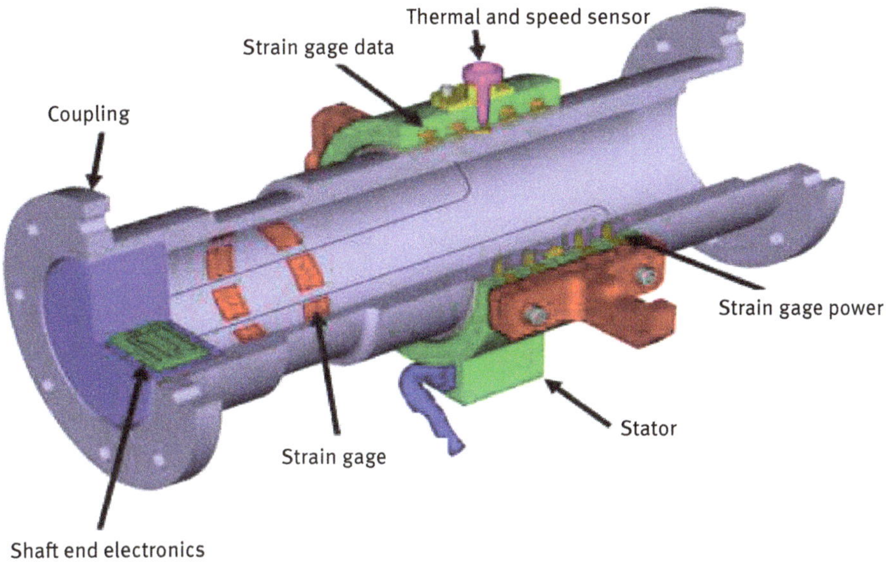

Figure 6.5: Indikon/Riverhawk torque meter components (source: *Indikon/Riverhawk Company, New Hartford, New York*).

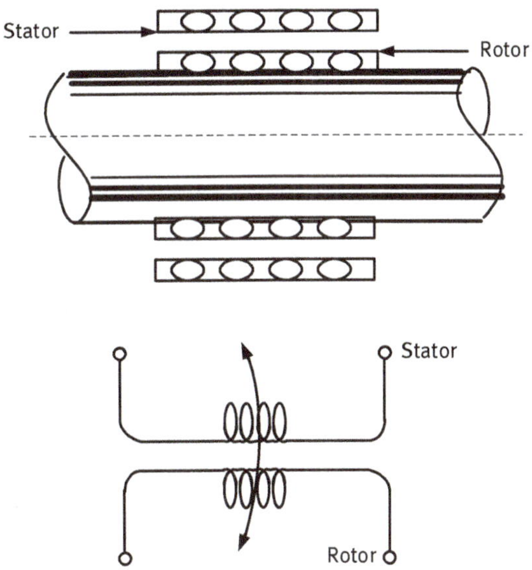

Figure 6.6: Electrical equivalent of rotary transformer (source: *Indikon/Riverhawk Company, New Hartford, New York*).

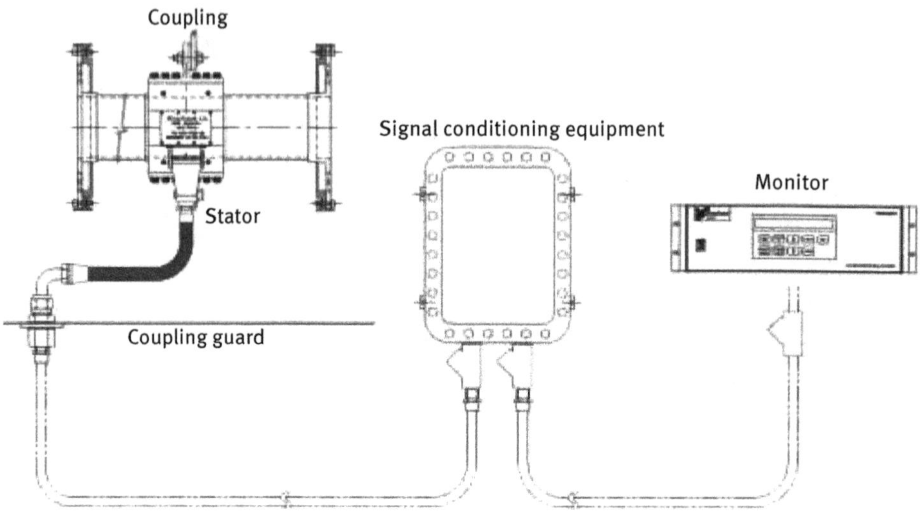

Figure 6.7: Indikon/Riverhawk torque-metering and alignment systems (source: *Indikon/Riverhawk Company, New Hartford, New York*).

The systems are fitted with technology that allows for thermal calibration. Accordingly, the output data are not affected by changes in temperature. Since both torque measurement and temperature compensation are done within a short span on the shaft, instrument or device accuracy is not affected by the temperature differentials typically experienced across couplings.

6.4 New optional features of torque meters

6.4.1 Hot alignment

Indikon/Riverhawk company's torque and alignment system (TAS) is currently (in 2020) the only system engineered to provide real-time coupling alignment data during normal operation. Like TMS, this provides torque, speed, and power data based on the most accurate sensing mechanism, the strain gage bridge. Similar to TMS, the TAS consists of three main components and two interconnecting cables. However, TAS also provides a separate analog output module to transmit data for higher level system monitoring. The totality of data consists of the TMS's torque, speed, and power and adds TAS's alignment data outputs (see Figure 6.7).

6.4.2 Transient/dynamic torque measurement

Using these devices to capture transient torsional data represent a novel approach. Knowing transient data allows specialists to increase equipment reliability. A potential use is to capture dynamic torque data during equipment start-up. Torque transients typically occur in motor start-ups where torque values can be as much as 8–10 times the normal running torque. The severity of transients can quite obviously affect coupling life and/or number of lifetime starts permitted.

6.4.2.1 Potential uses for dynamic torque monitoring

Dynamic torque monitoring can serve many purposes such as:

- Start-up data collection for evaluation of synchronous motor parameters and response.
- Continuous dynamic monitoring for potential excitation of torsional critical frequencies from low-amplitude sources. Dynamic torque monitoring can be used to sense and observe excitation caused by drift in certain process parameters or power supplies. Capturing such events at inception could point to specific needs for follow-up analyses and remedial action.
- Monitoring of operating machines for events and/or excursions that assist in maintenance optimization (predictive instead of time-based maintenance).
- Monitoring of efficiency degradation.

6.4.2.2 Torquetronics dynamic torque-metering system

Due to the large number of signal-generating teeth inherent in the Torquetronic™ system, the measurement of both steady-state and dynamic torque is possible. By applying a fast Fourier transform (FFT) to the torque data, the frequency and amplitude of torsional vibrations occurring in the compressor train can be obtained. Both can be viewed on the readout unit in real time (Figure 6.8), or further interrogated in a number of ways.

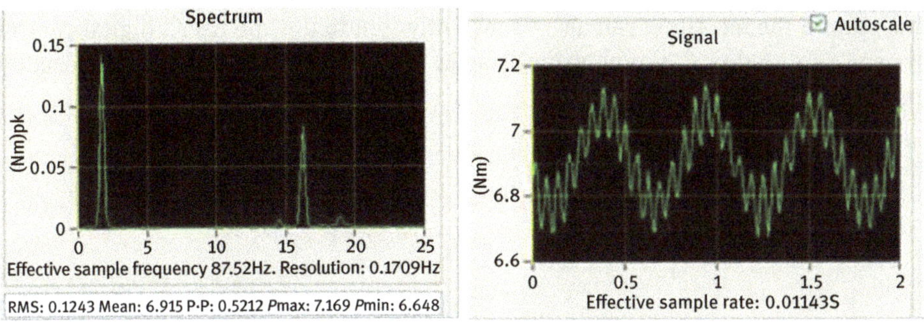

Figure 6.8: Torsional data showing FFT and raw signal from a machine train (source: *Torquemeters Ltd., Northampton, UK*).

Interrogation/analysis is judged particularly useful for variable frequency drive (VFD) electric motor-driven machines. In equipment trains receiving power from VFDs, actual torsional characteristics may be difficult to predict; their effects can be detrimental to the long-term health of a machine. The band width and amplitude resolution selected for Torquetronic™ systems depend on the needs of a particular application. Identification of torsional frequencies up to 4 kHz is possible with amplitude resolution down to less than 0.1% of full-load torque.

In situ calibration ensures that accuracy is maintained over the life of the equipment (typically 20 plus years). The commissioning or "baseline" values are stored in the readout unit; they are readily available for interrogation and review. Because there are no rotating electronics, high and time-stable system accuracy (better than 1% of full load) is inherent. Potential deviations from accuracy are derived from what fundamental science calls the "first principle properties" of all components in the signal chain. Readout units can be located in both safe and designated hazardous zones and areas; signals can be retransmitted to PLC (programmable logic controllers) and alarm trip systems needed for process control.

6.4.2.3 Indikon/Riverhawk dynamic torque-meter system

This system allows customers to capture torsional vibration during start-up (transients) as well as steady-state conditions. With the dynamic torque meter, users are able to monitor both instantaneous dynamic torque values and peak-to-peak torque values. Sampling is done at a rate of 5,000 samples per second.

Existing Riverhawk digital torque meters can be retrofitted easily by replacing old monitors with the new dynamic monitors.

The Indikon/Riverhawk dynamic torque meter can be used to validate equipment operation. It can also serve to obtain and visualize as a diagnostic tool wherein the data collected can be run through an FFT analysis, displayed as waterfall plots or Campbell diagrams. All can assist in obtaining and visualizing the data end users need for evaluation and analysis. Frequency response ranges from DC to 1,000 Hz.

Dynamic torque meters provide end users with instantaneous, recordable torque output; the recording can also be instantly sent to the end user's digital control system. Captured peak torque values stay in the system until the system is reset by an operator.

Chapter 7
Basic rotordynamics

7.1 Introduction to rotordynamics

To achieve the desired aerodynamic performance/efficiency, centrifugal compressors are designed with close radial clearances at their interstage seals, eye seals, and balance piston. Therefore, it is vital that the rotor's lateral vibrations are kept small enough during operation such that it does not contact these seals. Contact between the rotor and a seal increases the seal's clearance, resulting in larger leakage and reduced performance.

Keeping vibratory excursions low is important for the overall reliability of the compressor. Lateral vibrations can cause fatigue or wear of components such as the bearings and casing end seals. Torsional vibrations can lead to high-cycle and low-cycle fatigue failures of shafts and couplings as well as premature wearing of gear teeth.

Resonance and instability are the two phenomena that pose the highest threats to a centrifugal compressor's vibration performance. Familiar to many machinery engineers, resonance occurs whenever an excitation or forcing function's frequency coincides with a natural frequency, such as when a machine is operated near a critical speed.

Unlike resonances that are caused by external forcing mechanisms, instabilities are self-excited phenomena created by the presence of feedback loops. The most common examples of instability in machinery are shaft whip and oil whirl. The most famous rotor instability examples in compressors were encountered in natural gas injection service at the Kaybob plant [1] and Ekofisk platform [2].

Aerodynamic and rotodynamic design tendencies are usually in opposition. The ideal compressor design for aerodynamic efficiency would have the impellers mounted on a very small diameter shaft with large spans between impellers. Furthermore, all impeller stages would be fitted within a single casing. Such a utopian design would create significant vibration issues due to the resulting highly flexible shaft. Therefore, the compressor manufacturer must balance these tendencies with measures aimed toward low-risk vibration performance. Low-risk vibration is associated with large shaft diameters and short bearing spans.

Industrial standards, such as those provided by API [3], contain design and shop testing criteria to help avoid vibration problems in the field. Here, an overview will be provided on these requirements as well as the practical issues regarding lateral and torsional dynamics of centrifugal compressors. The overall subject of rotordynamics is too complex to be covered in great detail in this chapter. For a more comprehensive

---

Note: This chapter was contributed by (and full copyrights retained by) **BRG** Machinery Consulting–Charlottesville, Virginia.

https://doi.org/10.1515/9783110678765-007

and detailed treatment, the reader is encouraged to refer to many available textbooks on the subject as well as the API tutorial [4].

7.2 Lateral dynamics

7.2.1 Overview of performance requirements

Typically, the compressor's design must meet several lateral vibration performance requirements with respect to

- operating away from lightly damped natural frequencies,
- vibrations when operating between minimum and maximum continuous speed, that is, within the operating speed range,
- vibration amplitudes at critical clearances during start-up, shutdown, and overspeed trip, and
- Operating without high nonsynchronous vibrations.

The first of the earlier requirements is intended to avoid any resonance situations created by operating near rotor natural frequencies (critical speeds) or by structural natural frequencies. Since some residual unbalance is always present in the rotor assembly, this unbalance produces a forcing excitation at the same frequency as running speed. If the compressor operates near a lightly damped natural frequency, this unbalance excitation could be sufficient to cause damaging vibrations.

Whether or not a resonance situation leads to damaging vibrations is largely dependent on how well damped the natural frequency is during operation. Industrial specifications often recognize this fact by establishing separation margin regions that are dependent on a critical speed's amplification factor (AF). An example of these separation margin requirements is shown in Figure 7.1, relative to the operating speed range of a specific offshore gas injection compressor.

Several important points are illustrated in Figure 7.1. According to the requirements of API 617, no separation margin is required for AF less than 2.5, meaning a machine can operate on a natural frequency if it is sufficiently well damped. If a critical speed is more amplified (AF > 2.5), it must be located outside of the operating speed range. The amount of separation margin is dependent upon its AF as well as whether the critical speed is located below or above the operating range. As illustrated in Figure 7.1, API 617 requires greater separation margins for critical speeds located above the operating range than it does for ones below. This requirement stems from the need to limit vibrations up to the trip speed, which is typically 10% higher than the maximum continuous speed. An unintended benefit of this larger requirement is an improved robustness to handle any degradation, such as softening, which might occur in the bearing support system with time.

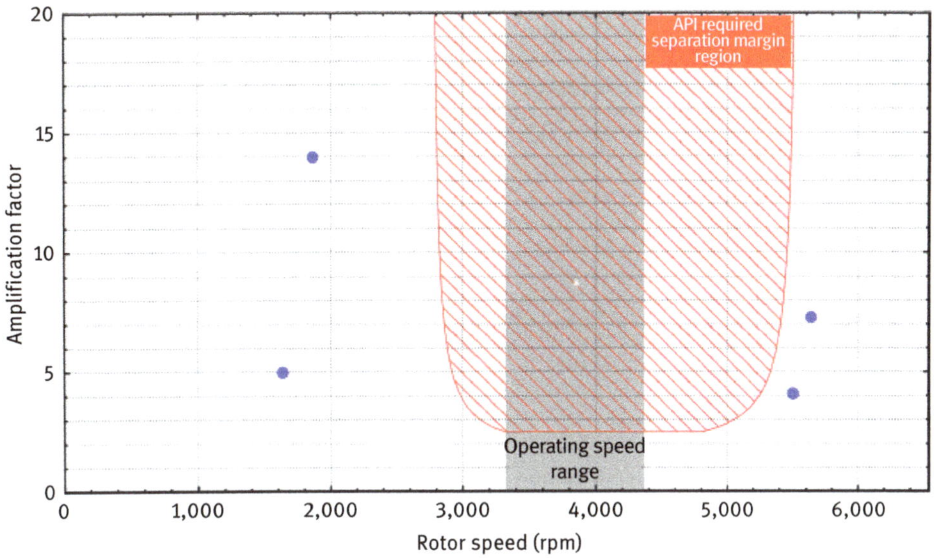

Figure 7.1: API separation margin region for a particular compressor (source: *BRG Machinery Consulting*).

Mitigating resonance problems through adequate separation margins help to ensure that the damaging vibrations do not occur in the operating speed range as well as trip speed. However, separation margin requirements do not specifically limit the experienced vibration amplitudes. Industry specifications typically address this by applying a second vibration performance criterion that ensures the machine runs smoothly, below a defined vibration limit A_{vl}, in the operating speed range.

Since many compressors' rotor vibrations are measured using noncontacting eddy current displacement probes mounted near the bearings, the limit applied is relative displacement amplitude based on industrial guidelines [3, 5]. Determined based on the compressor's maximum continuous speed, A_{vl} is applicable over the entire operating speed range and often represents the expected amplitude for a newly built machine undergoing a shop test.[1] Figure 7.2 illustrates the desired vibration performance, where the maximum vibration in operating range A_{max} is less than A_{vl}.

The compressor must be designed to remain below the vibration limit based on some balance quality for the rotor assembly. API [3] requires the design to handle twice the allowable residual balance *U*a limit for shop balancing. This 200% safety margin is meant to accommodate additional unbalance encountered post-balancing and during shop testing. These additional unbalance sources can be from components that are typically not mounted during a rotor assembly's balance process (couplings, dry gas seals,

1 A higher vibration limit may be specified once the machine is commissioned in the field.

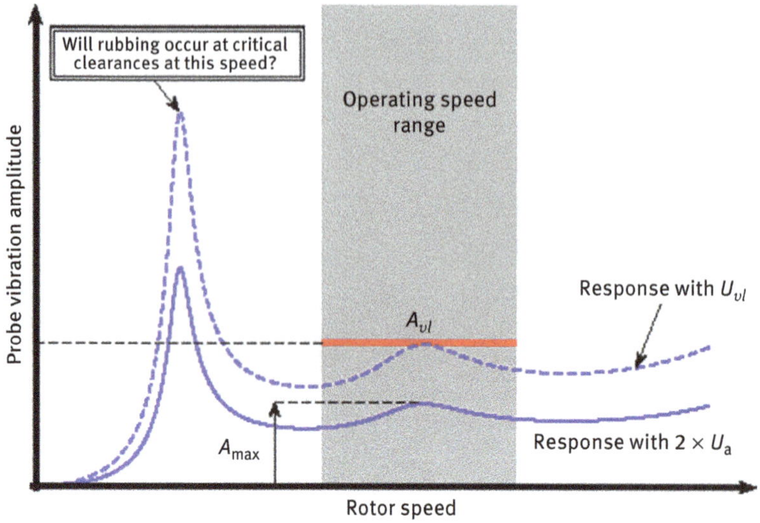

Figure 7.2: Lateral vibration response criteria (source: *BRG Machinery Consulting*).

etc.) as well as shrink fit adjustments that can occur under centrifugal loading during the shop test [6].

The third vibration performance criterion is intended to help ensure that critical clearance locations are not jeopardized during operation. The concern extends to internal seals that are vital to the machine's aerodynamic performance. These should not rub when the machine is started, shut down, or experiences an overspeed trip event. Since the normal operating speed range is typically surrounded by lightly damped critical speeds, high vibration may occur when the machine traverses or approaches these critical speeds during any of these transient events.

When accelerating through one of these critical speeds, vibration amplitudes along the rotor assembly are dependent on the underlying rotordynamics as well as the rotor's balance quality. For design purposes, the balance quality is typically assumed to be that which causes the probes to reach the vibration limit A_{vl} in the operating speed range [3]. Therefore, as shown in Figure 7.2, the compressor should not rub the close but important clearances when traversing critical speeds.

All three of the previous vibration criteria are primarily focused on limiting synchronous vibrations driven by residual unbalance in the compressor's rotor assembly. However, vibrations at other frequencies also must be considered due to the usual presence of nonsynchronous excitations. Nonsynchronous vibration can originate from misalignment between coupled shafts, flow pulses due to blade passing frequencies, gear tooth meshing, and so on. For centrifugal compressors, nonsynchronous vibrations need to be minimized because the presence of any large amplitudes usually indicates a faulty or abnormal condition that should be corrected. Therefore, the fourth vibration performance criterion applied typically limits the amplitude of any discrete, nonsynchronous frequency to a small percentage of the vibration limit A_{vl}.

From the rotordynamics perspective, the nonsynchronous vibration problem of greatest concern is rotor instability, such as shaft whip or oil whirl. Unlike many other nonsynchronous phenomena, rotor instabilities can be associated with violent and damaging vibration, causing failures that threaten both personnel and the physical environment. Avoiding these instabilities and their associated nonsynchronous vibrations becomes a central objective of the rotodynamic design process.

Because manufacturing and assembly tolerances exist, the resulting unbalance distributions are unknown. Moreover, operating conditions fluctuate and the rotordynamic design process must consider these variations; they can certainly influence each of the earlier mentioned performance requirements. For example, when assembled, the journal bearings' clearances will vary within their manufacturing tolerances. Such clearance variations may, in some instances, determine if a problematic resonance or instability occurs. Therefore, it is important that the compressor's rotordynamics be examined for the bearings' entire range of possible clearances and not just a single nominal value.

Whether or not a centrifugal compressor can reliably achieve these four vibration performance requirements depends greatly on the validity of the rotordynamic analyses performed during the design process. Various components within a compressor combine to work as a dynamic system that is assumed to be governed by the following linear equation:

$$\mathbf{M}\ddot{\mathrm{u}}(t) + (\mathbf{C} + \mathbf{G})\dot{\mathrm{u}}(t) + \mathbf{K}\mathrm{u}(t) \tag{7.1}$$

The mass **M**, gyroscopic **G**, damping **C**, and stiffness **K** matrices derive these properties from individual components or subassemblies. These will always include the rotor assembly, bearings, and annular seals. To obtain these properties, each component's physical characteristics must be converted into a dynamic model. Such component modeling is an important step to ensure the accuracy of the overall analysis.

7.2.2 Component modeling considerations

Each compressor shaft has portions of varying lengths and diameters; these yield different mass, inertia, and bending stiffness values. In order to accurately account for these differences, the rotor assembly is divided into sections to capture geometry changes and other rotordynamic locations, effectively creating a finite element mesh. Important locations that need to be included in the mesh are the centers of gravity of all mounted components; these include impellers, couplings, and thrust disks, as well as the centerline locations of bearings and seals.

Mounted components are typically modeled using their lumped mass and inertia (polar and transverse) values placed at their centers of gravity. Since the shaft sections are usually cylindrical, their individual mass and inertia are easily calculated. The bending stiffness of each shaft element is commonly determined from solid mechanics equations for cylindrical beams, in particular those developed by Timoshenko, which include shear deformations and rotary inertia [7, 8].

Shrink fits at impeller and sleeves may contribute additional bending stiffness to the rotor assembly [9]. Each equipment manufacturer uses their own experience-based methods to account for these contributions. These stiffening effects tend to be relatively small considering the size of other components' uncertainties. However, finite element analysis may be warranted to quantify the stiffening created by long fits or long components with fits at both ends. Frictional damping created by long shrink fits can cause shaft whip, which is one reason why short fits are preferred.

Process compressors are typically supported by either fluid film journal bearings or active magnetic bearings (AMBs); each type requires different modeling approaches to determine its dynamic characteristics. Since radial bearings are often the primary source of damping for controlling lateral vibrations, accurately modeling their dynamic characteristics is considered the central focus of a rotordynamics analysis. However, an API survey of manufacturers and consultants revealed that there continues to be a high degree of uncertainty in a journal bearing's predicted stiffness and damping [10].

Highly dependent on speed and static load, the stiffness and damping characteristics of fluid film journal bearings are often determined by solving the Reynolds equation for the steady-state and dynamic oil film forces. Depending on the application, additional physical effects may have to be considered in the calculations. Examples include turbulence in the oil film, heat conduction through bearing pads, (segments) and housings, as well as thermal and mechanical deformations of the pads and shaft. More advanced tilting-pad bearing designs, that is, ones that incorporate evacuated housings and direct lubrication devices require other modeling techniques. These techniques must account for oil flow characteristics and the possibility of lubricant starvation [11, 12]. Pivot deformations may also need to be included using Hertzian contact models [13]. The so-called FlexPad bearings are manufactured as a single piece with flexing pad support stems. Their details can be visualized from relevant websites.

The stiffness and damping of fluid film bearings may be very sensitive to clearance, preload, and inlet oil temperature. Since variations are inevitable in as-built and while-operating conditions, three combinations of these parameters' tolerances are typically calculated. The calculations reflect the designer's attempts to place limits on the expected range of a bearing's dynamic characteristics. Maximum bearing stiffness usually results from minimum clearance, maximum preload, and minimum inlet oil temperature. Minimum stiffness is usually the result of maximum clearance, minimum preload, and maximum inlet oil temperature. Figure 7.3 illustrates the variations in stiffness and damping that can occur in a compressor's journal bearing. For this particular bearing, the horizontal direct stiffness K_{xx} increases by over 400% when going from maximum to minimum clearance.

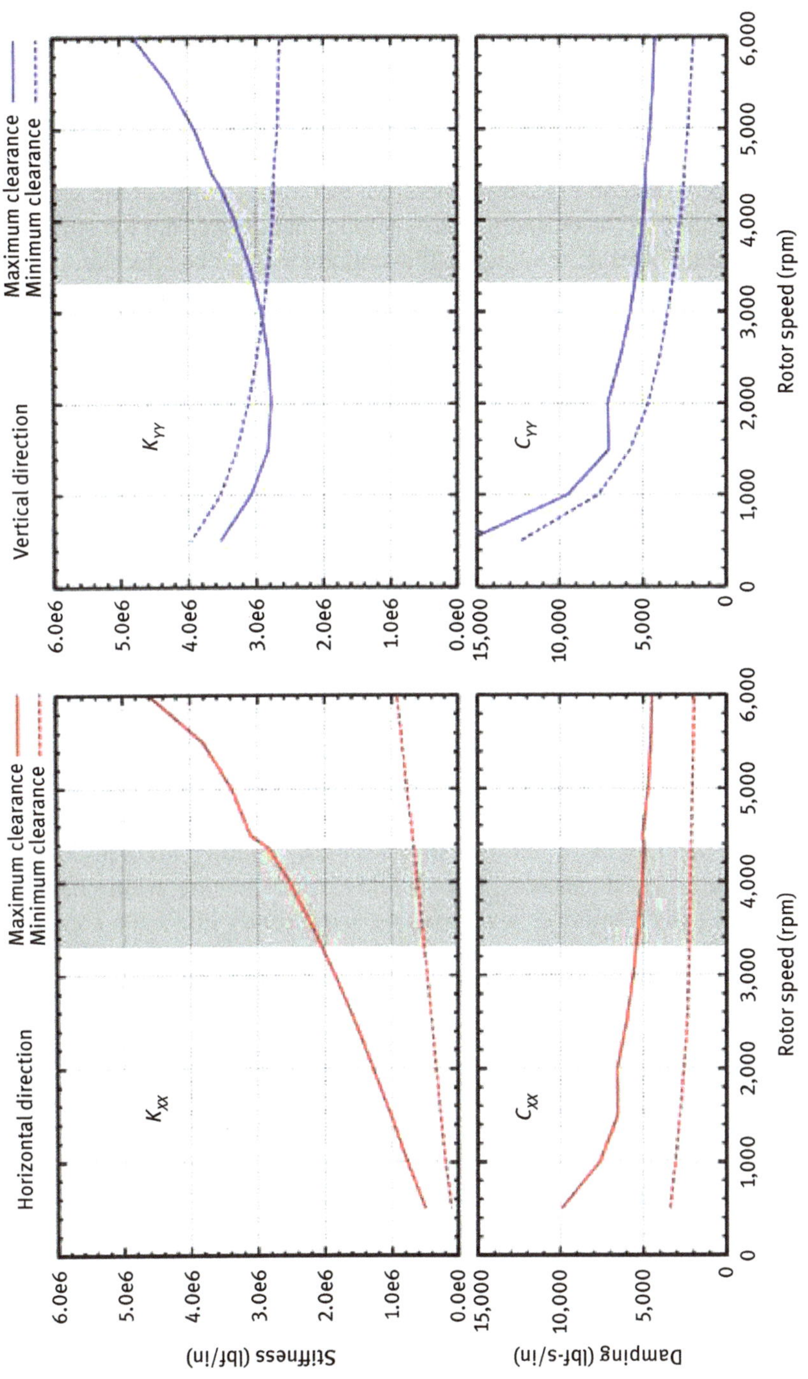

Figure 7.3: Dynamics versus speed and clearance condition for a particular tilting-pad journal bearing (source: *BRG Machinery Consulting*).

Although commonly used in aircraft jet engines, squeeze film dampers are sometimes incorporated into a centrifugal compressor's design when additional damping is needed. Each damper is usually integrated directly into the fluid film bearing's outer housing such that it utilizes the same oil supply. Damping generated by the oil film is typically calculated using the classic Reynolds equation [14]. The stiffness associated with the damper's centering and sealing devices, for example, O-rings or arc springs, must also be determined. If these sealing and/or supporting devices are too stiff relative to the inner journal bearing, the squeeze film damper will not be effective.

Well established as the preferred technology in cryogenic turboexpanders, AMBs are becoming more and more prevalent in the centrifugal compressor applications. The dynamic characteristics of the bearings are custom designed or tuned for the specific machine through software parameters in the control system. These dynamics are usually provided by the AMB manufacturer in the form of transfer functions that relate displacements measured at the radial sensors to bearing forces produced by the electromagnets. Such transfer functions cannot be modeled as the simple springs and dampers that replicate the responses of fluid film bearings. Furthermore, the AMBs' finite force capacity (static and dynamic) must be considered when examining unbalance response performance.

A compressor's casing end seals can also play an important role in the machine's lateral dynamics. Most centrifugal compressors that are designed and manufactured today (2020) incorporate dry gas seals to isolate the process gas from the outside atmosphere. Other than the mass and inertia of their rotating assembly, dry gas seals do not contribute any other dynamic characteristics.

The "benign" dynamic behavior of dry gas seals is in sharp contrast to the oil seals that preceded them for many decades. Oil seals and, in particular, their floating atmospheric-side bushings can generate strong stiffness and damping within their oil film. As mentioned earlier, stiffness and damping significantly affect the compressor's unbalance response and stability. The magnitude of these dynamics is very sensitive to the bushing's radial clearance, axial length, differential pressure, and even the bushing's operating eccentricity. To determine the resulting operating eccentricity and film dynamics, specialized tools that account for the bushing's weight and pressure balancing capability are necessary [15]. Suction pressure determines the pressure differential experienced by the oil seal's atmospheric-side bushing. Accordingly, the analysis should consider the full anticipated range of suction pressure conditions from routine start-up to emergency shutdown.

Since the 1990s, a compressor's internal seals (eye seals, shaft seals, and balance piston) have been recognized as significantly affecting lateral vibration performance. There are two predominant designs for these seals: labyrinth and damper types (hole pattern and honeycomb), each influencing dynamic properties in a different manner. The dynamics of labyrinth seals tend to be sensitive to the number of teeth, their location (at stator, rotor, or interlocking), operating clearance, pressure differential, and the circumferential velocity of the gas entering the seal, commonly known as preswirl or inlet swirl. Having small direct stiffness, labyrinth seals do not tend to have a significant impact on the machine's unbalance performance. However, their dynamics can cause shaft whip; they must be considered for reasons of stability.

Holes and honeycomb pattern type seals' dynamics tend to be significantly stronger than labyrinths. With direct stiffness magnitudes comparable to journal bearings, these seals should be included in the rotordynamics calculations. Their dynamics are sensitive to the geometry of the holes/cells; they are not unlike the parameters applicable to labyrinths. For balance pistons, where the high differential pressure can cause significant deflections of the seal's housing, it is important to maintain positive clearance convergence in the leakage flow direction of honeycomb or hole pattern seals during operation.

Predictions for both types of internal seals are typically based on compressible bulk flow theories using one, two, or three control volumes. An excellent overview of labyrinth seal modeling techniques can be found in [16]. A two-control volume approach is currently the industry standard for honeycomb and hole pattern-type seals [17]. Significant uncertainties remain in the accuracy of these bulk flow models' predictions, making seals the primary focus of current testing and computational flow dynamics (CFD) research.

A compressor's radial bearings are supported by multiple components that transfer the forces to ground. These components can include the bearing housing, machine pedestal, machine casing, baseplate, mezzanine, and foundation. While these structures are typically massive and stiff, their dynamics can greatly influence rotordynamics performance. This is particularly the case for large compression machinery and compressors with light-weight casings.

The importance of pedestal dynamics can be understood by recalling the well-known equation for the equivalent stiffness k_{eq} of two springs in series:

$$K_{eq} = \frac{K_{brg} \cdot K_{ped}}{K_{brg} + K_{ped}} \tag{7.2}$$

where K_{brg} is the bearing stiffness and K_{ped} is the effective stiffness of the combined pedestal structure. For two springs in a series, the equivalent stiffness is always dictated by the softer spring, that is, k_{eq}can never be larger than the softest spring. As shown in Figure 7.4, even when the pedestal stiffness is ten times the bearing stiffness ($K_{ped}/K_{brg} = 10$), the equivalent bearing stiffness is approximately 10% less. Such a reduction in stiffness can be important in situations where the critical speed locations are sensitive to bearing stiffness.

In rotodynamic analyses, it is commonly assumed that pedestal dynamics can be represented by a spring-mass-damper, single degree-of-freedom (SDoF) system whose parameters remain constant for all speeds. However, the dynamics of the pedestal system may be too complex to be accurately represented using such a simple SDoF model. In this case, methods that can accurately model pedestals with multiple natural frequencies in the operating speed range are available [18]. Regardless of the

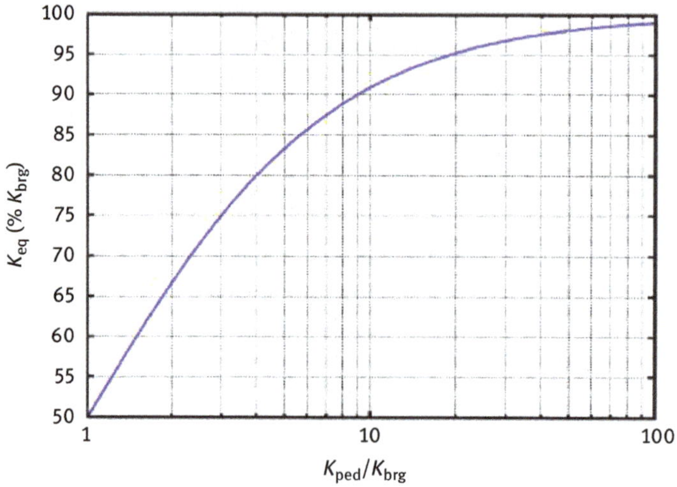

Figure 7.4: Pedestal stiffness influence on equivalent bearing stiffness (source: *BRG Machinery Consulting*).

pedestal model's complexity, its parameters should be based on measurements [19] or structural analysis [20] of the pedestal system.

The earlier components represent those typically considered in the rotordynamic design process. Debate continues on the contributions of other effects, such as the aerodynamic effects associated with the impellers [21] or blading [22]. Each manufacturer has their own philosophy on how each component is modeled and included in the analysis process.

7.2.3 Typical lateral analyses

There are three analyses commonly done during the design phase of centrifugal compressors:

1. undamped critical speed analysis
2. unbalance response analysis
3. stability analysis

Up until the 1980s, undamped critical speed analysis was often the only rotordynamic calculation performed to design a compressor. Today, while not associated with any design acceptance criteria, undamped analysis is still done for information purposes. Undamped analysis gives us insight into some of the machine's overall rotordynamic characteristics. Most notably, preliminary estimates of critical speed locations relative to the operating speed range are obtained from this well-established analysis, along with their expected degrees of severity or magnitude of amplification.

By incrementally increasing the stiffness at the bearing locations, undamped analysis determines the natural frequencies (eigenvalues) and their associated mode shapes (eigenvectors) by solving eq. (7.1) without damping and forcing in the system. The main products of these calculations are the critical speed map and the undamped mode shapes, Figure 7.5.

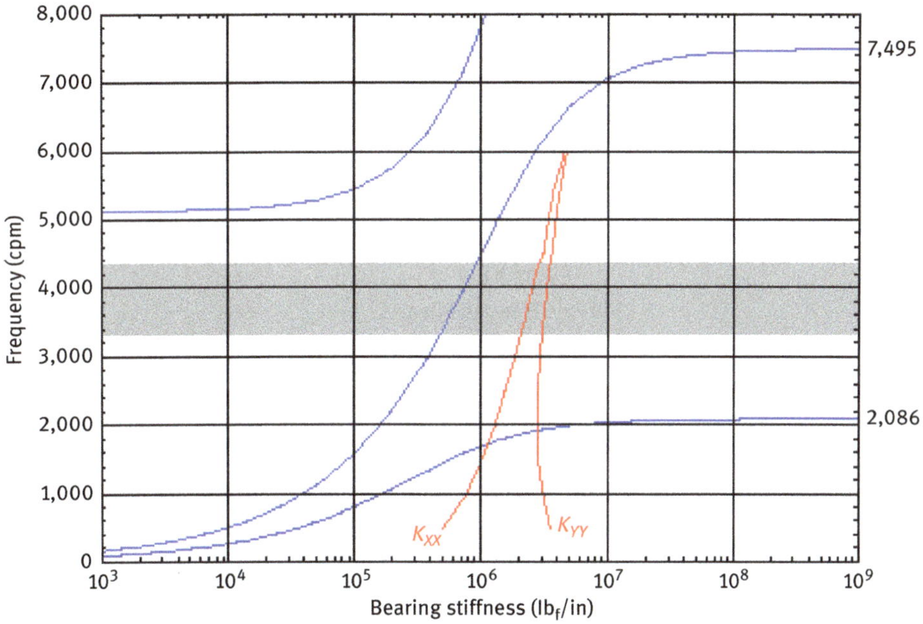

Figure 7.5: Undamped critical speed map with bearing stiffness curves (source: *BRG Machinery Consulting*).

Figure 7.5 presents an example critical speed map for a six stage, through-flow compressor in propylene refrigeration service. The map illustrates how each critical speed's natural frequency changes with bearing stiffness. At low (zero) bearing stiffness values, the rotor assembly is considered to be "free–free" such that the first two critical speeds are at zero frequency and their mode shapes, illustrated in Figure 7.6a, are associated with rigid motion (translation and rotation) of the rotor assembly with no bending. The first flexible rotor mode, where bending occurs in the free–free condition, is seen at 5,122 cpm for this compressor.

At high bearing stiffness values on the right side of the map in Figure 7.5, each critical speed line approaches a constant frequency value. For example, the rotor's first "rigid bearing" critical speed is at 2,086 cpm. The corresponding mode shapes in Figure 7.6b show that the rotor experiences significant bending at these high bearing stiffness conditions.

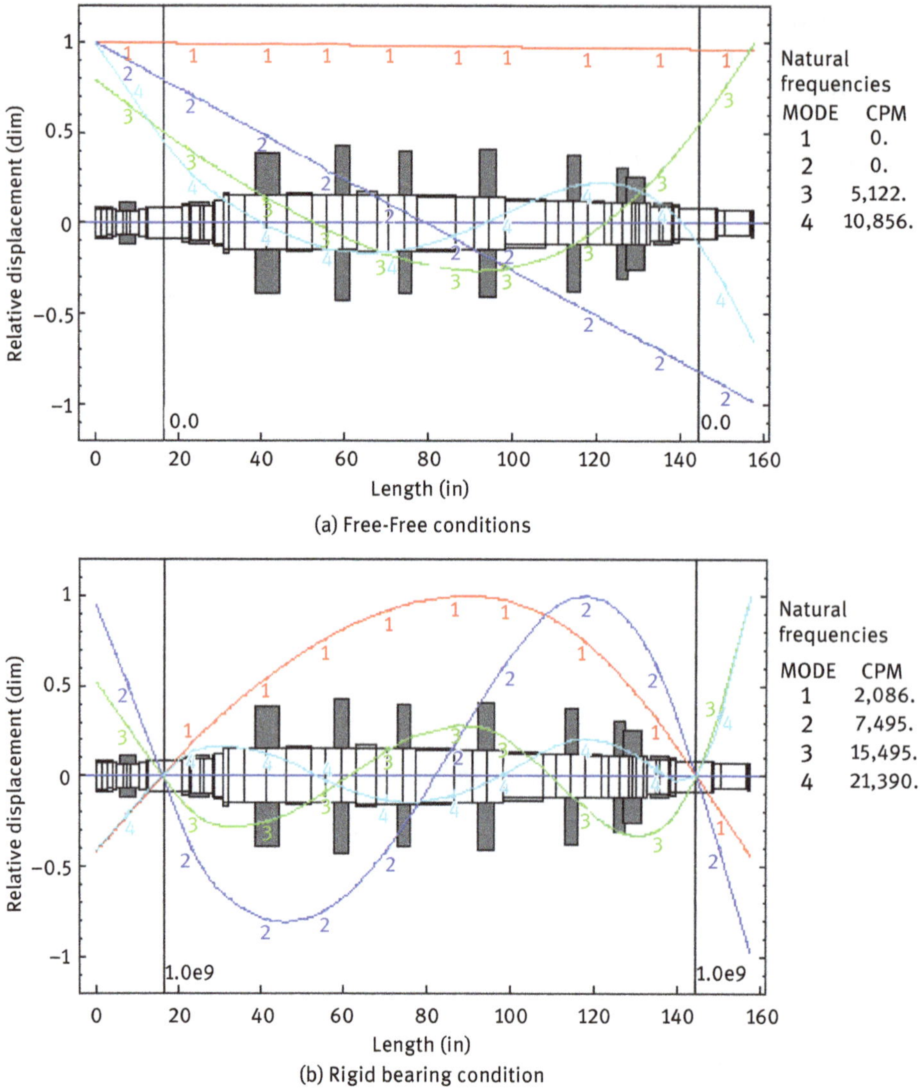

(a) Free-Free conditions

(b) Rigid bearing condition

Figure 7.6: Rotor free–free and rigid bearing undamped mode shapes (source: *BRG Machinery Consulting*).

Since the expected bearing stiffness values are known from their component models, it is common practice to overlay their predicted stiffness versus speed curves on the critical speed map, as illustrated in Figure 7.5. The locations of the machine's critical speeds are estimated based on the intersections of these bearing stiffness curves with the natural frequency lines. For instance, in Figure 7.5, the first critical speed is expected to occur at approximately 1,700–1,900 cpm.

The undamped mode shapes at the expected bearing stiffness levels help predict how effectively the critical speeds will be dampened and what unbalance locations will excite these critical speeds. Figure 7.7 shows that motions at the bearings are relatively small for the first critical speed, indicating this natural frequency will likely not be well damped. Unbalance near the midspan would effectively excite the first critical speed but coupling unbalance would not. Identifying the most effective locations for exciting the critical speeds of concern is an important consideration for the next analysis where unbalance is applied to the model.

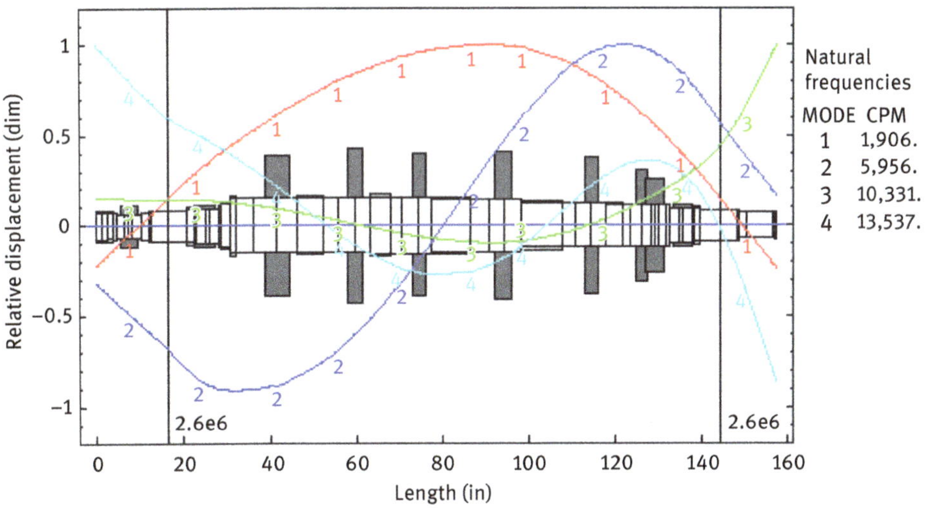

Figure 7.7: Undamped mode shapes at expected bearing stiffness levels (source: *BRG Machinery Consulting*).

Unbalance response analysis provides the best estimate of the compressor's critical speed locations and vibration amplitudes. Therefore, its results are used to determine whether the design meets the acceptance criteria associated with separation margins, vibration limits, and rubbing risks. Industrial standards require that these vibration performance criteria be evaluated by applying unbalance to the rotor system and calculating the resulting vibration response. In this case, all the component models with the exception of labyrinth seals are included in the system model. The analysis is done assuming steady-state conditions, where time transient and acceleration effects are not included. Instead, eq. (7.1) is transformed into the following steady-state equation:

$$\vec{U}(\Omega) = \left[-\Omega^2 M + j\Omega\{C(\Omega) + G(\Omega)\} + K(\Omega)\right]^{-1} \vec{F}_{\mathrm{UNB}}(\Omega) \tag{7.3}$$

where the unbalance distribution $\vec{F}_{\mathrm{UNB}}$ is applied along the rotor, and $\vec{U}$ contains the resulting displacement as a function of rotor speed Ω.

The analysis is usually done for several different unbalance distributions and bearing clearance conditions. Using the undamped mode shapes, the unbalance distributions are selected based on their likelihood and capability to excite the primary modes of interest; these, typically, are the first three critical speeds. For the example compressor, three unbalance distributions would likely be selected based on the mode shapes in Figure 7.7:

1. Midspan unbalance to excite the first critical speed.
2. Out-of-phase unbalance at the first and last stage impellers to excite the second critical speed.
3. Coupling unbalance to excite the third critical speed.

As discussed earlier, API [3] requires the compressor's probe readings to not exceed the vibration limit A_{vl} with twice the allowable residual unbalance Ua applied. This requirement establishes the amount of unbalance to be applied for each distribution. For a mounted component that is not balanced with the rotor assembly, perhaps couplings and dry gas seals, the level of unbalance applied should consider its individual balance quality and, in particular, its repeatability after disassembly and reassembly.

Separation margins and vibration limits are assessed by examining the calculated amplitude versus speed plots at the probe locations. Figure 7.8 presents the Bode plots for the four proximity probe locations of the example compressor. The calculated results should account for the angular orientation of the installed proximity probes or other vibration instrumentation.

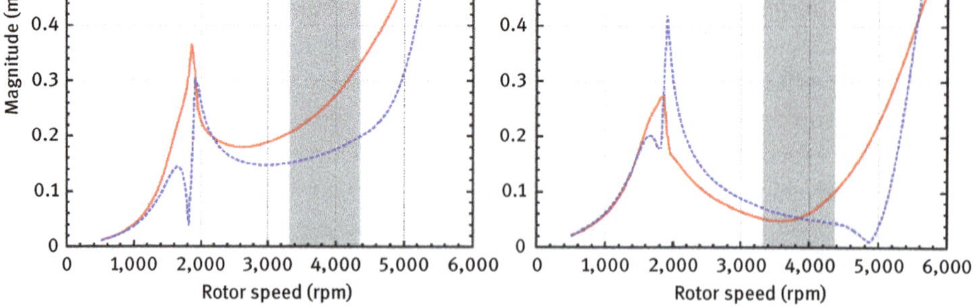

Figure 7.8: Proximity probe Bode plots for a specific unbalance distribution (2x*Ua*) and bearing/seal condition (source: *BRG Machinery Consulting*).

For each peak in the Bode plot, its AF is first calculated using the half-power method. A narrow peak produces a very high AF or "quality factor." Quality factor is the original term used by electrical circuit designers who developed the method to measure a resonance's sharpness. To assess the risk of resonance, the speed location and AF of each peak are then compared to determine whether its actual separation margin meets the separation margin requirements (see Figure 7.1). According to API [3], no separation margin is required for AF less than 2.5, which equates to 20% of critical damping ($\zeta = 0.2$).

Multiple peaks are observed around the first critical speed for this example compressor. This situation is common in compressors that use load-on-pad bearings or pedestals that have anisotropy or asymmetry in their direct stiffnesses (typically, $K_{xx} \ll K_{yy}$). Such a "split critical" can make AF calculations difficult, which is one reason to examine the results from damped Campbell diagrams [23].

In addition to assessing the risk of resonance, the Bode plot results also allow direct comparison with the vibration limit in the operating speed range. For this particular unbalance distribution and bearing condition, the example compressor's Bode plots indicate that the vibrations remain below the A_{vl} limit in the operating speed range.

Since the unbalance response analysis is linear, the amount of unbalance necessary for the probes to reach A_{vl} can be determined from the scale factor of A_{vl}/A_{max} (see Figure 7.2). With this degraded level of unbalance applied, the risk of rubbing critical clearances when going through critical speeds can be assessed using operating deflection shape (ODS) plots. An ODS for the example compressor at the first critical speed is displayed in Figure 7.9.

Unlike the mode shapes in Figure 7.7 where the amplitude of vibration is dimensionless, the ODS plot illustrates how much displacement is expected along the rotor relative to the clearances at critical locations. At the first critical speed of the example compressor, Figure 7.9 indicates that the greatest rubbing threat occurs near the midspan.

It is vitally important that the compressor design's separation margins, vibration levels, and rubbing risks be examined for each unbalance distribution and each bearing condition. When supported by AMBs, additional limits need to be considered; these include the bearings' finite static and dynamic force capacity. Ideally, bearing force capacity should never be the limiting factor, that is, exceeding a probe vibration limit should precede the loss of bearing capacity [6].

As discussed earlier, avoiding the nonsynchronous vibrations associated with oil whirl and shaft whip rotor instabilities is the central objective of the rotordynamics design process. Stability analysis is the analytical tool used by the industry for this purpose. Unlike undamped critical speed analysis, stability analysis is done with all components' models included in the assembled system as well as their damping characteristics. The system's governing equation of motion (eq. (7.1)) is then solved

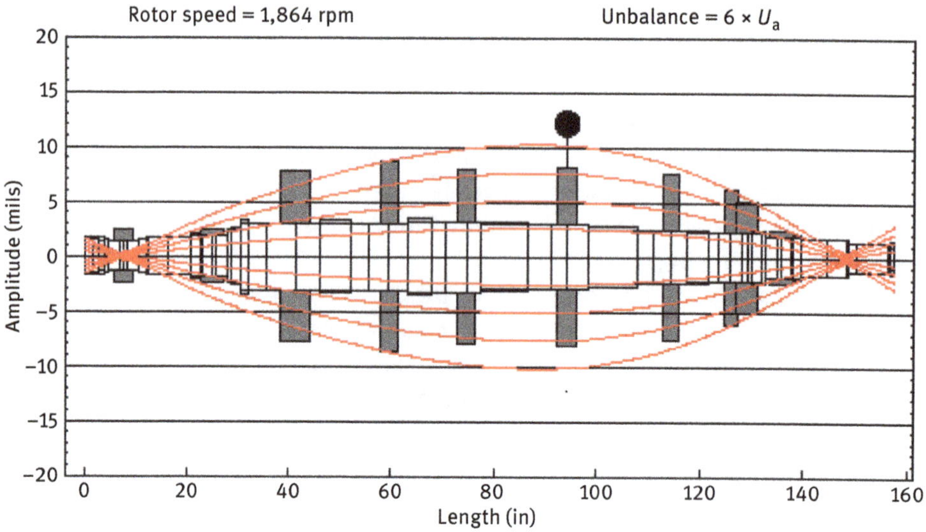

Figure 7.9: Operating deflection shape for example compressor (source: *BRG Machinery Consulting*).

without external forces to determine the "damped eigenvalues" and their associated mode shapes or "eigenvectors."

Each rotor mode of vibration typically possesses a pair of complex conjugate eigenvalues or poles of the form:

$$\begin{aligned}\lambda &= -\zeta\omega_n \pm j\omega_d \\ &= -\zeta\omega_n \pm j\omega_n\sqrt{1-\zeta^2}\end{aligned} \tag{7.4}$$

where ω_n is the mode's undamped natural frequency. A rotor instability occurs when a rotor mode's damping ratio ζ is negative ($\zeta < 0$), resulting in high nonsynchronous vibrations at the mode's damped natural frequency ω_d.

As an alternative to damping ratio, industry has adopted logarithmic decrement, or log decrement, as the indicator of a mode's stability level [4]. It can be calculated directly from a mode's eigenvalue or damping ratio:

$$\delta = \frac{-2\pi\mathrm{Re}(\lambda)}{|\mathrm{Im}(\lambda)|} = \frac{2\pi\zeta}{\sqrt{1-\zeta^2}} \tag{7.5}$$

For reference, a log decrement of 0.1 equates to a damping ratio of 0.0159, or 1.59% critical damping.

Oil whirl and shaft whip rotor instabilities typically are associated with the rotor's first forward whirling mode (1 F), where "forward" means the mode whirls in the direction of rotation. Therefore, the log decrement of the first forward mode (δ_{1F}) is the primary parameter of interest originating with the stability analysis. In particular, its value should not be negative (meaning unstable). When a machine is supported by AMBs, it is important to assess the stability of multiple rotor modes as well as casing/pedestal modes. This additional complexity arises from an AMB's ability to inject energy and destabilize any mode within their controller's bandwidth.

7.3 Lateral stability of rotors

Unlike unbalance response performance, only as of 2002 industry standards have stipulated analysis requirements and design criteria for stability concerns. For compressors incorporating oil film casing seals, or that are supported by fixed geometry AMBs, or are running above their first rigid bearing critical, API 617 requires a screening analysis of the machine's stability. This screening or "level 1" analysis is intended to determine whether the machine's dynamics warrant a more detailed "level 2" examination.

The main simplification during the screening analysis consists of an empirical relationship. This relationship adequately represents all of the compressor's internal dynamics, the ones associated with shaft seals, eye seals, balance piston, and so on. Instead of determining the stiffness and damping characteristics of all these internal components, the "aerodynamic cross-coupling" *qi* of each centrifugal or axial stage is calculated using the simple, empirical "modified Alford" equation [4]:

$$q_i = \frac{C \cdot \mathrm{HP} \cdot \beta}{D \cdot H \cdot N}\left(\frac{\rho_\mathrm{d}}{\rho_\mathrm{s}}\right) \tag{7.6}$$

where q_i is the estimated cross-coupled stiffness and C is a unit conversion constant. The stage efficiency coefficient β is a constant defined by the type of compressor stage, axial or centrifugal. For an axial stage, D and H are the blades' pitch diameter and height, respectively. For a centrifugal stage, D is the impeller's outer diameter and H is the minimum of either the diffuser's or impeller tip's width. Finally, the speed N, power HP, and suction ρs and discharge ρ_d densities for the stage are based on the compressor's normal or rated operating condition.

A stability sensitivity curve at maximum continuous speed, such as that shown in Figure 7.10, is the primary output from the screening analysis. At each extreme bearing condition, cross-coupling is varied from zero until the log decrement reaches zero. From these results, the minimum anticipated log decrement δ_A and stability margin Q_0/Q_A can be assessed relative to the anticipated cross-coupling Q_A, which is the sum of the individual stage cross-coupling values, that is, $Q_\mathrm{A} = \Sigma q_i$. If δ_A and Q_0/Q_A are not sufficiently high, the machine is required to undergo a detailed analysis where the individual stiffness and damping characteristics of the internal components should replace the modified Alford simplification.

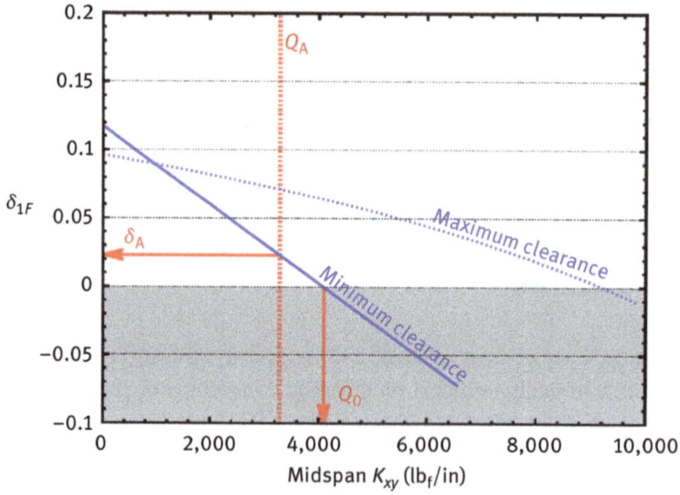

Figure 7.10: Example stability sensitivity curve (source: *BRG Machinery Consulting*).

A third screening criterion is to examine the machine's risk of instability based on industry experience. Unless the stability margin Q_0/Q_A is extremely high, a detailed analysis is required if the compressor's critical speed ratio (= $M\ C\ S/Nr_1$) and average gas density cause it to fall within region B in Figure 7.11. Many compressors have successfully operated in this region without oil whirl or shaft whip problems. However, industry experience has shown that the machines in this region should have their stability characteristics closely examined to avoid such instability problems.

When the compressor is supported by AMBs, it is recommended that all modes below twice the maximum continuous speed are examined during the screening process. As mentioned earlier, this additional requirement arises from AMBs' ability to destabilize any mode within half the controller's sampling rate, not just the first forward rotor mode. The screening criteria differentiate where the modes are relative to maximum speed. Modes below maximum continuous speed are expected to possess a higher stability level than modes above the operating speed range.

The "input sensitivity function" is another performance characteristic that is used to assess the stability margin of AMB-supported compressors. The input sensitivity's magnitude helps to define the feedback control loop's robustness against instability problems due to operating condition variations and other uncertainties in the rotor system [24]. ISO [25] sets a limit for the peak sensitivity of new AMB-supported machines.

A level 2 detailed analysis is intended to incorporate state-of-the-art tools and techniques for each component or mechanism contributing to the compressor's stability characteristics. For example, the details of the internal seals' construction, which are ignored by the empirical aerodynamic cross-coupling simplification in the screening analysis, should be modeled to determine their specific stiffness and

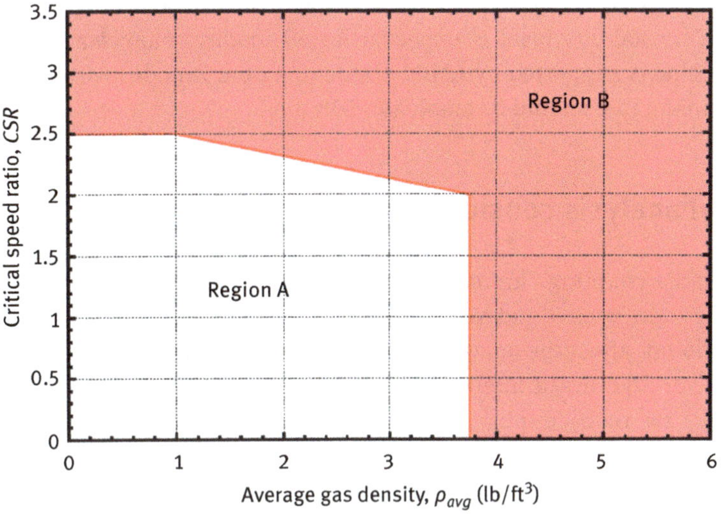

Figure 7.11: Screening regions to distinguish regions of stability risk (source: *BRG Machinery Consulting*).

damping properties. It should be noted, however, that "state of the art" is a relative term. Each compressor manufacturer will have their own tools, methods, and philosophies as to how components and mechanisms should be modeled.

API 617 requires that the detailed analysis include the stability results as components are progressively added to the system model, as shown in Table 7.1. In the example compressor's full model, the eye seals are shown to be destabilizing, reducing the log decrement from the rotor and bearings' "base stability" level. The balance piston, which for this example compressor incorporates a shunt bypass, is shown to add damping to this system and increases the log decrement.

Table 7.1: Example results from a detailed stability analysis.

Model	**Min clearance**		**Max clearance**	
	δ	**ω_d (cpm)**	**δ**	**ω_d (cpm)**
Rotor, bearings	0.123	1,967	0.098	1,892
+ Eye seals	0.076	1,895	0.085	1,826
+ Balance piston (full model)	0.087	1,894	0.105	1,825

With all the components included in the level 2 full model, API's design acceptance criterion is that the minimum predicted log decrement be greater than 0.1. Table 7.1

shows that the example compressor's full model does not achieve the 0.1 log decrement requirement. This stability level is expected for all modes below maximum speed when the machine is supported by AMBs. Higher frequency modes within the controller's half sampling rate should be stable ($\delta > 0$).

7.4 Other lateral analysis considerations

Typically, when flexible couplings are employed, the individual rotor assemblies within a compressor train behave independently, that is, the lateral dynamics of one rotor assembly do not affect the adjacent one. However, in some cases the coupling design may create significant interaction between machine bodies. This is of particular importance for rigid couplings as well as flexible couplings with relatively long spacers.

In these situations, the coupling spacers should be modeled as additional rotor assemblies, coupled together with the other machine bodies, and analyzed simultaneously as a train. Kirk et al. [26] provide guidelines on when such a train lateral analysis should be performed.

While the vibrations associated with oil whirl and shaft whip are a primary concern, the risk of another lateral instability phenomenon, thermal unbalance instability, should also be assessed during the design phase. Unlike oil whirl and shaft whip that cause high nonsynchronous vibrations, this phenomenon results in high synchronous (1×) vibrations that typically grow or oscillate with time.

Commonly known as the "Morton effect," the high vibrations are caused by the shaft differential heating that can occur when a journal synchronously whirls within a bearing's oil film [27]. The resulting thermal bow creates an unbalance feedback loop that can become unstable with time. A similar thermal unbalance instability phenomenon has been observed in AMB machines [28].

The most susceptible machines for thermal unbalance instability are those that have significant overhung mass, such as impellers, couplings, or thrust collars. Thermal unbalance is most commonly seen in overhung compressors; however, beam-style compressors can also experience the phenomenon. Regardless of the machine or bearing type, tools for prediction have been developed since about 2010. These tools should be utilized in the design phase to avoid testing and field problems with turbomachinery.

7.5 Rotordynamics design verification testing

The lateral vibration characteristics of a compressor are typically demonstrated during a variety of shop testing options, such as a four-hour mechanical run test, string test, or full load, full pressure test. However, these "vibration demonstration tests" cannot test every possible operating condition or combination of assembly/machining tolerance. Therefore, for this vast range of untested conditions, the rotordynamics model's accuracy must be relied upon to determine the acceptability of a compressor's design [29]. To help validate the accuracy of a compressor's lateral rotordynamics model, two types of "design verification test" options are available:

- unbalance response verification test (URVT)
- stability verification test (SVT)

The main objectives of the URVT are to assess the accuracy of the predicted critical speed locations and the predicted unbalance sensitivity in the operating speed range.

Since the number and location of known unbalances that can be applied during the URVT are usually very limited, significant excitation of the critical speed(s) of interest may not be possible. Nonetheless, the accuracy of the predicted vibration amplitudes due to the known unbalance can be examined.

SVT assesses the accuracy of the predicted log decrement. Unlike the URVT, an SVT typically requires additional hardware in the form of a temporarily mounted shaker device. This device is used to nonsynchronously excite the mode of interest for measurement of its damping ratio while the machine is operating. Thus, the SVT is a type of operational modal analysis.

API [3] provides a procedure and acceptance criteria for the URVT. Additional details about the testing methodology can be found in [29, 30]. While there is currently no industry standards in place to address SVT, such standards are likely forthcoming in the near future given the growing popularity of this type of testing. Pettinato et al. [31] outlined a methodology and criteria applied for acceptance testing, and additional details on SVT can be found in [29].

Several additional design verification testing options exist for machines supported by AMBs, since the bearings' control system can apply excitations and measure the resulting outputs before and after the controller. API 617 requires closed-loop transfer function measurements with the rotor levitated at standstill (0 rpm), and these measurements should correlate with predictions according to certain criteria. While not required, it is recommended to perform closed-loop measurements at maximum continuous speed as well. If closed-loop measurements correlate poorly with predictions, the bearings' transfer functions and the open-loop transfer functions should be measured to help identify the sources of discrepancies. Details on the physical meaning and measurement of these various transfer functions can be found in [24].

Regardless of the type of design verification test, if the correlation between the measurements and rotordynamics predictions is poor, the component models should be corrected or reconciled until they accurately predict the measurements. With this corrected model, the same rotordynamics analyses performed during the design stage should be repeated to confirm that the design meets the various acceptance criteria.

7.6 Torsional dynamics

7.6.1 Overview of performance requirements

Typically, the entire compressor train's design must meet several torsional vibration performance requirements with respect to

- operating away from natural frequencies that involve significant twisting of a coupling
- stresses when operating between minimum and maximum continuous speed, that is, within the operating speed range
- start-up life when driven by a synchronous motor
- survivability during an electrical fault, such as a short circuit

The first requirement is intended to avoid any resonance situations involving "coupling modes" and torsional excitations at synchronous 1×frequencies, electrical or mechanical, that tend to have relatively high excitation magnitudes. Figure 7.12 shows the coupling mode for a particular integral gear compressor driven by a synchronous motor. These are named so because they are governed by the coupling's stiffness, coupling modes are normally the lowest frequency torsional modes and involve significant twisting within the coupling and adjacent shaft sections. Resonance of such modes can cause high-cycle fatigue of these components, leading to catastrophic failure and ejection of the coupling. To avoid such failures, a 10% separation

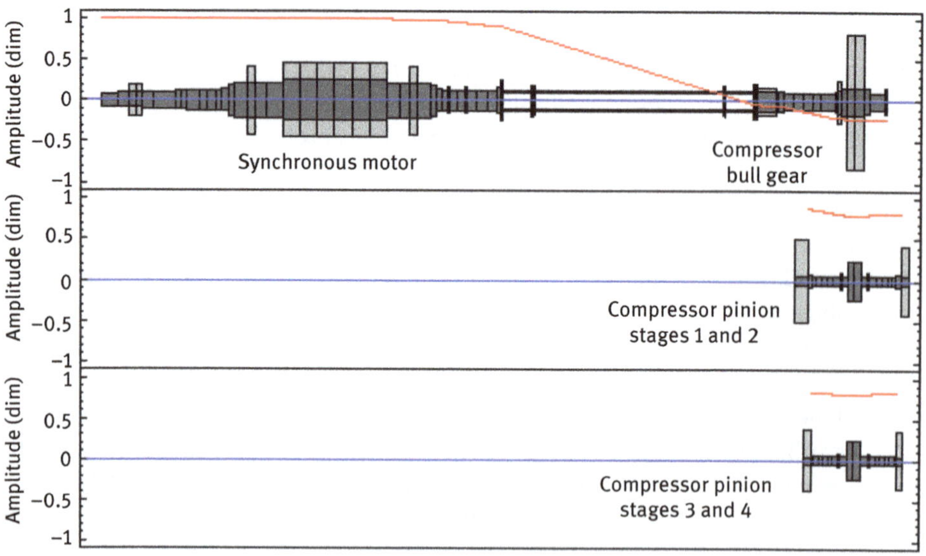

Figure 7.12: Example coupling torsional mode for an integral gear compressor train (source: *BRG Machinery Consulting*).

margin with synchronous frequency is typically required, at a minimum. Interferences without such margin should be resolved, often through a coupling redesign.

In addition to those at synchronous frequencies, there are many other torsional excitation frequencies that can occur during normal operation. These include harmonics of running speed and gear mesh frequency, as well as interharmonics associated with variable speed drives. It is desirable for all torsional modes to have similar separation margin levels with these other excitations. However, achieving these margins is practically impossible given the large number of natural frequencies and excitation mechanisms possible. Some torsional resonances are prone to exist during normal operation (within the operating speed range). Given this reality, the second torsional criterion wants to ensure that the stresses created by these resonances are low enough to avoid any high-cycle fatigue failure of the shafting or components.

Compressor trains driven by synchronous motors are the focus of the third performance criterion because these motors generate high oscillating torques during start-up. Occurring at twice-slip frequency, these large oscillating torques are illustrated in Figure 7.13 for the example integral gear compressor's synchronous motor. High stresses can be experienced during each start-up as this twice-slip frequency coincides with natural frequencies during acceleration up to synchronous speed. Over the course of many start-ups, these stresses can generate enough cumulative fatigue to cause shaft failures [32]. In addition, the high transient torques experienced may exceed the maximum torque capacity of the coupling. Given these

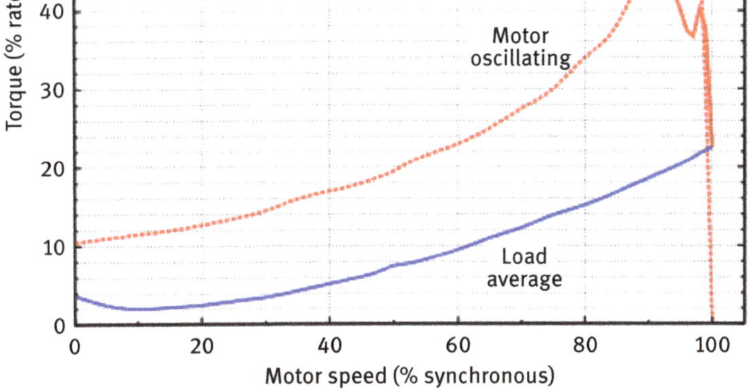

Figure 7.13: Start-up torques for an example synchronous motor (source: *BRG Machinery Consulting*).

threats in synchronous motor-driven trains, it is important to ensure that the coupling's torque capability is sufficient and that any fatigue damage experienced allows for a sufficient number of start-ups.

Electric motors and generators can also experience electrical faults. These faults include various phase–phase or phase–ground short circuits, as well as synchronization errors. Such electrical faults typically generate a transient torque of the following form:

$$T_{\text{fault}}(t) = A_0 e^{-\alpha_0 t} + A_1 e^{-\alpha_1 t} \text{Sin}\left(\omega_{\text{line}} t\right) + A_2 e^{-\alpha_2 t} \text{Sin}(2\omega_{\text{line}} t) \tag{7.7}$$

which consists of DC, line frequency ω_{line}, and twice line frequency components that decay with time. The machine train should be able to survive the stresses created by such faults without catastrophic failure. However, some components may need replacement due to the plastic deformations caused by the prevailing large stresses. Such deformations may not be physically visible.

As discussed earlier, the lateral vibration characteristics of the complete train only need to be analyzed for certain situations related to the specific coupling design. However, when assessing the torsional performance, the entire train must be analyzed as a single system. Various components within the train combine to work as a dynamic system that is assumed to be governed by the following linear equation:

$$\mathbf{Ip}\ddot{\theta}(t) + \mathbf{C}\dot{\theta}(t) + \mathbf{K}(t) = \mathbf{T}(t) \tag{7.8}$$

where $\theta(t)$ represents the angular deflection or twist at various locations along the train. The polar inertia **Ip**, damping **C**, and stiffness **K** matrices contain these properties from the individual components, such as the rotor assemblies, gears, and couplings. To obtain these properties, each component's physical characteristics much be converted into a dynamic model. Such component modeling is an important step to ensure the accuracy of the overall analysis.

7.6.2 Component modeling aspects

Rotors assemblies are modeled using a similar discretization approach as described for lateral modeling. In the past, it was common to develop a greatly simplified model of each rotor assembly consisting of only the largest inertias. However, today's computational power makes it easy to analyze rotor assembly models with the same high fidelity and complexity as lateral ones.

When determining the torsional stiffness of each shaft section, it is important to consider the diameter of adjacent shaft sections, as well as the presence of any keyed or shrunk-on components. Smaller diameter adjacent sections can "penetrate" a larger diameter section, effectively increasing the torsional length of the smaller diameter section. Such penetration factor effects are a strong function of the fillet radius between the two diameters [33]. Mounted components with shrink or keyed fits also create penetration factor effects, effectively reducing the length of the shaft section under the fit that can twist torsionally.

The torsional properties of couplings are provided by the coupling manufacturer. These properties usually consist of the coupling's overall torsional stiffness and polar inertia, which are often calculated according to an industrial standard such as in [34]. Elastomeric and torque-limited couplings require more complex dynamic models because of their nonlinear and torque-dependent characteristics [4].

In the train model, each coupling is often modeled as a single torsional spring that connects adjacent rotor assemblies. Given that the coupling manufacturer's torsional stiffness value usually includes the penetration factor effects of the two hubs, these effects must be accurately accounted for when developing the adjacent rotor assembly models.

Unless an elastomeric coupling or gearbox is present, there tends to be very little torsional damping within a compressor train. The sources of this damping, for example, material hysteresis within the shafting and internal friction, tend to be difficult to predict and have a high degree of uncertainty. Therefore, damping is usually only modeled by assuming a small damping ratio percentage ($\zeta \leq 2\%$) for each mode.

When a gearbox is present, additional torsional damping may be developed from the lateral–torsional interaction with the gearbox's journal bearings. Quantifying the level of torsional damping from the journal bearings can be assessed through a coupled lateral–torsional analysis. However, since this coupled analysis is rarely performed, any damping is also typically modeled using a modal damping approach. For nonspared, critical machinery trains incorporating gearboxes, a coupled lateral–torsional analysis should be strongly considered.

7.6.3 Typical torsional analyses

There are three torsional analyses commonly done during the design phase of centrifugal compressors:
1. undamped critical speed analysis
2. steady-state forced response analysis
3. time transient analysis

Unlike lateral dynamics, undamped critical speed analysis is the primary tool for assessing whether the torsional performance meets the acceptance criteria for separation margins between natural frequencies and excitation frequencies. Undamped analysis determines the natural frequencies (eigenvalues) and their associated mode shapes (eigenvectors) by solving eq. (7.4) without damping and external torques in the system. The main products of these calculations are the Campbell diagram and the undamped mode shapes.

Figure 7.14 presents an example Campbell diagram for an integral gear compressor train driven by a synchronous motor. Since the polar inertias and torsional stiffness characteristics are typically not speed dependent, each torsional natural frequency ω_n remains constant with speed. There is one natural frequency that is not shown on the Campbell diagram, the one associated with rigid body rotation of the train that occurs at 0 Hz. All the other natural frequencies involve twisting within the train. For example, the mode shape in Figure 7.12 illustrates the high degree of twisting that occurs within the coupling for the first natural frequency ($\omega_{n1} = 10$ Hz).

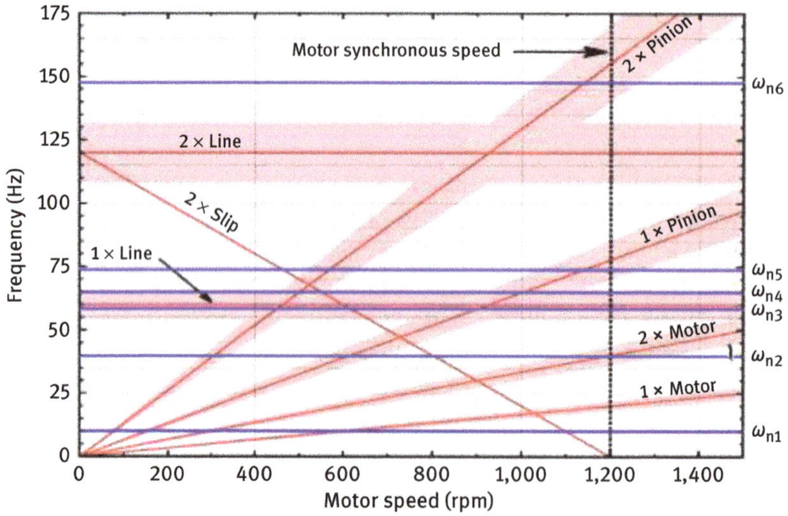

Figure 7.14: Campbell diagram showing torsional characteristics of synchronous motor-driven integral gear compressor (source: *BRG Machinery Consulting*).

Many intersections of natural frequency and excitation lines occur in Campbell diagrams. However, from the standpoint of the first two torsional vibration performance requirements, the intersections of concern or "interferences" are those that occur when the machine is within its operating speed range. When the example compressor's motor is at synchronous speed, Figure 7.14 indicates that the following natural frequencies violate the ±10% separation margin band with an excitation:

- ω_{n2} with 2 *x* motor speed
- ω_{n3} with 1 *x* electrical line frequency
- ω_{n4} with 1 *x* electrical line frequency
- ω_{n5} with 1 *x* pinion speed
- ω_{n6} with 2 *x* pinion speed

Once the interferences are identified, each mode shape is examined to determine the likelihood of the specific excitation that is manifested in the mode. If the excitation source occurs at a nodal location, the interference is not a concern. Otherwise, the acceptability of the resonance condition can only be determined through stress calculations.

Steady-state forced response analyses are performed to determine whether a resonance condition can lead to high-cycle fatigue or exceed the dynamic torque capabilities of components. In particular, this test is of interest to couplings. Industrial standards require that the stresses be evaluated by applying oscillating torque and calculating the resulting vibration response. The analysis is done assuming steady-state conditions, wherein time transient and acceleration effects are not included. Instead, eq. (7.8) is transformed into the following steady-state equation:

$$\vec{\theta}(\omega) = \left[-\omega^2 I_p + j\omega C + K\right]^{-1} \vec{T}\text{exc} \tag{7.9}$$

where $\vec{T}$exc is the excitation torque distribution applied along the train, and $\vec{\theta}$ contains the resulting angular displacements along the train. Unless a component such as an elastomeric coupling is present, the dynamic properties (inertia, damping, and stiffness) and excitation torque magnitude are constant; this means that only the angular displacements are a function of excitation frequency ω.

For each resonance or interference of concern, a steady-state forced response analysis is performed using the specific frequency range and torque distribution of interest. In each case, the resulting displacements, stresses, and dynamic torques are examined to determine their acceptability. Several engineering choices and parameters will vitally affect the accuracy of any analysis:
- magnitude of the excitation torque
- locations to examine stresses and dynamic torques
- stress concentration factors
- damping
- material properties

The excitation's torque magnitude as applied in the steady-state calculations is often based on industry experience, generally being a small percentage of the rated torque. Guidelines for magnitudes of various excitations can be found in [4]. In the case of a motor using a variable speed drive, harmonic and interharmonic torque magnitudes should be supplied by the manufacturer.

Typically, only a few of the elements within a train are examined in terms of their stresses and dynamic torques. At a minimum, these elements should include the shaft ends transmitting torque between bodies and the elements making up a coupling. However, additional locations, particularly those with high-stress concentrations (keyways, grooves, shoulders, etc.) or a high degree of twisting, may also need scrutiny.

Shear stress τ within a particular shaft element is directly related to the difference in angular displacement $\Delta\theta$ at its two ends. For a cylindrical shaft element, the maximum shear stress is calculated according to

$$\tau_{\max} = \text{SCF}\left(\frac{D_o G}{2L_e}\right)\Delta\theta \tag{7.10}$$

where D_o is the outer diameter, G is the material's shear modulus of elasticity, and L_e is the effective torsional length of the element that includes any penetration factor effects. SCF is the particular location's stress concentration factor, an important parameter that should be carefully determined using either well-accepted sources [35] or through performing finite element analysis.

To assess the risk of high-cycle fatigue, the calculated stresses are analyzed relative to the strength properties of the material. Ultimate and endurance strength values in the shear direction are often derived from those given for the tensile direction. Appropriate strength derating factors, such as those for surface finish and size, should be applied.

The final step of the steady-state forced response analysis is to assess whether all of the components possess infinite life for the particular resonance situation. API 617 specifies neither a particular high-cycle fatigue failure criterion nor safety factor that should be used in this assessment. One common approach is the use of Goodman diagram when considering the potential mean stress effects. Another typically applied criterion is that given in MIL-STD-167-2A [36], which focuses solely on the dynamic stresses.

When a compressor train includes an electric motor or generator, time transient analysis of the torsional dynamics is necessary to evaluate the train's survivability to various electrical faults. Furthermore, if the electric motor is a synchronous type, time transient analysis is required to assess the train's start-up life (number of safe starts) when subjected to the motor's large twice-slip frequency torques. In both situations, eq. (7.4) is solved to obtain the angular displacements and stresses as a function of time. The initial condition for an analysis of an electrical fault is typically at rated operating conditions as they relate to speed and torque, while the initial condition for a synchronous motor start-up is from standstill.

Figure 7.15 illustrates some results from a representative time transient analysis of the example compressor's start-up. For the expected torque versus speed characteristics given in Figure 7.13, the motor is predicted to reach synchronous speed at approximately 21 s. Just before reaching synchronous speed, speed oscillations occur as the train accelerates through 1,100 rpm. These speed oscillations are the result of the motor's twice-slip frequency torque exciting the train's first torsional natural frequency, as illustrated in Campbell diagram shown in Figure 7.14.

Dynamic stresses and torques are also the result of the time transient analysis. As in steady-state forced response analysis, it is important that these stresses and torques be examined for the critical locations within the train. Figure 7.15 shows the

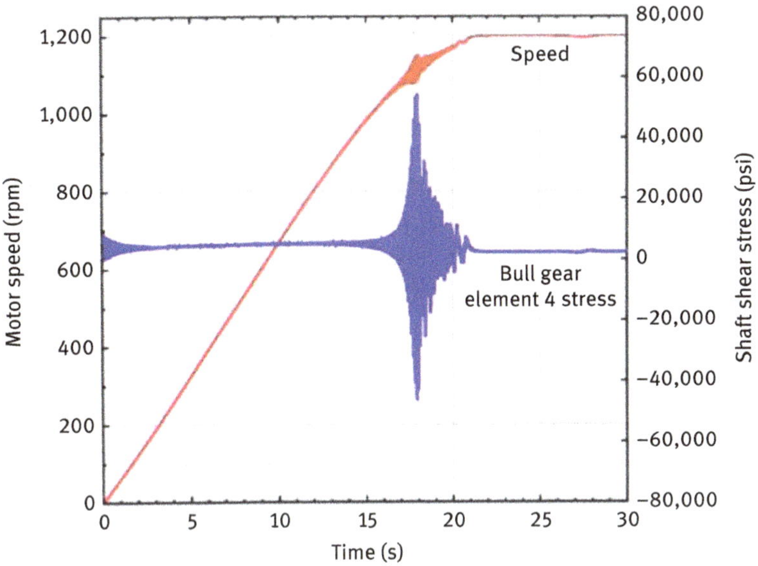

Figure 7.15: Transient start-up of synchronous motor driven integral gear compressor (source: *BRG Machinery Consulting*).

large stresses that occur within the compressor's bull gear shaft as the motor's oscillating torque component excites the train's first natural frequency.

Using the stress versus time history from the transient analysis, a cumulative fatigue damage analysis is performed as a post-processing calculation to determine how many starts a synchronous motor-driven train can survive, that is, the train's start-up life. Two competing theories are available for this fatigue analysis, stress-life theory or strain-life theory, with stress-life being the most commonly used [37]. Either theory can be used to establish a fatigue life curve (stress vs. number of cycles) for a particular shaft location. The fatigue life curve is then used in conjunction with the predicted stress history to determine the amount of damage during a single start-up. Through Miner's damage rule [38], the start-up life of that particular shaft location is subsequently predicted. Start-up life of the entire machine train is established by examining various critical locations and finding the minimum predicted.

API 617 has no specific requirement for the start-up life of a compressor train. To be agreed upon by the manufacturer and purchaser, the safe number of starts should be determined based on the desired life of the machine and the expected starts per year. A typical requirement is 1,000 safe starts. However, given the analytical uncertainties in the torsional damping, material properties, SCF, and others, some safety margin above the agreed-upon start-up life is recommended.

Survivability of electrical faults is determined through a much simpler assessment. Predicted dynamic stresses at critical locations must remain below the low-cycle fatigue limit of the material.

7.6.4 Other torsional analysis considerations

Torsional vibration problems can also be created in machinery trains that incorporate automatic speed control. Feedback of the speed signal to the controller creates a closed-loop system. If the speed controller's dynamics are not properly designed, a torsional mode can become unstable. Such torsional instabilities have been observed in motor-driven trains with variable frequency drives [39]. While there are currently no industrial standards to help mitigate this threat, combined electrical–mechanical dynamic models have proven effective in diagnosing and resolving such problems.

References

[1] Smith, K.J. An Operational History Of Fractional Frequency Whirl, in Proceedings of the Fourth Turbomachinery Symposium. Texas A&M University, 1975, 115–125.
[2] Geary, C.H., Damratowski, L.P., Seyer, C. Design and Operation of the World's Highest Pressure Gas Injection Centrifugal Compressors, in Eighth Annual Offshore Technology Conference, no. OTC-2485, 1976, 651–659.
[3] API. Axial and Centrifugal Compressors and Expander-compressors, 8th ed., American Petroleum Institute, 2014, no. STD 617.
[4] API. API Standard Paragraphs Rotordynamics Tutorial: Lateral Critical Speeds, Unbalance Response, Stability, Train Torsional, and Rotor Balancing, 2nd ed., American Petroleum Institute, August 2005, no. RP 684.
[5] ISO. Mechanical Vibration of Non-reciprocating Machines – Measurements on Rotating Shafts and Evaluation Criteria – Part 3: Coupled Industrial Machines, 1st ed., International Organization for Standardization, Geneva, Switzerland, 1996, no. 7919-3.
[6] Maslen, E.H., Cloud, C.H., Hauge, T., Kibsgaard, S.T., Lie, J.A. Unbalance Response Assessment of a Subsea Compressor, in Proceedings of the Thirteenth International Symposium on Magnetic Bearings. Arlington, Virginia: University of Virginia, August 2012.
[7] Timoshenko, S.P. On the Correction for Shear of the Differential Equation for Transverse Vibrations of Prismatic Bars, Philosophical Magazine, 1921, 41, 744–746.
[8] Timoshenko, S.P. On the Transverse Vibrations of Bars of Uniform Cross-Section, Philosophical Magazine, 1922, 43, 125–131.
[9] Smalley, A.J., Pantermuehl, P.J., Hollingsworth, J.R., Camatti, M. How interference fits stiffen the flexible rotors of centrifugal compressors, in Proceedings of the Sixth International Conference on Rotor Dynamics, Hahn, E.J., Randall, R.B., Eds., 2. Sydney, Australia: IFToMM, 2002, 928–935.
[10] Kocur, J.A. Jr., Nicholas, J.C., Lee, C.C., Surveying Tilting Pad Journal Bearing and Gas Labyrinth Seal Coefficients and Their Effect on Rotor Stability, in Proceedings of the Thirty-Sixth Turbomachinery Symposium. Texas A&M University, 2007, 1–10.
[11] He, M., Byrne, J.M., Cloud, C.H., V´azquez, J.A. Steady State Performance Prediction of Directly Lubricated Fluid Film Journal Bearings, in Proceedings of the Forty-First Turbomachinery Symposium. Houston, Texas: Texas A&M University, September 2012.
[12] He, M., Cloud, C.H., Byrne, J.M. Fundamentals of Fluid Film Journal Bearing Operation and Modeling, in Proceedings of the Thirty-Fourth Turbomachinery Symposium. Texas A&M University, 2005, 155–175.

[13] Kirk, R.G., Reedy, S.W. Evaluation of Pivot Stiffness for Typical Tilting-Pad Journal Bearing Design, ASME Journal of Vibration, Acoustics, Stress, and Reliability in Design, 1988, 110, 165–171.
[14] Gunter, E.J., Barrett, L.E., Allaire, P.E. Design and Application of Squeeze Film Dampers for Turbomachinery Stabilization, in Proceedings of the Fourth Turbomachinery Symposium. Texas A&M University, 1975, 127–141.
[15] Kirk, R.G. Oil Seal Dynamics: Considerations for Analysis of Centrifugal Compressors, in Proceedings of the Fifteenth Turbomachinery Symposium. Texas A&M University, 1986, 25–34
[16] Childs, D.W. Turbomachinery Rotordynamics: Phenomena, Modeling, and Analysis, John Wiley & Sons, Inc., 1993.
[17] Kleynhans, G.F., Childs, D.W. The acoustic influence of cell depth on the rotordynamic characteristics of smooth-rotor/honeycomb-stator annular gas seals, ASME Journal of Engineering for Gas Turbines and Power, 1996, 119, 949–956.
[18] V´azquez, J.A. Using Transfer Functions to Model Flexible Supports and Casings of Rotating Machinery, Ph.D. dissertation, University of Virginia, 1999.
[19] V´azquez, J.A., Barrett, L.E., Flack, R.D. Flexible Bearing Supports, Using Experimental Data, ASME Journal of Engineering for Gas Turbines and Power, April 2002, 124, 369–374.
[20] Moore, J.J., Vannini, G., Camatti, M., Bianchi, P. Rotordynamic Analysis of a Large Industrial Turbocompressor Including Finite Element Substructure Modeling, ASME Journal of Engineering for Gas Turbines and Power, 2010, 132(8), 082401.
[21] Moore, J.J., Ransom, D.L., Viana, F. Rotordynamic Force Prediction of Centrifugal Compressor Impellers Using Computational Fluid Dynamics, ASME Journal of Engineering for Gas Turbines and Power, 2010, 133(4), 042504.
[22] Ehrich, F.F. Rotor Whirl Forces Induced by the Tip Clearance Effect in Axial Flow Compressors, ASME Journal of Vibration and Acoustics, 1993, 115, 509–515.
[23] Lund, J.W. Stability and Damped Critical Speeds of a Flexible Rotor in Fluid-Film Bearings, ASME Journal of Engineering for Industry, 1974, 96(2), 509–517.
[24] Schweitzer, G., Maslen, E.H. Eds. Magnetic Bearings: Theory, Design, and Application to Rotating Machinery, Springer-Verlag, 2009.
[25] ISO, Mechanical Vibration – Vibration of Rotating Machinery Equipped with Active Magnetic Bearings – Part 3: Evaluation of Stability Margin. Geneva, Switzerland: International Organization for Standardization, 2006, no. 14839-3:2006.
[26] Kirk, R.G., Mondy, R.E., Murphy, R.C. Theory and Guidelines to Proper Coupling Design for Rotor Dynamics Considerations, ASME Journal of Vibration, Acoustics, Stress, and Reliability in Design, 1984, 106, 129–138.
[27] de Jongh, F.M., The Synchronous Rotor Instability Phenomenon – Morton Effect, in Proceedings of the Thirty-Seventh Turbomachinery Symposium. Texas A&M University, 2008, 159–167.
[28] Takahashi, N., Kaneko, S. Thermal Instability in a Magnetically Levitated Doubly Overhung Rotor, Journal of Sound and Vibration, 2013, 332, 1188–1203.
[29] Kocur, J.A., Cloud, C.H. Shop Rotordynamic Testing – Options, Objectives, Benefits and Practices, in Proceedings of the Forty-Second Turbomachinery Symposium. Houston, Texas: Texas A&M University, September 2013.
[30] Nicholas, J.C., Edney, S.L., Kocur, J.A., Hustak, J.F. Subtracting Residual Unbalance for Improved Test Stand Vibration Correlation, in Proceedings of the Twenty-Sixth Turbomachinery Symposium. Texas A&M University, 1997, 7–18.
[31] Pettinato, B.C., Cloud, C.H., Campos, R.S., Shop Acceptance Testing of Compressor Rotordynamic Stability and Theoretical Correlation, in Proceedings of the Thirty-Ninth Turbomachinery Symposium. Houston, Texas: Texas A&M University, October 2010, 31–42.

[32] Brown, R.N. A Torsional Vibration Problem as Associated with Synchronous Motor Driven Machines, ASME Journal of Engineering for Power, July 1960, 82, 215–220.
[33] Nestorides, E.J. A Handbook on Torsional Vibration, Cambridge University Press, 1958.
[34] ANSI/AGMA, Flexible Couplings – Mass Elastic Properties and Other Characteristics. American Gear Manufacturers Association, 2008, no. 9004–B08.
[35] Pilkey, W.D. Peterson's Stress Concentration Factors, 2nd ed., John Wiley & Sons, Inc, 1997.
[36] D.O.D., Mechanical Vibrations of Shipboard Equipment (Reciprocating Machinery and Propulsion System and Shafting) Types III, IV and V. US Department of Defense, May 1992, no. MIL- STD–167–2A (SH).
[37] Shigley, J.E., Mischke, C.R. Eds. Mechanical Engineering Design, 5th ed., McGraw-Hill, Inc, 1989.
[38] Miner, M.A. Cumulative Damage in Fatigue, Transactions of the ASME, Journal of Applied Mechanics, September 1945, 67, A159–A164.
[39] Kocur, J.A. Jr., Corcoran, J.P., Case study: VFD Induced Coupling Failure, in Proceedings of the Thirty-Seventh Turbomachinery Symposium. Texas A&M University, 2008.

Chapter 8
Approaches for sizing lube oil systems

8.1 General approach to equipment sizing

Lubrication systems are undoubtedly the lifeblood for all types of turbomachinery. Depending on the size and type of turbomachinery, the lubrication arrangement can come in many different forms and types as shown in Figure 8.1. However, in this section, we concentrate only on the typical circulating force-fed lubrication identified in the upper right-hand corner of Figure 8.1, and specifically system configurations that are defined as per API 614 for "special-purpose" equipment.

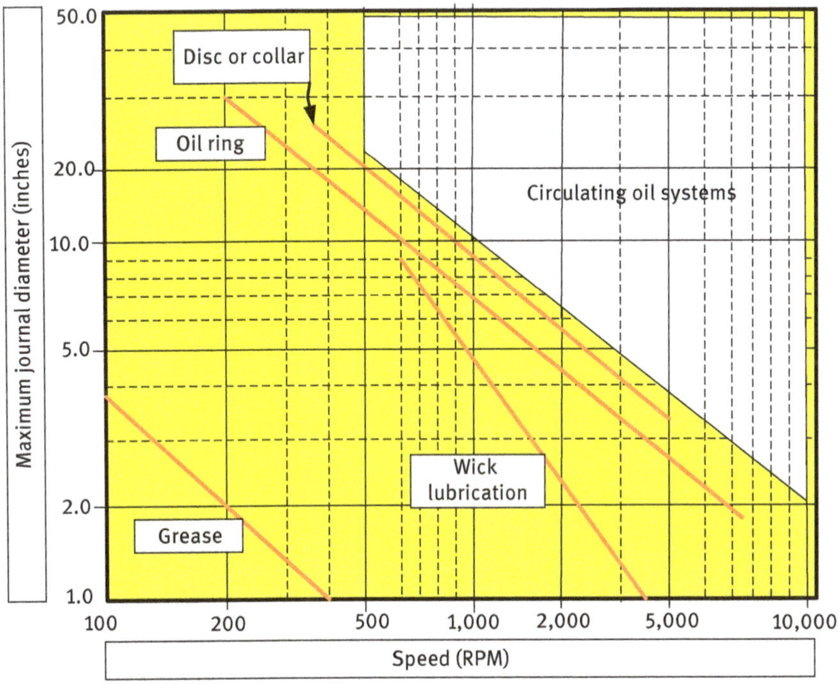

Figure 8.1: Upper limits of journal bearing lubrication methods [1].

API 614 standard sets guidelines for the designers of lubrication systems to follow; however, as these specifications are not binding, most vendors do not strictly observe them. These vendors do have in-house practices that sometimes deviate from the guidelines for various reasons, including cost constraints. Note that no matter how good the vendors' intentions might be, a faulty design lubrication system can have an impact on the scheduled startup of a grassroots plant. Moreover, and in

https://doi.org/10.1515/9783110678765-008

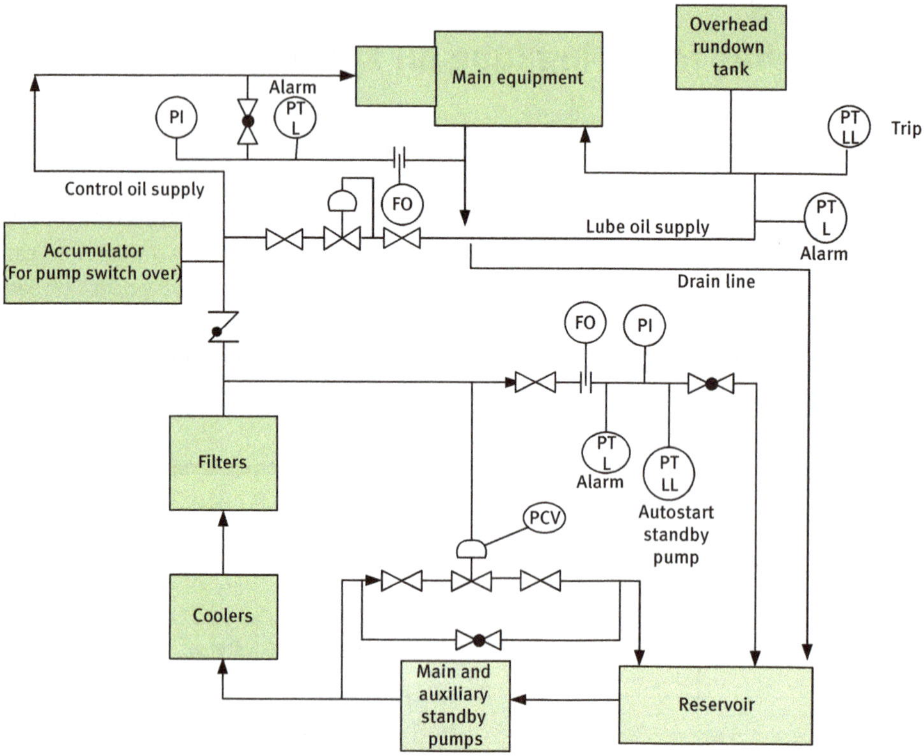

Figure 8.2: Block diagram of a typical API 614 system designed for special-purpose equipment.

some cases, for running plants, it could adversely affect its reliability. This possibility becomes more concerning after rerates are implemented.

The first intent of this section is to give the purchasers and end users a basic knowledge of a methodology on how some of the major components of these large "special-purpose" lubrication systems are sized. The second intent is to assist purchasers and end user in comparing the vendor's proposal with experience-based industry practices. These are spelled out in API 614 guidelines; a conscientious comparison allows buyers to ask proper questions whenever they are not comfortable with what is being proposed.

In order to understand the basic concepts outlined in API 614, our approach is to start with an example of how a particular vendor went about sizing some of the major components for a typical API 614 lubrication system after completing the rerate work. For this example, we will call the vendor doing the rerate Vendor-A. Our focus is on lubrication systems that are designed for "special-purpose machinery," as shown in Figure 8.2. We will review the sizing of some of the major components of these systems, namely:

- Main and auxiliary standby pump
- Oil coolers

- Oil reservoir
- Oil filters
- accumulator
- Overhead rundown tank (RDT)

8.2 Main and auxiliary standby pump

The first step is to calculate the lubrication requirements and heat load demanded by the equipment train, and then use the results to determine both the size of the pump and heat exchanger (cooler). The lubrication requirements and heat loads are obtained from bearing simulation software; the necessary oil throughputs are validated later during a mechanical run of the turbotrain. Each radial bearing is modeled at maximum continuous speed, and with nominal clearances and preloads. The thrust bearings are also modeled using different cases for compressor operation; oil flow rate and heat load are determined for the worst-case situation. We should state that the new or recent approach by most vendors is to standardize on bearing sizes. Standardization of bearing sizes allows the vendors to have a fair idea of the oil requirements and heat load demands. They can then validate these parameters during mechanical runs rather than modeling the bearings for every job. A general rule for oil distribution between journal and thrust bearings is to size the thrust bearing orifice based on a 2:1 ratio of oil going to the thrust bearing compared to that of the journal bearing.

8.2.1 Determining pump capacity

Knowing the oil flow and heat load required by each bearing, the oil flows are then added up to determine the minimum lubrication requirements. The total oil flow is then multiplied by 1.15 to determine the pump's capacity. Multiplying by 1.15 normally takes care of the maximum usage and includes system transients. However, maximum capacity should not exceed 125% of the total calculated flow due to driver sizing and peak efficiency considerations.

8.2.2 Pump's maximum pressure

Pressure for the pump is calculated by adding the following:

1. Highest oil pressure required for the driver and the driven equipment at the compressor deck (normal operation and shutdown conditions).
2. Static head – difference between the lubrication skid at grade level and the equipment centerline. Convert the linear measurement to pressure units.

3. Piping pressure loss from lubrication skid at grade to the (usually elevated) equipment location. This generally is in the range of 10–15 psi (0.7–1.0 bar).
4. Pressure drop across the reducing valve (backpressure controller). For estimation purposes, use approximately 20 psi (1.4 bar).
5. Piping loss for all lubrication skid piping at grade level. Generally approximated to be around 5 psi (0.35 bar).
6. Transfer valve pressure loss. Generally estimated to be around 5 psi (0.35 bar).
7. Pressure loss across the filter estimated at approximately 20 psi (1.4 bar).
8. Loss of pressure across the cooler can be approximated to around 10 psi (0.7 bar).

After the *capacity* and *discharge pressure* of the pump are calculated, the next parameter to be considered is the oil viscosity range from, say, 50 °F (10 °C) for a cold start, to approximately 150 °F (66 °C) for normal reservoir operating temperature.

Figure 8.3 shows some of the common grades of lubricating oils in use for turbomachinery and the equivalent viscosity units, including the viscosity change at different temperatures.

Knowing the *capacity*, *discharge pressure*, and *viscosity* range, the final step is to select the type of pumps for the system. Both centrifugal and positive displacement pumps are used in API 614 systems for "special-purpose machinery"; however, rotary positive displacement-type pumps are generally used in these large lubricating systems. The reason for using positive displacement pumps is explained by the performance characteristics of these pumps; they provide fairly constant flow rate against varying system pressures. For example, screw-type pumps can deliver flows as oil viscosity varies from cold start at 50 °F (10 °C), which is equivalent to a viscosity of 650 SUS, to 150 °F (66 °C), which is equivalent to a viscosity of 65 SUS. On the other hand, centrifugal pumps are generally used in low-pressure, high-flow applications, where only lubricating oil is required. There are many applications where centrifugal pumps are used in combined lube and seal oil systems. Here, the centrifugal pump supplies low-pressure lubricating oil to the system, while a portion of the discharge flow from the centrifugal pump is fed to a screw pump. The screw pump boosts the system pressure to that needed for control oil and/or seal oil.

Drivers for these lube oil pumps can be motors or steam turbines. In most cases, when there is a combination of steam turbine and motor drivers, the steam turbine driver is selected as the main and the motor as the standby. The reason is generally based on steam balance of the plant, and also on the fact that most motors will tend to auto start within 3–4 s. This is compared with steam turbines requiring continuous slow-rolling at about 200 rpm to maintain preheat and an adequate lubricant film in sleeve bearings. Turbines require in excess of 20 s to reach full speed from approximately 200 rpm at startup.

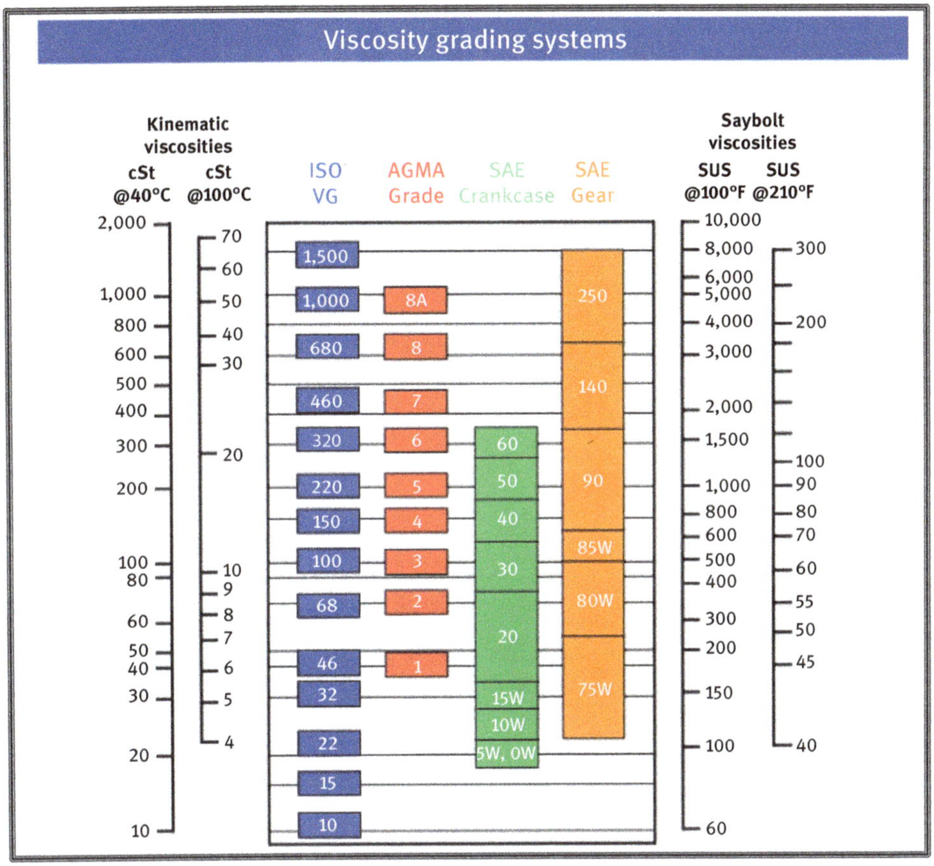

Figure 8.3: Common grades of lubrication oils showing equivalent viscosity units at different temperatures.

For calculation purposes, we will use an actual example of the evaluation exercise carried out by Vendor-A during the upgrade of an ethylene refrigeration compressor train driven by a steam turbine, as shown in Figure 8.4.

Table 8.1 represents the calculated values for the rerate of the ethylene refrigeration compressor train shown in Figure 8.4; it will be used to show how this experienced turbomachinery manufacturer went about sizing both the pump and the overall system:

Total normal oil flow = 31.50 m^3/h

Total flow including transient flow = 38.30 m^3/h

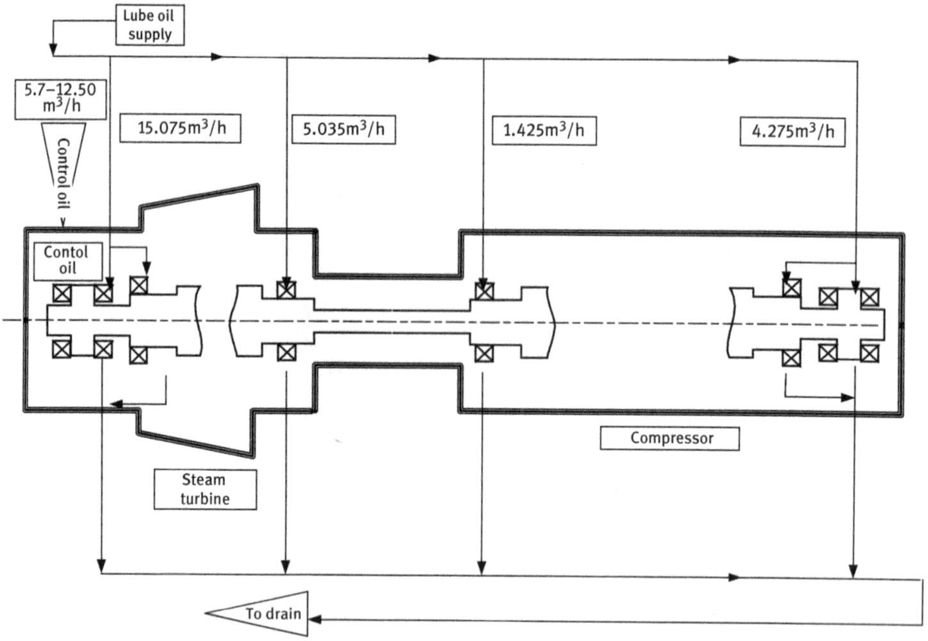

Figure 8.4: Lube oil distribution of an ethylene refrigeration compressor train driven by a steam turbine.

Table 8.1: Oil system requirements and calculated heat loads.

Ethylene refrigeration compressor train oil requirements					
Equipment	**Lubrication required (m^3/h)**	**Control oil (m^3/h)**		**Heat load (kJ/h)**	**Pressure at equipment (kPaG)**
		Normal	**Transient**		
Steam turbine					
– Lube oil	20.10				124.00
– Control oil		5.70	12.50		690.00
– Heat input				644,760	
Compressor					
– Lube oil	5.70				124.00
– Heat input				115,000	
Pump driver					
– Lube oil	0.00				
– Heat input					
Total	25.80	5.70	12.50	**759,760**	
		Normal	**Transient**		
Total oil flow required		**31.50**	**38.30**		

Guidelines given in API 614 for sizing positive displacement pumps recommend that the pump's capacity shall be at least 120% of the normal flow rate plus any transient flow. Therefore, the minimum pump capacity is calculated as follows:

$$\begin{aligned} Q \text{ minimum} &= (1.2 \times 31.5) + (38.3 - 31.5) \text{ m}^3/\text{h} \\ &= 44.6 \text{ m}^3/\text{h (196.4 gpm)} \end{aligned}$$

Using the eight steps outlined previously to calculate the pump's discharge pressure, the discharge pressure was calculated to be 2,035 kPa (20.35 bar). Therefore, the selected pump must be able to deliver 44.6 m^3/h of 60 SUS oil (cold start) at a discharge pressure of 2,035 kPa = 20.35 bar.

Vendor-A then proceeded to select a positive displacement lube oil pump from the catalog of a well-known pump supplier. Table 8.2 represents a section of the catalog showing the anticipated performance of a specific rotary positive displacement pump running at 1,800 rpm.

Table 8.2: Pump performance of rotary positive displacement pump from vendor's catalog.

Operating condition	Inlet temperature	Discharge pressure (kPa/bar)	Inlet pressure required (kPaa/bara)	Flow rate	Input power
Normal operation	65 °C	2,045/20.45	50/0.50	50.2 m^3/h	39.7 kW
Cold startup	10 °C	1,860 kPa	57/0.57	56.3 m^3/h	39.1 kW
Cold oil at relief valve	10 °C	3,014 kPa	57/0.57	55.4 m^3/h	58.1 kW

Based on the expected performance shown in Table 8.2 with the capacity range of 50–57 m^3/h, the vendor selected this particular pump, knowing that its actual capacity was more than adequate for the calculated capacity of 44.6 m^3/h.

8.3 Oil coolers

Most end users in the petrochemical industry require heat exchangers to be specified and built to ASME Code, Section VIII, and TEMA "C" (Tubular Exchanger Manufacturers Association) as was the case here for the example being worked out.

As per API 614 guidelines, an oil cooler must be sized to remove the total heat added to the oil from all sources, including bearings, seals, gears, and the heat contributed by the running pump during normal operation. For this example, the compressor had dry gas seals, also there was no gearbox; therefore, the heat load requirements are as shown in Table 8.1.

The heat load of the running pump is determined from the pump's shaft power converted to heat units and then added to the system as heat (Table 8.3).

Table 8.3: Comparison of vendor's design/sizing criteria versus API 614 for the rerate example.

Operating parameter	Vendor-A design criteria		API 614 design criteria	
	US customary units	**SI units**	**US customary units**	**SI units**
Maximum allowable working pressure (MAWP)	**90 psig**	**6 bars**	**≥100 psig**	**≥6.9 bar**
Oil				
Maximum oil outlet temperature	120 °F	50 °C	120°F	50 °C
Maximum oil pressure drop	10 psi	69 kPa		
Cooling water system				
Velocity in exchanger tubes	5–8 ft/s	1.5–2.5 m/s	5–8 ft/s	1.5–2.5 m/s
Maximum inlet water temperature	90 °F	32 °C	90 °F	30 °F
Maximum water outlet temperature	120 °F	49 °C	120 °F	50 °F
Water pressure drop	15 psi	103.42 kPa	15 psi	103.42 kPa
Maximum temperature rise	30 °F	17 °C	30 °F	20 °C
Minimum temperature rise	20 °F	11 °C	20 °F	10 °C
Fouling factor on water side				

Notes:
1. Ratio of maximum water to oil is 2:1.
2. Specified water coolers shall be suitable for 300 °F (149 °C) heating medium such as steam or a mixture of steam and water.

As part of the rerate, a new replacement bundle (heat exchanger) had to be supplied, since the existing cooler bundle was inadequate; it had been sized for a heat load of 527,900 kJ/h and oil flow of 22.34 m^3/h. Based on our previous calculations and the vendor guidelines given earlier, the new bundle had to be able to cool 31.5 m^3/h oil flow at 49 °C with a water inlet temperature of approximately 37 °C. Note that this inlet water temperature did exceed the vendor's criteria as well as API 614. Accordingly, a slightly larger cooler size had to be selected as follows:

Heat input from compressor train	759,800 kJ/h
Heat input from oil pump	142,900 kJ/h (39.7 kW)
Total heat duty	902,700 kJ/h

The size selection for the new replacement bundle was based on a rated heat load of 969,200 kJ/h, 31.5 m^3/h of oil flow, and 40.9 m^3/h of water flow at an inlet temperature of 36.7 °C. We also had to ascertain that the exchanger tubes were sized so as not to exceed the vendor or API 614 design velocity range.

8.4 Reservoir

Sizing the oil reservoir's capacity is based on the addition of the volumes between the system levels shown in Figure 8.5. These capacities are defined as per API 614 and are all referenced to the normal operating flow, which is defined as per API 614 as the total amount of oil flowing from the bearings, seals, gearbox, couplings, and steady-state controls. This does not include oil bypassed directly to the reservoir.

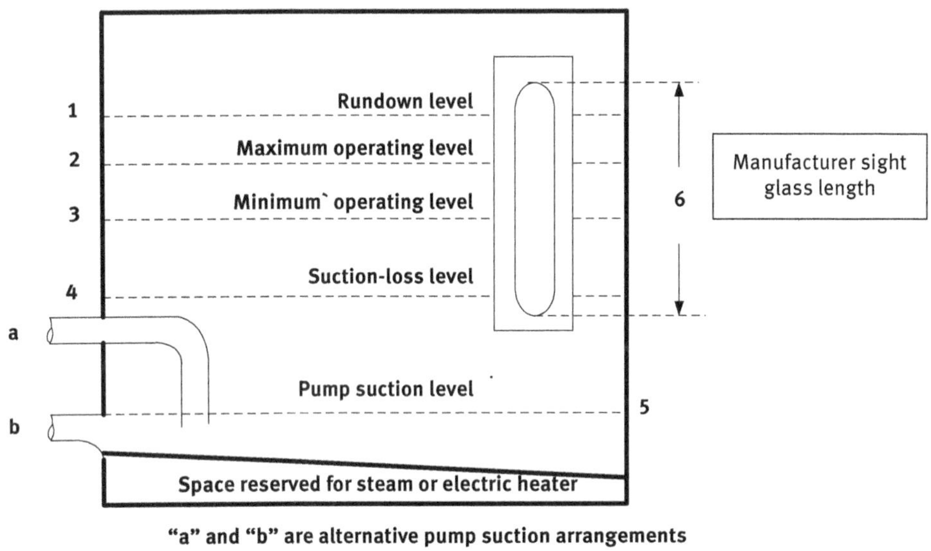

Figure 8.5: Reservoir levels defined as per API 614.

Figure 8.5 shows the reservoir levels as defined by API 614. The capacities/volumes between these levels are described as follows:

- The *working capacity* is defined as the capacity between levels 3 and 4 as shown in Figure 8.5, and shall be sufficient for 5 min of normal flow.
- The *retention capacity* is defined as the total volume below the minimum operation level or level 3 as shown in Figure 8.5, and shall be calculated to retain 8 min of normal flow.
- The level change (indicating volume/capacity change) between the *maximum* and the *minimum* operating levels shall be 50 mm (2 in) of the reservoir height.

- The *rundown capacity* is defined as all the oil drained from all of the components, such as bearings, gearbox, seals, accumulators, control elements, and the volume, of the RDT which will be addressed later. API 614 recommends adding 10% to the rundown capacity as an allowance for the volumes coming from the interconnecting piping.
- The free surface of the reservoir shall be sized based on a suggested minimum area of 60 mm^2 for each liter/min (0.25 ft^2 for each gal/min) of normal oil flow.

Having outlined the methodology used by API 614 to size reservoirs for "special-purpose" lubricating systems, we will continue to review how the rerate example fitted into the various sizing criteria outlined earlier.

The oil reservoir was a rectangular design with 2.47 m^3/h *working capacity* and a 3.53 m^3 *retention capacity*. Therefore, at a normal flow rate of 31.5 m^3/h to the equipment, the *working capacity* of the existing reservoir was calculated as 4.7 min and the *retention capacity* 6.7 min. Although these values were less than the guidelines outlined in API 614 of 5 min *working capacity* and 8 min *retention capacity* instead of 5 min, the equipment vendor was confident that the existing reservoir would perform adequately for the new operating conditions, which it did.

8.5 Oil filters

API 614 has an entire section on requirements for oil filters operating in "special-purpose" lubrication systems. However, for brevity and relevance to the rerate example we are dealing with, we will only examine the requirements outlined in Table 8.4.

A duplex set of filters was originally installed in the system being rerated with pleated paper cartridges. Vendor-A does not normally install pleated paper cartridges in their lubrication systems unless the cartridges are resin treated, which they were in this particular instance. Impregnated resin-pleated paper cartridges have a collapse pressure of approximately 100 psig (690 kPa, 6.90 bar). However, untreated pleated paper cartridges without resin treatment permit a lower collapse pressure only. Additionally, these untreated pleated paper cartridges cause the paper to swell if the lube oil contains small amounts of water, which effectively results in higher than the normal pressure drop.

The results of the rerate showed that the normal flow increased to 31.5 m^3/h (138.7 gpm). This new flow rate also showed that the reduction in pressure expected across the filter assembly with clean filter cartridges including the transfer valve would be approximately 55 kPa (8 psi). The vendor arrived at the conclusion based on the original design considerations highlighted later; also, they relied

Table 8.4: Allowable pressure drop and collapse pressure of filters.

Requirements	Pressure	Oil flow	Oil temperature	Oil viscosity
API 614 requirements				
Pressure drop for clean filters	30 kPa/5 psig	Normal	40 °C/100 °F	150 SUS
Cartridge/element collapsing differential pressure	500 kPa/70 psig			
Vendor-A requirements				
Pressure drop for clean filters	21 kPa/3 psig	Normal	49 °C/120 °F	100 SUS
Cartridge/element collapsing differential pressure	500 kPa/70 psig			

Notes:
1. Per **API 614** guidelines, whenever a specific cartridge element is required, the purchaser shall specify the type, maker, model, and construction.
2. **Vendor-A** requires that cartridge elements must be capable of passing normal oil flow at 10 °C/50 °F, which is equivalent to 650 SUS, without reaching collapse pressure.

on a subsequent analysis to determine the long-term capability of the installed cartridges:
- Design *normal flow* was 22.34 m^3/h (98.36 gpm) for the original system
- Number of cartridges installed was 22. Therefore, flow per cartridge was 1.015 m^3/h (4.47 gpm). This corresponded closely to one of the vendor's design criteria of not allowing more than 1.13 m^3/h (5 gpm) flow through individual filter cartridges.
- The system was designed with an adequate number of filter cartridges to allow an initial pressure drop of less than 5 psi per cartridge for normal flow with clean filters, as shown in Table 8.4.

From Figure 8.6, with a normal flow of 4.7 gpm for clean cartridges, the differential pressure is <5.0 psi with the original filter cartridge TYPE-X. However, after rerating, the flow per cartridge increased to 6.3 gpm resulting in a differential pressure of 8 psi. Although the existing filter cartridges TYPE-X could still be used under the new operating conditions, the time span between changes of filter elements would be reduced. Because of this, the vendor recommended using TYPE-Y instead, since this type of filter cartridges offered a lower pressure drop for the same flow and would be an exact drop-in without modification.

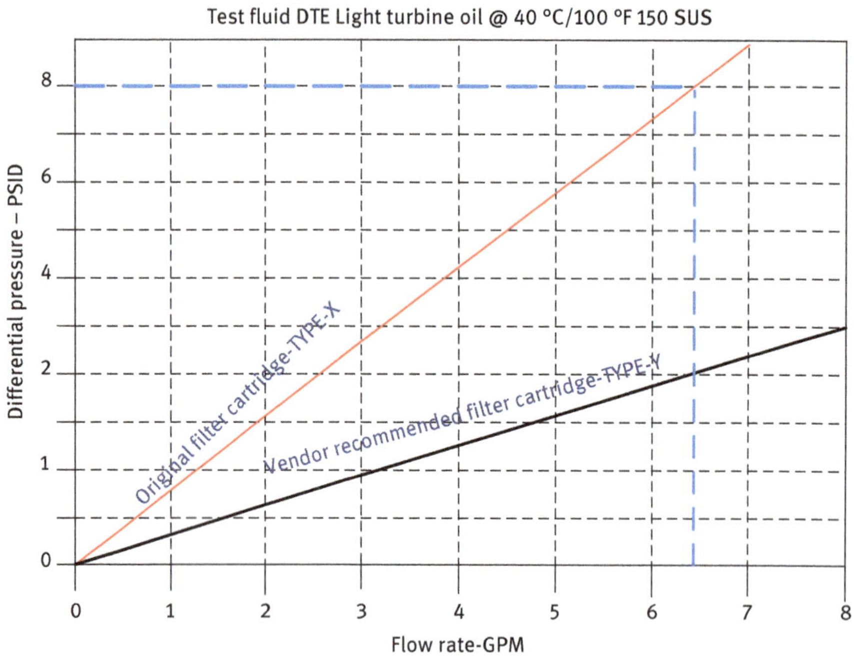

Figure 8.6: Flow through clean filter cartridge versus differential pressure (*for illustration purposes only*).

8.6 Oil accumulator

Using API 614 guidelines as we have done previously, it is recommended to have an accumulator installed, as shown in Figure 8.2, to promote stability during transients such as pump switchovers and control oil pressure fluctuations. Specifically, accumulator(s) are installed to accomplish the following:

1. To maintain the turbine control-oil pressure during servo control transients or maintain lube or seal-oil pressure while the standby pump accelerates from idle condition to operating speed. The control pressure shall be maintained above the equipment manufacturer's minimum specified supply pressure for all conditions (including transients). Typically, this is within 10% of the normal system pressure when a servomotor makes a full stroke in 1 s.
2. The system delivery pressure shall be maintained above the shutdown settings during standby pump acceleration or for at least 4 s for motor-driven pumps.

Figure 8.7 is an illustration of a transient situation when the main oil pump stops putting up enough pressure to sustain normal operation of the equipment. As the pressure of the system decreases, an alarm should be sounded and the accumulator(s) installed

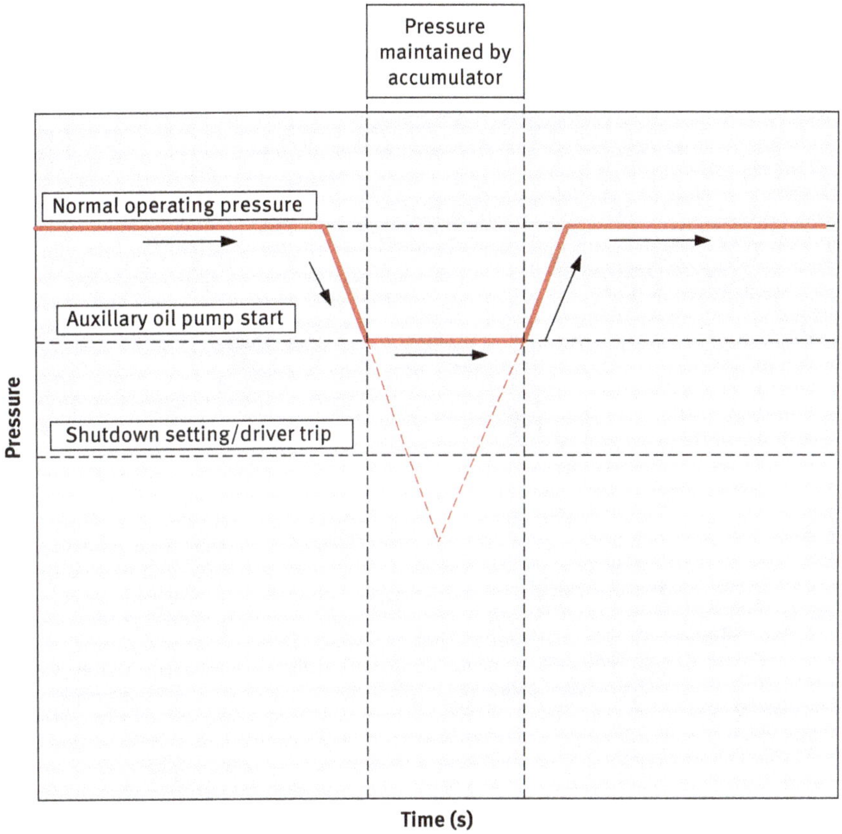

Figure 8.7: Auxiliary oil pump and accumulator reacting to system transient when main oil pump fails.

in the main oil header piping should be sized such that it/they deliver(s) enough oil at the minimum operating pressure level, while the auxiliary standby pump accelerates to design speed and maintains normal system header pressure. The required volume of oil discharged by the accumulator will be the normal flow rate required multiplied by the time allowed for the auxiliary pump to come up to design speed.

For large "special-purpose" lubrication systems, API 614 recommends using either *bladder*-type or *direct-contact*-type accumulators. These are shown in Figures 8.8 and 8.10.

Both types of accumulators can have piping and instrumentation configured to allow automatic charging of accumulator or they can be precharged. However, for simplicity of explanation we will discuss only precharged accumulators.

Figure 8.9 shows the sequence by which a typical bladder accumulator is placed in operation after draining and venting.

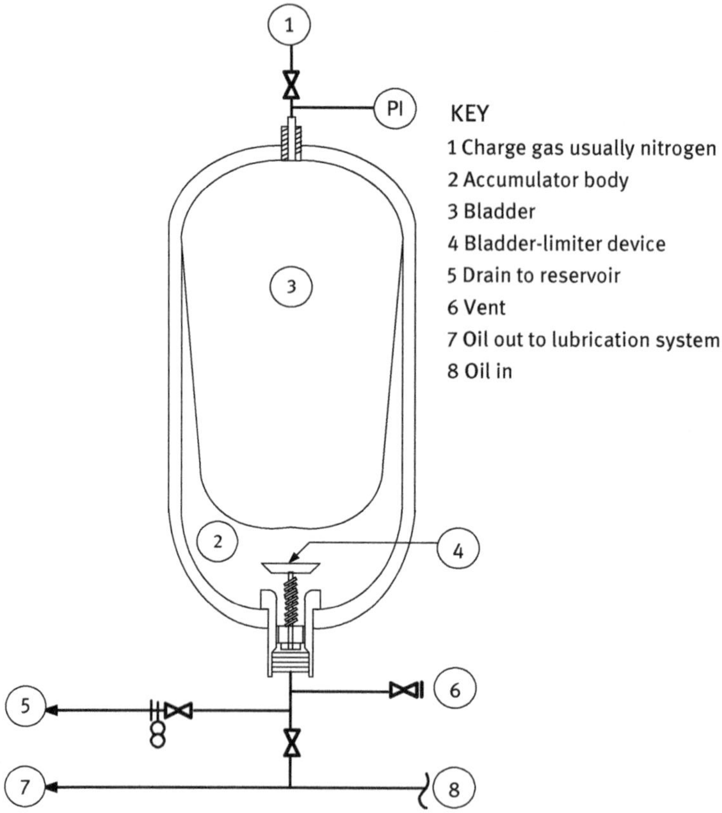

Figure 8.8: Bladder-type accumulator with bladder-limiter device.

- **Position A**: Precharging the bladder with nitrogen (N_2), causing the bladder to be extended, closing off the limiter device. The precharge pressure should be approximately 70–90% of the maximum working pressure.
- **Position B**: Bladder being compressed by the pressure of the incoming lube oil at the point of minimum operating pressure where bladder is set to function whenever the system loses the main lube oil pump, and the standby pump starts.
- **Position C**: Bladder is now compressed to the maximum operating pressure, and pressure within the bladder is in equilibrium with the lube oil system pressure.

The volume of oil available from a pre-charged accumulator can be calculated using the gas laws, knowing the accumulator volume as shown in Table 8.5, precharge pressure (typically between 70% and 90% of maximum system pressure), and the minimum system pressure at which the accumulator should function. The minimum system pressure should be at or near the set point pressure at which the standby pump starts:

$$P_0V_0 = P_1V_1 \tag{8.1}$$

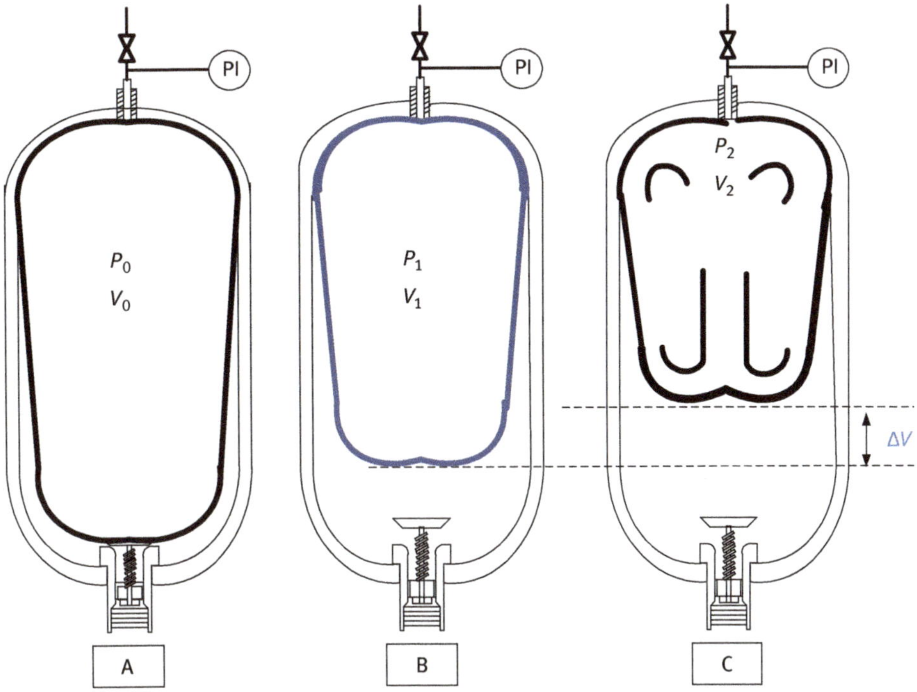

Figure 8.9: Commissioning sequence and/or operation for typical bladder accumulator.

Transposing eq. (8.1)

$$V_1 = P_0 \frac{V_0}{P_1} \tag{8.2}$$

$$P_1 V_1 = P_2 V_2 \tag{8.3}$$

where P_0 is the gas precharge absolute pressure, V_0 is the total accumulator volume, P_1 is the absolute system pressure at which accumulator functions (near or at standby pump start pressure), V_1 is the compressed gas volume inside bladder at P_1 pressure, P_2 is the final operating pressure in absolute units, V_2 is the compressed volume inside bladder at P_2 pressure, ΔV is the volume of oil that enters the system during pressure reduction caused by loss of main oil pump.

Next, we find the volume of oil entering the system from the accumulator when system pressure drops to P_1

$$\Delta V = V_1 - P_1 \frac{V_1}{P_2} = V_1 \left(1 - \frac{P_1}{P_2}\right)$$

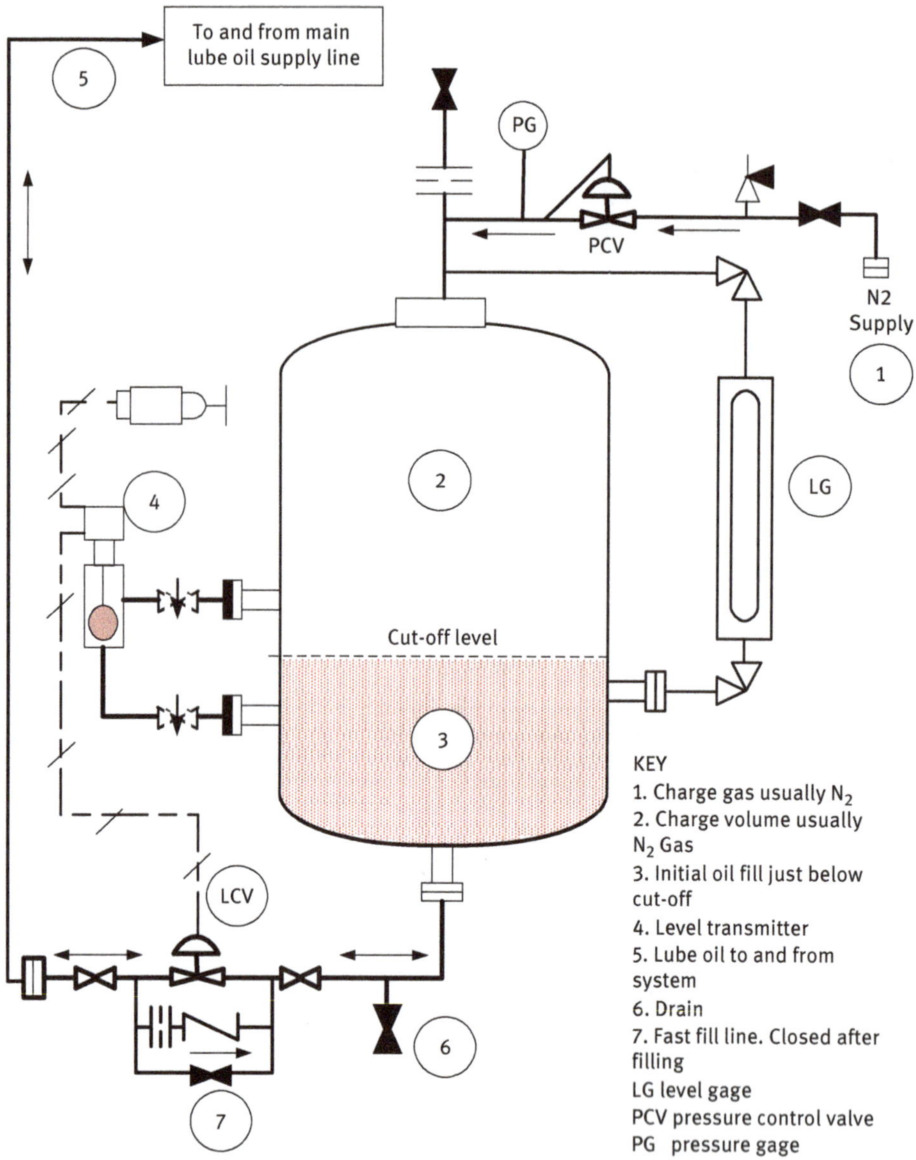

Figure 8.10: Direct contact-type accumulator with limiting level control valve.

Table 8.5: Typical accumulator sizes showing actual volumes.

Nominal size (gal)	Actual volume (gal)	Actual volumes (L)	Actual volumes (m^3)
10	9.0	34.07	0.034
25	25.0	94.63	0.095
40	40.0	151.4	0.151
80	80.0	302.8	0.303

Example 1: A 10 gal accumulator is installed in an oil system operating at a pressure of 120 psig normal, and with the standby set to auto start at 105 psig. If the precharge is 80% of the normal system pressure, calculate the volume of oil delivered by the accumulator and period of time of the delivery when the main lube oil pump trips if the normal flow to the system was 31.5 m^3/h for the revamp example:

$$P_0 = 110.7 \text{ psia}$$
$$V_0 = 10 \text{ gal}$$
$$P_1 = 119.7 \text{ psia}$$
$$P_2 = 134.7 \text{ psia}$$
$$V_1 = 110.7 \times \frac{10}{119.7}$$
$$= 9.25 \text{ gal}$$
$$\Delta V = 9.25 \times (1\text{–}119.7/134.7)$$
$$= 1.03 \text{ gal } (3.9 \text{ L}) = 0.0039 \text{ m}^3$$

Normal oil flow

$$= (31.5/3{,}600) \text{ m}^3/\text{s}$$
$$= 0.00875 \text{ m}^3/\text{s}$$

Time of delivery

$$= 0.446 \text{ s (less than desired)}$$

Example 2: By using a larger accumulator size, say 80 gal with the same operating conditions as in Example 1, calculate the volume and time of delivery during a trip of the main lube oil pump:

$$P_0 = 110.7 \text{ psia}$$
$$V_0 = 80 \text{ gal}$$
$$P_1 = 119.9 \text{ psia}$$
$$P_2 = 134.7 \text{ psia}$$
$$V_1 = 110.7 \times \frac{80}{119.5}$$
$$= 73.98 \text{ gal}$$
$$\Delta V = 73.98 \times (1\text{–}119.7/134.7)$$
$$= 8.24 \text{ gal } (31.23 \text{ L}) = 0.0312 \text{ m}^3$$

Normal oil flow

$$= (31.5/3{,}600) \text{ m}^3/\text{s}$$
$$= 0.00875 \text{ m}^3/\text{s}$$

Time of delivery

$$= 3.6 \text{ s (less than API 714 minimum requirement)}$$

Reviewing both examples, the point to make is that for large "special-purpose" lube systems using bladder-type accumulator to assist with transient situations during pump swaps or from the loss of main lube oil would require very large bladder-type accumulators or several accumulators installed in parallel. It should be noted also that bladder-type accumulators do have their shortcomings, namely:

- Bladders do have limited lives, as a result of low-cycle fatigue failure (sometimes as low as 40–50 cycles).
- Space is normally very tight around or on top of the lube oil skid assembly platform; hence, having a large bank of accumulators in parallel is not preferred.

Considering these shortcomings of bladder-type accumulators, large "special-purpose" lubricating systems tend to install *direct contact*-type accumulators to handle system transients. These are generally installed at grade level near or on top of the lube oil skid platform.

Charging and placing a *direct contact*-type accumulator in service are done in a similar manner to bladder types. The main difference is that precharge is done with a small quantity of oil blocked in just below the cut-off level, where the level control valve (LCV) closes and shuts off oil from the main system. Precharge pressure ranges from 70% to 80% of the main operating pressure. Once precharge is accomplished, oil from the system is allowed to enter the accumulator by opening the fast-fill valve #7 with the LCV opening as well. Oil pressure from the system compresses the N_2 precharge until the N_2 pressure is in equilibrium with the system pressure. Once pressure equilibrium is achieved, fast-fill valve #7 is closed, and a small amount of oil will enter and leave the accumulator through the restriction orifice installed next to the check valve.

This frequent cycling of oil back and forth is normal and depends on the responsiveness of the system to small transients. Sizing of these direct contact accumulators can be done using eqs. (8.1)–(8.3), as shown earlier.

In the previous revamp example, a large direct acting accumulator was installed originally; however, after calculations by the vendor and using the increased normal flow of 31.5 m^3/h, it was determined that the existing accumulator would still provide enough flow during transients; therefore, it did not require a change out.

8.7 Optimizing coast-down time for RDTs

The main purpose of lube oil RDT is to act as the "*last-resort*" lubricating source whenever the main lubricating system is no longer functioning, as in the case of an outright trip on major rotating equipment trains.

The RDT provides lube oil not only for lubrication during coast down but also for cooling the rotor, especially in the case of steam- and gas turbine-driven trains, owing to the fact that bearings act as heat sinks. Therefore, integrating these RDTs into the lubricating system is considered critical to the overall reliability of major machinery trains. Criticality notwithstanding, not much attention is paid to these tanks by an EPC (Engineering-Procurement-Construction) purchasing department or client engineers. Even the presumed reliability-focused API 614 Standard has RDTs as optional for large "*special-purpose*" lubrication systems.

RDTs can be divided into two categories: pressurized and nonpressurized. A nonpressurized tank is open to atmosphere. Pressurized RDTs are usually mounted at grade level; they can be very complicated considering the amount of instrumentation to be installed and the logic that is applied for them to operate as designed. This section therefore deals with atmospheric RDTs similar to the one shown in Figure 8.11. We will also explain how these can be optimized for extended coast-down time with little or no extra investment in hardware.

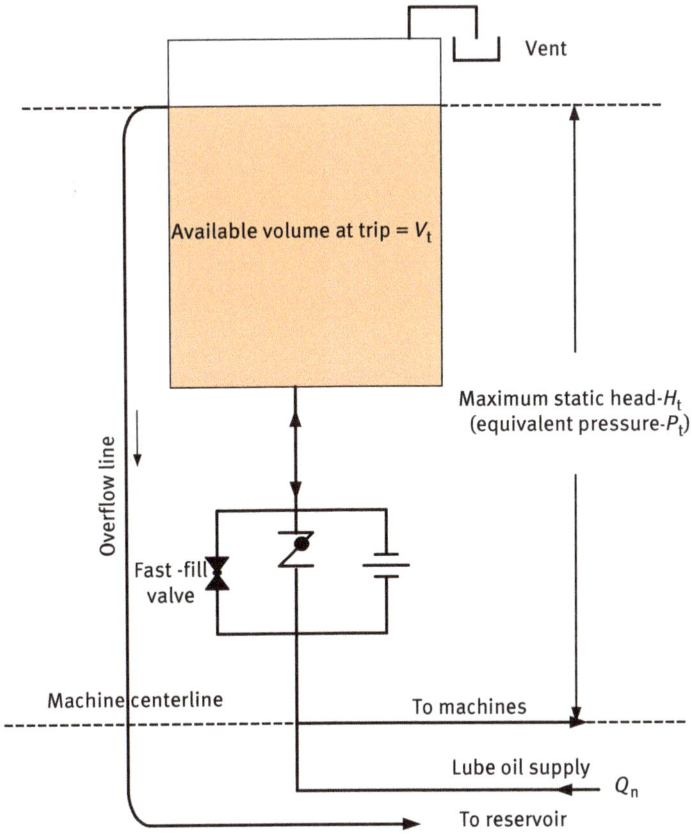

Figure 8.11: Basic concept for atmospheric overhead rundown tank.

These atmospheric RDTs are installed at specific elevations to provide sufficient static head for the required lubrication and cooling during coast down. The ability of these atmospheric RDTs to meet key requirements specified in API 614 depends on a fair amount of design engineering effort by the machinery vendors. Regrettably, the need for thoughtful design is not always shared with the purchaser–owner or client's engineers unless requested. Therefore, the authors recommend that purchasing departments and client engineers request calculations from machine vendors to ensure proper functioning of these RDT systems. Alternatively, the client's EPC should perform the calculations. Validation should be done by examining a certain key design parameter outlined in API 614, which will ensure the proper functioning of the RDT, and possibly offset equipment damage during coast-down situations.

Two case histories will be studied to compare a typical installation with a more modern, highly recommended engineered approach to achieve longer and more effective coast-down time.

8.7.1 Key design parameters to be examined

1) *Maximum static head in tank during normal operation*: as per API 614–5 paragraph 4.9.2.2, *the maximum static head shall be less than the lube oil trip pressure but not less than 30 kPa (0.3 bar; 5 psig) at the beginning of a cost down.* Note: Adhering to this criterion allows the lube oil system to trip at low–low pressure without interference from the static head imposed on the system by the RDT.

   Salisbury et al. suggest that the equivalent static pressure should not be more than 2–3 psi less than the low–low lube oil pressure trip setting [2].
2) *Tank sizing*: As per API 614–5 paragraph 4.9.2.1, *the tank shall be sized for not less than 3 min of normal operating lube oil flow.* Note: *The minimum sizing criteria require more than 3 min of coast-down oil flow due to supply pressure decay.* This becomes extremely important if the train is driven by steam or gas turbines. Making sure that there is adequate oil capacity to cool the shaft and lubricate the bearings during coast down will avert serious damage to the train. The primary concern is whether or not there is sufficient capacity designed into the RDT to satisfy different shutdown scenarios. Visualize a sudden depressurization of the compressor train or a simple shutdown. The coast-down time can be much more than 3 min. In all instances, the designer must consider not only the train inertia but also gas inertia effects. Obviously, the two scenarios will have very different coast-down times, and so the vendor designing the RDTs must always err on the conservative side of the equation. For example, a simple way to size the capacity would be

   Tank capacity = Normal operating flow (gpm) × Coast-down time (min)

However, because the coast-down time is not exactly known, nor is there an exact quantity of oil established to feed the bearings as the train coasts down, vendor and purchaser shall mutually agree on a coast-down time for the machine train. This can be done during the design audit phase of a project. For steam and gas turbine trains, 8–20 min is not uncommon for coast-down times depending on the train size.

8.7.2 Methodology used to validate key parameters

Validation of the stated key parameters will be done by a series of checks as follows:

Check 1: Static head in tank during normal operation
Let the lube oil trip pressure given by vendor = P_{vt} (psig)
Maximum static head in tank during normal operation = H_t (ft) (see Figure 8.11)
Convert H_t to pressure head P_t and compare it with P_{vt} (psi)

$$P_{vt} > P_t = < 5 \text{ psi (per API 614 guidelines)}$$

Preferably $P_{vt} > P_t$ by 2–3 psi as suggested by Salisbury et al. [2]

Check 2: Tank sizing
Let the normal oil flow to bearings = Q_n (gpm)
Let the available volume of oil in RDT to support coast down = V_t (see Figure 8.11)
Convert V_t to gallons Q_{RDT} and compare it to 3 min × Q_n

$$Q_{RDT} = > 3 \times Q_n \text{ (gallons)}$$

Check 3: Time supplying oil from the tank during coast down
Let $Q_t = \sqrt{\frac{P_t}{P_n}}(Q_n)$ [3]
Q_t = Oil flow at trip (gpm)
Q_n = Flow to bearings at normal pump pressure (gpm)
P_t = Oil pressure at trip (psi)
P_n = Normal oil pressure at header (psi)
Let the time supplying oil from tank during coast down = T
Then $T = (Q_{RDT} + Q_{Piping})/Q_t$

Case study 1: Figure 8.12 shows the typical RDT installation arrangement for a large 50 MW three-body compressor train driven by a steam turbine, with dimensional details of the RDT shown in Figure 8.13:

Normal Lube oil system pressure	P_n	= 20 psig
Height of overflow line to machine centerline	H_t	= 165.35 in
Static height in tank	H_s	= 143.7 in

Static height in piping to machine centerline	H_p	= 165.35–143.7 in
		= 21.65 in
Normal flow to machine train	Q_n	= 669 US gpm
Lube oil system trip pressure (given by vendor)	P_{vt}	= 8 psig
Diameter of RDTs	D	= 66.93 in
Diameter of pipe from RDT to system	D	= 6 in
Lube oil		= ISO 32
Density of ISO 32 oil (at 60 °F)	ρ	= 7.10 lb/gal or 0.0308 lb/in^3
Note: 1 US gallon		= 231 in^3

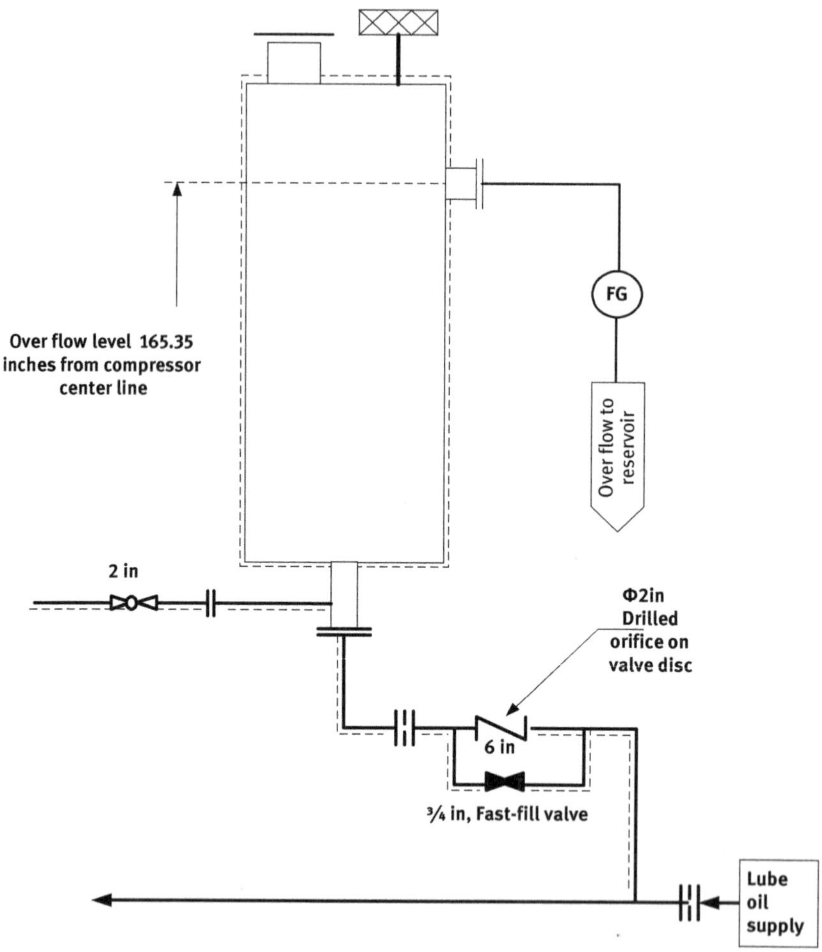

Figure 8.12: Typical atmospheric rundown arrangement for trains mentioned in Case 1.

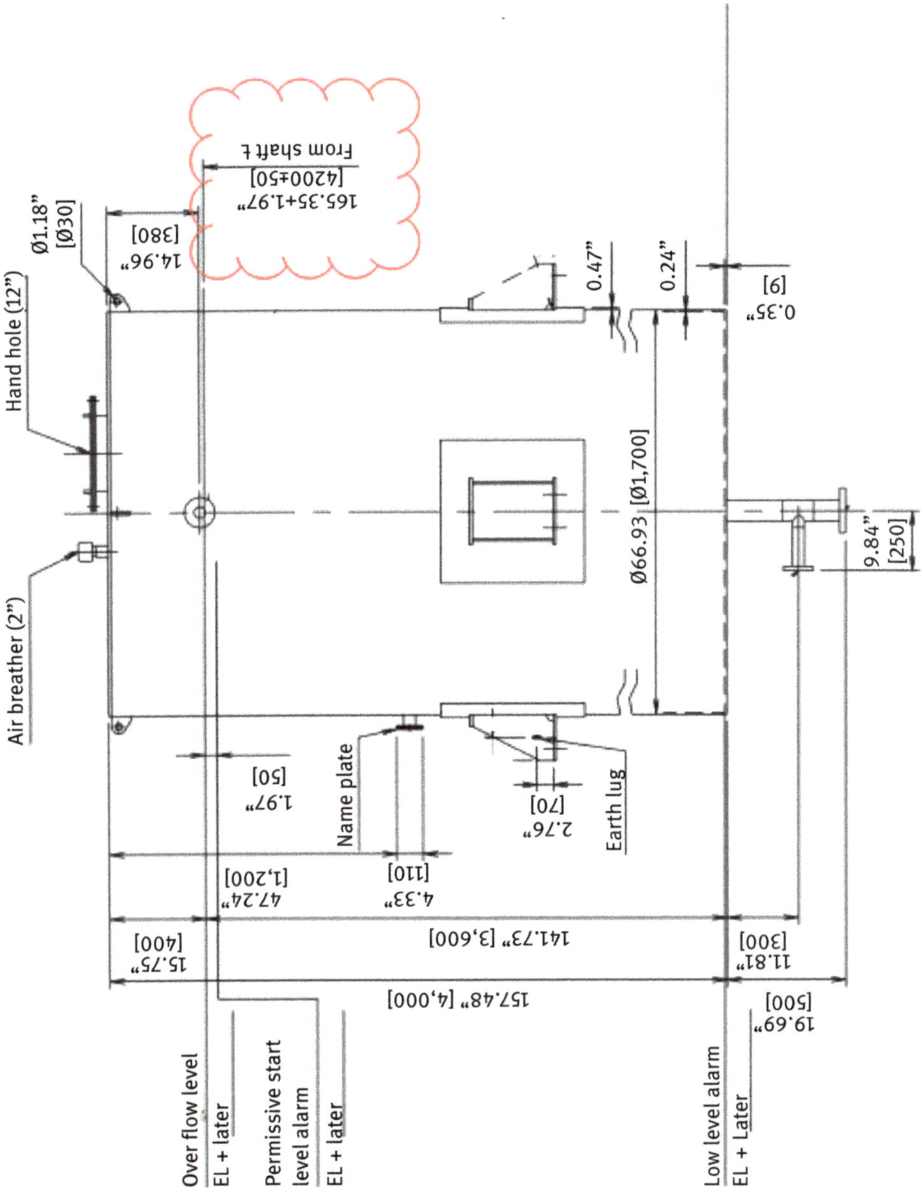

Figure 8.13: Dimensional details of the rundown tanks for Case 1.

Check 1:

Static head in RDT during normal operation $P_t = 0.0308 \times 165.35$ lb/in^2

$= 5.09$ psig

Since $P_{vt} = 8$ psi, the height of the RDT meets API guidelines

Therefore, $P_{vt} > P_t > 5$ psi (meets API 614 guidelines)

Also $P_{vt} > P_t$ by 2.91 psi; well within the 2–3 psi range suggested by Salisbury et al. [2].

Check 2:

Normal flow to bearings $Q_n = 669$ gal/min

As per API 614, tank must be sized for at least 3 min of normal operating flow

$= 669$ gal/min $\times$ 3 min

$= 2{,}007$ gal or 463,617 in^3

Volume of oil available in tank before coast down $Q_{RTD} = 143.7 \text{ in} \times \pi D^2/4$

$= 505{,}578.2$ in^3 or 2,188.7 gal

Therefore $Q_{RTD} > 3 \times Q_n$

The size of the tank meets and exceeds API 614 guidelines.

Check 3:

We know that the flow from the RDT will be less than the normal flow and is required only for cooling and lubrication until the train comes to a stop. Therefore, let us calculate the approximate flow rate from the tank and the time the tank will be supplying oil to the train.

Total oil available before the trip $= Q_{RTD}$ + oil in piping from RDT to machine

$= 505{,}578.2 + (21.65 \times \pi d^2/4)$ in^3

$= 506{,}190.33$ in^3 or 2,191.3 gal

Oil flow at trip from RDT $= (8/20)^{0.5} \times 669$ gpm

$= 423.11$ gpm

Time supplying oil from tank during coast down $T = 2{,}191.3/423.11$ min

$= 5.17$ min

Therefore, the tank meets API 614 guidelines for $T \geq 3$ min

Case study 2: This case evaluates a large 32 MW axial air compressors train driven by a gas turbine through a reduction gearbox. The RDT had to be designed specially to match the coast-down flow requirements given by the gas turbine vendor as follows. The compressor vendor had the responsibility to design the lube oil system including the RDT.

Gas turbine lubrication requirements	
Oil type	Mineral oil
Viscosity	ISO VG32 or ISO VG46
Temperature	Turbine start permissive 60 °F
Normal supply	113–130 °F
High alarm	150 °F
High–high shutdown	155 °F

Pressure	Normal supply 20 psig
Maximum supply	30 psig
Low alarm	17 psig
Low–low shutdown	12 psig
Normal lube oil flow	65 US Gal/min
Normal speed	5,050 rpm
Coast down:	
At 100% speed	100% of Normal oil flow required
At 60% speed	40% of normal oil flow required
At 20%	10% of normal oil flow required
Cooldown	0.5 US gallon/min for 90 min

Based on these lubrication requirements by the gas turbine vendor during coast down, the compressor vendor designed the RDT with an internal stilling-pipe arrangement with orifices shown in Figure 8.14. This tank design was able to deliver different flow rates at specific times during the coast down, matching the gas turbine lubrication requirements.

Overall system design parameters (*including gas turbine, gearbox, and compressor*)

Normal lube oil system pressure	P_n	= 20 psig
Height of overflow line to machine centerline	H_t	= 336.17 in
Normal flow to machine train	Q_n	= 376 US gpm
Lube oil system trip pressure (given by vendor)	P_{vt}	= 12.0 psig
Diameter of RDTs	D	= 60 in
Lube oil		= ISO 32
Density of ISO 32 oil (at 60 °F)	ρ	= 7.21 lb/gal or 0.0312 lb/in^3
Note: 1 US gallon		= 231 in^3

Check 1:

Static head in RDT during normal operation $P_t = 0.0308 \times 336.17$ lb/in^2 = 10.35 psi

Since P_{vt} = 12.0 psi, the height of the RDT meets API guidelines

Note: For cases where P_{vt}–P_t is less than the 2 psi as in this particular case, the authors recommend raising the trip setting by at least 1 psi to get within the range suggested by Salisbury et al. [2].

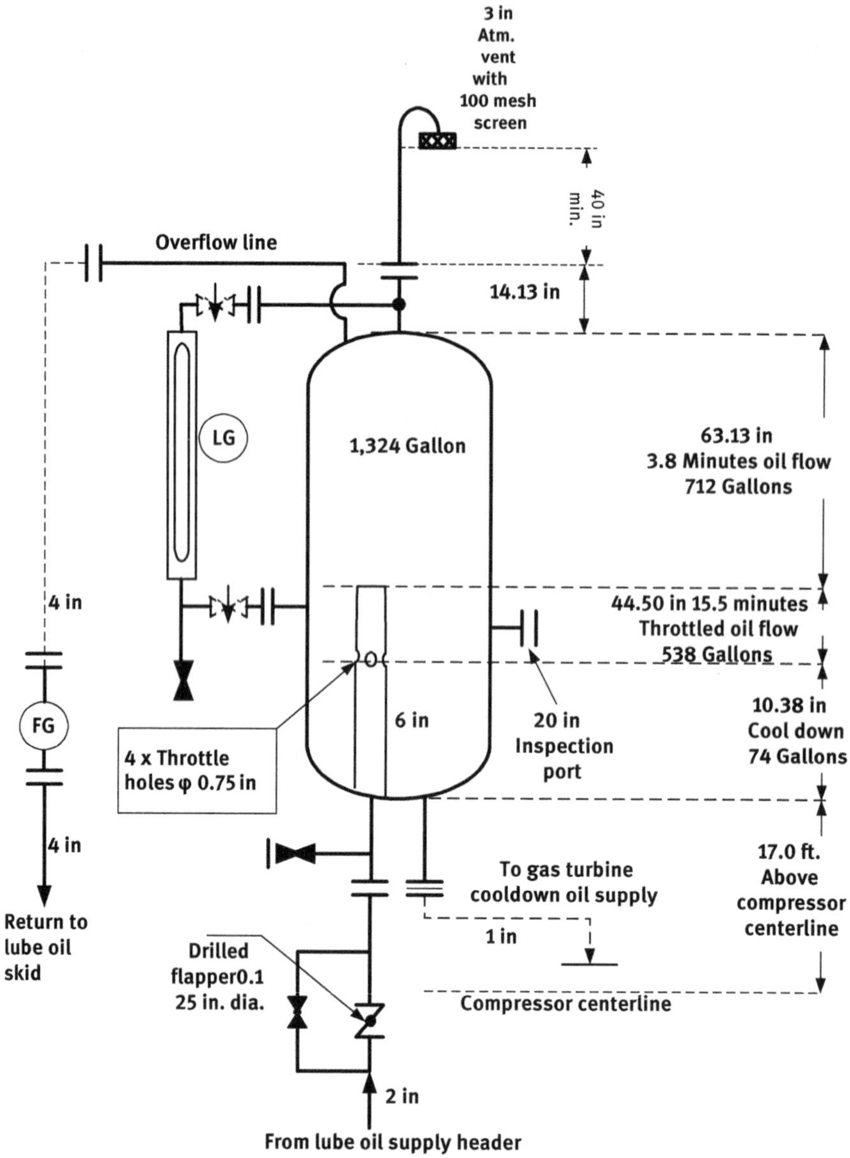

Figure 8.14: Atmospheric rundown with internal pipe arrangement.

Check 2:

Normal flow to bearings $Q_n = 376$ gal/min

As per API 614, tank must be sized for at least 3 min of normal operating flow

$$= 376 \text{ gal/min} \times 3 \text{ min}$$
$$= 1{,}128 \text{ gal}$$

Volume of oil available in tank before coast down $Q_{RTD} = 1{,}324$ gal

$$Q_{RTD} > 3 \times Q_n$$

Therefore, the size of the tank meets and exceeds API 614 guidelines.

Check 3:
In order to match the supply flow requirements of the gas turbine during coast down, the RDT was designed to modulate the supply flow into three different flow regimes:
- Unthrottled flow design volume was 712 gal
- Throttled flow design volume was 538 gal
- Cooldown flow for the gas turbine 74 gal

Figure 8.14 illustrates how this was done.

To achieve these three flow regimes, the RDT was designed with a 6 in diameter internal piping arrangement with four 0.75 in diameter drilled holes, located at a specific height along the length of the pipe, and the pipe protruded internally inside the tank, also at a specific height.

This internal piping design made it possible to achieve the following flow rates with the RDT installed at a height of 17 feet above the machine train centerline:
- Unthrottled flow rate during coast down was 187 gal/min with the estimated supply time of 3.8 min
- Throttled flow rate during coast down was 33.8 gal/min with the estimated supply time of 16 min
- Cooldown flow rate was 0.5 gal/min, estimated to be in excess of 90 min

The calculations on how the compressor vendor was able to design and achieve the extended time supplying oil to the machine train during coast down are outside the scope of this book; however, the point to make is that RDTs can be engineered and optimized to give lengthy times supplying oil during coast down when needed.

References

[1] Khonsari, M., Booser, E.R. Matching Lube Oil systems to Machinery Requirements, Machinery Lubrication, Noria Corporation, 1328 E. 43rd Court, Tulsa, OK 74105, USA, November 2004.

[2] Salisbury, R., Slack, R., Sassos, M.J. Lubrication and Seal Oil Systems – Proceedings of the Thirteenth Turbomachinery Symposium, Turbomachinery Laboratory, Texas A&M University, College Station, Texas, 151–175.

[3] Brown, R.N. Turbomachinery Problem: Lube and Seal Systems Revisited – Proceedings of the Seventh Turbomachinery Symposium, Turbomachinery Laboratory, Texas A&M University, College Station, Texas, 119–123.

Chapter 9
Dry Gas Seals, Auxiliaries, and Support Systems

9.1 Dry gas seals (DGS)

Dry gas seals (DGSs) are the seals of choice for compressors today. Although they can be found in a few pumping services, such as light hydrocarbons, the majority are applied in compressors. The popularity of these seals in compressors is attributed to their frequently higher reliability compared to traditional wet seals. Generally speaking, they have been proven to represent a reliable sealing option at reduced overall cost and lower process gas leakage loss compared to many wet seals. Figure 9.1 shows a typical DGS cartridge.

Figure 9.1: Typical dry gas seal cartridge arrangement (source: *Eagle-Burgmann*).

To support the use of these seals, the American Petroleum Institute (API) developed the API 692 standard. The standard provides users with guidance in selecting the appropriate seal configurations for different process application. It also recommends the design of the supporting systems. This standard was developed with input from

Note: Contributed by EagleBurgmann.

https://doi.org/10.1515/9783110678765-009

experienced users, engineering and design providers (also known as engineering/procurement/construction companies), seal manufactures, DGS panel manufacturers, and compressor manufacturers. API standards give guidance on seal system configurations and associated requirements.

The design intent for DGS-related standards is to assist end-users with better understanding on how these seals should be applied in a variety of available options. The selected option must be best for a specific process or service. In particular, the selection must be influenced by the vendor–manufacturer's favorable field experience. This experience must be verified by the owner–purchaser's reliability engineers in the pursuit of their machinery quality assessment (MQA) activities. Compliance with API standards and the owner–purchaser's supplementary specifications must be established during MQA.

This chapter summarizes typical DGS technologies and systems, along with some newer technologies designed to improve DGS reliability. These newer technologies allow end-users a wider range of applications; the wider range is now being demanded by the oil and gas industry. Our aim is twofold: First, to assist readers in becoming more familiar with the API 692 standard; next, to equip readers to most effectively select and specify reliable sealing systems.

9.1.1 Principle of operation

DGSs work on the principle of creating a gap between the faces of two rings with gas as a sealing medium between the seal faces. One ring rotates while the other ring remains stationary. The rotating ring is designed with spiral groves etched into its face at a depth of approximately 0.0003 in/0.007 mm. This seal is considered an axial end-face seal as opposed to a radial seal. (Radial seals have a clearance gap between a rotating shaft and a stationary labyrinth or floating ring.)

With a radial seal, the gap will vary due to temperature gradients and different types of materials having different coefficients of thermal expansion. With radial seals, a fixed distance from the shaft or shaft sleeve must be maintained at all operating conditions. Doing so will prevent contact between a rotating shaft or sleeve and stationary component. This results in a large gap and high leakage rates for radial seals compared to axial end-face seals. Axial end-face seals are designed to have a 3- to 4-µm gap during operation. These seals will effectively produce a more consistently smaller leakage rate compared to radial seals.

Note: *To understand what 3–4 µm equates to, the average human hair is 50–70 µm.* *Reference:* https://willrich.com/metrology-education/understanding-millionths-inch/

In producing the gap between the sealing rings, O-ring positions, ring geometry, and grooves in the rotating seal ring face perform critical functions. Forces applied to the rings affect gap width. As the seal ring attached to the shaft rotates, the grooves in the face pump additional gas between the rings. A stiffer gas film and slightly increased gap will result. The stiffer film prevents contact between the rotating and stationary faces as the compressor rotor moves both axially and radially. Hydrodynamic forces similar to the ones acting in fluid film bearings are created; this ensures minimum seal leakage and prevents contact between the faces. A dynamic O-ring or similar secondary sealing element allows the stationary component to float axially to effectively maintain a 3- to 4-µm gap between the rings. Loss of free axial movement of the stationary ring, known in the industry as O-ring hang-up, will either cause the seal gap to increase and produce high seal leakage or compress the gas in the gap. O-ring extrusion can cause the seal ring faces to contact. The dynamic sealing system comprises a dynamic secondary sealing element, a thrust ring, springs, and balance sleeve. The dynamic sealing system ensures that the stationary seal ring moves when required to effectively maintain the sealing gap.

Another important factor for producing a consistent gap between the seal rings is the seal ring face flatness. Flatness and gap parallelism are of extreme importance. Early DGS designs used tungsten for the rotating ring and carbon graphite for the stationary ring. Carbon graphite is a forgiving material in the event of face contact. However, there are limitations when using this material in larger seal rings, in hot temperatures, and at high pressures. These can combine to cause deformation of the stationary ring, and larger gaps will be required to prevent face contact. This increases leakage and limits applications where a carbon graphite seal ring can be effectively applied.

Rotating rings made of tungsten carbide are advantageous for maintaining flatness. However, as seal size and speeds increased, rotating weight became a limiting factor. Contact or liquid entering the seal gap introduces heat. This produces heat checking or stress cracks; it weakens the material and cause hoop stress on the rotating seal ring resulting in fracture. However, when operated under ideal conditions, the material can provide a reliable seal. On the other hand, if the right conditions are not met, then end-users might experience seal failures, resulting in extensive damage to the seal and compressor.

Due to the limitations of tungsten and carbon graphite, newer seals now generally incorporate silicon carbide for both rotating and stationary rings. By utilizing this material, higher pressures and speeds can be achieved along with lower leakage rates. Although tungsten material is still used today for manufacturing some DGSs, many experienced end-users have discontinued its use.

It is obvious that in rotating seal ring, stationary seal ring, O-rings, as well as secondary seals (elastomers), the dynamic "sealing-in-motion" components are the critical parts in a basic DGS. Design features and materials for these components have been refined by seal manufactures to achieve their increasing capabilities. As a result, seal reliability increased and leakage was reduced.

To manage the wide variety of applications and produce safe and reliable sealing solution, additional components are constantly being designed to optimize a complete sealing solution. A total sealing solution comprises DGS cartridge(s) with

a single set of seal rings or dual sets of seal rings oriented in tandem or apposing configurations. The cartridge comprises the following components:
- Primary process seal (stationary and rotating rings),
- Intermediate labyrinth,
- Backup seal/secondary seal (stationary and rotating rings), and
- Separation seal (acts as a barrier, preventing oil intrusion).

Additional components depend on the type of compressor and hazardous level of the process gas being sealed. The other critical component of a sealing solution is the seal support system; it will be discussed later.

9.1.2 Merits of using different seal configurations

9.1.2.1 Tandem seal arrangement

Tandem DGSs (Figure 9.2a) were the original (and most commonly used) configuration in pipeline applications. This type of seal consists of a process labyrinth, primary

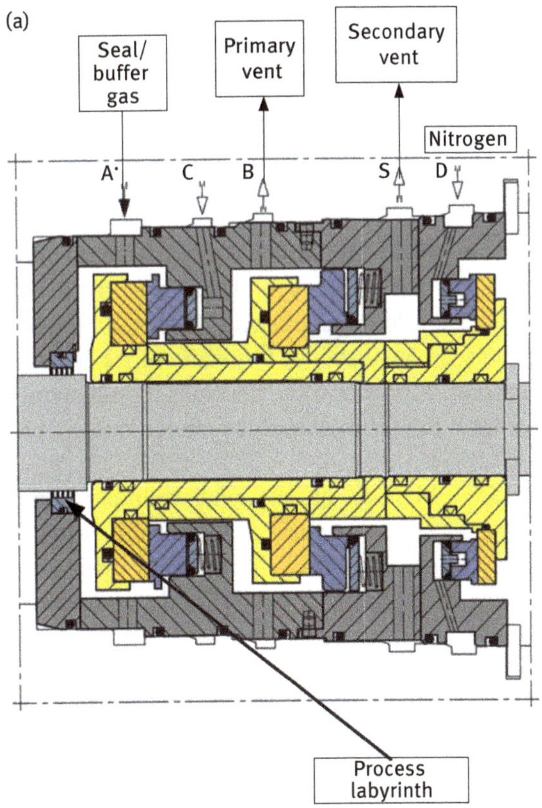

Figure 9.2a: Tandem seal arrangement (source: *Eagle-Burgmann*).

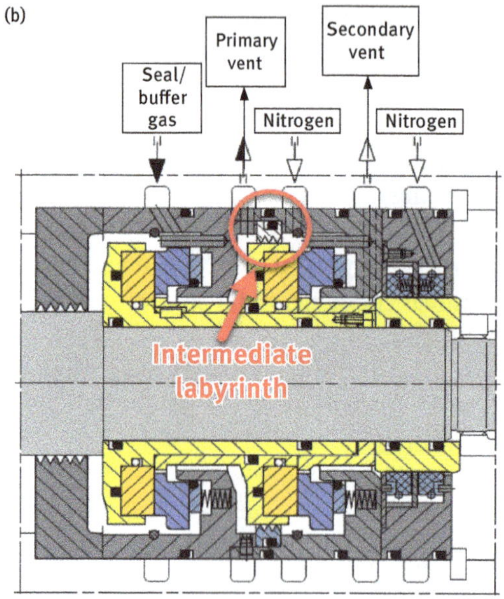

Figure 9.2b: Tandem seal with intermediate labyrinth arrangement (source: *Eagle-Burgmann*).

seal mating rings, secondary seal mating rings, and a separation seal assembly. For this seal, the secondary seal medium is the process gas leakage from the primary seal. Both the primary and secondary leakages are vented to atmosphere for pipeline applications because most compressor stations have neither flares nor vapor recovery systems. Wherever a flare or vapor recovery system is available, the primary leakage is directed into these systems; the secondary vent is taken to a safe location, typically to atmosphere.

For hazardous process, gases such as those containing H_2S, an intermediate labyrinth, and secondary seal gas are being used. This type of seal is called a tandem seal with intermediate labyrinth (Figure 9.2b). The labyrinth is placed between the primary and secondary seals with a secondary seal gas (typically nitrogen) supplied between labyrinth and secondary seal. Nitrogen, thus, becomes the secondary seal medium and ensures any leakage from the primary seal is directed out the primary vent.

The intermediate labyrinth provides additional safety even when a secondary seal gas is not used. If a primary seal failure occurs, the labyrinth restricts debris and gas coming from the primary seal; hopefully, it prevents these contaminants from entering the secondary seal. Should a catastrophic seal failure occur whereby both the primary and secondary seals fail, the intermediate labyrinth will act as a restriction, reducing the volume of process gas passing through the seal. API 692 recommends using an intermediate labyrinth with all tandem seals.

9.1.2.2 Double-seal arrangements

In refineries and chemical plants, tandem seals are still being used. The leakage from the primary vent is typically directed to a flare system. However, since these flare systems often have pressures higher than the sealing pressure for these compressors, there exists a potential problem: Tandem seals allow back pressuring of the primary seal, which can cause the seal to fail. Such failure events have indeed been experienced and reported by owner–operators. When using this configuration, a seal gas must be introduced at elevated pressure. Usually, this is achieved by injecting compressor discharge gas into the seal as the primary seal gas, or by using a clean external buffer gas. Sealing systems designed in this manner are a bit more complicated and must be carefully designed.

A double-seal arrangement designed for low-pressure applications is shown in Figure 9.3. Double seals are often the most suitable option in situations where back pressuring could occur. Double seals have two sets of seal-mating rings in an opposed (mirror image) assembly; seal gas is injected between the two sets of seal rings. Double seals are typically used for pressures below 20 bar (290 psi) but can accommodate pressures of 60 bar (870 psi) or higher, depending on seal size and operating speeds. In essence, double seals provide a simpler, more robust sealing solution for lower pressures.

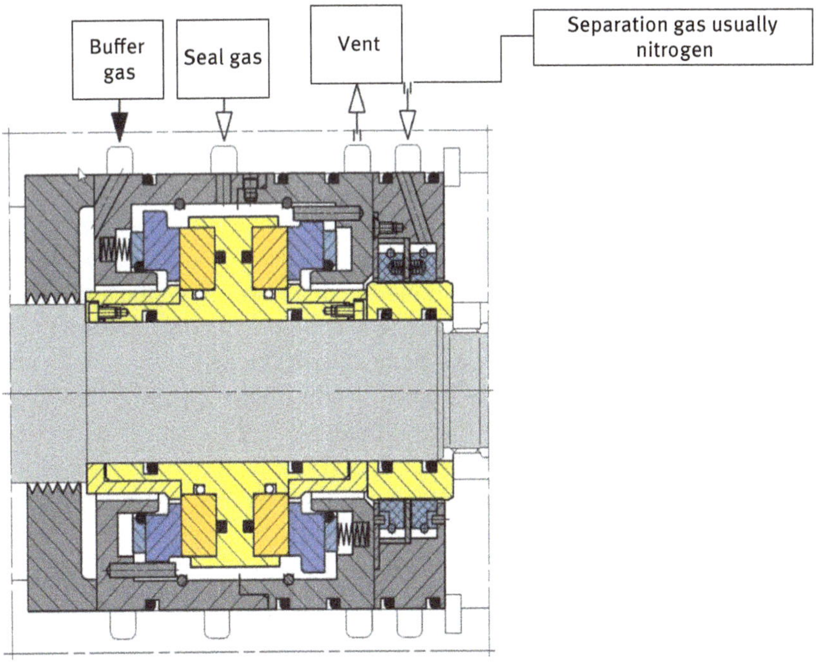

Figure 9.3: Double-seal arrangement (source: *Eagle-Burgmann*).

The two arguments most often used against a double seal are:

1. It requires the use of an external sealing gas, and is therefore susceptible to contaminating the process
2. There is no backup/safety seal similar to the tandem seal arrangement. This lack of backup is a consequence of how the seal is configured.

In the author's opinion, these are often unfounded arguments. The seal gas can come from the process, which then eliminates the concern of contaminating the process. When using process gas, an external source or means of increasing the seal gas pressure is usually required during startup conditions or when there is not enough differential pressure available to produce the gap between the seal mating rings. Increasing the seal gas pressure during startup conditions will be addressed later. Generally, the amount of seal gas leaking into the process is very low, typically less than 0.25 scfm (0.42 m^3/h). Therefore, the end-user must assess if this poses an issue with potentially contaminating the process.

It can be reasoned that radially segmented abradable stationary labyrinth seals can effectively serve as backup emergency seals for double DGS styles. These seals have extremely small radial clearances and can control leakage until shutdown and repair are scheduled and effected.

9.1.3 Dry gas seal support systems

DGS support systems (Figure 9.4) are used to provide an adequate quality seal gas supply for the process. Well-designed systems supply clean, dry seal gas to the primary mating rings and intermediate labyrinths. These systems also send gas to the separation seals where it is applicable and designed/selected; the system controls the seal gas flow, measures seal leakage, provides warnings when issues are identified, and shuts down the compressor when a seal has failed. A well-designed seal gas system warns the operator when conditions occur that could damage the seal. Well-designed seal gas systems can also shut down a compressor should a seal failure be imminent or if a hazardous condition is about to develop.

Compressor manufacturers sometimes wish to add their own shutdown logic for added compressor protection and to ward off liabilities. Experience shows that these add-ons are not always justified. Well-managed facilities conduct hazardous and operability studies. Citing multiple failure modes will define the shutdown requirements.

Figure 9.4: Dry gas seal support system (source: *Eagle-Burgmann*).

9.1.3.1 Seal gas supply

On many original seal gas supply systems, only switches and orifices were used for simplicity. The problem with switches is that they are not helpful at troubleshooting seal failures. It is conceivable that the measuring orifice restricted flow when more flow was needed, which in turn contributed to seal failures. Today, most systems use transmitters and control valves with programing logic. These control the delivery of an adequate gas supply through valve stroke movement when supply gas pressure changes.

Unfortunately, these changes still do not fully address the need for adequate pressurization of the seals during startup/standstill conditions. Seal gas boosters have been developed and are available to provide these needs at their proper times, which will be discussed in the DGS innovations section later.

A common discussion regarding seal gas supply is whether to use flow control or differential pressure control. As per API 692, the main criterion for effectively delivering seal gas to the system is the minimum flow needed to satisfy the velocity requirement across the process labyrinth inside the compressor. The gas velocity requirement is achieved by either of these two methods: (a) Measuring differential pressure using an orifice or (b) flow control using a control valve. If there is a restriction orifice in a differential pressure-controlled system, then, logically, it is a flow control system. The orifice size is determined by identifying the flow required to achieve a specified differential pressure. Most beam style compressor manufacturers incorporate restriction orifices in their seal gas systems at both ends of the compressor. Often, the sealing pressure will be different on one end of the compressor compared to the other; hence, the reason for having independent orifices. These orifices are sized to manage a fair amount of sealing pressure differences, even when the compressor balance piston is worn or when the compressor is being operated at its upper limits. Either of the two conditions produces higher sealing pressures on the discharge end of the compressor.

The downside of using a true differential pressure control system is how it reacts to an inrush of high flowrates through the system, which can increase substantially under certain operating conditions. It will require a system with much larger than "normal" flow capability to handle these types of transients. API 692 defines the seal gas system capability: *The seal gas system shall be designed to handle three times the normal flow in order to manage excessive flows caused by transient conditions.* However, with a flow control system, excessive flow conditions caused by transient situations are rare, and so the system is usually oversized unnecessarily.

9.1.3.2 Monitoring primary seal

The health of the primary DGS is determined by the volume of process gas escaping into the primary vent. The volume of gas is measured to confirm the seal is operating effectively, and the seal gas system is programmed to provide warning when flow increases above a set point. A signal impulse initiates shutdown when flow is excessive. API 692 recommends that the HIGH-alarm limit for the gas flow coming from the primary seal be set at 300% of normal flow, and the HIGH–HIGH alarm be set at 500% of normal flow to initiate shutdown. These functions can be achieved with a flow or pressure transmitter, depending on the location where the primary vent/seal leakage is sent.

Originally, primary seal monitoring systems were designed with a pressure switch and orifice located in the primary vent line to atmosphere. This made it easy to determine a change in gas flow because the pressure increased upstream of the orifice. However, as environmental rules became more stringent, the primary gas leakage was directed to a collection system. Collection systems such as flare systems do have varying pressure conditions, and so the sealing system required better means of determining flow changes. Today's best practice for installing vent lines to these collections systems is to install a flow meter or differential pressure transmitter along with the inline orifices.

The system should also have a relief valve or rupture disk to bypass the orifice when gas flow becomes excessive. By configuring the system in this manner, damage to the flow meter and excessive pressure on the secondary seal are prevented. Earlier designs focused their attention on monitoring only the primary seal, with no means provided to monitor the secondary backup seal. This led to many dangerous situations where the secondary backup seal would fail before the primary seal, and the end-user was not warned of the impending seal failure. The danger came when the excessive leakage resulting from the failure of secondary backup seal was not properly contained. Excessive leakage of process gas would imperil the bearings, eventually getting to the lube oil tank, ballooning the tank, and even causing fires and explosions. User industries now incorporate instrumentation and logic to monitor the health of the secondary seal.

9.1.3.3 Secondary seal monitoring

As identified earlier, original seal system designs were not effective at monitoring the secondary seal or detecting a secondary seal failure. However, a number of hazardous incidents that occurred in the petrochemical industry pointed to the need for secondary seal monitoring. It is not easy to effectively monitor the secondary seal, since most of the methods used over the years do have their pros and cons. However, close cooperation with the seal and system manufacturers will ensure that best available sealing arrangements are identified and that means are provided to fully monitor the secondary seal.

Monitoring the secondary seal can be difficult because secondary seal vent flow is a combination of secondary seal leakage and flow from the separation seal. Since the volume of gas coming from the secondary seal leakage is only a fraction of the flow coming from the separation seal, it is very difficult to set an effective alarm identifying a secondary seal failure in relation to a change in separation seal gas flow. In pipeline applications, monitoring the lower explosive limit (LEL) of gas leaving via the primary vent line is likely the most common method of tracking the condition of a secondary seal. Again, the volume of separation gas can dilute the increased leakage from the secondary seal, so even with a seal failure the LEL may not be reached. It is also common to have erroneous alarms due to oil vapors migrating through the separation seal. Monitoring face temperature via an resistance temperature detector (RTD) on the secondary stationary seal ring is another viable method. However, it can be a difficult task to ascertain the fine wires from an RTD are not damaged while installing the seal.

The most common method used today for monitoring secondary seal failure is measuring the back pressure in the primary seal vent. An increase in back pressure of the primary vent without a corresponding increase in either primary or secondary seal gas flow is an indication of the secondary seal having failed. Monitoring the primary vent back pressure is extremely important for low-pressure applications. If the primary vent pressure becomes higher than the sealing pressure, a reverse pressure is acting on the primary seal and the seal will fail. As previously indicated, for low-pressure applications, one would consider double seal as the preferred choice. This seal arrangement will be more reliable and will address the secondary seal monitoring concerns.

9.1.3.4 Separation seal supply system and monitoring

There are many types of separation seals and separation seal support systems. The design of each seal manufacturer varies from its competitors and each will likely use different control systems to support their DGS. Some utilize a pressure control system, while others use a flow control system. Working with the compressor and seal manufacturers will ensure that the appropriate system is selected and an appropriate monitoring strategy is implemented for the separation seals.

One of the most important considerations for the separation seal is ensuring that separation seal gas is being supplied to these seals before starting the lube oil pump. This is usually done by incorporating a permissive feature or interlock between the start switch of the lube oil pump and the pressure transmitter of the separation seal gas supply. The aim is to prevent oil contamination of the DGS. New technology is now available; we will describe it later and examine how it eliminates the need for interlocks.

As in the case of separation seals, the DGS support system can have many variants. This is generally attributable to varying applications and the experiences of different stake holders; these include DGS vendor/manufacturers, panel suppliers, users, and compressor manufacturers. API 692 has tried to assemble best practices for these systems by obtaining input from all legitimate sources and participants. However, because the individual stake holders tend to firmly defend and adhere to their normal practices, obtaining commonality of approach has proved elusive. Accordingly, the standard currently defines what is being used in industry, which may not be best practice. Also, what is defined in the present standard applies only to perhaps 80% of typical process compression services.

Covering every process application would require an even more extensive standard. The other issue with these systems is cost. Ascertaining and then following best practice usually costs more than bare minimum. Therefore, the end-user must assess what is truly needed for the specific application to ensure reliable seal operation and monitoring, but be cost-effective at the same time.

Unfortunately, a project sometimes purports a *fit-for-purpose* philosophy and ends up not looking at life-cycle cost. This occasionally leads to methodologies unable to produce the most reliable seal system. Selecting the best seal and support system should be entrusted to the end-user's rotating equipment engineer. This engineer should work only with a seal manufacturer who has thorough knowledge of their DGS and seal support systems. One should closely interrogate every step in the selection and manufacture of DGS systems. The seal manufacturer must have turnkey responsibility and should *never ever* be allowed to take the position that controls or related panels were the responsibility of a subvendor.

Contracts for entire compression systems must make accountabilities absolutely clear. These should be spelled out for the purpose of establishing actionable pursuits that can have grave consequences in isolated situations. We know of instances where having to pay for engineering errors and wildly optimistic promises forced the purveyors of nothing but fast-talk and promises into bankruptcy. Only this type of contract will ensure that the best DGS technology is selected and implemented for the specific service. A contract that obligates the vendor/manufacturer to pay for parts and, occasionally, labor, makes the owner–purchaser the test bed for design extrapolations and lower cost components. Parts and labor reimbursements are generally far less than 1% of what plant unavailability will truly cost the owner.

9.1.4 Dry gas seal innovations

As indicated at the beginning of this chapter, users continue to push DGS limits to extremes; this then is very often the driver for technology refinement and/or improvement. Along with pushing limits, improvements were required in process applications where DGSs were not as reliable as their more traditional predecessor seals. Through failure analysis and better understanding of DGSs, improvements have been made to both seals and support system. In turn, these led to increased DGS reliability and to pushing their limits even further. Through these improvements, systems have become more complicated.

System complexity, on the one hand, is not ideal for operators and is more costly to produce. On the other hand, properly implemented complex systems have resulted in enhanced overall reliability. When comparing the total cost of lost production, seal refurbishment, or replacement, against the added expense for a more elaborate but safer and more reliable DGS system, the incremental cost of superior DGS is truly negligible. With that in mind, this section provides information on technology improvements for:
- Increased sealing pressure,
- Expanding temperature limits,
- Reducing seal leakage,
- Slow speed (turning and ratcheting) situations,
- The ideal separation seals, and
- Increasing seal reliability while handling dual phase gases.

9.1.4.1 Increased sealing pressure capabilities

In the early 2000s, typical DGS applications ranged from low pressures to ~1,500 psi (100 bar). Currently, DGS products are being developed for ~9,500 psi (650 bar) services. This requires detailed engineering for all seal components and the interaction between components. One of the bigger challenges to overcome is deformation of the seal rings when these are exposed to higher pressures. Material selection, precise seal ring design, and optimized geometries are imperative to keep deformation in check. Seals with deformed faces will not achieve the lowest possible seal leakage rate. Deformed seal faces risk contacting the opposing face. Face contact implies that no gas film separates the faces. Operating a DGS without such a film will not be feasible for very long; it should not be allowed as a matter of principle. Adherence to this principle is absolutely necessary and we should not be swayed by misleading advertisements to the contrary.

While it is feasible at slow roll speeds and with special coatings for short times and with potentially high wear, disaster is imminent if we allow operation at high speeds without a gas film. Not to be overlooked are secondary sealing elements; these are important for enabling operation at higher pressures. With extreme heat possibly being generated at seal faces without gas film, certain Viton elastomers will have been cooked to a crisp.

DGSs use O-rings as secondary sealing elements to contain the pressurized process gas within the seal cavities of compressors with rotating shafts. At pressures above ~800 psi, gas tends to seep into the natural voids present in most elastomeric O-rings. When the compressor is depressurized, the gas trapped in the O-ring expands and damages the O-ring. Depressurization occurs at rates greater than 300 psi/min. The depressurization rate and pressure at which damage occurs in O-rings depend on O-ring material, grade and hardness of the elastomer, type of process gas, and its maximum temperature. Therefore, disclosure or general assistance from the O-ring or seal supplier may be required to identify the true limits of an O-ring.

Rapid expansion of tiny gas bubbles in an elastomer is known as explosive decompression. Such decompression prevents the O-ring from later sealing effectively when the compressor is repressurized. To alleviate the just concerns associated with explosive decompression, polytetrafluoroethylene (PTFE) secondary sealing elements became the component of choice for higher pressure applications. PTFE seals have no voids that might allow gas penetration; the explosive decompression concern is thus eliminated. The downside to PTFE seals is that they have a fine edge where the sealing contact is made, and this edge can be easily damaged during installation. It simply means that above-average care is needed when installing DGSs containing PTFE secondary sealing elements.

Using carbon graphite for the stationary rings can pose difficulties when handling higher pressure applications. Carbon graphite deforms considerably under higher pressures; silicon carbide materials have been qualified and proven to replace carbon graphite for these applications. For operation at higher pressures, tapered faces and special features can be implemented. These will maintain a parallel gap between the sealing rings faces; the special features are usually incorporated in the design of the rings. The owner's reliability professional should insist on disclosure of these special features. A purchaser is entitled to know what he/she is buying.

Another consideration for high and ultrahigh-pressure sealing is the design of the secondary dynamic seal configuration. The functional gap of a tandem DGS shown in Figure 9.5 is the radial gap between the balancing sleeve and the support

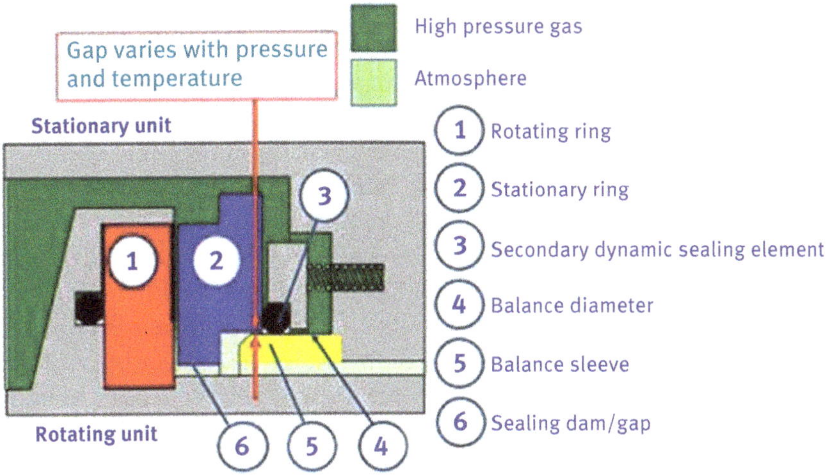

Figure 9.5: Dry gas seal showing functional gap (source: *Eagle-Burgmann*).

ring of the dynamic secondary seal. This is designed with only a few hundredths of a millimeter to prevent the sealing element material from being extruded. If the secondary sealing element is extruded, this will change the seal balance diameter and affect the balancing forces of the seal. The problem to be addressed and overcome, therefore, is to allow axial movement of the mating ring and dynamic seal ring under all operating conditions without the secondary sealing element being extruded. This is a difficult design and manufacturing step because the operational gap width will probably be influenced by both temperature and pressure. Obviously, this influence must be minimized. Tungsten carbide (WC) material seems to be the best choice. It must offset the combined influence of high pressure and high temperature while maintaining a functional gap. Finite element calculations are usually carried out by the seal manufacturer before finalizing the design.

9.1.4.2 Expanding temperature capabilities

To increase the temperature range of DGSs, two main factors must be accomplished.

1. **Selecting the secondary sealing element.** Viton O-rings with 90 durometer hardness are typical for DGS applications but have limited temperature capabilities; these range from −40 °F/°C up to 380 °F/193 °C; the allowables vary, depending on the Viton grade. DGS applications have been documented as temperatures ranging from a low of −94 °F/−70 °C to a high of 608 °F/320 °C.

   When seal temperatures go from ambient to extremely hot or cold, the biggest design concern is associated with the dynamic secondary sealing element. This sealing element rides on the balance diameter ring (sleeve), which is typically made from the same material as the seal housing. One side of the sealing element is sealing against the backside of the stationary ring, which is a carbide material. The two different materials (balance diameter ring and the stationary mating ring) can pose a design problem, inasmuch as maintaining the correct film gap (face gap) is necessary, regardless of temperature-induced component growth.

2. The stationary ring inside diameter has a minimum clearance between it and the balance sleeve to prevent extrusion of the O-ring. However, for hot temperatures, this gap is increased. With the difference in coefficients of linear expansion for both materials, one of these two material will expand more than the other when exposed to elevated temperatures. Therefore, this needs to be factored into the design. If the clearance is too tight, it will prevent the seal ring from moving axially. If the clearance is high, the O-ring or sealing element might experience extrusion under pressure. The reverse occurs for low temperatures; the gap between the ring and balance sleeve will open with the lower temperature.
3. **Managing the functional gap**: To effectively manage the gap between the seal ring and balance diameter the balance sleeve and seal ring should be made of similar materials. The best approach is to manufacture the balance diameter sleeve from a tungsten carbide material. This material has an almost identical thermal expansion coefficient as that of a silicon carbide seal ring. Having an additional ring supporting the dynamic sealing element made from tungsten carbide will allow maintaining this gap regardless of rates of change and temperatures involved. This design has been used for sealing services ranging from −93 °F/−70 °C to +482 °F/+250 °C.

9.1.4.3 Reducing seal leakage

As indicated previously, grooves are etched into the face of the rotating seal ring (Figure 9.6). This grooving assists in generating the seal gap and providing seal gap stability (stiffness). Every seal manufacturer has their own groove design and claims of why theirs is better than the others. Some designs probably originated with efforts to prevent violating another party's patent. Most, if not all, designs are based on initial testing of different groove shapes, followed by practical operation to verify a chosen groove geometry. An infinite variety of these groove shapes is possible if the designer varies the groove depth from one end to the other. In general, the cost of machining grooves is not a problem. Spark erosion and chemical milling have been used since the 1950s.

Groove designs can differ. There exist unidirectional groove designs, meaning the seal is designed to rotate only in one direction; these would require different seal assemblies for each end of a compressor. There are also bidirectional grooves, which can rotate in either direction; they allow the same seal to be used on either end of the compressor. The industry has always identified unidirectional seals as having the

Figure 9.6: Photo showing groves in rotating ring (source: *Eagle-Burgmann*).

better film stiffness. The downside of a unidirectional seal is its inability to safely cope with reverse rotation. Since many compressors are subject to occasional reverse rotation, unidirectionality versus bidirectional capability can be made a matter of hot debate. Whether or not obtaining somewhat greater film stiffness is sufficient to warrant the use of unidirectional seals is, at best, anybody's guess.

Seal manufacturers have developed their own proprietary software to determine the effect of groove geometry on seal gap. They have generally concluded that unidirectional seals do have better film stiffness than their bidirectional counterparts. They thus argue that unidirectional groove geometries should be preferred. However, while this comparison still holds true among two-dimensional (2D) grove geometry seals, certain three-dimensional (3D) geometry grove seal designs often show a different result.

Tests have shown that 3D bidirectional groove geometry seals are able to produce a greater film stiffness than the typical 2D design. The risk of faces making contact is thus reduced and seal leakage is less than what is typically experienced by 2D seal. Moreover, 2D groove seals are designed with an edge all the way around the groove. This produces a region where small debris can become trapped inside the groove. As debris collects over time, the groove slowly becomes less effective in generating the needed seal gap. The advantageous 3D groove face design is considered self-cleaning by having the bottom of the groove slope up toward the seal face surface. Since this design does not allow debris to collect, the seal gap is thought to generate the desired face distance throughout the life of the seal. The vendor–manufacturers of 3D bidirectional DGS products believe they will provide the best film stiffness and deliver the most robust seal design.

One vendor–manufacturer offers seal faces with diamond coatings to better survive operating conditions detrimental to seal reliability. Taking their cues from pumping services where occasional dry running had been experienced, the DGS manufacturer developed diamond-coated hard faces for compressor seal applications. Due to the hardness of diamond, the seal life was greatly extended, which appeared to prove its value for many sealing applications. The high hardness, low friction coefficient, and excellent heat dissipation all proved to be of great benefit for seals. With this experience and generally impressive background, diamond face seals were then applied to DGSs.

The advantage claimed for diamond faced seals is the ability to reduce the seal gap even further. The 3–4 μm typically seen for the gap in traditional 2D designs seals was now reduced to 1 μm. With a thin diamond layer on the seal face, there might be fewer concerns with faces making short-term contact. The marketing advantage is largely due to the implied benefits one associates with diamonds. That diamonds are a girl's best friend is widely known. That diamonds are brittle and can be cut with a diamond dust-infused blade or rotor is not usually recalled by the noninquisitive.

However, hardness, low friction coefficient, and good heat dissipation come to mind when we get into technical details. An objective observer may also ask if high cost is counterbalanced by someone else's well-engineered DGS design where thoughtfully engineered DGS support systems ascertain that only clean, dry gas reaches the seal face gap, and where seal leakage disappears into the compressed gas stream. Thorough comparisons and relevant experience checks must always trump advertising and marketing bulletins.

Combining the ideal 3D groove geometry and reduced seal gap made possible by diamond face seals cannot be faulted. The leakage for DGSs could effectively be reduced to a fraction of the leakage typically seen in the most common seal designs. Still, if leakage gas is returned to the compressor's suction, the entire issue becomes moot. Understanding the different product designs, and pondering over the fullest possible meaning of the saying "trust, but verify" should be at the very top of the reliability professionals' DGS strategies.

9.1.4.4 Slow roll turning and ratcheting

Slow roll turning and ratcheting refer to conditions when the compressor is rotated at speeds below 500 rpm. Many design factors are involved as we contemplate this mode of preoperation, such as:

- Seal size
- Effect on the process gas
- Speed than must be reached to produce an effective seal gap

End users should work closely with the seal manufacturer to prevent issues associated with these operating conditions. And indeed, in slurry pump applications, the diamond face seal was able to defer wear-out under highly abrasive and/or dry running conditions. Dry running occurs on DGSs when a compressor is placed on turning gear or on a ratcheting cycle. Under these slow turn conditions, there is insufficient speed to produce an effective seal gap when no pressure is present. Moreover, even with low seal cavity pressure, the secondary seal is the seal at most risk. This is explained by the fact that this seal has little or no gas fed to it from the primary seal. Accordingly, it is good practice to supply purge gas (usually nitrogen) during slow roll. Many seal failures have occurred whenever proper preoperating procedures were disregarded.

Using diamond-like carbon coated silicon carbide on both seal faces (rotating and stationary ring) will minimize wear during slow roll. Minimizing or, better yet, avoiding any wear during contact of the seal faces will extend the seals life. A seal with a carbon graphite face would also tolerate wear, but the carbon face would produce carbon dust. This dust would fill the grooves if they are not self-cleaning. Dust-laden grooves will eventually prevent the seal rings from generating an effective seal gap.

Unfortunately, for large steam turbine-driven compressor trains the time period of exposure to slow roll can be very long, depending on how long it takes to return a steam turbine rotor from its natural or "set" or bow condition to almost perfect straightness. The problem is compounded if dry nitrogen is used as the seal gas during slow roll for tandem seals with intermediate labyrinth and carbon rings. Dry nitrogen is not a good lubricant for carbon, and so the carbon ring wear rate may increase exponentially with long-term dry nitrogen usage.

Again, to address concerns during rotor slow roll, diamond coating (Figure 9.7), similar to the coatings used in slurry pump seals, has been successfully applied to DGSs.

In summary, using diamond-coated seal faces for DGS applications has benefits which include:
- Reasonable wear at lower turning speeds
- Reduced seal leakage due no narrower face gap
- Tolerating small amounts of liquid

9.1.4.5 Separation seals

The separation seal is an important part of a DGS cartridge. Liquids are detrimental to DGSs. Since lubricating oil exists adjacent to the DGS, it is important to prevent the lubricant from entering the DGS. Similarly, if process gas passes through the DGS cavity to the bearing housing, it can create unsafe conditions. Occasionally, this happens when seals do not have an intermediate labyrinth and a secondary seal gas supply. The purpose of a separation seal is to prevent process gas from migrating into the bearing cavity and to prevent lube oil from reaching the DGS.

Split labyrinth seals were originally used as separation seals. Separation gas would be injected between two sets of labyrinth teeth, with the separation gas split into two streams. One stream would flush process gas out through a vent, the other would prevent migration of oil from the bearing cavity into the DGS. Split labyrinths

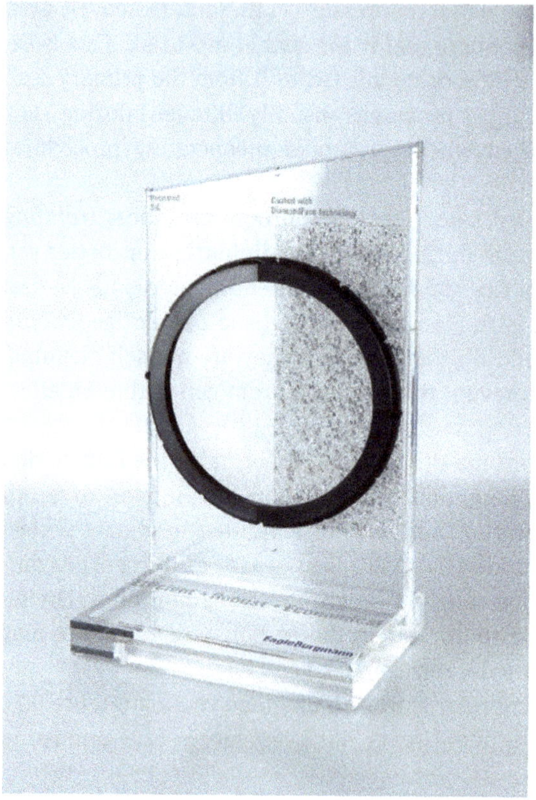

Figure 9.7: Diamond-coated seal ring (source: *Eagle-Burgmann*).

were not ideal. DGS were often located close to bearings and flooding the bearing cavity tended to contaminate the DGS. Also, when catastrophic seal failures occurred and both the primary and secondary seals gave way, high volumes of process gas entered into the bearing cavity. At times, the lube oil tank ballooned, and fires and even explosions were not unheard of. Carbon ring seals were later applied to produce more restrictions and reduce consumption of the separation gas, compared to labyrinth seals. Unfortunately, carbon ring seals were still not the ideal solution as seal failures still occurred from lube oil contamination. Limited seal life due to carbon ring wear was experienced by many users.

Safety issues, limited seal life, oil migration and high separation gas consumption caused by wear were the most common issues with carbon ring separation seals. To resolve these issues, the coaxial barrier seal (CobaSeal) shown in Figure 9.8 was developed by a prominent manufacturer utilizing DGS technology. This seal incorporates an axial end face with gas injected into the center of the stationary seal face. At standstill, the seal is completely oil tight, even without a supply of separation gas.

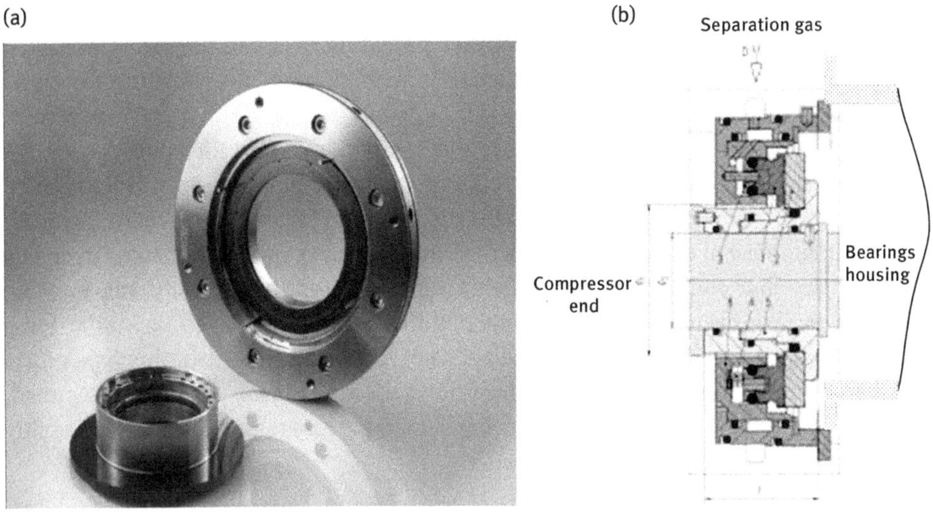

Figure 9.8: Coaxial barrier seal (source: *Eagle-Burgmann*).

Other features of the CobaSeal include:
- Toleration of flooded bearing housing
- Bidirectional
- Self-cleaning 3D gas grooves
- Extremely low leakage
- Slow roll capability
- Insensitivity to radial vibration
- Operating pressures up to 2.7 bar (39.2 psi)
- Seal faces are silicon carbide against ductile stainless steel

Reportedly, this seal worked so well that it is now used for low-pressure applications.

9.1.4.6 Increasing seal reliability

One of the key factors for the seal supply system is to reliably supply an adequate amount of clean gas to the seals. While running, compressor process gas supply to the seal systems generally comes from a higher-pressure source inside the compressor. The gas is filtered, ensuring effective protection against contamination that might be entrained in the process gas. However, during slow roll or transient operation of the compressor, there could be an interruption of high-pressure flow to the seal system. Lack of flow to the seal system can result in damage to the seals. As a consequence, industry had to find solutions to the problem.

A seal gas booster unit was developed to address the issue; it proved effective and thus enhanced seal reliability. However, the concern with the original seal booster designs was their own inherent reliability. The original seal gas boosters were air-driven reciprocating piston style miniature compressors that incorporated many moving parts. Although they were reciprocating designs, they were expected to somehow meet the seal gas flow demands with smooth nonpulsating flows. At their limits of design flow, these reciprocating boosters had to run at or above limits to deliver seal gas flow. The result was low reliability and short life span.

In response to this experience, a seal manufacturer developed a vertical sealed (canned) motor-cum-regenerative impeller compressor (Figure 9.9) to meet the need. Its centrifugal design eliminates flow pulsations. The basic model is shown; impellers differ in size and configuration to accommodate different gas density and/or flowrates. Of course, there are other approaches that produce seal gas flow at varying pressures and gas densities to meet the needs of DGS products. Also, this type of gas booster is more expensive than the traditional booster types. It can be assumed, though, that the effectiveness of producing satisfactory gas flow and asset reliability easily justify the cost.

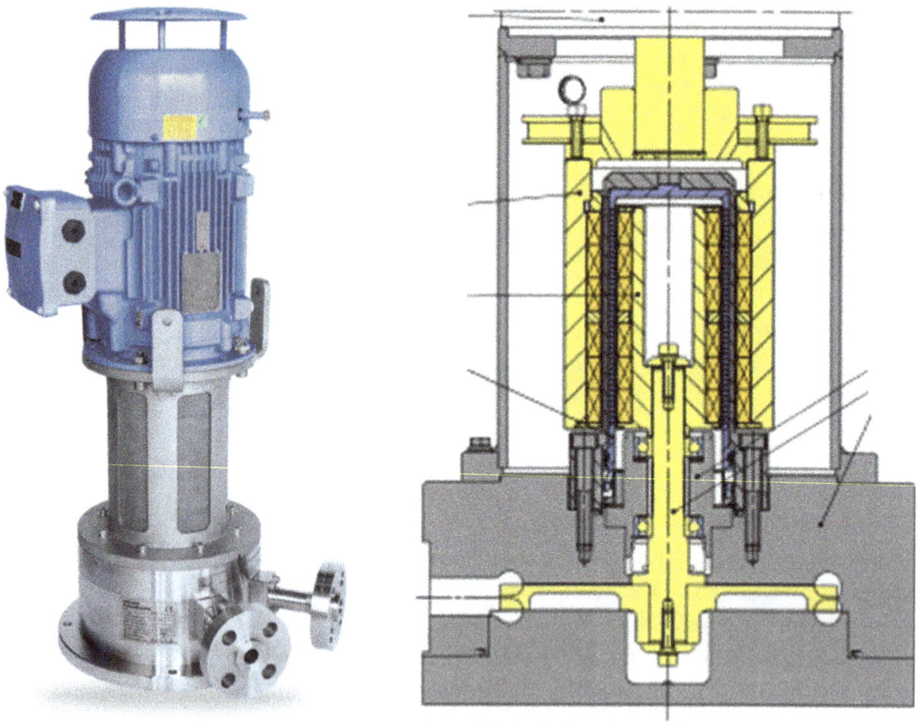

Figure 9.9: Vertical canned motor regenerative impeller gas booster compressor (source: *Eagle-Burgmann*).

Nevertheless, and on the subject of reliability, there is an important admonition that must be heeded beyond DGS technology:

> A rigorous engineering assessment of the true cost of DGS products versus true cost of traditional oil seals and their oil reservoirs, rundown tank, and fully spared pump/filter/heat exchangers offered by experienced compressor manufacturers may be in the best interest of the owner–purchaser.

While the DGS approach is widely favored for new compressors, we can be certain that it will not be best for every conceivable compression service. Likewise, abandoning an existing, relatively trouble-free conventional lube oil console with its fully spared pumps, filters, and heat exchangers during a compressor upgrading project may also not be a smart idea. However, it is never a mistake to engage in a rigorous and entirely fact-based comparison of the two existing and widely used sealing solutions: Conventional "wet" seals and DGSs.

9.1.4.7 Dual-phase gas

Another concern regarding DGS failures centered on wet seal gas or dual-phase gases. Typical process gases consist of many different components. When the gas temperature and/or pressure changes, some of these components change into the liquid phase. Seal gas typically branches off from the compressor discharge line. Its pressure is reduced to the sealing pressure – typically suction pressure. The seal gas pressure then decreases to atmospheric pressure as it moves across the seal faces. These changes in pressure and temperature can cause component fractions in the gas to change into liquid, which is detrimental to DGSs.

Liquid trapped between the seal faces will cause failure during operation or when starting a compressor. During operation, liquids have a much higher shearing force than gases. This shearing force generates heat, which causes seal face deformation, resulting in face contact. The contact of the seal faces generates more heat, which is usually enough to cause a seal ring failure and certainly endangers elastomers. Due to the flatness of the seal face, liquid present between the seal faces, when starting a compressor, requires high torques to break the bond. Indeed, occasionally, one of the seal rings will break. These are some of the reasons why DGSs cannot tolerate liquids.

To prevent failures caused by these conditions requires seal gas conditioning, such as:
- Removing liquids, if present
- Heating the seal gas, thereby removing the conditions that promote liquification of some of the gas constituents

Seal gas conditioning should consist of:
- Coolers
- Liquid knockout vessels
- Heater and heat traced supply lines

Coolers (heat exchangers) are typically only used when an analysis of the seal gas suggests that a lower temperature is required. Most conditioning systems consist of a liquid knockout drum, heater, and heat tracing.

Conditioning systems become quite expensive, especially when duplicating equipment to guarantee uninterrupted seal gas flow during seal operation, along with additional utility costs for operating such a system. Duplication also makes the system more complicated, which is not what end users desire.

To overcome the failure risk of liquid ingestion, one seal manufacturer set his/her sights on developing a DGS capable of handling dual-phase gases. The critical factors considered in the design were:

- High torque requirement to accomplish liquid shear
- Dissipating the heat generated by the shearing force

Again, diamond coating became one of the key elements in developing these seals. Diamond can dissipate considerable heat and has a very low friction coefficient; should the faces make contact, only a moderate amount of heat would be generated. Designing in 3D grooves was another key element to accommodate moderate amounts of liquid contamination. This type of 3D groove was found to substantially reduce the torsional force needed to break the bond between rotating and stationary rings in the presence of liquids.

A somewhat counter-intuitive feature was incorporated to accommodate small amounts of liquid; it consisted of soft torque transmission drive pins and antirotation pins used on the seal rings. Soft torque transmission prevents seal ring breakage, which often occurs with solid torque transmission and antirotation pins. Combining all these features produced a seal capable of accepting a liquid in dual or gas phase sealing media. This allowed users to dispense with extensive seal gas conditioning systems. Unless extrapolated beyond good practice, the approach described here tends to result in simpler systems, reduced maintenance, and lower energy cost.

9.1.4.8 Material selection

The final section of this chapter represents an overview of material selection and brief highlights of advancements in materials of construction. Material selection for compressor DGS products tends to be simpler than having to deal with pump seals; compression services are far fewer. The seal housing and other metal components are typically made from ANSI 316 and 410 stainless steel or equivalent materials. For harsher environments, these are upgraded to duplex stainless steel, super duplex, Inconel, and even Hastelloy. High concentrations of H2S, wet H2S or acids are typical where an upgraded material is required. In general, close to 80% of dry gas compressor seals are constructed with 316 and 410 stainless steel series metals.

Years ago, materials for secondary DGS sealing elements started out as Viton 70 Durometer O-rings, but most were later upgraded to Viton 90 Durometer hardness as a customary standard. DGS manufacturers use proprietary grades of Viton to accommodate various seal designs and operating conditions. Therefore, to ensure that sealing elements are replaced with the proper material, the end-user should go through the seal manufacturer. Doing so advances the hope that the seal will operate without problems. A standard 90 Durometer O-ring from an O-ring supplier may not meet all the requirements for the application, although common sense tells us that not everything called a requirement is always needed for success. The opposite holds true as well: Even a reputable manufacturer is sometimes shocked by the reality of customers seeing fit to upgrade and/or improve parts designed and fabricated by said reputable manufacturer.

As there is a pressure limitation for O-rings, seal manufacturers use various grades of PTFE, some with impregnated materials for higher pressure applications. Chemrez, Kalrez, or other O-ring materials are used when dealing with special applications. It is reasonable to work with one's seal vendor for defining the best secondary sealing elements based on their experience and engineering practices. However, it is equally reasonable for a highly motivated reliability professional to ask his/her supplier why a worthy competitor has, for some time, gained fame for DGS products that are innovative and successful in spite of using "Material X" versus our supplier holding on to traditions that are – mercifully – labeled as just that: Traditions.

There are several materials that are used for seal rings; these include metal, tungsten carbide, silicon carbide and various grades of carbon. Associated with each material are advantages and disadvantages. Seal manufacturers use what has worked for them, although it may not necessarily be the most suitable material combination for some special services or applications. Take, for instance, tungsten carbide (WC) and carbon, which DGS manufacturers usually identify as a hard-soft combination. This is the original material combination used for DGSs. This combination has worked well in many applications and can be more economical to manufacture than hard–hard material combinations. Accordingly, and given the successful operation of the seals for many years in a variety of applications, users are willing to specify or accept the use of WC.

However, many experienced users will not accept this combination. They base their stance on having seen consequential damage when WC failed. They have had access to documentation and litigation following exorbitant downtimes and/or repair frequencies. Over the years, reliability professionals have studied these matters and have realized that the damage often reached way beyond the seal itself. Understanding the pros and cons of each material is critical when making a material choice. Making informed choices again invokes the axiom: "Trust, but verify." Yes, it takes knowledge to make informed choices. Translation: The most popular DGS may not be the best choice for the reliability focused.

Below is a table of pros and cons as a general reference guide. There is also an informative annex in API 692 that can provide additional guidance. Keep these in mind when working with your seal manufacturer for defining the best material combination for your application.

Table 9.1: Pros and cons for selecting stationary face materials.

Material	Pros	Cons
Carbon graphite	Tolerates some contact	Carbon dust
	Seal manufacturer's experience	Low thermal dissipation coefficient
		Excessive wear with dry nitrogen
		Produces higher seal leakage
		Increased face deformation with temperature, heat, pressure, and size
Silicon carbide with DLC	Resists deformation with temperature, heat, pressure, and size	Wear during extended contact periods
	Produce lower and consistent leakage	
	Good thermal dissipation coefficient	
Silicon carbide with diamond	Excellent thermal dissipation coefficient	
	Low friction coefficient	
	Reduced leakage No damage from abrasives	
Silicon nitride	Increased tensile strength to manage high speed applications	Cannot be coated

9.1.4.9 Stationary face

Table 9.2: Pros and cons for selecting rotating face materials.

Material	Pros	Cons
Metal/ductile	No brittle failure mode	Limited pressure and temperature capabilities
Tungsten carbide	Tensile strength	Heat checking
	Seal manufacturers experience	High hoop stress
		Damage caused by fractured material
Silicon carbide with DLC	Preventing damage to seal faces under limited contact conditions	
Silicon carbide with diamond	Not damaged from debris	
	Extends length of time seal faces can operate in contact	
	Effectively dissipates heat	

9.1.4.10 Rotating face

A final comment: Principles governing material selection rarely – if ever – change. The basic operating principle of DGSs remains the same as the original concept many years ago. Over time and as with any mechanical device, there are many improvements to the hardware. Improvements make components and subassemblies into items or systems that make them more effective and efficient for the intended task or function. Items such as sealing element materials, face materials, torque transmission features, antirotation mechanisms, and coatings are among the many that were improved over the decades.

These developments allow the seals to operate under more severe conditions than the original design, providing capabilities for high pressure, a second or two added to the time allowed for actual face contact, extended temperature range, lower leakage rates, and a wider range of available sealing media. Just because a compressor manufacturer prefers working with seal manufacturer "X" does not mean that "X" makes the best DGS for your particular application. Understanding the potential consequences of material selection, having a firm grasp of one's application requirements, and knowing best available technology are of immense value. Having that knowledge will ensure access to the best sealing technology. That knowledge-based technology is what provides the most reliable solution and lowest cost of ownership. Reach out to that knowledge.

Chapter 10
Equipment purchasing

10.1 Consider single-point responsibility for major fluid machines

Equipment purchasers are always interested in meeting the cost and schedule targets for projects. There could, however, be conflicting goals when purchasing compressor drivers and/or auxiliaries from vendors other than the compressor manufacturer. While cost targets might initially be met, missed schedules and future warranty disputes present risks that must be considered. It is worth noting that how experienced owner–purchasers avoid these conflicts:

Agreement on "train responsibility"

Whenever best-of-class companies separate the procurement of a major fluid machine from the procurement of its driver, one of the two suppliers is given train responsibility. The owner–purchaser then pays the responsible vendor – usually the compressor manufacturer – a modest but equitable sum for accepting single-point or equipment train responsibility.

When this rule was disregarded by an overzealous project person at a company (usually) priding itself as a sophisticated best-in-class performer, a lot of unproductive finger pointing resulted. A little later, the issue received renewed attention during a performance appraisal. The project person's employer–manager considered it a career-limiting oversight to not assign single-point responsibility for compressor trains.

Exceptions to the rule

Of course, giving a vendor single-point responsibility does not mean that all risk has been removed. As a case in point, reciprocating compressors had been purchased from a manufacturer who had just acquired another company's product line. Unfortunately, the owner–purchasers quickly encountered glitches in spare parts identification and timely availability of spares. In a second instance, a manufacturer of reciprocating compressor frames and main running gear tried to adapt another manufacturer's pistons and cylinders. Part of the assembled machine thus originated as subassemblies for high-speed compressors; the rest of the machine consisted of API 618-compliant slow-speed subassemblies. The "marriage" had a real rough time early on. Initially, there was much consternation as to who was at fault when the compressors suffered from insufficient capacity and different components began to fail.

https://doi.org/10.1515/9783110678765-010

Soon after these hybrid machines were installed on a floating production/storage/offloading unit (commonly called FPSO) things really went wrong. The owner–purchaser initiated litigation against as many parties as possible and years were wasted in tedious and contentious arguments. In fairness, the problems were ultimately solved; so, keep these machines on your approved bid invitation list for future consideration.

Regrettably, both of the above experiences and many other case histories remain hidden from our view. The parties sign protective nondisclosure agreements because neither side wants their reputation tarnished. All involved are reluctant to publish detailed articles or papers about these events, and lessons learned turn into lessons forgotten.

Focusing on the bright side

In contrast, much less risk exists when a purchaser opts for entire compressor systems from manufacturers with a proven record of long-term quality and as-promised performance. A case in point is modern process compressor packages (Figure 10.1) that are built in accordance with applicable portions of the different API standards covering integrally geared centrifugal to meet the demands of the petroleum, chemical, power, and gas industries (API-617). To date, integrally geared compressors have been built for services including gas turbine fuel gas (methane), hydrocarbon refrigeration gases, ammonia, carbon monoxide, carbon dioxide, syngas, and gas field gathering.

Figure 10.1: Cutaway view of a six-stage process gas compressor (source: *Cameron, Buffalo, NY*).

Conventional wisdom used to suggest these are somewhat smallish machines. Today, however, we see flow rates of 400,000 m^3/h, mole weights from 6.5 to 58 MW, stage pressure ratios as high as 2.6:1 (Figure 10.2), inlet pressures from ambient to 45 bar, and discharge pressures to 200 bar.

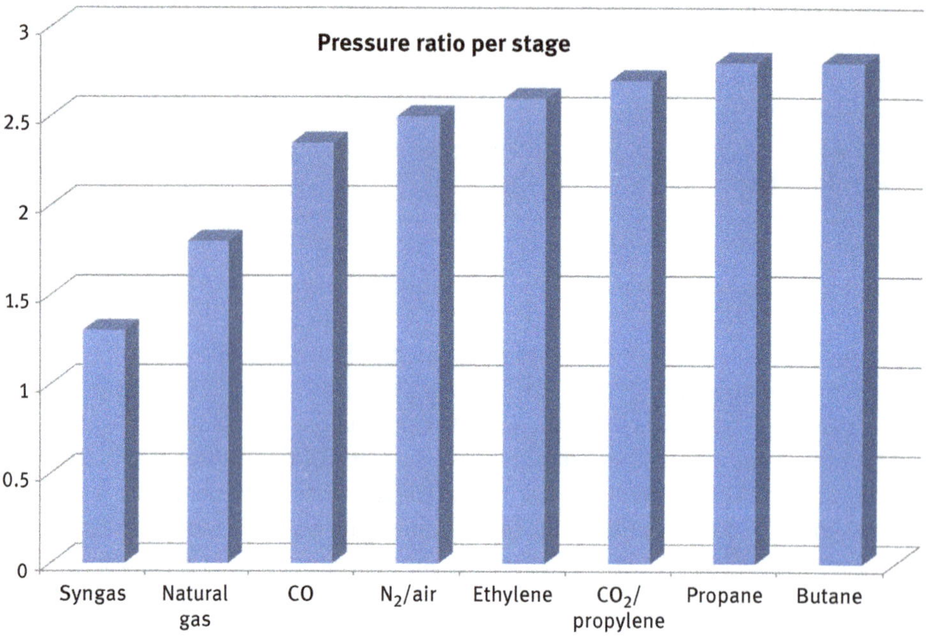

Figure 10.2: Achievable per-stage pressure ratios are mole-weight dependent (source: *Cameron, Buffalo, NY*).

One of the most experienced US manufacturers of integrally geared compressors has designed these for drivers up to 20,000 kW. This manufacturer and/or its precursor companies have produced 12,000 integrally geared machines; up to six high-speed stages are available per gearbox from this manufacturer. However, as many as 12 stages have been incorporated in an integrally geared compressor designed by a European manufacturer and drivers have, on occasion, approached 60,000 kW.

Today's efficiencies can approach 90%. Some manufacturers have designed and built their own gearboxes for decades. Many producer factories have undergone expansions over the past 50 years. Their machines are extensively shop tested and the overall designs are always fully up-to-date. Seal designs have kept pace with modern dry gas sealing technology, as have their well-proven intercoolers.

In any event, there are many opportunities to use multistage integrally geared machines where mere tradition may have steered us in the direction of positive displacement machines. The design of these integrally geared late-generation dynamic compressors overwhelmingly favors single-point responsibility. As an additional feature, modern integral gear-driven compressors often represent the best possible blend of standardization and customized design (Figure 10.3). These machines certainly merit consideration and invite comparisons with every other type of compression machinery. We are highlighting them here because they are the prototypical single-source machine. Drivers, sealing, and lubrication systems are perhaps provided by different vendors, but whenever the responsibility stayed with the compressor manufacturer and the package was "single sourced," value was amplified and missed performance became increasingly rare.

Figure 10.3: A modern integrally driven compressor (source: *Cameron, Buffalo, NY*).

Finally, multistage integrally geared 10-stage compressors for carbon sequestration, large (6,000 kW) water-filled twin-screw compressors for coal gas, and hermetically sealed integral direct motor-driven multistage centrifugal compressors for a multiplicity of services are among compressor types that have been used with great success since the late 1990s and early 2000s. Upon closer examination, we may have to adjust our thinking about packaged machines and about single-point responsibility fluid machines for process plants. Where "packaged" often meant lack of easy maintenance access and design compromises, the new meaning may well be proven performance and desirable single-point (single-source) responsibility. Keeping informed on best-in-class practices will have its own rewards.

Chapter 11
Design basics of tabletop foundations for machinery engineers

11.1 Introduction

The detailed design of large concrete foundations for centrifugal compressor trains can be very complex and is well beyond the scope of this book. Designing these foundations remains an important task for structural and geotechnical specialists. They must accurately determine the static and dynamic responses of these structures.

However, these designs can influence the overall performance of compressors and drivers, for which concrete foundations provide the underpinnings. Major compression machinery can be surprisingly sensitive to foundation designs and today's mechanical engineering specialists are asked to accept the overall responsibility for long trouble-free performance of multimillion-dollar compressor installations. Taking responsibility does not mean that they must immerse themselves deeply in the design detail. It does mean, however, that they must be able to converse with the designers. The reliability-focused machinery engineer must become the gatekeeper, seeing to it that no risky shortcuts are taken. Shortcuts that could potentially impair the performance of critical machinery must never be allowed to creep into project philosophies.

It follows that the compressor specialist assigned to a project must be able and equipped to become the said gatekeeper. The first step in enabling someone involves imparting an understanding of what is involved in "safe passage through the gate." In this instance, the essential elements of obtaining a satisfactory tabletop design need to be conveyed to all involved parties. Accordingly, the intent of this chapter is threefold:

1. Summarize our understanding of how to engage with the team that designs the foundations and/or structures for major machines.
2. Highlight some basic design parameters, enable the reader interpret the building blocks that make up a suitable analysis, and facilitate comparing the results of both static and dynamic studies with experience-based industrial practices.
3. Encourage rotating equipment engineers to ask questions whenever they are not comfortable with how foundation designers explain the results of their analysis.

This chapter gives an overview of how machinery foundations are designed. It recommends an approach for rotating equipment engineers to adopt during the design phase. Later, in Appendix D, some detailed tools and outputs are described. Even if these tools will not be used by rotating equipment engineers, the machinery engineer on a project team must be in a position to recommend these to the person doing the analysis. Tools that are ultimately chosen by the analyst should facilitate gaining a reasonable understanding of machine behavior. Getting to know the dynamic response of

https://doi.org/10.1515/9783110678765-011

the foundation structure during transients and while operating under steady-state conditions is not just important; it is mandatory.

It is intuitively evident that, as process plants are getting larger, the risk of structural crack development and even failures of large foundation increases. Early involvement during the design phase by all stakeholders should include rotating equipment engineers; the goal should be to "get it right the first time." The value of their timely guidance is very high and usually acknowledged by good managers.

For example, typical designs of these structures narrowly focus on how the foundation structure reacts within the normal operating speed range, often from minimum governor speed to maximum continuous speed when driven by turbines. However, rotating equipment engineers will want to know how these structures behave during the warmup stage of the turbine. Therefore, the rotating equipment engineer should understand and relay to others the results of the dynamic analysis by ensuring that:

a) The amplitudes associated with the vibratory modes will not affect the structural integrity of the foundation structure:

   > Check to see that the foundation limits its motion to amplitudes that will not endanger the satisfactory operation of the machine or disturb people working in the immediate vicinity [1].

b) Work with the project team to mitigate any potential vibration excitation during design; ensuring avoidance of dynamic resonance within at least 20% separation margin from the operating speed range.
c) Check to see that transmissibility to neighboring structures is avoided.

The various input requirements for the dynamic analysis should come from the equipment manufacturer; they should have been reviewed and clarified by the rotating machinery engineer before the data reach the structural engineer. However, the structural engineer leads the team and performs several iterative analyses of the structure.

11.2 Common types of compressor foundations

Large centrifugal compressor trains in petrochemical plants are considered very critical equipment (VCE), and therefore vital in maintaining stable and cost-effective operation. Large compressor trains are therefore installed on well-designed foundation structures, structures that can withstand perturbations not only from the machines themselves, but also from associated piping and occasional process upsets. VCE machinery foundations must be designed with these factors in mind:

- Resist warpage from soil settlement
- Remain unaffected by thermal stresses

- Safely support static loading
- Tolerate anticipated vibration, vibration excursions, and dynamic loading

To achieve every one of these factors in the foundation design, many different types of reinforced concrete (RC) structures offer themselves to the designers. Available elements include block type, frame type, or frame type with vibration isolators. The elevated frame type is commonly called a tabletop and is usually the foundation of choice for large rotating equipment in petrochemical plants.

The block-type foundations in Figures 11.1 and 11.2 are among the simplest turbomachinery foundation constructed with RC. These foundations are generally used for pumps, reciprocating compressors, and small centrifugal compressor trains with upcomer piping arrangement (*process piping installed on the upper half of compressor casings*). The mat thickness generally ranges from approximately 1.2 to −1.7 m (3.9–5.6 ft). The focus of this chapter and its relevant Appendix D is on elevated foundation structures for large centrifugal compressor trains.

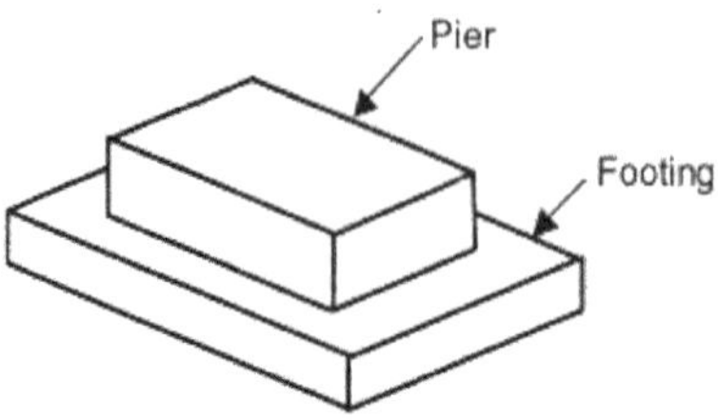

Figure 11.1: Block-type foundation with base footing.

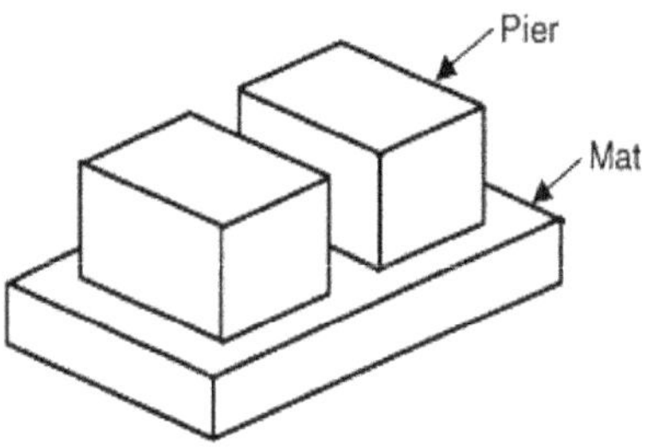

Figure 11.2: Block-type foundation with base mat.

Compressor “trains” sometimes consist of one, two, or even three compressor bodies coupled in tandem, and driven by a steam turbine or gas turbine. Bodies with downcomer pipe arrangement or upconnected (*process piping installed on the lower half of compressor casings*) are mounted on elevated structures while radially split (“barrel”) compressors are commonly mounted at grade levels.

11.3 Predimensioning of foundation tabletop

Predimensioning is the initial process used by the responsible civil or structural engineer to size the components of the structure. Sizing is done by a trial based on the equipment manufacturer's outline drawings, local codes, and practices, as shown later.

A tabletop structure consists of a footing or pile cap, top deck, beams, walls, and/or columns, as shown in Figure 11.3.

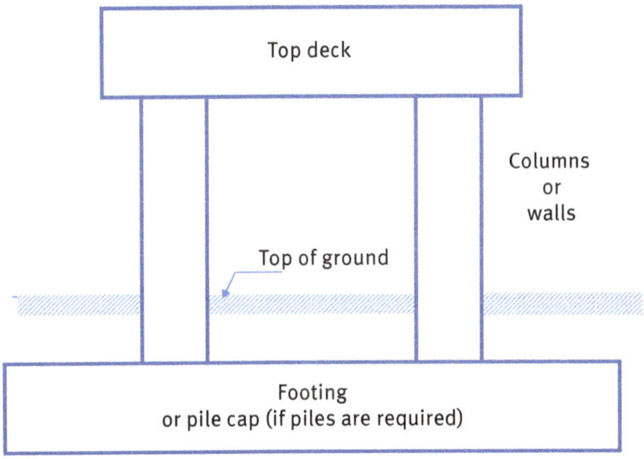

Figure 11.3: A tabletop structure.

The dimensions of the top deck are governed by the baseplate drawing; it is normally provided on the machine manufacturer's outline drawing. Sizing of the structural members is based on the industrial guidelines and practices shown below:

- **The total mass of the structure shall be:**
  - At least three times the weight of the supported equipment for compressor trains <50,000 HP (37,285 kW).
  - At least six times the weight of the supported equipment for compressor trains >50,000 HP (37,285 kW).
  - At least five times the weight of the supported equipment for reciprocating compressor trains.
- **Soil:**
  - The geotechnical report should recommend if the structure may rest on the soil, or if piles are necessary. The report should also indicate the reasons for the recommendation as follows: soil is weak, anticipated soil settlement is too large, and so on.

- The maximum static soil bearing pressure should be less than 50% of the allowable soil bearing capacity "q_{all}." By adopting this criterion, the assumption is that the stress and the strain of the soil (or pile) will have a linear and elastic relationship during vibratory excursions. A second assumption is the need to avoid fatigue failure due to the anticipated cycles of loading and unloading. Designs that minimize fatigue failure permit the soil to absorb the additional pressure from dynamic loads and will minimize structural deflections:

  $$\text{Maximum static soil bearing pressure} = \left(W_{mach} + W_{fou}\right) / \left(L_f \times B_f\right) \tag{11.1}$$

  where W_{mach} is the weight of machine assembly including baseplate, W_{fou} is the weight of foundation, L_f is the horizontal length of foundation raft/footing perpendicular to the axis of rocking, and B_f is the breadth/width of foundation.

- **Footing:**
  - The bottom of the footing should be below the grade level, and its thickness "t" shall not be less than
    $t = 0.0104\ L^{4/3}$ where L is in mm
    $t = 0.07\ L^{4/3}$ where L is in ft
    L is the average distance between two adjacent columns
  - The center of soil pressure should be within 300 mm (~1 ft) of all superimposed loads.
- **Piles and pile cap**: These are part of a design whenever the geotechnical report indicates that piles are necessary.
  - Typically, the piles are verticals; however, the edge piles can be modified if required to control lateral deflections.
  - The maximum load on the piles should be less than 50% of their allowable capacity.
  - The centroid of the pile arrangement should be within 300 mm (~1 ft) of all superimposed loads.
  - The weight bearing capacity of the pile cap should be 1.5–2.5 times the weight of centrifugal compressor trains and 2.5–4 times the weight of reciprocating machine trains. For pile distribution, the designer should consider the geotechnical recommendations, the static loads, and the dynamic deflections.
  - The pile group width is important to provide resistance against rocking vibration modes. The width should be 1–1.5 times the vertical distance from the base to the machine centerline.
- **Columns or walls, beams, and top deck**
  - Locate intermediate columns below the coupling or gear box of the machine train.

- The top deck should have enough width and length to accommodate the skid of the equipment, including the required edge distance for the anchor bolts plus piping and other equipment.
- The top deck thickness should be one-fifth of the clear span. Verify that its vertical deflection under static loads is less than 0.50 mm (0.02 in).
- The weight of the top deck plus the weight of top half-height of the columns should be greater than the weight of the equipment.
- The center of gravity of the column or wall should be coincident (±5%) with the center of gravity of the equipment plus the top deck and the top half-height of the columns. This verification is needed to spot and avoid excessive torsional vibration of the top deck.
- Analyze the structure under the assumed static equivalent loads of 50%, 30%, and 10% of the equipment weight for the vertical, transverse, and longitudinal directions, respectively. The loads are assumed to act independent of each other but will be combined with the vertical dead load designated as "DL." The DL includes the weight of the foundation, dead weight of the machine, and its auxiliaries:
  DL + 50% of the equipment load applied vertically
  DL + 30% of the equipment load applied transversely
  DL + 10% of the equipment load applied longitudinally
  Verify that the maximum horizontal deflection for these equivalent loads is less than 0.5 mm (0.02 in).
  Verify that the columns deflect equally under these load combinations.
  These equivalent loads and their deflection limits may change according to the equipment manufacturer specifications.
- The columns and beams shall have natural frequencies with at least ±20% separation margins away from the machine's operation speed range. To obtain the natural frequency of each beam and columns, the engineer may perform a quick analysis using simplified structural models.

11.4 Materials

The tabletop is constructed using RC. These structures are considered massive because of the large dimensional requirements outlined in the predimensioning guidelines.

RC comprises two important materials: reinforcing steel (rebar) and RC. The rebar follows ASTM specifications A615 or A706. The yield strength of the rebar is approximately 400 MPa (60 ksi) with a modulus of elasticity of 200 GPa (29,000 ksi). Stronger rebars are unnecessary because the design is typically governed by the minimum steel ratio required by local construction codes or engineering practices.

Avoid cracks in the concrete mass by ensuring proper steel distribution. To accomplish this, invoking the following criteria is recommended:

- During the construction process, avoid steel concentration in areas that can promote superposition of stirrups in the same layer, or hooks at the end of columns intersecting with the beam reinforcement.
- Install rebar close enough to avoid cracks due to thermal movement of the structure, but with enough spacing to permit easy concrete pouring.

The concrete quality is very important; it must ensure that the vibrations produced by dynamic loads are within the allowable range and below the expected (and calculated) results.

For weight, the concrete typically used is categorized as normal weight concrete, with a density of approximately 2,300 kg/m^3 (144 pcf).

- The cement, aggregates, and additives are selected according to the durability required for the structure. The durability of the structure is defined by the permeability of the concrete. The permeability factor of the concrete is the property that indicates how much it prevents intrusion of corrosive elements, like sulfates and chlorides of the soil, or chlorides from the air when the structure is situated close to the seashore.
- The concrete strength f'_c shall be ≥21 MPa (3,000 psi). Minimum concrete strength is typically $f'_c = 28$ MPa (4,000 psi).
- Control of deflections: Concrete mixtures with higher strength can be useful to control deflection. A larger modulus of elasticity helps to reduce deflections.
- Temperature control during pouring: Concrete releases heat during curing; this heat must be controlled to avoid premature concrete cracking. Using fly ash Class F to replace an equal amount of cement helps to reduce heating and cracks.
- Curing: Good planning is necessary for the curing process. ACI 207.1R-05 [2] provides information about the best curing methods for mass concrete.
- The amount and distribution of the reinforcing steel depends basically on the minimum requirements to avoid cracking. Frequently, the stresses from external loads do not control the RC design of these massive structures.

11.5 Structural modeling of the tabletop main objectives of modeling

- Derive the structural behavior of the foundation imposed by static and dynamic loads:
  - Amplitude of vibration should be well within the specified boundaries during steady-state and transient operations.
  - The foundation should be structurally safe in terms of stress levels.
  - The design of the foundation system should be such that the soil dampens the dynamic forces of the machine.

  - Investigate the resulting vibrating modes with high energy components and associated frequencies; these frequencies should be ≥20% outside the operating speed range.
  - Review the graphical representation of each vibratory mode.
- Perform a comparison between the results derived from the model and the machine manufacturer's requirements. Determine if refinement to the model is required:
  - The precision of the results is highly dependent on the properties of the soil, and concrete. Consider also the interaction of the machine with its attached piping and with adjacent machines. It is common practice to isolate the tabletop from adjacent structures, such as platforms, shelters, and pipe supports; realize that different pipes are connected to the machine.
  - Ascertain that the structural model output permits adjustments to pile spacing and to change dimensions of the structural elements.

References

[1] American Concrete Institute, ACI 351-3-3R-04. Foundations for Dynamic Equipments.
[2] American Concrete Institute, ACI 207.1R-05, 2012. Guide to Mass Concrete.

Chapter 12
Special-purpose couplings

12.1 Introduction

The growth and technological evolution of industrial gas turbines was accompanied by the need for high-torque and/or higher speed flexible couplings. These, of course, are needed to connect the shafts of driving and driven fluid machines. In short, the design of flexible couplings for large gas turbines and other rotating equipment had to keep pace with enhancements made to the machines. Improved material selection resulted in greater power density; however, this is just one of many innovations contributed by flexible coupling designers and manufacturers.

Increased power density of a coupling is best explained as the optimized combination of lighter weight and higher capacity. Gear couplings have high power density but require lubrication and are unable to operate at ever-higher speeds. While still in use in slow-speed machines and demanding ruggedness, gear couplings no longer predominate. Nevertheless, with good maintenance practices and keeping gear couplings well lubricated, they can perform well.

Nonlubricated coupling designs represent highly suitable alternatives to gear couplings. Disk pack couplings and flexible diaphragm couplings are much preferred by today's reliability-focused users. While they differ from each other, both offer advantages in the applications for which they were designed.

12.2 Diaphragm couplings

Diaphragm couplings were first used in modern aircraft in the 1950s. After years of reliable service and after perfecting the design, diaphragm couplings found acceptance in industrial gas turbines in the late 1960s. Over the years, this style of coupling (Figures 12.1 and 12.2) has become known for its high reliability, lightweight, and extremely low maintenance requirements.

The flexing element of a diaphragm coupling is a plate, or a series of plates, called the "diaphragm(s)." Diaphragms accommodate misalignment and transmit torque using the metal between the outside and inside diameters (OD and ID) of the diaphragm. Its cross section is contoured to keep stresses well within acceptable limits. These stresses result from the slight but unavoidable misalignment (or skew) between driving and driven shafts. All coupling designs must incorporate a measure of flexibility to accommodate slightly out-of-parallel and/or skewed alignment between equipment shafts.

https://doi.org/10.1515/9783110678765-012

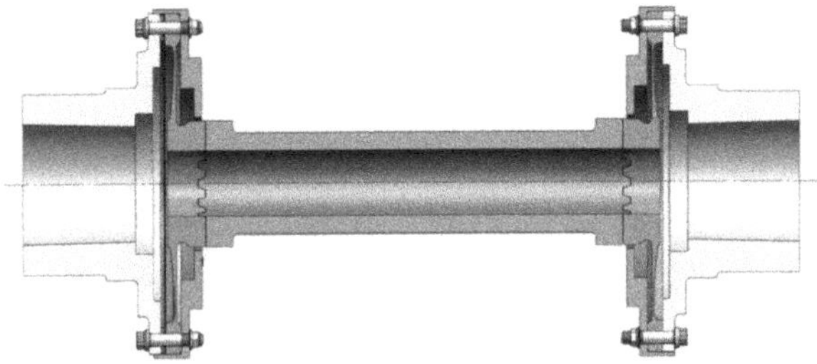

Figure 12.1: Contoured diaphragm coupling (source: *Riverhawk Co.*).

There are two types of diaphragm couplings, the contoured diaphragm and the convoluted diaphragm. A contoured diaphragm coupling typically uses a single diaphragm per end as the flexible member. Contoured diaphragm profiles have a thickness that varies from the OD to the ID; this design provides optimized stress distribution (Figure 12.1). As a general rule, an inherently thicker dimension exists at the hub (i.e., the ID) and a thinner dimension is located near the rim (the OD). The real purpose of the contoured profile is to allow the diaphragm to be thin enough to transmit the required torque, yet still be able to flex and keep the stresses within allowable levels. Design stresses for these diaphragms are based on alternating and mean stresses resulting from the transmitted torque and shaft misalignment. The combined stresses are low enough to give the diaphragms infinite life; designers will consult one of the widely used modified Goodman allowable stress diagrams found in mechanical engineering texts [1].

The profile shape, thickness, and diameter of a diaphragm are tradeoffs between a specified flexibility and torque transmission capability. Thicker diaphragms can handle more torque but are not as flexible. Thinner diaphragms will have greater flexibility at the expense of somewhat reduced torque transmission capability.

Parallel misalignment, axial push–pull, and angular misalignments of a diaphragm coupling are taken up by the deflection of the diaphragm. Under an axial misalignment, the diaphragm's OD will deflect away from the ID as shown in Figure 12.2a. This deflection is elongating/stretching the coupling or reducing/compressing the coupling, which is commonly associated with thermal growths that a turbine will typically experience during operation. For angular misalignment, the diaphragm will deflect in one direction and, simultaneously, in the opposite direction 180° away, (see Figure 12.2b). This deflection can occur with a parallel offset from initial misalignments of the shaft ends; it can also result from thermal growth of the drive turbine and its driven equipment.

There also are diaphragm couplings with wavy or "convoluted" diaphragms. Such couplings typically incorporate multiple diaphragms, each with a wavy or modified profile instead of the single diaphragms shown earlier in Figures 12.1 and 12.2.

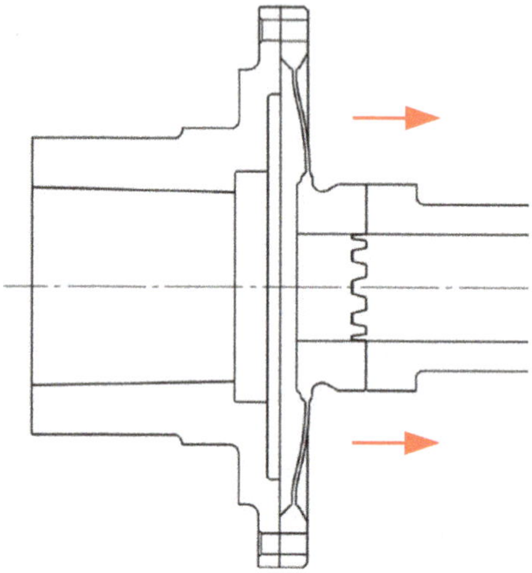

Figure 12.2a: Coupling showing axial deflection (source: *Riverhawk Co.*).

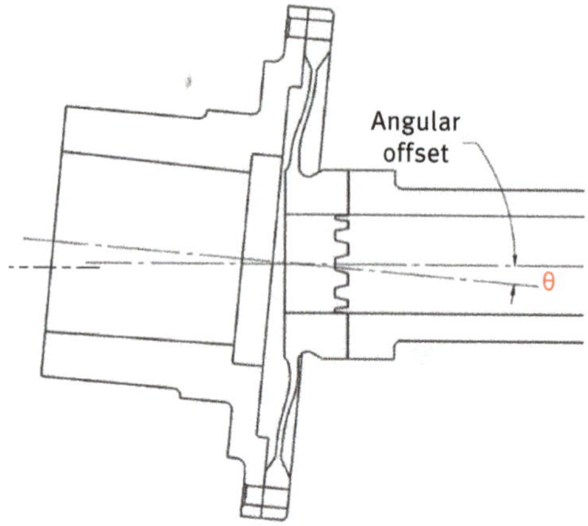

Figure 12.2b: Coupling showing angular deflection (source: *Riverhawk Co.*).

This modified design was developed years after the contoured diaphragm, and brought with it an improvement to flexibility. However, an inherent difficulty existed with multiple diaphragm designs. Fatigue failure could result from fretting contact as multiple diaphragms rubbed against each other. Such failures led to a temporary setback until the design was improved. Filler rings were devised that kept the diaphragms from rubbing against each other, thus eliminating fretting. Using different materials, attention to thicknesses and a more thoughtfully selected number of diaphragms allowed designers to achieve optimized performance. Combining torque and misalignment capabilities in the right proportions allowed diaphragm couplings to satisfy all reasonable application requirements.

One of the benefits of diaphragm couplings is their ability to accommodate bending moments and axial forces; both are functions of the profile's stiffness. This is a property not matched by gear couplings. In gear couplings, bending moments and forces can vary greatly because sliding coefficients of friction between mating teeth are neither constant nor predictable. In gear couplings, this coefficient depends on lubricant effectiveness, tooth materials, and tooth surface finish quality. In contrast, stress analysis methods and the designer's understanding of stress magnitudes in diaphragm profiles allow these components to be easily modified or "tuned." Tuning accommodates an application that may need increased torque capability and allows reduced misalignment capability or vice versa. These favorable design options greatly contribute to the acceptance of, and preference for, diaphragm couplings in modern turbomachinery.

Diaphragm couplings must incorporate a guard (Figure 12.3). This guard serves several essential roles in the coupling's performance. The guard is piloted to locate the diaphragm and the hub, or flange end, and to allow for excellent balance repeatability. Another important feature of the guard is to protect the thin profile shape from scratches, nicks, or dents that would become stress risers on the diaphragm. The guard is also a good antishatter device that will keep the center section of the coupling contained in the unlikely event of a diaphragm failure. Note again overall simplicity of the single diaphragm coupling fitted with a guard, as depicted in Figure 12.3.

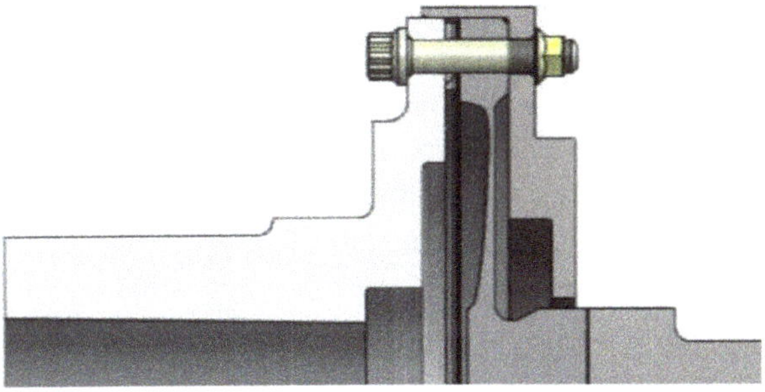

Figure 12.3: Coupling cross section showing guard assembled around the diaphragm (source: *Riverhawk Co.*).

Regardless of diaphragm type, contoured or convoluted, diaphragms (the flexing members which are an integral part of a rim) are attached to the other components with bolts, splines, or welds depending on manufacturers and their design preferences. All transmit torque in the same manner, that is, torque transmission through the diaphragm is in the radial direction from the OD to the inner or vice versa, depending on the direction of torque.

Some advantages of the diaphragm coupling are worth recalling:

- Known to be stronger than the disk coupling and can be made in diameters larger than disk couplings. They therefore accommodate higher torques.
- Contoured diaphragm couplings eliminate the fretting corrosion sometimes associated with multiple disk couplings
- Known to have exceptional balance repeatability.

12.3 Disk couplings

The history of disk couplings goes back even further than that of the diaphragm couplings. The invention of flexible disk couplings, dated around 1919, pointed coupling manufacturers in a new direction. Over the many decades of design and performance evolution, disk couplings led to them being considered both reliable and cost-effective. They have joined diaphragm couplings in popularity and are rightly considered among the most reliable flexible coupling commonly used in today's industrial turbomachines.

The basic flexing elements of such couplings are flat sheet metal rings assembled into stacks, also known as disk packs, and occasionally called a "sandwich." The entire sandwich is replaced every 6–8 years during major turnarounds, also known as maintenance repair outage. However, unlike a diaphragm coupling that connects the driving and driven components from the OD to the ID, the disk pack coupling does this by using bolts, all positioned on the same bolt circle. The bolts are connected in an alternating sequence, one fastened to the driver, the next one connected to the driven components. For example, if there are six bolts in the disk pack, three will connect to the driver's hub flange and three will connect to the driven machine's hub flange (see Figure 12.4). The bolts in the disks are commonly referred to as the "links." These links allow for some shaft misalignment while maintaining significant power (torque) transmission capabilities. It is entirely possible to examine disk pack integrity while the machine is operating at its normal speed. If observation under a stroboscopic light shows serious bowing of the links, the sandwich should be replaced.

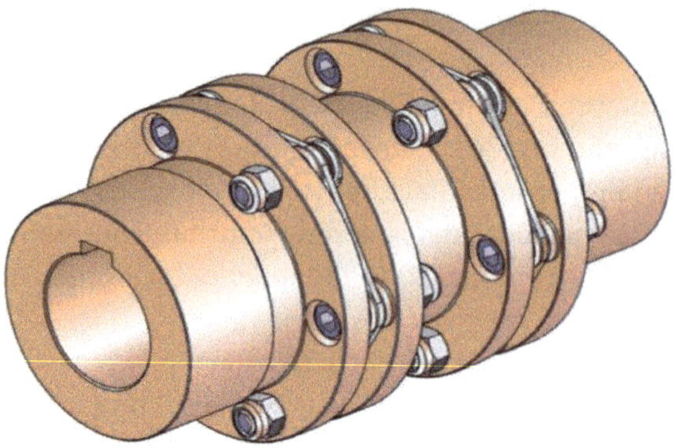

Figure 12.4: Disk pack coupling.

Torque transmission will put tension through half of the links of the disks and the remaining links will operate in compression. The directional sense of the coupling torque action determines which links are in tension versus which are in compression. The diameter of a disk pack is a good indicator of the amount of torque it can transmit; the larger the diameter the more torque capability it will have.

The flexibility of a disk pack comes from the length of the links between adjacent bolts: The longer the distance between the bolts, the more flexible the coupling. Flexibility depends also on the thickness of the disks. Multiple thin disks have an

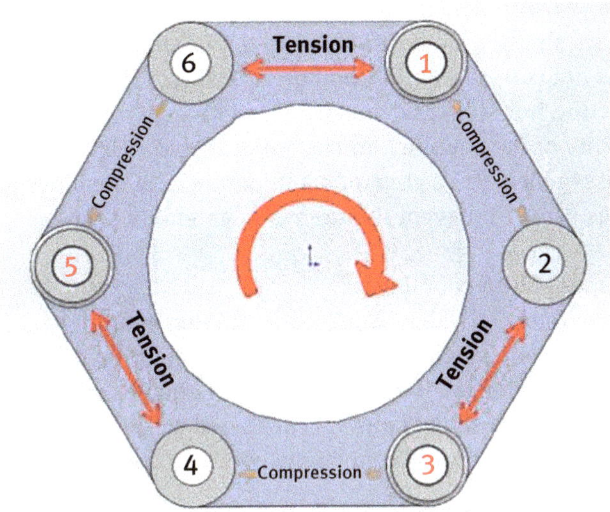

Figure 12.5: Disk showing the segments of the disk that operates tension and compression.

advantage over one thick disk when flexibility is required; multiple disks also have the advantage of adding redundancy. If one or more disks fail, the remaining disks can carry the load and will thus allow planned shutdowns before a major failure takes its toll. Tensile and compressive loading alternate, as shown in Figure 12.5.

There are three common disk designs: circular, hexagonal, and scalloped. The circular disk design was the original design; it is still in use today and stands out as the easiest to manufacture. It can accommodate low misalignments but has higher stresses than other disk designs. The hexagonal and scalloped designs have been developed to further enhance the capabilities of disk pack couplings; as of 2020, they represent the design most frequently used. Hexagonal disks offer improved torsional stiffness; the scalloped design is lighter and has a lower stiffness (Figure 12.6).

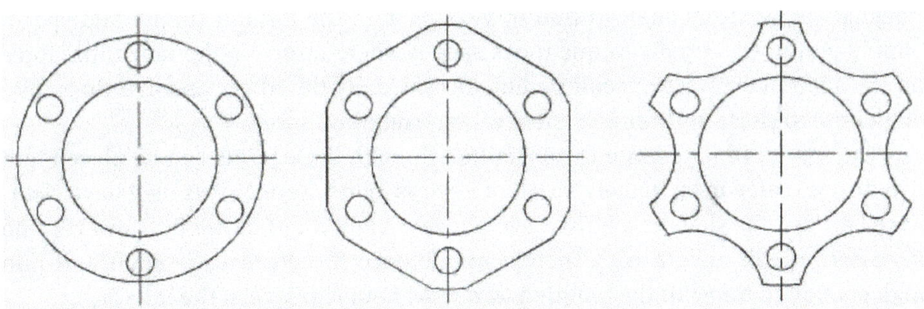

Figure 12.6: Three common types of disks.

A common misunderstanding with disk pack couplings is that the disks can be viewed in the same manner as shims and can thus be used to adjust the overall length of the coupling. This is not correct and this kind of “adjusting” should never be done. Adding to or subtracting from the disks in a pack will change both operating behavior and load capability of the coupling. In custom designed disk packs, it is common to see the packs assembled at the point of manufacture with bushings in the bolt holes. These bushings serve to prevent the sandwich assembly from being turned into individual slices.

Summarizing the advantages of disk couplings:

- Usually smaller in diameter than equivalent capacity diaphragm couplings
- Generally, less expensive than equivalent capacity diaphragm couplings
- Offer a measure of built-in redundancy. If an individual disk should fail, it will not disconnect the coupling so long as the bolts are capable of carrying the load in shear. This feature will usually allow the coupling to continue operating until safe shutdowns can be effected.

Overall, both disk and diaphragm couplings are nonlubricated, have low maintenance, and infinite life shaft connectors. Disks and diaphragms superimpose lower loads than gear couplings when misaligned; they will accept greater misalignments at higher speeds than gear couplings. Disk and diaphragm couplings can be balanced more precisely than gear couplings. But again, any of the coupling types examined in this text and operating within an approved design envelope will function for many years if given proper care and maintenance.

12.4 Technological improvement to shaft–hub interface joint for low moment couplings

The torque capacity for transmitting power across machines is often governed by the mating connection between the shaft and hub. This critical connecting joint transmits power from shaft to hub or vice versa in the form of torque and speed. Simply put, power equals torque times speed, where torque is the tangential force on the shaft acting at a specific radius (occasionally described as force times moment arm) to cause a turning moment at some required speed.

The way in which torque is transmitted through these joints can be placed into two distinct categories, namely keyed or keyless joints. *Keyed joints* may have clearance between the shaft and hub or even have a slight interference fit. However, the *keyless joint* will have a high friction interference fit between the shaft and hub high enough to transmit the required torque without slippage at the interface.

Generally, *keyed joints* are designed for low-power applications, up to a few thousand horsepower (or kW). In contrast, *keyless joints* are interference-fitted joints capable of transmitting much greater power. The *keyless joint* uses friction between the shaft and hub to transmit torque at a certain speed to produce power. The friction set up within the joint is caused by an interference fit between the shaft and hub. Assembling an interference fit means that the bore of the hub is slightly less than the shaft diameter; therefore, one must have the means of expanding the hub during installation. In large turbomachinery, these *keyless joints* are virtually always used on tapered shafts.

Next, we will examine the torque-strength limits of *keyless joints*; improving them beyond the present industrial norms will add value. We know that higher interference fits can result in considerably greater torque transmission capability but we also know that excessive interference fits can cause damage to the shaft or hub.

Friction-fitted *keyless joints* are the preferred choice of equipment manufacturers and end users for high-performance turbomachinery. This type of fit does not have the stress concentration and balance issues normally associated with keyed hubs. However, to achieve the needed nonslip gripping action required between shaft and coupling hub, a friction-fitted hub must be mounted on the shaft with a tremendous clamping force. High gripping or clamping forces are produced by high contact pressures at the shaft OD. In simpler terms: The contact pressure results from dimensional interference at the shaft–hub interface.

Installation of the hub is done either by heating or hydraulically dilating the hub. Although experiments have shown that the dry-friction coefficient from a heated hub is around 0.20 compared to 0.12 for hydraulically installed hubs, the latter is the preferred choice. The downside for heated hubs is the handling of the hub while hot, and the fact that it requires an onsite heating oven. Heating by open flame lacks uniformity and control; it is not an acceptable hub-installation practice. Hydraulic dilation of the hub is considered best installation practice within the hydrocarbon processing industry for interference-fitted joints.

The term "hydraulic" refers to fluid being injected at the high dilation pressure needed to assemble and disassemble a friction-fitted *keyless joint*. As mentioned above, these joints consist of tapered mating hub ID bores and shaft OD surfaces. Interference and contact pressure are generated by axially pulling or pushing the hub up the taper of the shaft. Figure 12.7 depicts a typical hydraulically fitted *keyless joint* hub being installed.

To facilitate proper loading of the joint, the male and female tapers are precisely machined. Their surface finishes must be no rougher than 16 RMS and their mating profiles verified to make perfect contact over at least 80% of the entire surface. During installation, the mating surfaces form a seal that can retain oil pressure. Although in some instances sealing is done with O-rings between the two surfaces to trap the oil as the space between the hub and shaft is pressurized, an O-ring presents vulnerability if it should burst. Should that happen, hub removal will be costly and time consuming; moreover, shaft surfaces may have to be refinished.

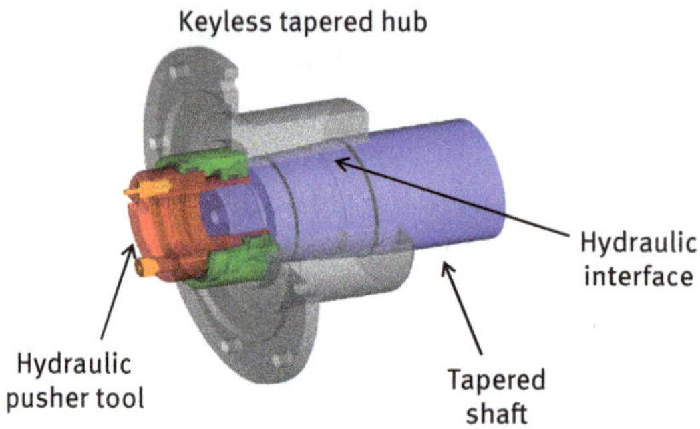

Figure 12.7: Hydraulic installation of a keyless tapered hub (source: *Riverhawk Co.*).

At installation, the hub is driven up the shaft taper by using a combination of axial thrust force from a pusher tool while simultaneously injecting hydraulic pressure into mating space. Hydraulic pressures are never allowed to reach the yield point of the hub material. Yielding means permanent deformation and the hub's ability to clamp itself to the shaft will be lost. Using pressures in the vicinity of 17,000–21,000 psi (120,000–145,000 kPa) expands the hub ever so slightly, and the lubricating properties of the hydraulic fluid now greatly facilitate axial sliding (pull-up) along the taper. Once the required axial pull-up is achieved and an end nut is in place, the dilation pressure is relieved. This allows the oil or similar fluid to drain. Sufficient strain energy will be trapped in the hub and shaft materials to retain the gripping forces.

In practice, hydraulic *keyless joints* are designed to mate with about 16,700 psi (115,000 kPa) grip pressure. However, in some challenging applications this grip pressure can be as high as 21,000 psi (145,000 kPa). Typically, hubs for turbomachinery have an OD/ID ratio of 1.5, as shown in Figure 12.8. Unlike the exaggerated taper shown in this illustration, the actual angles use in modern industry can be as little as 0.5 degrees included (total) angle. This combination of diameter ratio and pressure requires interference levels of 0.0020 to 0.0025 in/in.

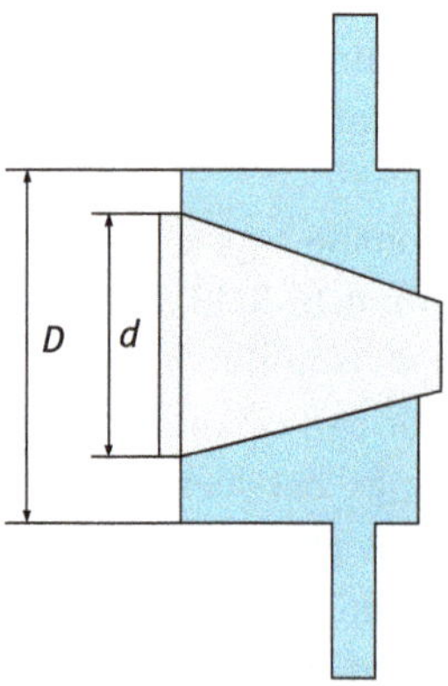

Figure 12.8: Typical cross section of tapered shaft and hub.

While contact pressure is the factor that relates to joint strength, interference is the parameter that field personnel can measure and control. A series of relationships between pressure, interference and torque are presented below to provide clarity as to how these parameters interact.

For a given shaft size, pressure is determined by the magnitude of interference as follows [1]:

$$P_i = E \times (i \times d) \times \frac{(D^2 - d^2)}{2 \times D^2 \times d} \tag{12.1}$$

where P_i is the contact pressure, psi (kPa); E is Poisson's ratio, usually steel, psi (kPa); D is the hub OD in or mm, assumed uniform profile; d is the shaft joint OD in or mm; i is the relative interference in/in (mm/mm).

For turbomachinery, the most common D/d ratio for hubs is 1:5. Applying this here allows simplification of the following:

$$P_i = \frac{E \times i}{3.6} \tag{12.2}$$

The torque required to slip the joint is the product of pressure, friction coefficient, contact area, and shaft radius:

$$T_s = \frac{P_i \times \mu \times \pi \times d^2 \times L}{2} \tag{12.3}$$

where T_s is the slip torque in-lb, (N-m), a limit of joint strength; μ is the friction coefficient, a value of 0.15 is used most commonly; L is the length of contact between hub and shaft.

The L/d ratio affects machinery operation in various ways; equipment manufacturers must weigh these influencing factors when making design decisions. Most commonly, hydraulic joints are designed for the customary L/d ratio of 1:0. Using this ratio here will allow further simplification as follows:

$$T_s = \frac{E \times i \times \mu \times \pi \times d^3}{7.2} \tag{12.4}$$

Equation (12.4) is realized after rationalizing eqs. (12.2) or Equation (12.3) and substituting $L/d = 1.0$ in eq. (12.4).

These equations reveal a number of important relationships.

Equation (12.3) shows that the torque capacity of the joint is proportional to the contact pressure for a given shaft diameter and contact length "L."

Equation (12.4) shows that the torque capacity of a hydraulic joint is roughly proportional to the cube of the diameter when designed according to normal industrial practices. It also shows that the slip torque is proportional to the amount of interference designed into the joint. Comparing eqs. (12.3) and (12.4), it is evident that pressure and achievable interference are proportional to each other. This is true when the outer profile of the hub is unchanged, that is, when the hub does not have multiple ODs.

For designers of shaft joints, contact pressure is the key variable because it relates to the total clamping force. In the field; however, interference is far easier to measure and control. The angle of taper and axial pull-up can be measured. Simple trigonometry allows accurate calculation of the resulting interference. Accordingly, interference became the most commonly used expression to describe pressure within a friction joint.

Equations (12.1) and (12.3) can be useful for detailed shaft-end design work, while eq. (12.4) may be more useful for initial sizing of shafts for equipment where friction joints are to be used.

Over the decades, hydraulic shaft connections have been in highly successful service. Indeed, this connection technique has proven to be generally reliable and effective at performing the required functions. However, as with most technologies, time has also allowed this technology to be pushed to its limits. Limitations have become evident as engineers dealt with the need to transmit ever more torque and power through these joints. Most limitations fall into two categories:

1) Dilation pressure control
   a. Leakage of high pressure during installation and removal
   b. Incomplete draining. There are retained pockets of pressure when installation is done
2) Surface damage during installation and removal
   a. Galling and shaving due to sliding and localized high contact pressure

These shortcomings increase dramatically with increased contact pressures. The factors that determine just how much interference a surface pair can tolerate are difficult to quantify. Most experts limit interferences to values ranging between 0.0020 and 0.0025 in/in (mm/mm). Where special cases need more clamping pressure, an upper limit of 0.0030 in/in (mm/mm) can be achieved. It comes at the expense of reduced installation life along with the need for extreme care in design and usage. The API 671 standard [2] limits interference levels to 0.003 in/in (mm/mm), making this the absolute upper limit for hydraulic *keyless joint* designs.

The objectives and a 2020 solution for improving the design of keyless joints are highlighted and summarized as follows:

1. Transmit more torque across the shaft–hub interface to accommodate increased power demand while considering the present limitations.
2. Eliminate the potential for installation damage to the shaft end while transmitting the full design power.

In 2020, a noted manufacturer of couplings and specialty engineered equipment released information on their recently developed clamping method; the design pertains to low moment coupling hubs. The development did accomplish the above objectives.

The new design was shown to boost maximum torque capacity and eliminates high pressure sliding on the shaft-to-hub contact surfaces. As of this writing, the manufacturer's patent for the novel clamping method (shown in Figure 12.9) is pending.

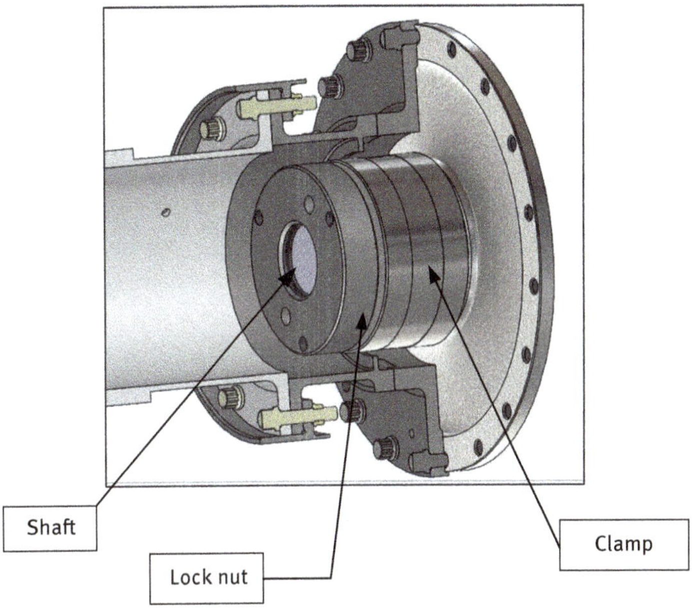

Figure 12.9: New design of the hub–shaft interface keyless-joint for enhanced torque (source: *Riverhawk Co.*).

Presently aimed at low moment coupling hubs, it addresses the needs of industry's highest power machines. The new design utilizes hydraulic hub technology but applies it in a new way. The high-pressure sliding elements are removed from the shaft surface and located at the OD of the hub. This approach proved advantageous to designers, manufacturers, and end users for a number of reasons:

- It allows the highest interference levels for low-moment couplings with reduced risk of damage to the shaft–hub interface.
  - Higher contact pressures between shaft and hub in excess of 30,000 psi (207,000 kPa) can now be achieved. The needed clamping forces are now applied at the OD of the hub.
- The new approach eliminates trapped pressurized oil pressure from the torque-transmitting interface.
  - Torque capacity is less affected by procedure, making the coupling's power transmission capability more predictable.
  - Higher effective friction coefficient.

- The hub can now be installed and removed without risk of surface damage to the shaft end.
  - The designers demonstrated no surface damage after several installation and removal cycles. Figure 12.10b shows no damage to the shaft end after more than 20 installation and removals. This exercise simulated the typical life cycle of most turbomachinery assets. For example, if we assume 4 years between turnaround cycles and the coupling hub is removed during each turnaround, then this exercise simulates a 40-year span.
- The design is tolerant of small amounts of surface damage on existing shafts.
  - Allows hub assembly on damaged or marginal shaft ends to operate with little or no loss in torque capacity.
- It was developed specifically for low-moment couplings.
- This new design is easily adapted to existing tapered shafts.

For new machines and uprates:
- The new design can provide a cost-effective solution where traditional couplings will have insufficient coefficients of friction to resist slipping on the shaft.
- Brings many potential turbomachine uprates within reach of feasibility by increasing the power capabilities of an existing shaft size
- Keeping shaft, bearing, and seals size smaller keeps capital and operating costs down.
- Keeping shaft size smaller allows lighter couplings for more control over dynamic design factors.

Figure 12.10 a and b shows a sample shaft end before and after testing. Note that grooves were machined into the shaft taper to replicate the mounting encountered on an existing shaft. No high-pressure oil or axial pushing forces were needed (or employed) on these surfaces.

(a) (b)

Figure 12.10: (a) New shaft before installation and (b) shaft after 20 removals of the hub (source: *Riverhawk Co.*).

12.5 Torsional peak shaver safety hub

Machinery designers and operators are occasionally faced with the need to run through short-duration high-torque transients. Machinery known to experience such transient torques must be designed to protect their shaft systems against dangerous torque spikes; their often very high magnitude can result in damage to the equipment. An example of such damage is slippage of a coupling hub on a shaft end.

Torque spikes can originate from within the process or from the power source. Examples of potentially damaging sources of torque spikes are:

1. Missed synchronization timing of generators upon connection with the power grid
2. Switch "bouncing" in the case of induction motor drives
3. Ingestion of fluid slugs during compressor operation
4. Short circuit in the local power network
5. Mechanical failure of another member of a machine train

Torque spikes ramp up instantaneously and are followed by near-instantaneous cessation or decay of the transient. All but the last two cases listed (4 and 5) might benefit from allowing the equipment to keep running

Usually, the first objective with respect to torque spikes is survival of the machinery. A relatively frequent and competing objective is continuation of the process or "running through the undesirable deviation." Although these two are conflicting objectives, both must often be weighed and considered.

Thoughtfully engineered peak shaver safety hub (PSSH) designs can satisfy both objectives by absorbing an energy spike while allowing the process to continue. One particular PSSH design incorporates internal slip surfaces dimensioned and manufactured to tolerate short-duration slippage without damage. A PSSH can be integral to a flexible coupling; it can either be mounted on a shaft (in a configuration that replaces the hub) or as a spacer located between driving and driven shafts.

Figure 12.11 shows the cross-sectional view of a novel PSSH (PSSH). The sleeves are clamped together by a hydraulic torque coupler. Hydraulic torque application allows frictional contact between the two surfaces, which then operate as a solid assembly. Whenever a torque spike occurs, the sleeve design allows a predefined amount of slippage to prevent overloading the driven machine. The definable slippage between the sleeves allows driving and driven machines to stay online (i.e., continue to "run through") until the transient situation has cleared. The PSSH then regains grip without slipping and continues to transmit normal torque to the driven (loaded) machine.

Figure 12.12 simulates normal torque action and a superimposed transient torque spike. The PSSH serves to clip or shave the spike during the "run through" period. It continually transmits allowable torque to satisfy process demands.

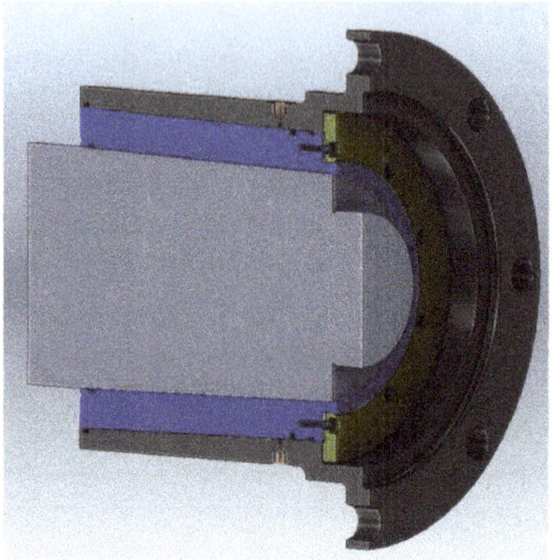

Figure 12.11: Cross section of a shaft-mounted peak shaver hub (source: *Riverhawk Co.*).

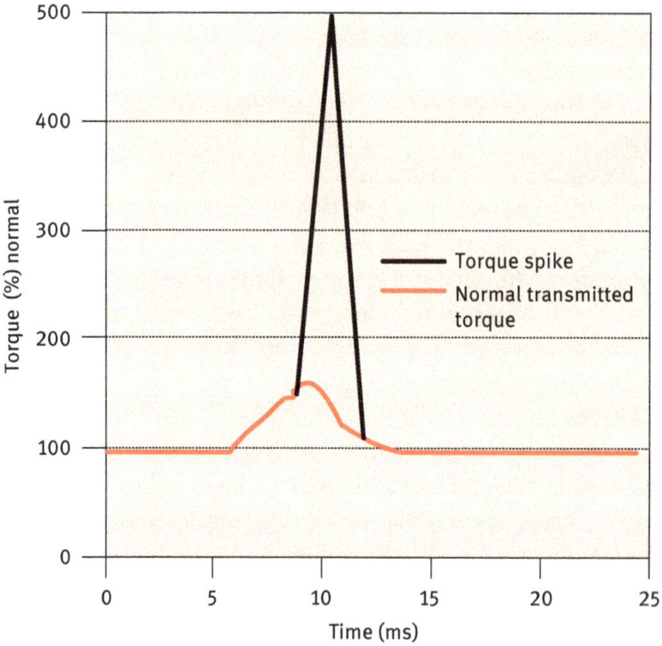

Figure 12.12: Peak shaver representative torque transmission.

However, friction surfaces do wear with repeated slippage. The units should be periodically examined to determine the extent of damage due to slippage. Wear rates depend on the magnitude of energy being dissipated and on event frequency. Although some PSSH units have demonstrated 1,000 successful slip cycles, not all experience the same rate of wear. Accordingly, and to allow estimation of prudent operating time before precautionary replacement or repair, it is deemed appropriate to check slip frequency and wear rate, even if upsets are infrequent. Such an examination takes on added importance when high-value assets must be protected.
In cases where extremely high or long-duration energy dissipation is needed, the spike energy absorbed in friction surfaces could cause overheating and permanent damage to the PSSH. On such relatively infrequent occurrences, a releasing device should be considered in the design.

Unlike old-style shear section assemblies which are often limited by material fatigue, PSSH designs will not cause unjustified releases, that is, "uncoupling." An additional advantage of the modern PSSH development is its "drop-in" replacement design; in other words, it can be a substitute for conventional coupling hubs. This, of course, greatly simplifies the protection of major machinery from transient torque spikes.

12.5.1 Data requirements and example

To apply this solution, the following information are needed:
- Shaft-end dimensions and coupling flange dimensions
- Normal operating power and speed
- The maximum torque that the shaft can tolerate, or the torque that must not be exceeded
- Description of the expected transient
  - Expected magnitude of torque spike
  - Time duration of transient

An example of a common case is given to show how the PSSH handles heat. It involves an 8,100 hp (6,040 kW) synchronous motor–compressor train operating at 1,800 rpm with direct (across the line) start. The designers had to be established that the safety hub could meet the API 546 requirement of three starts in 1 h at 100% voltage. Start events were modeled as one start for every 15 min, with conductive heat dissipation only.

For the thermal analysis, a lumped system time-step model of the start sequence was created and run with a PSSH. The results of the analysis are represented as the torque curve shown in Figure 12.14.

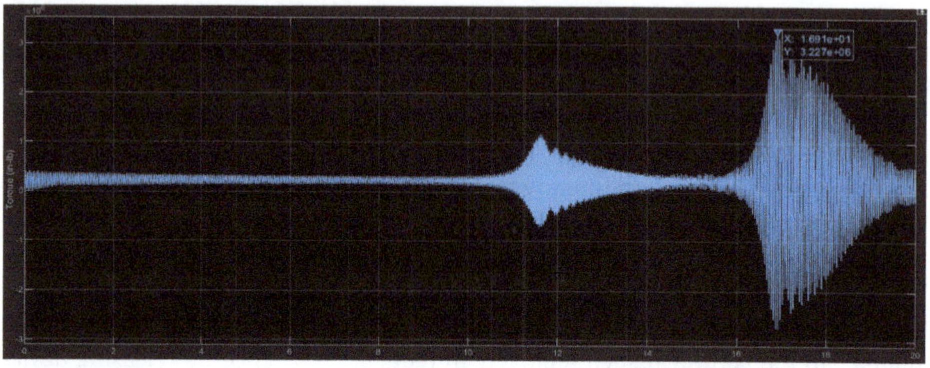

Figure 12.13: Torque versus time for time-based modeling of the synchronous motor–compressor train without damping protection during line start. Peak amplification ~11.3 p.u (source: *Riverhawk Co.*).

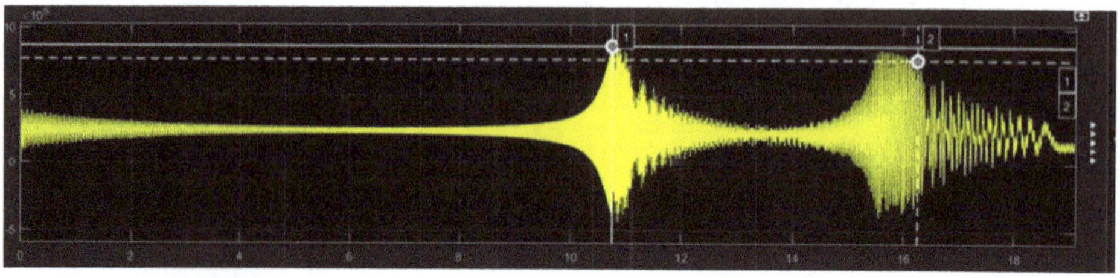

Figure 12.14: Torque versus time results for time-based modeling the synchronous motor–compressor train, line start equipped with a peak shaver safety hub (source: *Riverhawk Co.*).

The dotted line in Figure 12.14 represents the protection torque setting of the unit. Note how the torque spikes are truncated to this level. Also, the duration of the dynamic interaction is reduced compared to Figure 12.13.

This model included modeling the slip characteristics of the peak shaver. Figure 12.15 shows the heat generation at the safety hub slip interface for one start; 28 slips can be identified. These data were used for input to a time-stepping finite element analysis (FEA) thermal model of the unit. The thermal FEA was run for three starts and two rest periods. Table 12.1 shows the approximate average temperature of the interface; Figure 12.16 shows the temperature distribution immediately after the final start. In this case, the maximum allowable interface temperature was 350 °F. The design met the application programming interface requirement with a significant unused margin of safety left.

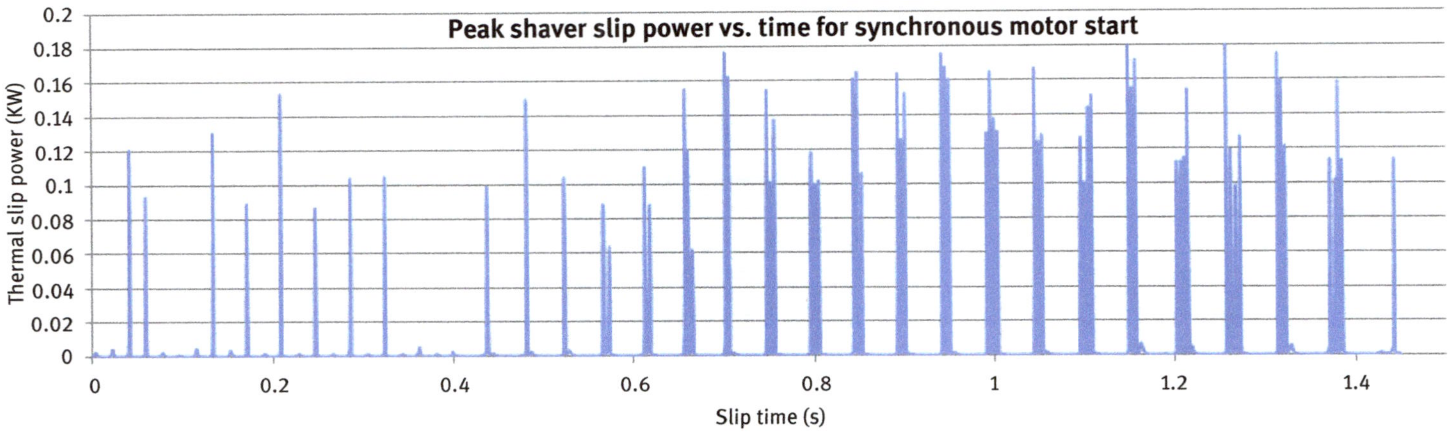

Figure 12.15: Time versus slip power absorption for peak shaver during synchronous motor line start. Integration of these spikes yields total absorbed spike energy of a start event (source: *Riverhawk Co.*).

Table 12.1: Interface temperature within PSSH after three starts in 45 min.

Time	Comments	Interface temperature (°F)
0 s	Initial start	70
1.45 s	End of first start	180
15 min and 1.45 s	End of first rest and beginning of second start	82.7
15 min and 2.90 s	End of second start and beginning of second rest	192.7
30 min and 2.9 s	End of second rest and beginning of third start	95.4
30 min and 4.35 s	End of third start	205.4

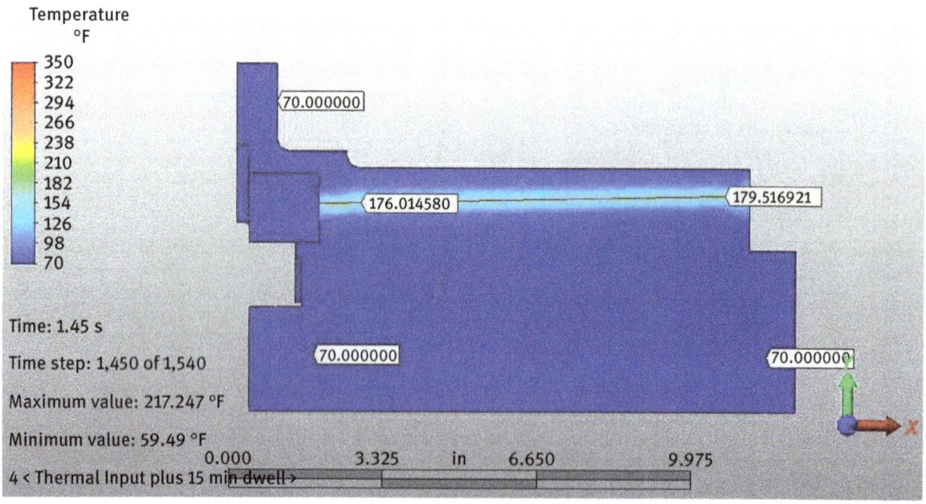

Figure 12.16: Thermal model of safety hub after 3 starts in 1 h (source: *Riverhawk Co.*).

Figure 12.17: Peak shaver safety hub mounted on shaft (source: *Riverhawk Co.*).

References

[1] Shigley, J.E. Mechanical Engineering Design, 3rd, McGraw-Hill, ISBN 978-0-07339820-4.

[2] American Petroleum Institute (API) 671, 4th, Special Purpose Couplings for Petroleum, Chemical and Gas Industry Services.

Chapter 13
Compressor controls and protection systems

13.1 Turbomachinery control functions in the distributed control system

In modern plants, distributed control systems (DCS) dominate the landscape for regulatory systems. DCS provide reliable hardware with fault-tolerant architecture, proven software solutions for many processes, and convenient operator and engineering interfaces. In many cases, DCS vendors also supply the plant safety systems, taking responsibility for the implementation of the majority of the process control and safety scope.

Historically, due to more stringent performance requirements and specialized functions, DCS had only limited penetration into the area of turbomachinery control (TMC). However, over the last two decades, as the cost and footprint of microprocessor-based system decreased, while computing power increased exponentially, DCS vendors have begun to penetrate into the TMC space. A few DCS manufacturers added specialized high-performance hardware to their offering, intended to meet the requirements of TMC for process-critical machines.

Today, DCSs have a leading share of the overall TMC market, followed by the OEM systems, which are mostly PLC based, and then by systems supplied by the companies specializing in TMC. The latter vendors typically dominate the niche for large, process-critical machines, such as main air blower in the FCCU (Fluid Catalytic Cracking Unit) in a refinery, or a 60 MW refrigeration compressor in a baseload LNG (Liquefied Natural Gas) plant.

Despite the advances in the performance of general-purpose control systems, purpose-built systems in many cases provide several advantages for TMC. Yes, it is possible to use general-purpose instrumentation; however, achieving the same level of performance as a purpose-built system requires obtaining high-end processors or controllers. This includes adapting available modules for handling the types of I/O required for TMC systems, as well as fast and deterministic networking, for which they were not originally designed. As a result, the cost of implementing and maintaining such systems, quite attractive at a first glance, may in fact be inferior to purpose-build system. Moreover, the know-how built into the system and expertise of the personnel engaged in engineering and commissioning such system are crucial to the success of the installation – often more so than the hardware platform used.

https://doi.org/10.1515/9783110678765-013

13.2 TMC requirements

A simplified control system diagram (CSD) for motor-driven compressors is shown in Figure 13.1a; a CSD for turbine-driven compressors is shown in Figure 13.1b. TMC includes the following main areas:

1. Regulation of the rotating speed of steam and gas turbines
2. Antisurge control
3. Capacity control of turbomachinery train

13.2.1 Overview of TMC systems

In Figure 13.1a and b, the main objective is to maintain the suction pressure of the machine. In Figure 13.1a, suction pressure controller PIC-1 varies the flow rate through the machine by manipulating a suction throttling valve, whereas in Figure 13.1b, flow is the control objective and is achieved by adjusting the speed of the train. The antisurge

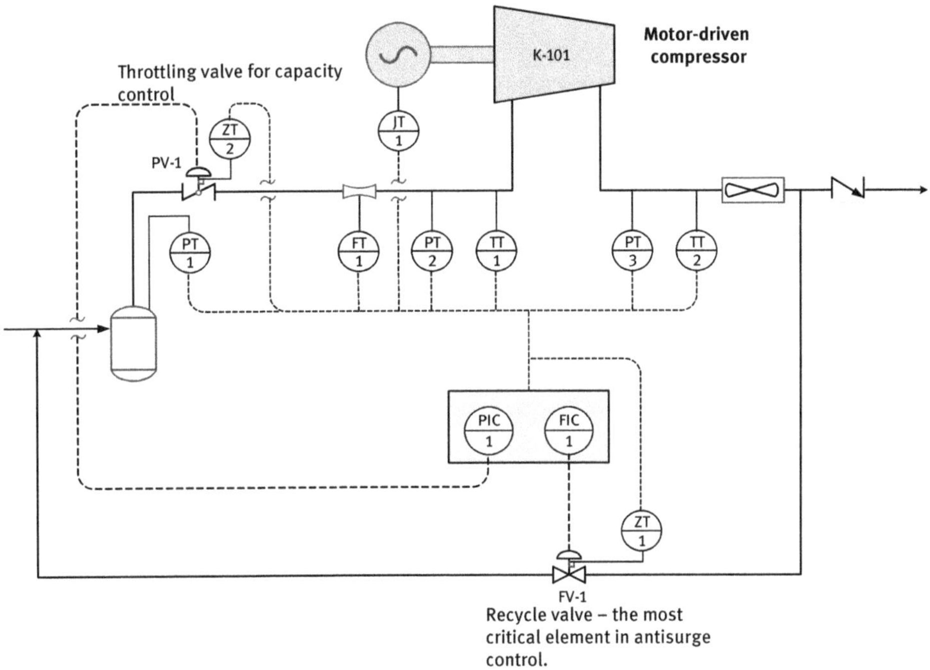

Figure 13.1a: Motor-driven compressor (source: *Comp. Controls Corp.*).

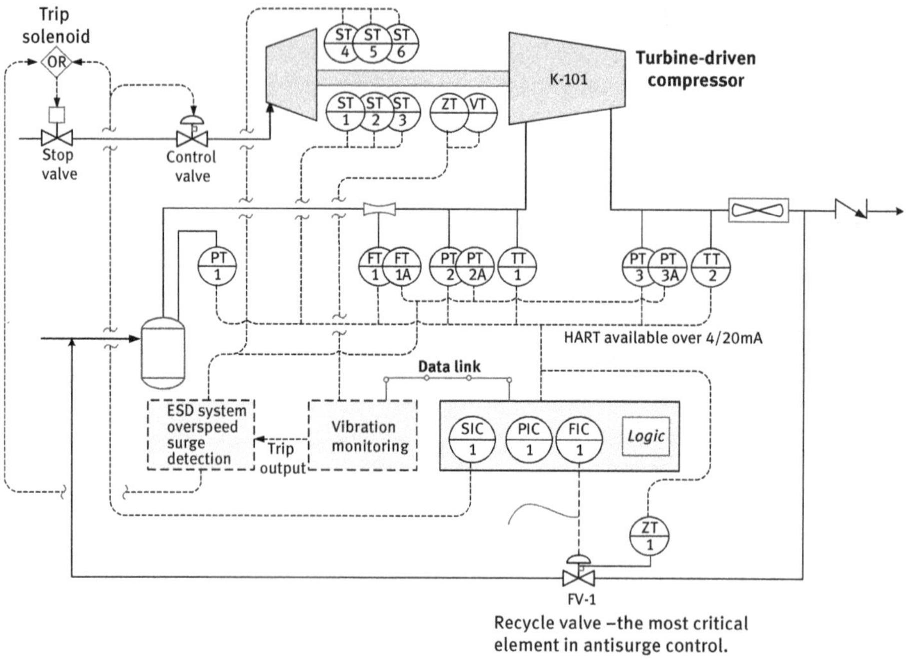

Figure 13.1b: turbine-driven compressor (source: *Comp. Controls Corp.*).

control typically receives the process measurements shown in the CSDs. Temperature measurements may not always be used for antisurge control as such but are very useful for monitoring purposes. Note that Figure 13.1b shows redundant measurements, which is a common practice for process-critical units. The antisurge controller manipulates the recycle valve around the unit, so as to always maintain unit flow above the onset of surge.

Other important aspects are control, sequencing, and monitoring of the auxiliary systems, such as lube oil system, isolation valves, and seal gas system are not considered here.

13.2.2 Speed control

Rotating speed regulation requires response times that are significantly below those necessary for general-purpose control loops. For example, fuel control systems for some aeroderivative gas turbines have control loops executing at 10 ms. Industrial gas and steam turbines from several of the top equipment manufacturers use 40 ms control loops or faster. Here, loop time refers to the total execution cycle of the system, including I/O processing. On the other hand, general-purpose loops, such as level control, can be executed at around 250 ms or higher without any significant degradation in loop performance. For overspeed protection systems, API 670 fifth edition (pp. 8.4.4.1) mandates that the overall system response from overspeed detection to the change of state of the trip relays shall be within 40 ms. In order to provide desired redundancy for overspeed prevention, the primary speed control system should have similar response times.

Power generator control applications have generally higher risk of overspeed, when sudden loss of electric load results in high acceleration rates for the rotor. However, although not as drastic as the breaker opening, a rapid decrease in power consumption during compressor surge can also result in high rotor acceleration rates that can lead to an overspeed condition. The flow through the compressor during surge can decrease to zero within 150 ms, requiring corresponding reaction times from the speed control system. Even without the safety considerations for prevention of overspeed, the rotor time constants can be such that controlling the speed of the machine within a few rpm may be only achievable with loop times of 100 ms or less.

The electronic speed control loop requires specialized hardware; it must be able to process frequency inputs from various types of sensors installed on the unit. The actuators for steam and gas turbines may use hydraulics, requiring precise positioning of the pilot-ram actuation systems, utilizing position feedback, typically from linear variable differential transformer sensors. In this case, the control system must position servo valves with loop times of less than 10 ms, and be capable of interfacing with the special sensors.

13.2.3 Antisurge control

Antisurge control also requires relatively fast execution times. Historically, antisurge system vendors have used 40 ms loop execution time. The requirement arises mainly from the high rates of change of flow during surge and in the region approaching surge. The slopes of the performance curves of a centrifugal compressor generally tend to "flatten" near the surge limit, as shown in Figure 13.2. Even a small pressure perturbation causes a significant change in flow, relative to the flow at surge. The stability margin of a control system decreases as the slope of the curve tends to approach zero, which may cause a system with a slower loop time either to become unstable, or to be detuned to the extent that it can no longer prevent surge. If the magnitude of the surge control margin is of little importance, then the margin can be set sufficiently large, so that the antisurge control system always operates in the relatively steep region of the curve. In that case, even a slower system may do an adequate job of preventing surge. However, in most cases the magnitude of the surge margin does matter. The machine turndown without recycle is reduced with a large margin; accordingly, using a system with a loop time of around 40 ms represents a proven and safer choice.

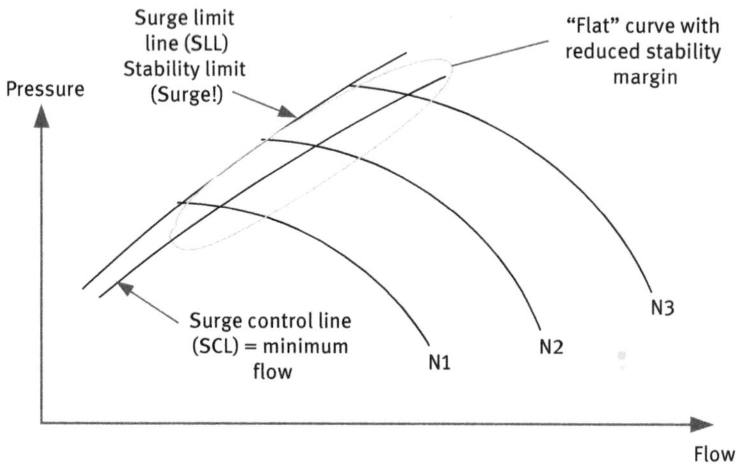

Figure 13.2: Surge limit and control lines, performance curves (source: *Comp. Controls Corp.*).

Proponents of general-purpose control systems for TMC applications question the validity of requiring loop times <40 ms for antisurge control. The reasoning goes something like this: "I know that my process does not change very rapidly, my recycle lines are long, and there are large volumes in the system – if my process time constants are >10 sec, why do I need a control loop executing at <40 msec?"

As long as the compressor operates in a steady state far away from the surge limit, the argument is indeed correct. In fact, one could even go further and say that if the compressor would always operate in a steady state and away from the surge limit, there would be no need for any control system at all. The problem is that process upsets do happen, and the ability of the control system to respond in a timely manner will translate into the difference of either losing the process, tripping the unit and, at times, preventing machine damage.

Antisurge systems typically use standard off-the-shelf instrumentation; however, there are important criteria for response times, performance characteristics, and installation practices, which will be covered in more detail further.

13.2.4 Capacity control

Compressor capacity control does not necessarily require the same speed of execution as the speed of antisurge control. However, in cases when the unit is operating with recycle or blow-off, the integration between capacity, antisurge, and speed control becomes important. For example, a reduction in the rotating speed of the unit in the region approaching the surge limit can result in surge. This occurs because, due to process volumes, the discharge pressure decreases more slowly than the ability of the machine to deliver the required pressure ratio as speed is reduced. Therefore, it is beneficial for the capacity control system to send a so-called feed-forward or decoupling signal to the antisurge system for initiating opening of the antisurge valve even before the operating point reaches the control line. In order for the decoupling action to be effective, the capacity control system should not be an order of magnitude slower than the antisurge system. Values of around 100 ms have proven adequate for capacity control.

13.2.5 Other consideration

Another important aspect of the TMC systems is determinism, where the control system ensures that the execution time will be within a defined interval in each cycle. Variable execution times can result in delays in response when the control system microprocessor is under high load, therefore compromising the effectiveness of the control loop. This may occur during critical situations, such as overspeed or surge.

13.3 Various approaches to TMC architecture and interfaces to DCS

Historically, the architecture shown in Figure 13.3a has been widely used for TMC systems. The overall unit control system (UCS) comprises the TMC, the control and monitoring of the auxiliaries, and the protection systems. The UCS may be installed in a local control room near the unit, or in the same location as the main plant control room. A local operator panel on the deck of the process unit may be provided, since operating procedures at many plants require that operators perform unit startup and shutdown while located next to the unit. It is anticipated that with progressively higher reliability of communications and video equipment, these requirements may be relaxed in the near future. Today, ruggedized PC-based workstations and smart displays have been replacing the analog displays and pushbuttons. With the exception of the emergency shutdown (ESD) buttons, these workstations communicate with the UCS via a digital communication bus, which may be configured for redundancy.

In Figure 13.3b, the TMC is located in the field, next to the unit. Such installation saves on cabling costs, but requires an extended environmental operating range and, in many instances, a hazardous area certification, at least for Class 1 Div. 2, or ATEX Zone 2. Many modern control systems today do have the hazardous area certification, but the extended operating temperature range without climate control may prevent installation of many systems in the field, or on the skid.

Figure 13.3c shows a compromise solution, where ruggedized I/O is placed in the field, but the processor modules and other system equipment are located in the control room environment. A number of high-speed industrial communications protocols can be used for communications between the I/O subsystem and the TMC; EtherNet IP,

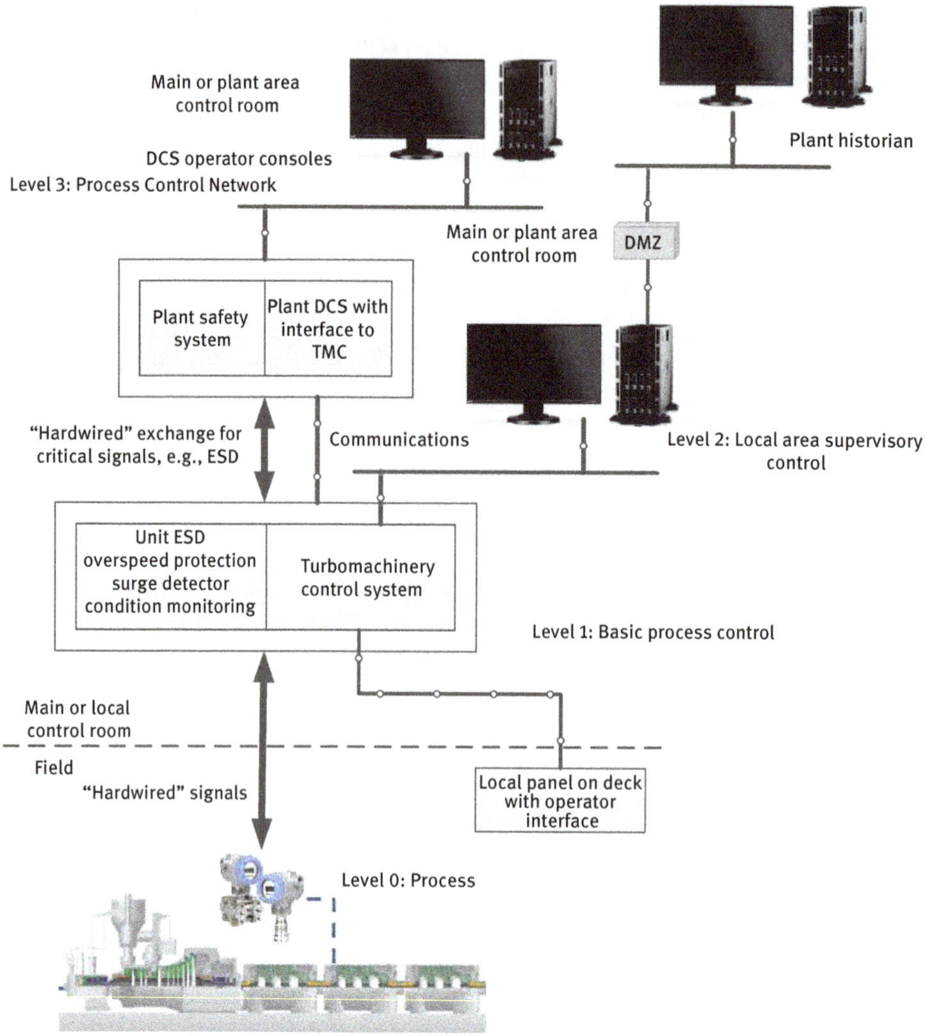

Figure 13.3a: Architecture of turbomachinery control system and DCS interfaces, TMC located in main or local control room building (source: *Comp. Controls Corp.*).

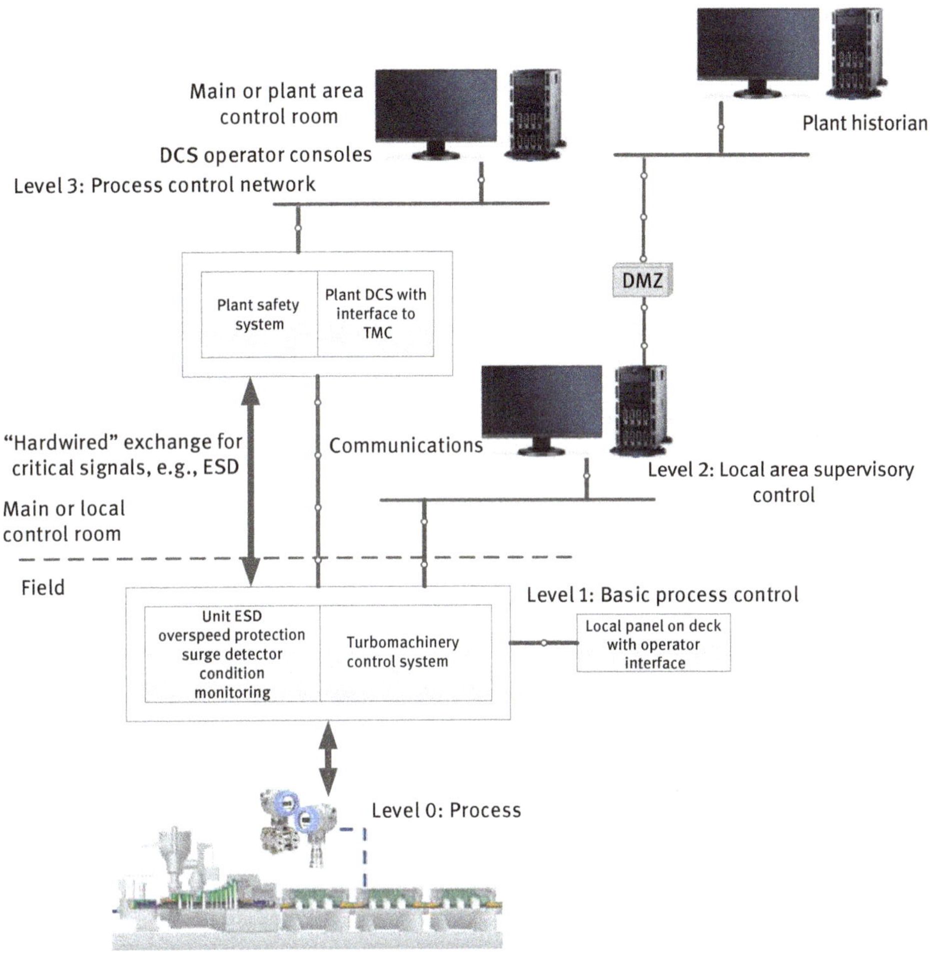

Figure 13.3b: Architecture of turbomachinery controls and protection and DCS interfaces, TMC located next to the unit in the field (source: *Comp. Controls Corp.*).

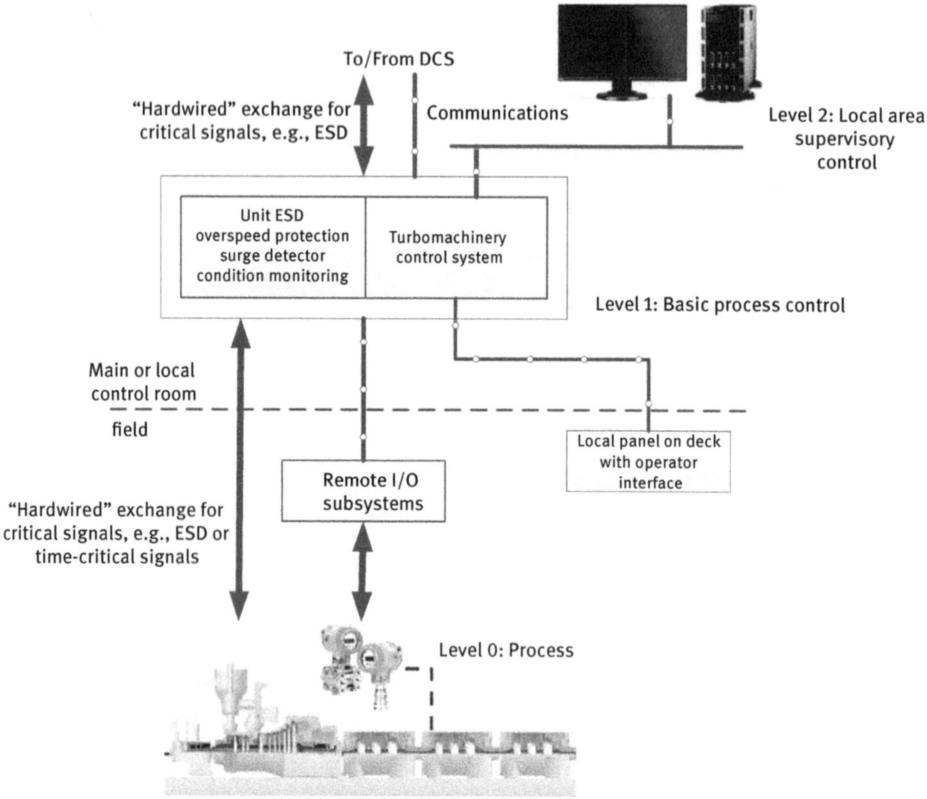

Figure 13.3c: Architecture of turbomachinery controls and protection and DCS interfaces, TMC with remote I/O (source: *Comp. Controls Corp.*).

ProfiNet, and others come to mind. However, not all protocols can provide sufficient communication speed. Critical signals, such as rotating frequency and differential pressure from the flow-measuring device for antisurge control may still need to be "hardwired."

Due to simplification of wiring and reduction of cabling costs industry has been moving in the direction of replacing analog signals with various fieldbuses. In fieldbus systems, a transducer, such as a flow transmitter, communicates with the control system digitally instead of via analog signal. However, most fieldbuses today do not provide sufficiently fast response times for TMC systems and should not be used for TMC.

13.3.1 Communications

The entire system may be divided into formal levels, as shown in Figure 13.3 a and b. Operating personnel would like to perform majority of their functions at level 3, where the main DCS consoles are located. Such architecture necessitates reliable communications with the TMC system, located at level 1; therefore, communication schemes typically have redundancy. application layer protocols with Ethernet as the physical layer such as MODBUS TCP/IP, ProfiNet, and EtherNet IP are typically used for communications between the TMC and the DCS. In some cases, open platform communication data access (OPC DA) communicating from the TMC workstation to the DCS has been implemented. Less frequently, open platform communication alarm and events (OPC AE) communication has also been employed. OPC DA and AE provide important features, such as time stamping of data and sharing alarms native to TMC with the DCS. However, OPC DA and AE communications are only available from the workstation, thereby introducing another element with much lower reliability into the communication scheme. In addition, OPC DA and AE have major cyber security flaws in network communications.

The introduction of open platform communication unified architecture (OPC UA) holds much promise. This platform is not limited to the Windows-based platforms and can be implemented on the actual control systems, eliminating the need for any devices between TMC and DCS. Moreover, OPC UA provides adequate cyber security features for complying with developing cyber security standards. OPC UA integrates data and alarms, thus providing much higher degree of integration between TMC and DCS.

In most cases, the communications between the DCS and TMC contain process and unit information for the operating personnel, diagnostics, and operator commands, such as set points from the DCS to TMC. The exchange does not usually involve information needed to calculate the responses of control loops. Therefore, communications are not time-critical, provided the update time does not cause delays in operator response time.

1. The TMC system may have its own interface, although today such an interface may carry out more of a diagnostics and engineering tasks, rather than operator interface functions. Generally, DCS update times are much slower than TMC. This means that the DCS historian does not have the necessary resolution for analyzing process signals related to turbomachinery, with data points less than 1 s apart. Therefore, TMC workstation serves an important purpose for analyzing critical events, such as occurrences of ESD or surging. The communications between TMC and the Workstations allow for data sampling and archiving at less than 100 ms rate. This facilitates identifying the root cause of various malfunctions. In addition, several vendors provide data archiving at the loop time of the controller (e.g., 40 ms) onboard the controller itself. The critical event files can be uploaded from the controller to the workstation for analysis. Using specially developed software, high-resolution data can also be passed on to the OSI PI-based plant historian.

13.4 Fault-tolerant architectures

Turbomachinery inevitably represents some of the most critical assets of a plant. Therefore, the reliability of both TMC hardware and software significantly affects its entire operation. For a large plant, catastrophic failure of a unit and subsequent lost production, not even considering repair costs, can result in severe financial losses. In refining and petrochemical plants, TMC systems for main machines are typically either dual redundant or triple modular redundant (TMR).

In upstream or midstream applications, which include gas and oil production and general hydrocarbon transportation, simplex systems are sometimes used. This is the case where plant designs include an entire standby unit. However, the costs for dual redundancy systems have generally come down in the decades after 2010 and the cost may not significantly exceed that of a simplex approach. Many of today's redundant systems include online or hot-swappable modules, improving the availability of a redundant system even further.

Today's TMC systems generally are modules with central processing units (CPUs). These modules execute application software; communicate with various types of I/O modules; and have termination assemblies for connecting field wiring. Many systems support some degree of fault tolerance and online replacement of the redundant modules in case of failure. The TMC systems utilize either off-the-shelf high-performance PLCs, DCS hardware, or purpose-built control platforms. Some of the purpose-built systems have an architecture similar to a PLC, with processor and I/O modules.

Some of the more popular brands of TMCs are shown at the end of this chapter.

The years 1990s saw the advent of TMR systems. These can be sourced from Triconex (now Schneider Electric), GE (their Mark VI), and Woodward (MicroNet), among several others that also have demonstrated solid experience. Due to the advantages of 2oo3 (two out of three) architecture, these systems initially had superior reliability compared to the dual-redundant systems available at the time. However, with the development of more extensive diagnostics coverage, modern dual-redundant systems have proven to have sufficiently high reliability numbers for TMC systems.

Today, many DCS and PLC vendors provide dual-redundant 1oo2D systems that are widely accepted for nonsafety applications.

A typical TMR system has an architecture similar to the one shown in Figure 13.4.

Each CPU votes based on 2oo3 on the three signals received from the input modules that process the signals, and then votes on the value sent to the output modules. In turn, the output modules vote on the received values and then vote on the final value of the output.

The dual-redundant systems shown in Figure 13.5 operate in a 1oo2D mode. The signals are similarly exchanged between the modules, only instead of the 2oo3 voting diagnostics used to eliminate a value in case of any failures in its channel, some systems use the module active or duty concept. With this concept, the voting

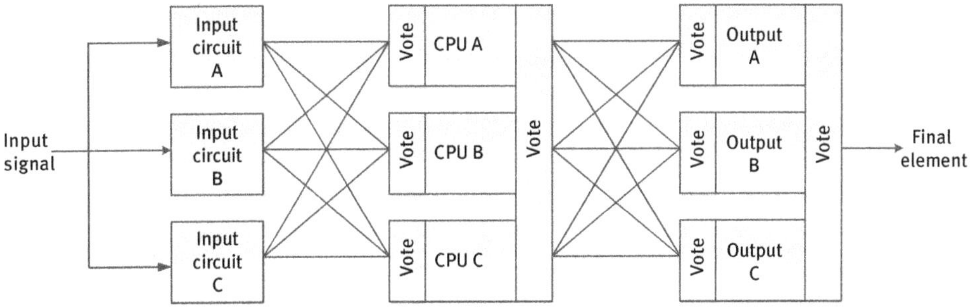

Figure 13.4: TMR architecture (source: *Comp. Controls Corp.*).

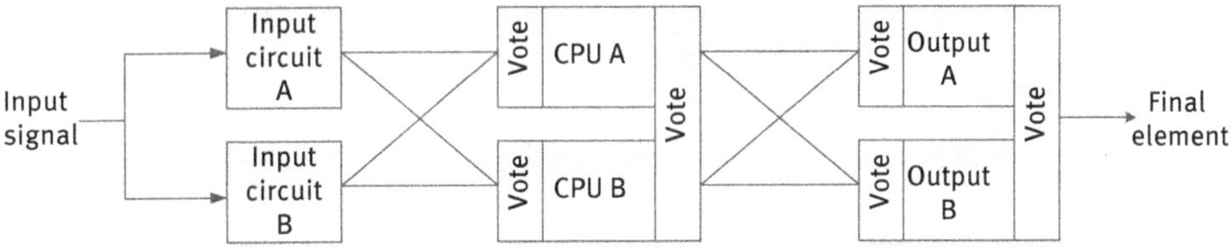

Figure 13.5: Dual-redundant architecture (source: *Comp. Controls Corp.*).

mechanism allows the active modules to determine the output of the system, as long as the system is healthy. If there are active module faults, the redundant peer module becomes active.

In many applications, the actuator accepts a single position demand signal. Accordingly, a redundant system must supply a single output, which is determined either via a switching relay scheme, or, in case of the current output, each module produces its share of the output current. In many cases off-the-shelf PLCs do not have the capability of producing a single output from redundant output modules. The situation often results in the scheme shown in Figure 13.6, with simplex I/O and redundant processors.

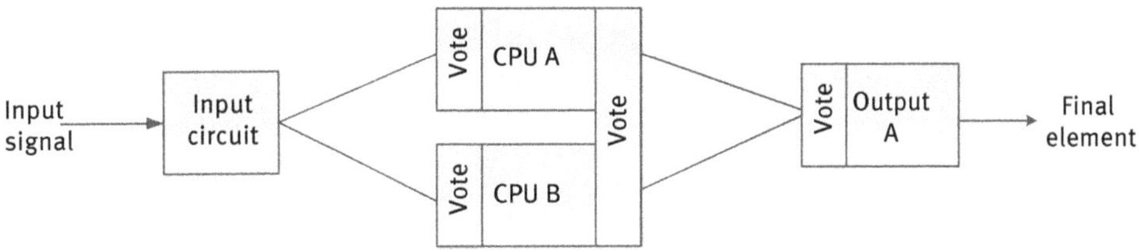

Figure 13.6: Simplex I/O and redundant CPUs (source: *Comp. Controls Corp.*).

In some systems, particularly those initially designed for safety applications, the CPUs are synchronized at each step of execution, ensuring voting with data acquired at the same time for each execution cycle. However, such synchronization requires significant complexity and overhead. For nonsafety applications and when the execution cycles are an order of magnitude faster than the process time constants, the processor synchronization does not noticeably improve the behavior of the system and may not be necessary. Determinism, that is, the ability of a control system to guarantee that a control loop executes within a prescribed time interval, is extremely important for TMC systems. Determinism guarantees a certain response time, regardless of CPU loading or other factors.

Many of today's systems are "plug-and-play," meaning that they do not require any configuration prior to, or after replacing of a failed module. Instead, the healthy CPU module remaining in the system automatically recognizes that a new module has joined the system and downloads the required firmware, application software and configuration to the new module.

13.4.1 Software standards

Adoption of the IEC 6-1131 software standard for programmable logic controllers has been increasing dramatically within the automating industry. Many PLC and DCS vendors today provide software development packages based on the IEC 6-1131 concepts. Purpose-built TMC systems have also been incorporating IEC 61131, increasing the reliability and maintainability of the application code.

13.5 Surge detection and API-670, fifth edition

Surge in centrifugal and axial compressors often manifests itself as oscillations of flow and pressure. This also relates to other compressor parameters, such as rotating speed. Many of the devices, which incorporate surge detection function, use either the location of the operating point relative to the surge limit line or physical manifestations of surge, such as high rates of changes of selected process signals for identifying surge. Using both methods increases the reliability of detection. For example, the rate-of-change methods can identify surge when the SLL is incorrectly configured, instrumentation malfunctions, or in case compressor characteristics change due to some types of failure, such as a partially blocked inlet.

API 670, fifth edition, lists compressor flow, discharge pressure, and inlet temperature as the signals that should be used for surge detection. The surge detection function should be able to identify each surge cycle. An accepted approach in the industry is to issue an alarm, or actually trip the unit if the number of cycles exceeds a predefined threshold within a given time period, for example, three cycles in 10s.

The API 670 fifth edition also prescribes using a segregated surge detector for axial compressors and recommends segregated detection for critical centrifugal machines.

The standard recommends detecting surge when flow decreases below lowest possible surge flow, or on a rapid decrease of the signal. For the inlet temperature method of detection, surge would be assumed when "the compressor inlet temperature rises considerably above the upstream gas temperature." Using temperature for surge detection requires probes with fast response times, which are seldom used in process applications. There are no specific recommendations for discharge pressure, other than detecting rapid decrease and recovery. Detection of rapid decrease in the signal means that the surge detector should be able to calculate the rate of change of the signal.

In practice, reliable implementation of surge detection based on the rate of change presents a number of difficulties. First of all, a detection function should be able to differentiate between process changes and surging. The surge signature of the compressor may be quite different in the field compared to shop testing. Therefore, surge testing should be conducted in the field in order to identify rates of change during surge. However, in many cases this is not done, particularly in new installations with large and critical machines with high potential for damage from forces acting on machine components. A typical case where this is not done is in base-load LNG plants. Field surge testing is often not conducted, due to the risk of a machine being damaged. Therefore, the surge signature remains unknown.

Furthermore, the instrumentation used for measurement, such as differential pressure transmitters used for flow measurement, may limit the sensitivity of the rate of change determination. This is due to poor speed of response and excessive damping used for noise suppression.

In Figure 13.7 the flow exhibits a typical behavior during the surge cycle, while discharge pressure experiences only a relatively small decrease. This occurs quite often in process machines during the first surge cycle, which is when the discharge pressure time constant can be quite large. This can be attributed to process volumes and occurs when the discharge pressure transmitter is located at a significant distance from the actual machine discharge. In Figure 13.8, both pressure and flow experience significant decrease in magnitude during surge. In order to minimize nuisance detection, at least two parameters should be used for detecting surge. The practical choice is often reduced to flow measurement because in a particular process application the pressures do not experience sufficient fluctuations during surge. Reliable detection is also difficult when sensor locations and process flows work against us.

The location of the instrumentation plays an important role. Flow and pressure measurements distant from the actual discharge and suction of the machine may not exhibit significant changes during first several surge cycles. In addition, the detector should also be able to discern between high rates of change during signal

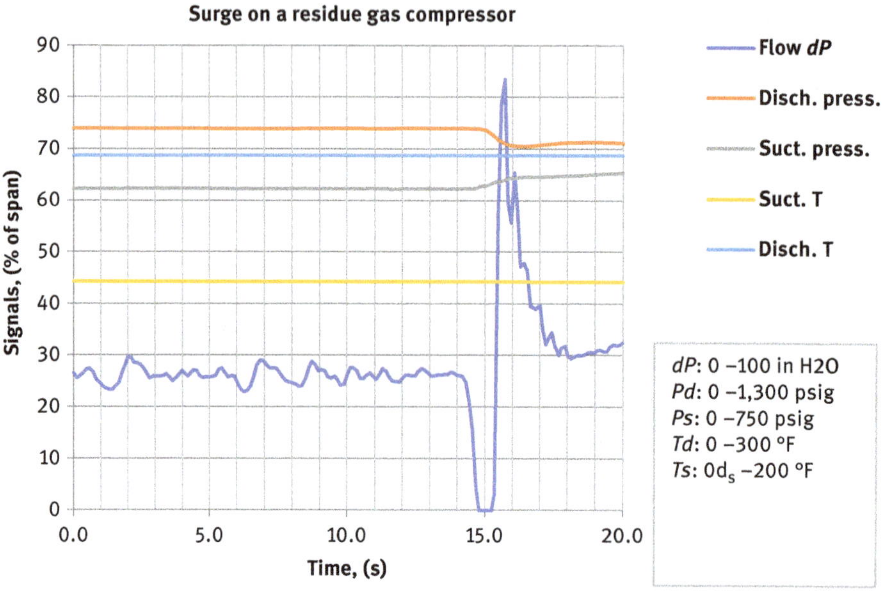

Figure 13.7: Surge signature of a residue gas compressor (source: *Comp. Controls Corp.*).

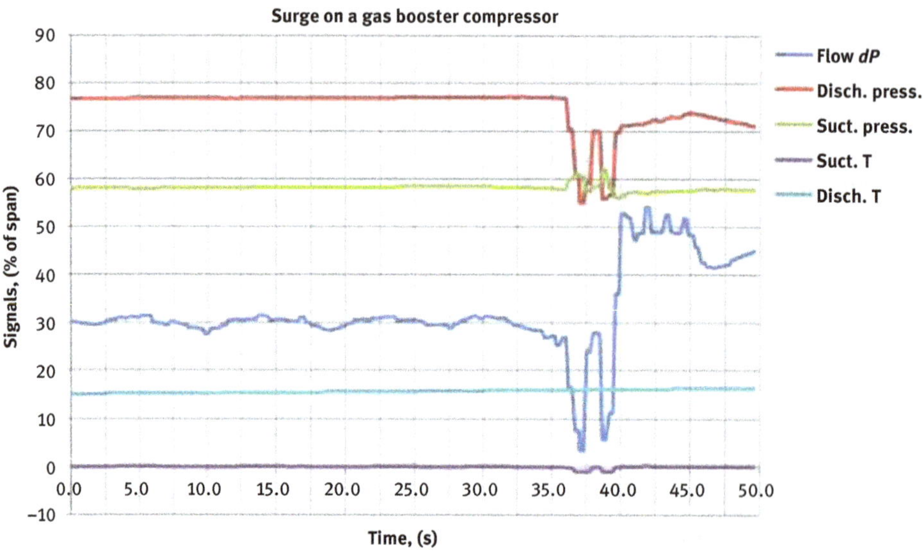

Figure 13.8: Surge signature of a gas booster compressor (source: *Comp. Controls Corp.*).

failures and surging. Considering that flow can decrease to zero within 150 ms during a surge cycle, this is one more reason why the control system should have loop execution times of around 40 ms.

13.6 New development for TMC systems

Faulty measurements from signals used by the TMC systems often result in a unit's going off-line. There can then also be nuisance trips and, consequently, a major process disturbance. Deployment of smart instruments has improved the diagnostic coverage; however, issues such as blocked impulse lines, condensation, and processor failures in the instrument still cause a significant number of incidents. Differential pressure measurements are most susceptible to faulty readings, generally attributable to the relatively low level of the signal and its sensitivity to distortions due to issues in the impulse lines. While many of the smart transducers available on the market exhibit very high reliability, there have been cases where the device processor malfunctioned and stopped updating its output value; basically, it just "froze." The question is, then, how to identify faulty measurements when the transmitter itself does not provide diagnostics and the signal is electrically "healthy?"

13.6.1 Signal drift detection

One method for failure detection in such cases is to use correlation with the other measurements coming from the turbocompressor unit, that is, we can use physical characteristics of the machine to reject behavior that does not correlate with the expected machine behavior.

Figure 13.9 contains a simplified logic diagram for determination of a failure of one signal in a redundant pair. Please note that many of the logic details are not shown for the sake of simplicity. The concept can be extended to nonredundant transmitters as well.

For a compressor unit, signals, such as flow, suction, discharge pressure and temperature, motor power or current for machines with electric drives, and rotating speed, have certain physical relationships to each other. Flow through the compressor cannot change without a corresponding change in pressure ratio, consumed power, or rotating speed. In Figure 13.9, a signal is considered in steady-state condition when its derivatives over multiple time intervals are within a configurable dead-band. A system is considered in steady state when all of the participating signals are in steady state. One of the signals in a redundant pair (signals 1A and 1B in Figure 13.9) is rejected when:

1. The spread (difference) between redundant signals is above a threshold, and
2. The steady-state condition of the system does not match the signal steady-state condition.

The signal is restored when its condition matches the system condition and the spread between redundant signals is within tolerance.

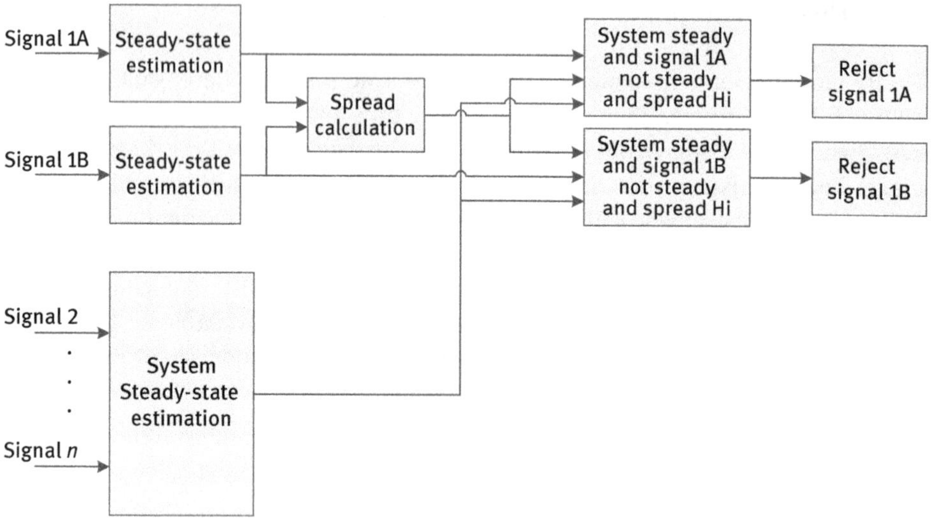

Figure 13.9: Simplified diagram for rejecting a drifting signal (source: *Comp. Controls Corp.*).

13.6.2 Detecting a "frozen" signal

When a transmitter does not update its output value, or when the isolation valve(s) on a pressure transmitter is closed, the characteristics of the signal usually change, and the amplitude of the noise or pulsations present in the signal decreases. The algorithm for identifying a "frozen" signal detects when a signal remains within a certain tolerance. The tolerance to be chosen must have values less than the normally observed values; the excursion must be observed over an extended period of time.

13.6.3 Reducing the risk of antisurge valve not opening

Filed experience shows that it is possible for a recycle valve to become "stuck" in one position due, most likely, to increased stiction or debris buildup in or around the seat of the valve. The longer the valve remains in one position, the higher is the risk of the valve becoming stuck. Figure 13.10 shows a recycle valve "stuck" at around 30% opening although the position demand is at 0% of fully closed position. After some time, the valve finally moves, causing a large disturbance to the process, and approaching the surge limit of the compressor.

Periodic exercising of the valve can mitigate the risk of the valve being stuck. The algorithm, shown in a simplified schematic in Figure 13.11, periodically sends a test command signal to the valve. The test signal can only be generated when the system is in steady-state condition and in the presence of a permissive response

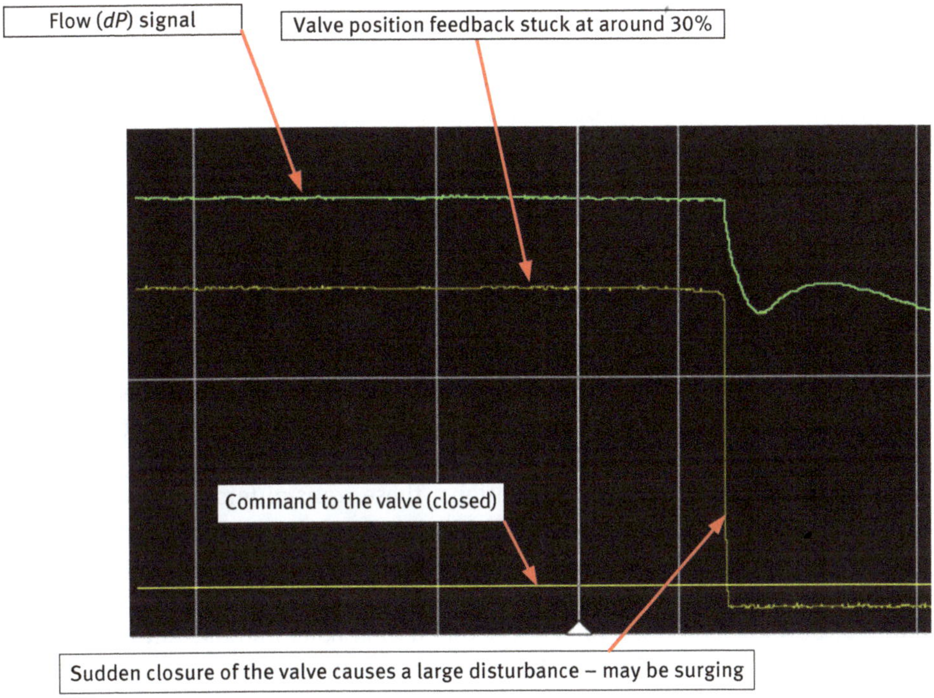

Figure 13.10: Stuck antisurge valve (source: *Comp. Controls Corp.*).

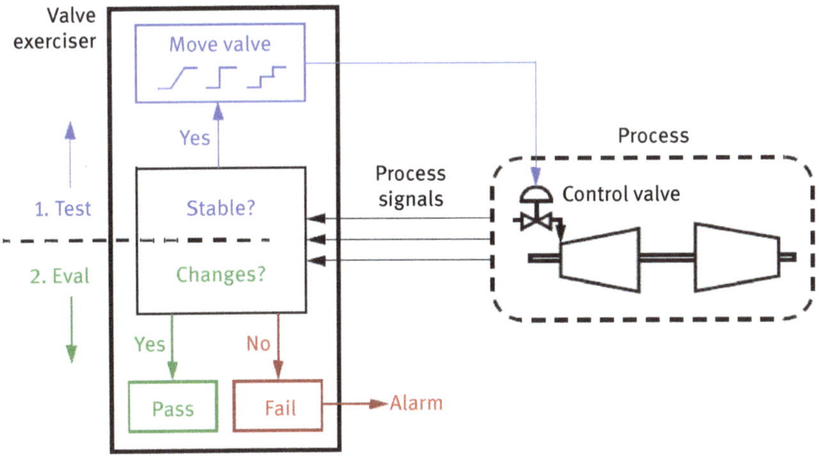

Figure 13.11: Partial stroke testing (valve exerciser) (source: *Comp. Controls Corp.*).

from the operator. The test signal may consist of a step, a series of steps, or a ramp. The test criteria may be based not only on the valve position feedback but also on other signals, such as flow and proximity to surge parameter. The test passes if the steady-state condition, based on the selected signals, is no longer true; that is, anyone of the selected signals began to change within the configured time period.

13.7 Instrumentation selection

13.7.1 Flow measurement

The most common instrumentation for TMC systems includes flow measurement using obstruction-type flow meters. These flow meters require the installation of differential pressure transmitters, and pressure measurements with associated pressure transmitters. For gas applications, these transmitters must be installed above the process piping. Impulse tubing length must not exceed 3 m with a rising slope of at least 30° to allow for free drainage of any condensate that may form within the tubing. Bends and elbows should be minimized in order to avoid leaks that may develop as a result of bending the tubing. Long impulse lines contribute to signal noise, as pulsations in the pipe can be amplified by the gas resonance phenomena in the impulse tubing.

Figure 13.12 is an example of severe signal noise. The signal from the flow transmitter is represented by the green line, while the red line represents the deviation from surge calculated by the controller. The signal noise in this example was so severe that surge testing and continual operation of this compressor could not be accomplished until the problem was rectified.

13.7.2 Antisurge valve

Transients in the region close to surge can be extremely fast due to the relative flatness of the compressor performance curve in this region. For example, the operating point may shift from the safe area into the surge region in less than 1 s, and the duration of a typical surge cycle can be less than one second. Therefore, to adequately protect the compressor from surge, the valve's speed of response must be similarly fast. At the same time, the valve positioning system must provide stable and precise control while operating on the surge control line for extended periods of time. Poor performance of the position system may result in larger antisurge control margins and increases the risk of surging. Therefore, the requirements for dynamic performance of the valve positioning system must reflect both the speed of response and precise control.

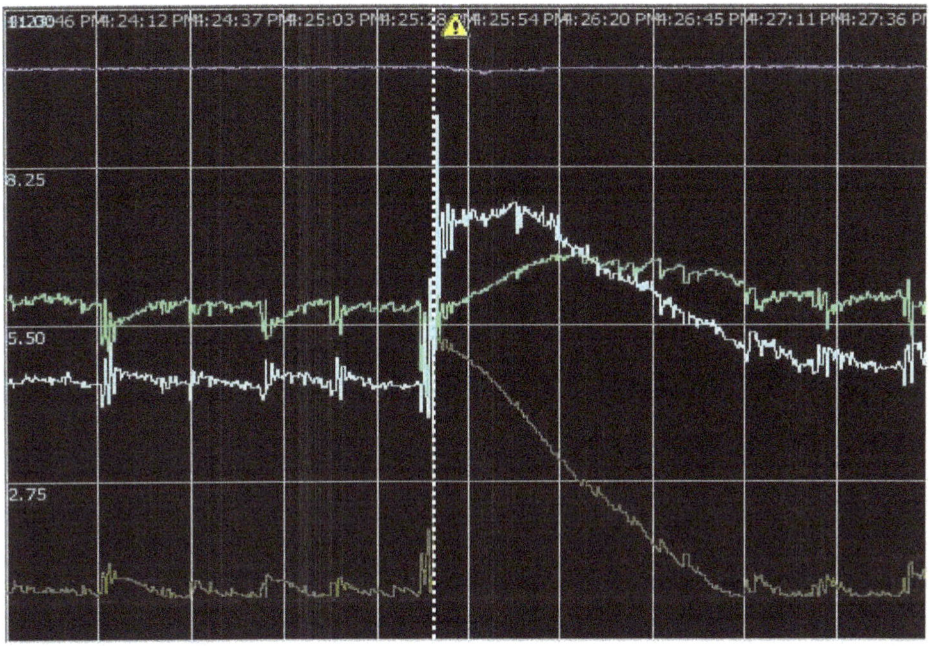

Figure 13.12: Effects of noisy flow signal on antisurge control (source: *Comp. Controls Corp.*).

Table 13.1: General antisurge valve characteristics.

Valve characteristic	Recommended behavior
Flow direction	Flow to open
Positioning/actuation failure mode	Fail open
Flow characteristic	Linear, equal percentage accepted in certain cases

Table 13.2: Full stroke test requirements.

	Description	Criteria	Notes
1	Stroking speed: full opening by automatic control in response to the control signal step from 20 to 4 mA.	*T*95 ≤ 2.0 s preferred 2.5 s max	Dead time included. Stroke time measured at 95% of the opening for cases with slow endpoint movement
2	Closure by automatic control in response to the control signal step from 4 to 20 mA.	Equal to opening stroke time preferred *(10 s max)*	*T*95 from 100% to 0%

Table 13.2 (continued)

	Description	Criteria	Notes
3	Opening dead time, starting from full thrust on seat and control signal step to 100% open.	≤0.5 s	Measure to the point when the valve starts moving (0.5% opening)
4	Full opening time by de-energizing the solenoid valve	*As per project requirements*	

Table 13.3: Dynamic response test requirements.

	Description	Criteria	Notes
1	Large step response (overshoot)	single overshoot ≤3% of span	Started from 10% open, step tests in both directions with minimum 15 s hold period between steps. Recommended steps: 10%, 20%,30%, 40%,. . ., 80% (see Figure 13.8)
2	Stiction Ramp test	No stick/slip motion excluding the end points $dx/dt > 0$	Control signal ramp from 4 to 20 mA and 20 to 4 mA. Full stroke time = 5 min, (ramp rate ≈ 0.3%/s). The response should be a continuous ramp – no "stair steps"
3	Terminal-based linearity	Within ±1%	Test at 0%, 25%, 50%, 75%, and 100% based on actual stem motion, not position feedback
4	Dead band test (dead band + hysteresis)	≤1.0% of command signal	Measure the amount of change in the position command that initiates reverse movement of the position signal. Test at 10% and 50% of the stroke: 0.25, 0.5. 1, 2, 5 steps (see Figure 13.9)
5	Maximum control signal to initiate movement off the valve seat from the full thrust position	≤3% of stroke	Initiate step of 0.5% starting from fully closed position. Check for valve movement. If none occurs within 10 s, initiate another step starting from closed position. Continue until valve begins to move. Steps 0.5%, 1%, 1.5%, 2%, 2.5%, . . .

Appendix A Physical properties of various gases

A.1 Tables and charts for properties of common gases

Table A.1: Individual characteristic gas constants of some typical gases.

Gas		Molecular weight	Individual/characteristic gas constant – *R*				
Name	**Formula**	**(g/mol), (kg/kmol)**	**(J/kg K)**	**(kJ/ kg K)**	**(kcal/(kg K)), (Btu.(IT)/ lb °F)**	**(ft lbf./ lb °R)**	**(ft lbf/ slug °R)**
Acetylene	C_2H_2	26.038	**319.32**	0.3193	0.07627	59.35	**1,910**
Air	A mixture	28.9647	**287.05**	0.2871	0.06856	53.35	**1,717**
Ammonia	NH_3	17.031	**488.21**	0.4882	0.11661	90.74	**2,919**
Argon	Ar	39.948	**208.13**	0.2081	0.04971	38.68	**1,245**
Butane	C_4H_{10}	58.122	**143.05**	0.1431	0.03417	26.59	**855**
Butene	C_4H_8	56.106	**148.19**	0.1482	0.03539	27.54	**886**
Carbon dioxide	CO_2	44.01	**188.92**	0.1889	0.04512	35.11	**1,130**
Carbon monoxide	CO	28.01	**296.84**	0.2968	0.0709	55.17	**1,775**
Chlorine	Cl_2	70.906	**117.26**	0.1173	0.02801	21.79	**701**
Chloromethane	CH_3Cl	50.488	**164.68**	0.1647	0.03933	30.61	**985**
Ethane	C_2H_6	30.069	**276.51**	0.2765	0.06604	51.39	**1,654**
Ethene	C_2H_4	28.053	**296.38**	0.2964	0.07079	55.09	**1,772**
Fluorine	F_2	37.997	**218.82**	0.2188	0.05226	40.67	**1,309**
Helium	He	4.003	**2,077.1**	2.0771	0.4961	386.05	**12,421**
Hydrogen	H_2	2.016	**4,124.2**	4.1242	0.98506	766.54	**24,663**
Hydrogen chloride	HCl	36.461	**228.04**	0.228	0.05447	42.38	**1,364**
Hydrogen sulfide	H_2S	34.081	**243.96**	0.244	0.05827	45.34	**1,459**
Methane (natural gas)	CH_4	16.042	**518.28**	0.5183	0.12379	96.33	**3,099**
Nitrogen	N_2	28.013	**296.8**	0.2968	0.07089	55.17	**1,775**

https://doi.org/10.1515/9783110678765-014

Table A.1 (continued)

Gas		Molecular weight	Individual/characteristic gas constant – *R*				
Name	**Formula**	**(g/mol), (kg/kmol)**	**(J/kg K)**	**(kJ/ kg K)**	**(kcal/(kg K)), (Btu.(IT)/ lb °F)**	**(ft lb$_{f.}$/ lb °R)**	**(ft lb$_f$/ slug °R)**
Nitrogen dioxide	NO_2	46.006	**180.73**	0.1807	0.04317	33.59	**1,081**
Nitrogen trifluoride	NF_3	71.002	**117.1**	0.1171	0.02797	21.77	**700**
Nitrous oxide	N_2O	44.012	**188.91**	0.1889	0.04512	35.11	**1,130**
Oxygen	O_2	31.999	**259.84**	0.2598	0.06206	48.29	**1,554**
Propane	C_3H_8	44.096	**188.56**	0.1886	0.04504	35.05	**1,128**

Table A.2: Critical properties for common gases.

Substance	Critical temperature		Critical pressure		Boiling temperature (1 atm)	
	(°F)	**(°C)**	**lb/in² [kPa]**	**(bar)**	**(°F)**	**(°C)**
Air	−220.94	−140.52	549.08 [3,786]	37.858	–	–
Ammonia (NH_3)	270	132.4	1,636	112.8	−27.4	−33
Argon	−188	−122	705.6	48.7	−302.5	−85.8
Butane	305.6	152	550.4	38	32	0
Carbon dioxide (CO_2)	87.8	31.2	1,071.6	73.8	−110	−79
Carbon monoxide (CO)	−220.5	−140.3	507.5 [3,499]	35	−310	−190
Chlorine	291	144	1,118.7 [7,713]	77.1	−29.3	−34
Ethane	90.0	32.2	708 [4,882]	48.9	−127.4	−88.5
Ethanol (alcohol)	467	242	914 [6,302]	63	173	78.4
Ethyl ether	381	194	522 [3,599]	36	94.2	34.6
Ethylene	48.9	9.4	735 [5,068]	50.7	−272.6	−169.2
Fluorine	−200	−129	808.5 [5,575]	55.8	−307	−188
Helium	−456	−271	33.2 [229]	2.3	−452	−269
Hydrogen (H)	−400	−240	188.2 [1,298]	13.0	−423	−253

Table A.2 (continued)

Substance	Critical temperature		Critical pressure		Boiling temperature (1 atm)	
	(°F)	(°C)	lb/in² [kPa]	(bar)	(°F)	(°C)
Hydrogen chloride	125	51.6	1198 [8,260]	82.7	−121.1	−85.1
Isobutane	274	135	529.2 [3,649]	36.5	10.9	−11.7
Isobutylene	293	145	580 [3,999]	40.0	19.6	−6.9
Isononane	590	310	335.1 [2,311]	23.1	303	151
Methane	−117	−82.6	673.3 [4,642]	46.5	−259	−162
Nitrogen (N)	−232.6	−147	492.4 [3,395]	34.0	−321	−195
Nitrous oxide (N_2O)	97.4	36.4	1,047.6 [7,223]	72.3	−127	−88.5
Oxygen (O_2)	−181.5	−118.6	732	50.5	−297	−183
Phosgene			823.2	56.8	46.9	8.3
Propane	206.1	96.7	617.4	42.6	−44	−42
Propyl benzene	689	365	464.2	32	319	159
Propylene	198	92.4	670.3	46.3	−54	−48
Refrigerants						
Undecane	691	366	287.2	19.8	385	196
Water	705	374	3,206.2	220.5	212	100

Note: Pressures are in absolute units

Table A.3: Critical constants and acentric factors for selected fluids.

	T_c (K)	P_c (bar)	υ_c (cm³/mol)	ω
Elements				
Argon	150.8	48.7	74.9	0.001
Bromine	588.	103.	127.2	0.108
Chlorine	416.9	79.8	123.8	0.090
Fluorine	144.3	52.2	66.3	0.054
Helium-4	5.19	2.27	57.4	−0.365
Hydrogen	33.0	12.9	64.3	−0.216
Iodine	819.	116.5	155	0.229
Krypton	209.4	55.0	91.2	0.005
Neon	44.4	27.6	41.6	−0.029
Nitrogen	126.2	33.9	89.8	0.039

Table A.3 (continued)

	T_c (K)	P_c (bar)	υ_c (cm^3/mol)	ω
Oxygen	154.6	50.4	73.4	0.025
Xenon	289.7	58.4	118.4	0.008
Hydrocarbons				
Acetylene	308.3	61.4	112.7	0.190
Benzene	562.1	48.9	259.	0.212
n-Butane	425.2	38.0	255.	0.199
1-Butene	419.6	40.2	240.	0.191
Cyclobutane	460.0	49.9	210.	0.181
Cyclohexane	553.8	40.7	308.	0.212
Cyclopropane	397.8	54.9	163.	0.130
Ethane	305.4	48.8	148.3	0.099
Ethylene	282.4	50.4	130.4	0.089
n-Heptane	540.3	27.4	432.	0.349
n-Hexane	507.5	30.1	370.	0.299
Isobutane	408.2	36.5	263.	0.183
Isobutylene	417.9	40.0	239.	0.194
Isopentane	460.4	33.9	306.	0.227
Methane	190.4	46.0	99.2	0.011
Naphthalene	748.4	40.5	413.	0.302
n-Octane	568.8	24.9	492.	0.398
n-Pentane	469.7	33.7	304.	0.251
Propadiene	393.	54.7	162.	0.313
Propane	369.8	42.5	203.	0.153
Propylene	364.9	46.	181.	0.144
Toluene	591.8	41.0	316.	0.263
m-Xylene	617.1	35.4	376.	0.325
o-Xylene	630.3	37.3	369.	0.310
p-Xylene	616.2	35.1	379.	0.320
Miscellaneous inorganic compounds				
Ammonia	405.5	113.5	72.5	0.250
Carbon dioxide	304.1	73.8	93.9	0.239
Carbon disulfide	552.	79.0	160.	0.109
Carbon monoxide	132.9	35.0	93.2	0.066
Carbon tetrachloride	556.4	45.6	275.9	0.193
Carbon tetrafluoride	227.6	37.4	139.6	0.177
Chloroform	536.4	53.7	238.9	0.218
Hydrazine	653.	147.	96.1	0.316
Hydrogen chloride	324.7	83.1	80.9	0.133
Hydrogen fluoride	461.	64.8	69.2	0.329
Hydrogen sulfide	373.2	89.4	98.6	0.081
Nitric oxide	180.	64.8	57.7	0.588
Nitrous oxide	309.6	72.4	97.4	0.165
Sulfur dioxide	430.8	78.8	122.2	0.256

Table A.3 (continued)

	T_c (K)	P_c (bar)	υ_c (cm^3/mol)	ω
Sulfur trioxide	491.0	82.1	127.3	0.481
Water	647.3	221.2	57.1	0.344
Miscellaneous organic compounds				
Acetaldehyde	461.	55.7	154.	0.303
Acetic acid	592.7	57.9	171.	0.447
Acetone	508.1	47.0	209.	0.304
Acetonitrile	545.5	48.3	173.	0.278
Aniline	699.	53.1	274.	0.384
n-Butanol	563.1	44.2	275.	0.593
Chlorobenzene	632.4	45.2	308.	0.249
Dichlorodifluoromethane (freon 12)	385.0	41.4	216.7	0.204
Diethyl ether	466.7	36.4	280.	0.281
Dimethyl ether	400.0	52.4	178.	0.200
Ethanol	513.9	61.4	167.1	0.644
Ethylene oxide	469.	71.9	140.	0.202
Isobutanol	547.8	43.0	273.	0.592
Isopropyl alcohol	508.3	47.6	220.	0.665
Methanol	512.6	80.9	118.0	0.556
Methyl chloride	416.3	67.0	138.9	0.153
Methyl ethyl ketone	536.8	42.1	267.	0.320
Phenol	694.2	61.3	229.	0.438
1-Propanol	536.8	51.7	219.	0.623

Table A.4: Benedict–Webb–Rubin constants.

Gas	Formula	R (kJ/kg K)	A_0 (Nm/kg^2)	B_0 (m^3/kg) × 10^{-3}	C_0 (Nm^4 K^2/kg^2) × 10^7	a (Nm^7/kg^3)
Methane	CH_4	0.518	731.19	2.65735	0.88964	1.21466
Ethylene	C_2H_4	0.296	430.55	1.98649	1.69071	1.19119
Ethane	C_2H_6	0.276	466.27	2.08914	2.01509	1.28892
Propylene	C_3H_6	0.198	350.22	2.02308	2.51642	1.05482
Propane	C_3H_8	0.189	358.58	2.20855	2.65194	1.12224
i-Butylene	C_4H_8	0.148	288.57	2.06958	2.98871	0.97316
i-Butane	C_4H_{10}	0.143	307.31	2.36826	2.55256	1.00195

Table A.4 (continued)

Gas	Formula	R (kJ/kg K)	A_0 (Nm/kg^2)	B_0 (m^3/kg) × 10^{-3}	C_0 (Nm^4 K^2/kg^2) × 10^7	a (Nm^7/kg^3)
n-Butane	C_4H_{10}	0.143	302.87	2.14127	2.98168	0.97334
i-Pentane	C_5H_{12}	0.115	249.39	2.22006	3.40357	1.01546
n-Pentane	C_5H_{12}	0.115	237.38	2.17426	4.13424	1.10159
n-Hexane	C_6H_{14}	0.0965	197.24	2.06498	4.53487	1.12913
n-Heptane	C_7H_{16}	0.0830	177.04	1.98756	4.79543	1.0402

Gas	Formula	b (m^6/kg^2) × 10^{-5}	c (N-m^7 K^2/kg^3) × 10^5	α (m^9/kg^3) × 10^{-9}	γ (m^6/kg^2) x 10^{-6}
Methane	CH_4	1.31523	0.62577	30.1853	23.3469
Ethylene	C_2H_4	1.09451	0.97139	8.08173	11.7469
Ethane	C_2H_6	1.23191	1.22361	8.97220	13.0701
Propylene	C_3H_6	1.05806	1.39829	6.13014	10.3453
Propane	C_3H_8	1.15892	1.52759	7.09776	11.3317
i-Butylene	C_4H_8	1.10774	1.58056	5.16963	9.41616
i-Butane	C_4H_{10}	1.25806	1.47891	5.48279	10.0799
n-Butane	C_4H_{10}	1.18582	1.63610	5.62184	10.0799
i-Pentane	C_5H_{12}	1.28545	1.87887	4.53682	8.90805
n-Pentane	C_5H_{12}	1.28545	2.22807	4.83038	9.13893
n-Hexane	C_6H_{14}	1.47181	2.40013	4.40244	8.99353
n-Heptane	C_7H_{16}	1.51575	2.49275	4.33982	8.97754

Table A.5: Molal heat capacity of common gases.

Molal heat capacity MCp (ideal gas state) Btu/lb-mol °R										
Gas	**Chemical formula**	**Mol. weight**	**MCp**							
			0 °F	**50 °F**	**60 °F**	**100 °F**	**150 °F**	**200 °F**	**250 °F**	**300 °F**
Methane	CH_4	16.043	8.23	8.42	8.46	8.65	8.95	9.28	9.64	10.01
Ethyne (acetylene)	C_2H_2	26.038	9.68	10.22	10.33	10.71	11.15	11.55	11.90	12.22
Ethene (ethylene)	C_2H_4	28.054	9.33	10.02	10.16	10.72	11.41	12.09	12.76	13.41
Ethane	C_2H_6	30.070	11.44	12.17	12.32	12.95	13.78	14.63	15.49	16.34
Propene (propylene)	C_3H_6	42.081	13.63	14.69	14.90	15.75	16.80	17.85	18.88	19.89
Propane	C_3H_8	44.097	15.65	16.88	17.13	18.17	19.52	20.89	22.25	23.56
1-Butene (butylene)	C_4H_8	56.108	17.96	19.59	19.91	21.18	22.74	24.26	25.73	27.16
cis-2-Butene	C_4H_8	56.108	16.54	18.04	18.34	19.54	21.04	22.53	24.01	25.47
trans-2-Butene	C_4H_8	56.108	18.84	20.23	20.50	21.61	23.00	24.37	25.73	27.07
Isobutane	C_4H_{10}	58.124	20.40	22.15	22.51	23.95	25.77	27.59	29.39	31.11
n-Butane	C_4H_{10}	58.124	20.80	22.38	22.72	24.08	25.81	27.55	29.23	30.90
Isopentane	C_5H_{12}	72.151	24.94	27.17	27.61	29.42	31.66	33.87	36.03	38.14
n-Pentane	C_5H_{12}	72.151	25.64	27.61	28.02	29.71	31.86	33.99	36.08	38.13
Benzene	C_6H_6	78.114	16.41	18.41	18.78	20.46	22.45	24.46	26.34	28.15
n-Hexane	C_6H_{14}	86.178	30.17	32.78	33.30	35.37	37.93	40.45	42.94	45.36
n-Heptane	C_7H_{16}	100.205	34.96	38.00	38.61	41.01	44.00	46.94	49.81	52.61
Ammonia	NH_3	17.031	8.52	8.52	8.52	8.52	8.52	8.53	8.53	8.53
Air	Air	28.964	6.94	6.95	6.95	6.96	6.97	6.99	7.01	7.03
Water	H_2O	18.015	7.98	8.00	8.01	8.03	8.07	8.12	8.17	8.23
Oxygen	O_2	31.999	6.97	6.99	7.00	7.03	7.07	7.12	7.17	7.23
Nitrogen	N_2	28.013	6.95	6.95	6.95	6.96	6.96	6.97	6.98	7.00
Hydrogen	H_2	2.016	6.78	6.86	6.87	6.91	6.94	6.95	6.97	6.98

Table A.5 (continued)

Molal heat capacity MCp (ideal gas state) Btu/lb-mol °R										
Gas	**Chemical formula**	**Mol. weight**	**MCp**							
			0 °F	**50 °F**	**60 °F**	**100 °F**	**150 °F**	**200 °F**	**250 °F**	**300 °F**
Hydrogen sulfide	H_2S	34.076	8.00	8.09	8.11	8.18	8.27	8.36	8.46	8.55
Carbon monoxide	CO	28.010	6.95	6.96	6.96	6.96	6.97	6.99	7.01	7.03
Carbon dioxide	CO_2	44.010	8.38	8.70	8.76	9.00	9.29	9.56	9.81	10.05

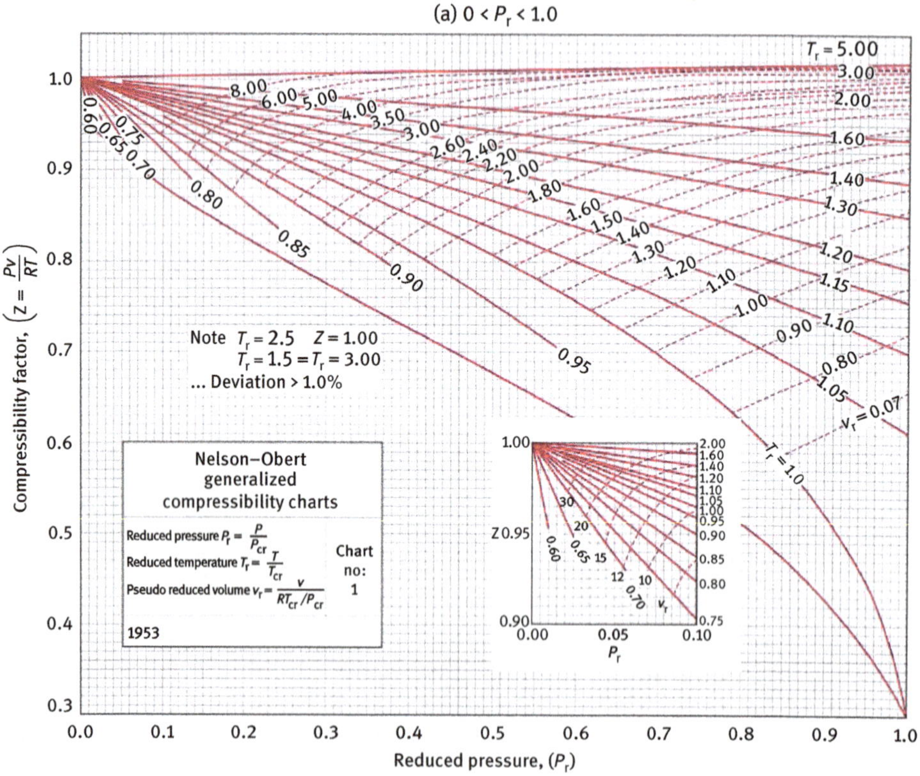

Chart A.1: Nelson–Obert generalized compressibility factor charts.

Reduced pressure, (P_r)

(b) $0 < P_r < 7$

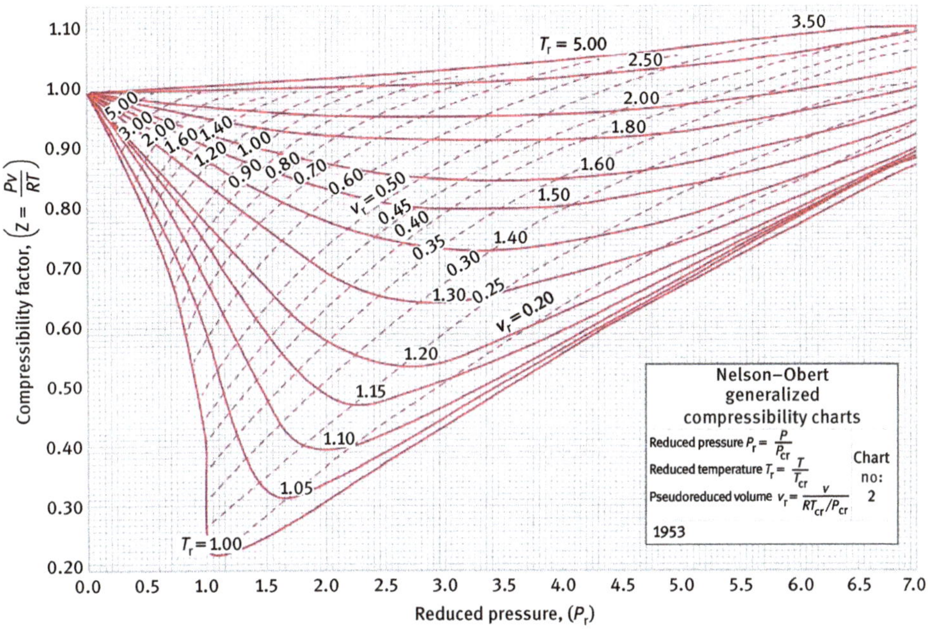

Chart A.2: Nelson–Obert generalized compressibility factor charts.

Nelson–Obert generalized (averaged) compressibility chart

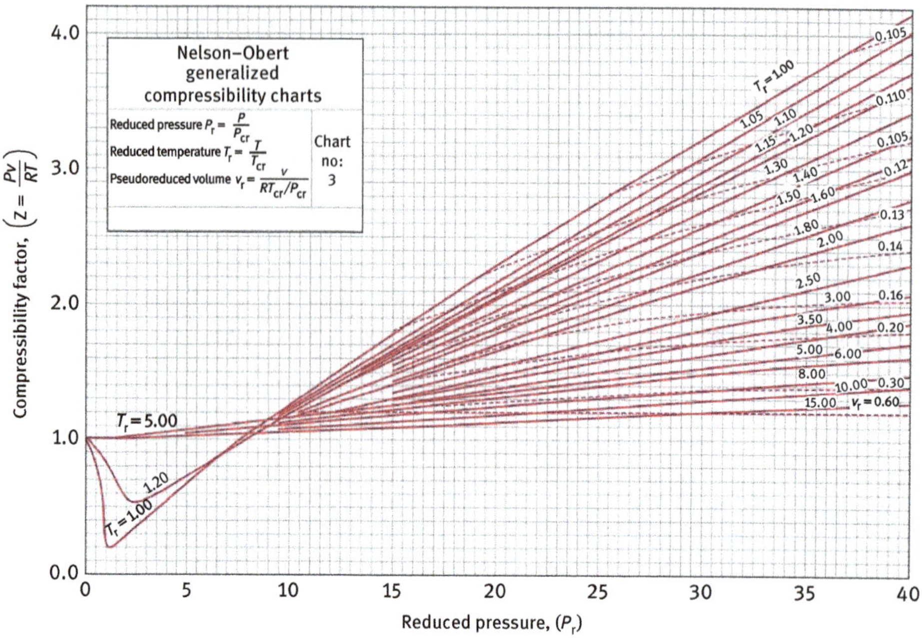

Chart A.3: Nelson–Obert generalized compressibility factor charts.

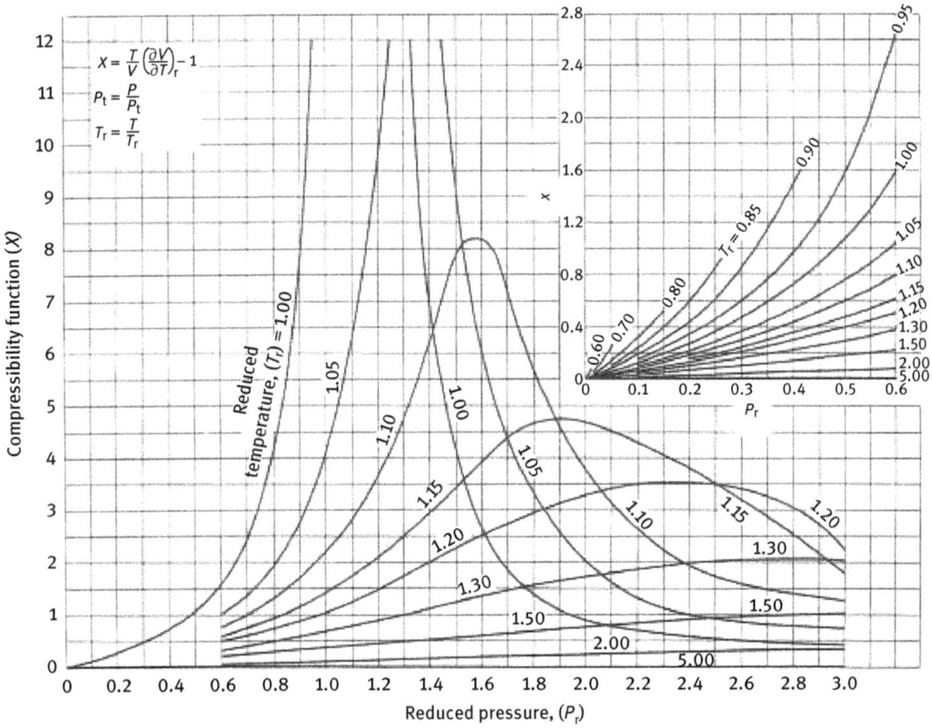

Chart A.4: Generalized compressibility function *X*.

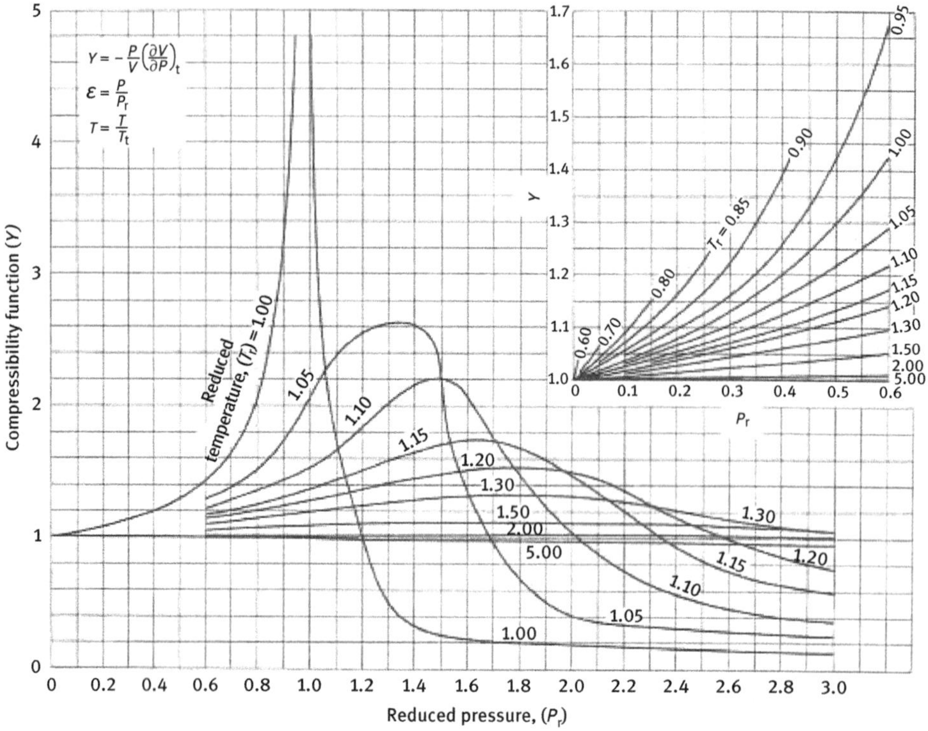

Chart A.5: Generalized compressibility function *Y*.

A.2 Conversions factors for commonly used units in the process industry

Working with engineering procurement and construction personnel (and even our engineering colleagues), we find that mistakes are sometimes made when converting process units from the US customary to metric and/or vice versa.

Since both the US Customary and the metric systems coexist in engineering, we intend here to give engineers and process designers a quick reminder of important conversions factors used for common engineering parameters.

Meaningful conversion can be a bit of a challenge. Engineers tend to think quantitatively in the units to which they are accustomed. They learned to express these measurements before fully appreciating the importance of rapidly grasping unit quantities as they are known to others. In one such event, the water injection rate into a dry screw compressor was calibrated in US customary units, although it should have been calibrated in SI units. This simple oversight resulted in wrecking the compressor soon after initial startup.

There is even confusion in the way measurements are best expressed within the family of US customary units. As an example, the American Petroleum Institute Standards API 617, 618, 619, 672, and 681 use the following units to express volume flow:

- Million standard cubic feet per day (MMSCFD). MMSCFD is a unit commonly used in the United States for measuring the volume flow of natural gas, liquefied petroleum gas, and compressed gases that are extracted, processed, and transported in large quantities. The volume is calculated based on 14.7 psia and 60 °F.
- Standard cubic feet per minute (SCFM). SCFM is similar to MMSCFD, and used in the same way, except for smaller volume flows. It is also based on 14.7 psia and 60 °F. For comparison:

$$1\,\text{SCFM} = 0.00144\,\text{MMSCFD}$$

- Inlet cubic feet per minute/actual cubic feet per minute (ICFM/ACFM). ICFM and ACFM are used interchangeably since both are defined as the volume of the gas entering a compressor section or stage (the intake volume) and calculated based on the actual temperature and pressure of the gas.

Conversion of ACFM to SCFM is done using the ideal gas equation of state:

$$\text{ACFM} = \text{SCFM}\left(\frac{P_s}{P_1}\right)\left(\frac{T_1}{T_s}\right)$$

where $P_s = 14.7$ psia, $T_s = (60\ °\text{F} + 460\ °\text{R}) = 520\ °\text{R}$, $T_1 =$ inlet temperature in °R of the gas, $P_2 =$ inlet absolute pressure of the gas.

In the metric system of units, volume flows are expressed as follows:

- Normal cubic meters per hour (Nm^3/h). The unit of volume is calculated based on 1.013 bar and 0 °C (273.15 °K).
- Standard cubic meters per hour (Sm^3/h). The unit of volume is calculated based on 1.013 bar and 15 °C (288.15 °K).
- Actual cubic meters per hour (Am^3/h). This is the gas volume calculated at the actual condition of temperature and pressure.

Conversion is done using the ideal equation of state.

$$\frac{P_1 V_1}{T_1} = \frac{P_2 V_2}{T_2}$$

Table A.6: Gas volume conversion table – for transporting large quantities.

Gas flow units	Million std. cubic feet per day (gas) MMSCFD	Std. cubic feet of gas per minute SCFM	Std. cubic meter per hour @ 15 °C Sm³/h	Normal cubic meter per hour @ 0 °C Nm³/h
Million standard cubic feet of gas per day MMSCFD	= 1 MMSCFD	694.44 SCFM	1,177.17 Sm³/h	1,115.89 Nm³/h
Standard cubic feet per minute (gas) SCFM	0.00144 MMSCFD	= 1 SCFM	1.69512815 Sm³/h	1.6068862 Nm³/h
Standard cubic meter per hour at **15 °C** Sm³/h	0.000849493 MMSCFD	0.589926 SCFM	= 1 Sm³/h	0.94794378 Nm³/h
Normal cubic meter per hour at **0 °C** Nm³/h	0.00089643 MMSCFD	0.622321 SCFM	1.054914882 Sm³/h	= 1 Nm³/h
Mega standard cubic meter per day at **15 °C** MSm³/h	35.39555 MMSCFD	24,580.24586 SCFM	41,666.6667 Sm³/h	39,497.65747007 Nm³/h

Table A.7: Conversion factors for other commonly used engineering units.

Parameter	US customary units (a)	Convert to metric	
		Multiplication factor (b)	Metric units (*c*) = (*a*) × (*b*)
Length	Inches (in)	2.54	centimeter (cm)
	"	0.0254	meter (m)
	"	0.0000254	kilometer (km)
	Foot (ft)	30.48	centimeter (cm)
	"	0.3048	meter (m)
	"	0.0003048	kilometer (km)
Force	Pound (lbf)	453.59	gram force (gmf)
	"	0.4535924	kilogram force (kgf)
	"	4.44822	newton (N)

Table A.7 (continued)

Parameter	US customary units (a)	Convert to metric	
		Multiplication factor (b)	**Metric units (c) = (a) × (b)**
Mass	Pound (lbm)	453.59	gram force (gm)
	"	0.4535924	kilogram force (kg)
	"	0.000453592	metric ton (t)
Pressure	Pound/square inch (psi)	0.06804596	atmosphere
	"	0.06804596	bar
	"	6894.757	Pascal
	"	6.894757	kilopascal (kPa)
	"	0.006894757	megapascal
	Inches of mercury	0.03386388	bar
	"	33.86388	millibar
	"	25.4	millimeters of mercury
Energy	British thermal unit (BTU)	252.0	Calorie
	"	0.2519958	kilocalorie
	"	1055.06	joule (newton–meter)
	"	0.000293	kilowatt–hour
	"	0.293	watt–hour
	Pound force–foot (lbf-ft)	0.3238316	calorie
	"	0.000323832	kilocalorie
	"	1.355818	joule (newton–meter)
	"	0.000376616	watt–hour
Power	BTU/min	17.56863	watt
	"	0.01756863	kilowatt
	"	1.76E-05	megawatt
	Foot–pound/min	0.02259697	watt
		2.26E-05	kilowatt
		2.26E-08	megawatt
	Horsepower (hp)	746	watt
		0.746	kilowatt
		0.0007457	megawatt

A.3 Calculating key properties of gas mixtures

Gas mixtures are very common in compressor applications. Therefore, understanding how to calculate properties of mixtures is extremely important to engineers. We will use the example in Table A.8 to show a method for calculating certain properties of a mixtures, including specific heat ratio "k" and compressibility factor "Z."

By applying several of the equations previously presented in Chapter 1, Sections 1.2 and 1.7, we can calculate the following properties of the mixture for the example shown in Table A.8:

- Molecular weight = 14.751 lb/lb mol
- Molal specific heat = 8.274 Btu/lb mol °R
- Critical pressure = 628.11 psi
- Critical temperature = 316.75 °R

From eq. (1.53):

$$\text{Specific heat ratio } k = \frac{MC_p}{MC_p - 1.99} = \frac{8.274}{8.274 - 1.99} = 1.32$$

$$\text{Reduced pressure } Pr = \frac{88}{628} = 0.14$$

$$\text{Reduced temperature } Tr = \frac{(50 + 460)}{316.75} = 1.61$$

$$\text{Compressibility factor } Z = 0.99 \text{ [From Chart A} - 3]$$

Table A.8: Tail gas mixture at 50 °F and 88 psia.

Gas composition	Mole fraction Y	Mol. weight of individual gas M(1)	Mol. weight of mixture $\sum Y \times M$	Individual molal specific heat MCp (2)	Molal Specific Heat of Mixture $\sum Y \times$ MCp	Component critical Pressure (psi) P_c (3)	Critical pressure of mixture $\sum Y \times P_c$	Component critical temperature (°R) T_c (3)	Critical temperature of mixture $\sum Y \times T_c$
Methane	0.901	16.042	14.452	8.42	7.585578	673.3	606.576	343	309.0087
Hydrogen	0.095	2.016	0.1921	6.86	0.653758	188.2	17.93546	60	5.718
Ethylene	0.001	28.052	0.0393	10.02	0.014028	735.0	1.029	508.9	0.71246
Carbon dioxide	0.002	28.01	0.0672	8.70	0.02088	1071.6	2.57184	547.8	1.31472
Mixture properties	**1.000**		**14.751**		**8.274244**		**628.1123**		**316.75388**

Notes:

(1) Taken from Table A1

(2) Taken from Table A5

(3) Taken from Table A2

Appendix B Sizing different types of compressors

B.1 Aerothermodynamic fundamentals of different types of compressors

B.1.1 Centrifugal compressor impeller

Fundamentally, we can apply Newton's third law of motion to any centrifugal impeller to determine the change in angular momentum of the fluid entering and leaving the impeller. Figure B.1 illustrates a typical cross-section of a centrifugal impeller with impeller-eye radius r1 and outside diameter hub radius r2. Simply stated: *The rate of change in angular momentum of the fluid will be equal to the net torque imposed on the fluid by the rotor.*

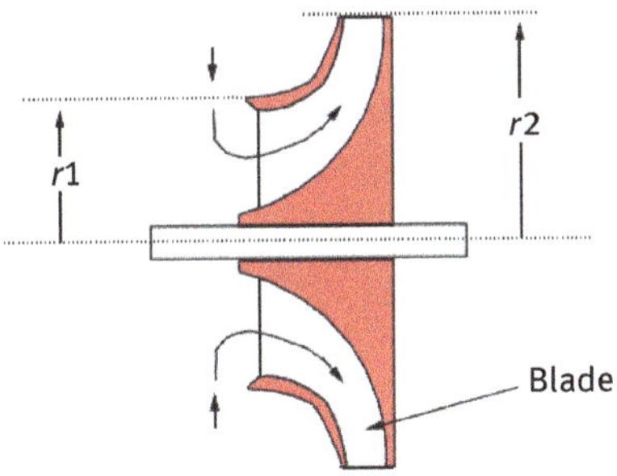

Figure B.1: Outline of typical centrifugal impeller.

$$\tau = \frac{m}{g}(r_1 Vu_1 - r_2 Vu_2) \tag{B.1}$$

where τ represents torque, m is the mass flow, Vu is the tangential component of fluid velocity, and g is the gravitational constant.

The rate of energy/power transferred from the rotor to the fluid "*Eu*" = $\tau\omega$, where ω is the angular velocity.

Therefore:

$$Eu = \frac{m}{g}\omega\ (r_1 Vu_1 - r_2 Vu_2) \tag{B.2}$$

Substituting $U = \omega r$ into eq. (B.2)

$$\frac{Eu}{m} = \frac{1}{g}(U_1 Vu_1 - U_2 Vu_2) = \text{Energy } \textit{per unit mass} \text{ “}H\text{”} \tag{B.3}$$

where U_1 is the inlet tangential velocity at the eye of the impeller and U_2 is the exit tangential velocity at the impeller outside diameter.

https://doi.org/10.1515/9783110678765-015

Equation (B.3) is the basic equation for all forms of dynamic turbomachinery, developed by Leonhard Euler, and is known as the Euler equation.

Euler's equation simply states that the energy transfer between the rotor and the fluid must be accounted for by the difference between the product U_1Vu_1 at entry and product of U_2Vu_2 at the exit of the impeller.

Assuming simple two-dimensional (2D) impeller blade/vane; and straightening vanes upstream of the impeller; then there is no angular momentum in the incoming fluid. The Euler equation reduces to

$$H = \frac{U_g Vu_2}{g} \tag{B.4}$$

where H is the head of the impeller (energy/unit mass).

Equation (B.4) implies that for ideal flow with no inlet swirl designed to give the fluid at any inlet angular momentum, the discharge conditions determine the energy transfer.

The exit angle of the blade/vane is shown in Figure B.2, and is represented by $\beta 2$. Centrifugal impeller blade designs are sometimes radial, in which case, $\beta 2$ will be 90°. However, the blade angle at exit can be less than or greater than 90°. If $\beta 2 < 90°$, the impeller is referred to as a "*backward leaning*" impeller. If, on the other hand, $\beta 2 > 90°$, the impeller is referred to as a "*forward leaning*" impeller.

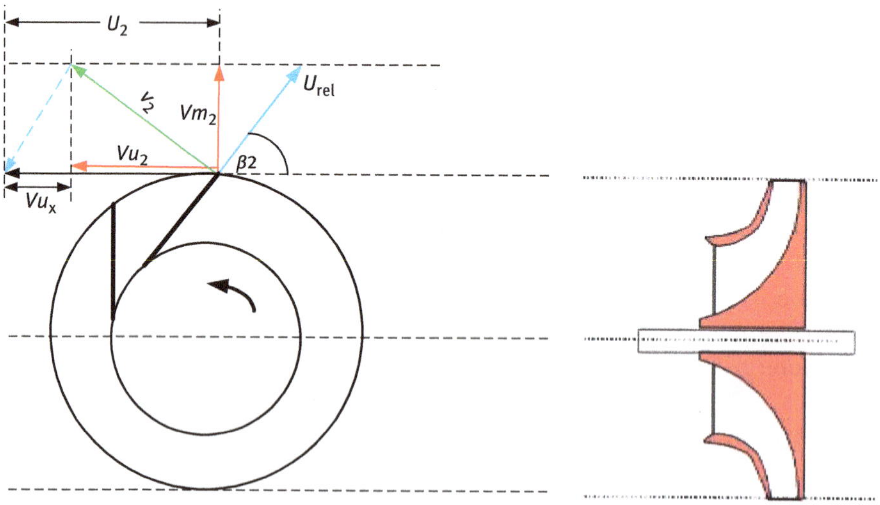

Figure B.2: Flow vector at impeller discharge.

By examining the flow vector at the discharge of the impeller and using eq. (B.4), we can relate the capacity to the head of the impeller.

$$Vu_2 = U_2 - Vu_x$$

But

$$Vu_x = Vm_2 Cot\beta_2$$

Therefore,

$$Vu_2 = U_2 - Vm_2 Cot\beta_2$$

Substituting Vu_2 in eq. (B.4), we get

$$H = \frac{U_2}{g}[U_2 - Vm_2 Cot\beta 2] \tag{B.5}$$

However, since Vm_2 is related to the volume flow "Q" of the impeller, that is, $Vm_2 = \frac{Q_2}{A_2}$ then we can rewrite eq. B.5 in terms of volume flow:

$$H = \frac{U_2}{g}\left[U_2 - \frac{Q_2}{A_2} Cot\beta 2\right] \tag{B.6}$$

Equation (B.6) relates flow to head and shows that for constant speed the Euler head versus volume flow is essentially a linear function.

Equation (B.6) can be rewritten as

$$H = C + MQ_2 \text{ [equation of a straight line]} \tag{B.7}$$

where

$$C = \text{Intercept of the straight line} = \frac{U_2^2}{g}$$

$$M = \text{Slope of the line} = -\frac{U_2 Cot(\beta_2)}{gA_2}$$

The slope "M" depends on the value of β_2. If β_2 is between 0° and 90° the cotangent is positive, and the slope is negative. For $\beta_2 = 90°$, the cotangent is zero, and therefore, the head = C = Constant. Figure B.3 shows how the slope varies with different blade exit angles.

Compressors with backward leaning blades are the machines of choice in the process industry. The reason being their characteristic curves are more appropriate to process applications, since the negative slope provides better differential head for process control.

The H–Q lines derived from the Euler equation assume that the fluid velocity is uniform across the impeller cross section and exit along a path guided by the blade's exit angle. However, the fluid does not exit perfectly along the blade's exit angle, since flow separation between the fluid and the surface of the blade starts to take place long before the fluid exits.

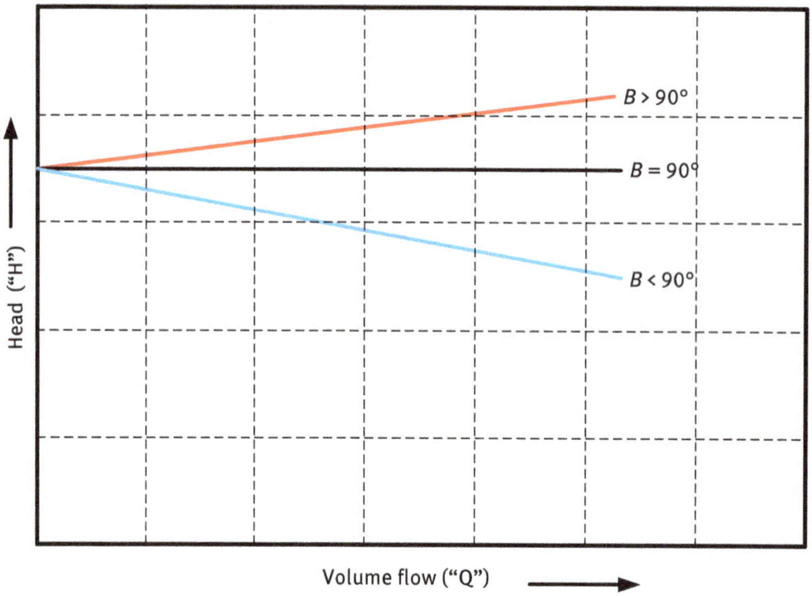

Figure B.3: *H–Q* lines derived from Euler's equation for different blade angles at exit.

This separation of the fluid from the blade's exit angle is called *slip*. Several studies on *slip* have been well documented by Stodola, Stanitz, Balje, and others. Each of the authors mentioned have derived different equations for calculating slip factors. However, we mentioned slip to give the reader a sense of the ideal *H–Q* curve and why the real curve is different.

From Figure B.4 we determine that $Vu_{2\ \mathrm{Actual}} < Vu_{2\ \mathrm{Ideal}}$

From eq. (B.4),

$$H_{Ideal} = \frac{U_2 V_{u2\ Ideal}}{g}$$

$$H_{Actual} = \frac{U_2 V_{u2\ Actual}}{g}$$

$$\text{Therefore, } H_{Actual} < H_{Ideal}$$

The actual head accounts for losses from the following:
- Fluid separation from the blade surface causing a decrease in exit whirl velocity or slip
- Nonuniform velocity distribution across the impeller cross section
- Fluid friction
- Turbulence

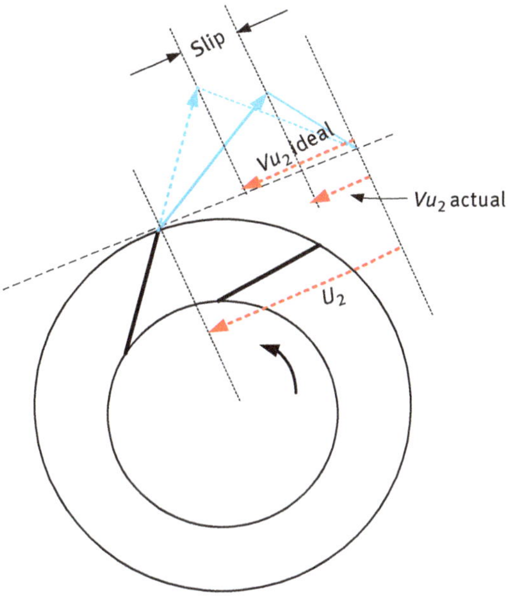

Figure B.4: Velocity triangle with flow separation at exit.

Figure B.5 shows the mathematical corrections required to go from the Euler derivation to an actual *H*–*Q* curve at constant speed. The actual curve has a basic slope as expected from our previous derivation in eq. (B.7) but trails off at points "*A*" and "*B*" due to the fluid losses previously mentioned.

As flow increases, the impeller continues to guide flow through its passages until *Q* reaches a point where no more flow increase is possible. At this point, the impeller passages are in a state of choke. In theory, the fluid velocity is approximately equal to the speed of sound at the inlet operating conditions.

The speed of sound at inlet operating condition is given as

$$a_{0i} = \sqrt{gKZRT_1}$$

where a_{0i} represents the speed of sound, g is the gravitational constant, k is the isentropic specific heat ratio, Z is the compressibility factor, T_1 is the inlet temperature.

However, as flow decreases, *Q* gets to a point where turbulence of the fluid causes flow instability and the stage is said to be in surge.

Choke and surge accounts for the instability regions of the *H*–*Q* curve defining the operating range *A*–*B*. The operating range and slope depend much on the blade's exit angle β_2, and the molecular weight/type of gas being compressed. Figures B.6 and B.7 show how these two factors affect the head and the operating range.

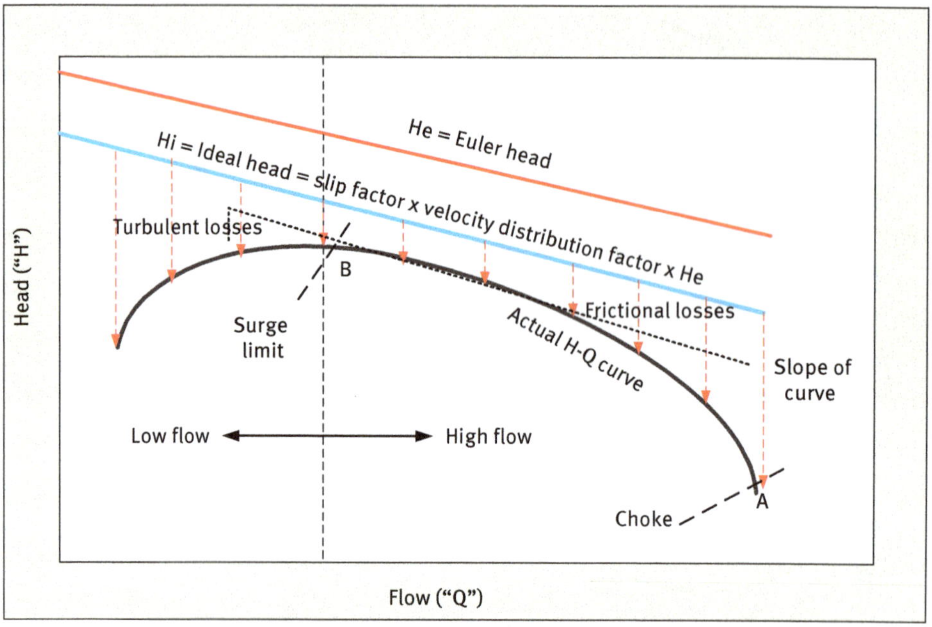

Figure B.5: Mathematical conversion required to go from the Euler derivation to an actual *H–Q* curve at constant speed.

Impeller design fundamentally involves aspects of science; it also involves compromises between science and manufacturing tolerances. Accurate predictions of impeller performance involve R&D followed by detailed testing.

Stage testing is done within the manufacturer's shop to assure accurate predictions. The test is usually done with air and the results correlated with high fidelity computational flow dynamics.

Measurements of pressures and temperatures are performed at several locations along the flow path of the compressor stage.

The correlated results are generally based on the polytropic process. In order to arrive at the calculated head and efficiency values, the following basic relationships previously derived in Chapter 1 are used. These include:

I. Temperature ratio $\frac{T_2}{T_1} = \left(\frac{P_2}{P_1}\right)^{\frac{n-1}{n}}$

II. Polytropic efficiency $\eta_p = \left(\frac{n}{n-1}\right)\left(\frac{k-1}{k}\right)$

III. Polytropic head $Hp = ZRT_1 \frac{n}{n-1} \left[\left(\frac{P_2}{P_1}\right)^{\frac{n-1}{n}} - 1 \right]$

Each impeller or stage is tested by varying the flow and speed, resulting in a characteristic performance curve specific or unique for just that impeller; it is based on the impeller's dimensional design. The impeller dimensions will include inlet and exit blade angles, impeller diameter, tip width, solidity, and so on.

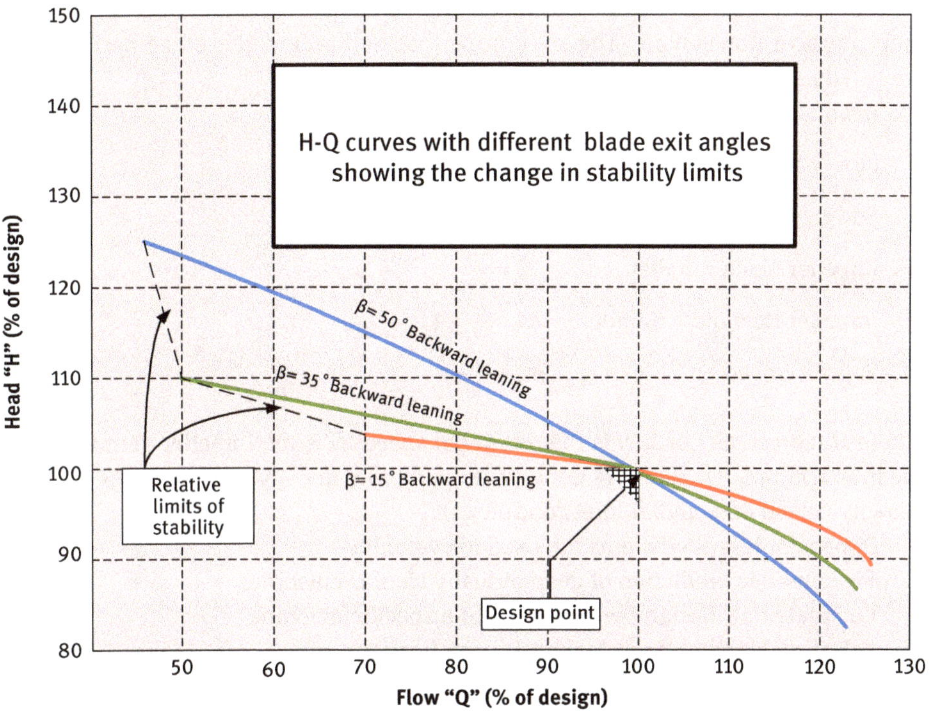

Figure B.6: Effects of blade exit angles on actual *H–Q* curves.

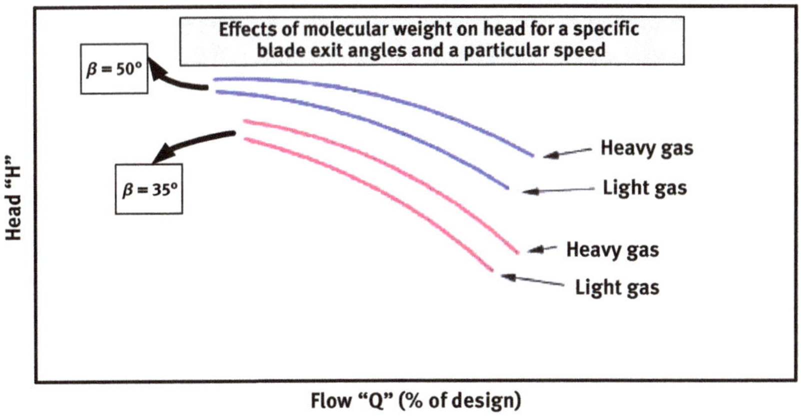

Figure B.7: Effects of molecular weight on actual *H–Q* curves.

Dimensionless quantities can be used to nameplate impeller stages based on their physical dimensions. The six groups that define impeller stage performance are listed below:

I.	Head coefficient	$\mu = \frac{H}{N^2D^2}$
II.	Flow coefficient	$\varphi = \frac{Q}{ND^3}$
III.	Power/work coefficient	$\psi = \frac{P}{\rho N^3D^5}$
IV.	Impeller Mach number	$M = \frac{ND}{a_{01}}$
V.	Impeller Reynold's number	$R_E = \frac{\rho ND^2}{\text{Dynamic viscosity}}$
VI.	Specific speed	$Ns = \frac{NQ^{1/2}}{H^{3/4}}$

where H represents head, N is the rotational speed, D is the impeller diameter, Q is the inlet volume, P is the work done on fluid, ρ, fluid density, and a_{01} is the acoustic velocity (speed of sound at inlet conditions).

Dimensionless coefficients are used to evaluate:

- Performance prediction of geometrically similar machines
- Calculated off-design performance for a specific machine
- Selecting impeller types for specific applications
- Anticipated impeller efficiencies based on impeller type

Following testing of a compressor stage or section, dimensionless curves are plotted; they will be similar to those shown in Figure B.8. If a section is tested, the average diameter of the wheels is used to calculate the impeller coefficients previously defined.

Theoretically, from the derivation of Euler's equation, we know that for a given impeller operating on a given gas, the head–flow relationship is a function of speed.

Variables are based on the following relationships:

Volume flow	$Q\alpha N$
Head	$H\alpha N^2$
Power	$P\,\alpha N^3$

Having produced these dimensionless curves for a specific dimensionally designed impeller, a family of curves can be generated for geometrically similar impellers. This can be done based on speed and type of gas.

In order to combine the effects of speed and type of gas, the equivalent tip speed method can be used.

$$\text{Tip speed of gas being investigated "}U_i\text{"} = \text{Tip speed of test impeller} \times \frac{a_{\text{ref.gas}}}{a_{\text{act.gas}}}$$

where U_i is the equivalent tip speed of gas being investigated, $a_{\text{ref.gas}}$ is the acoustic velocity of reference gas [example air at 80 °F (27 °C)], $a_{\text{act.gas}}$ is the acoustic velocity of actual gas

The single dimensionless curves can now be expanded into a family of curves with respect to speed and molecular weight, as shown in Figure B.9.

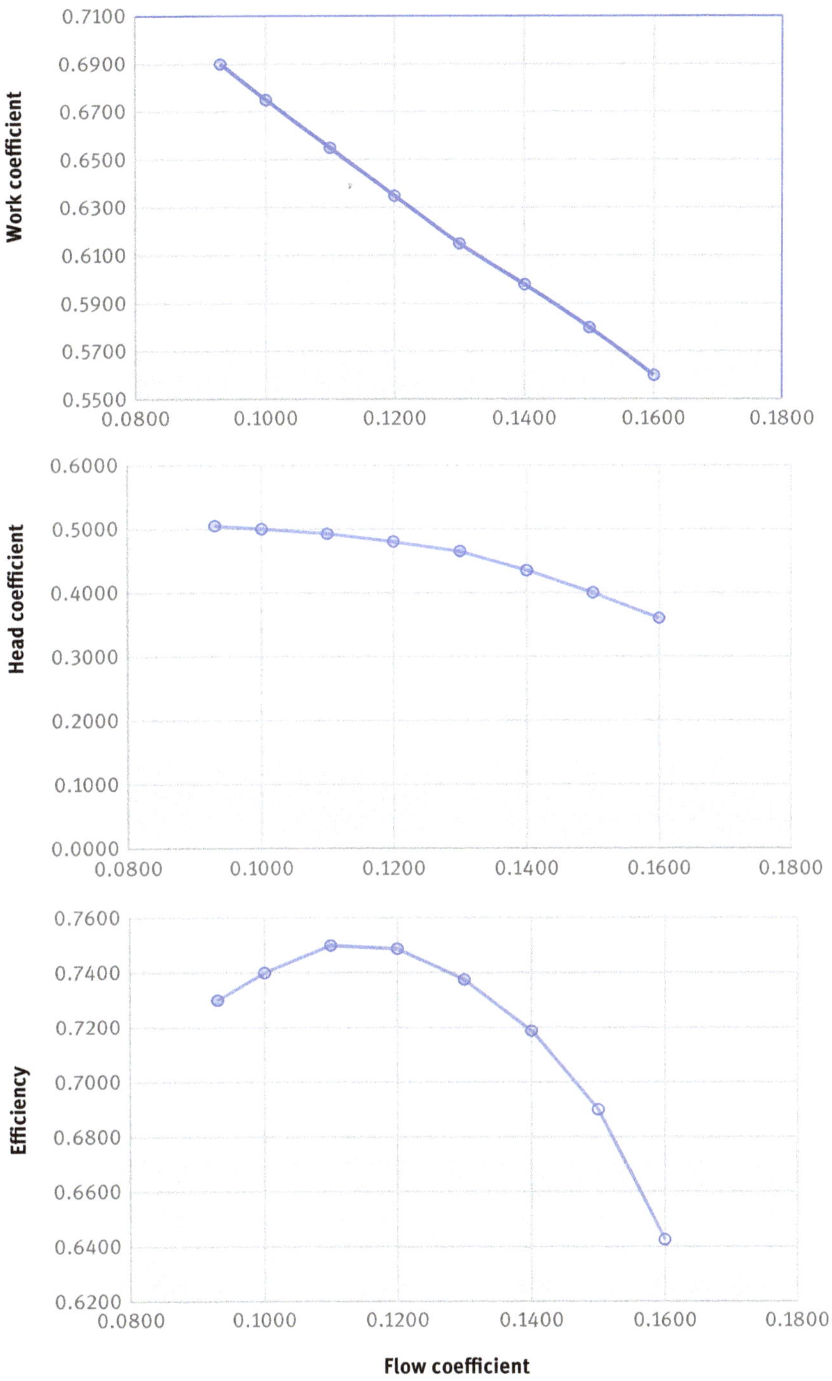

Figure B.8: Typical dimensionless curves for a centrifugal compressor stage.

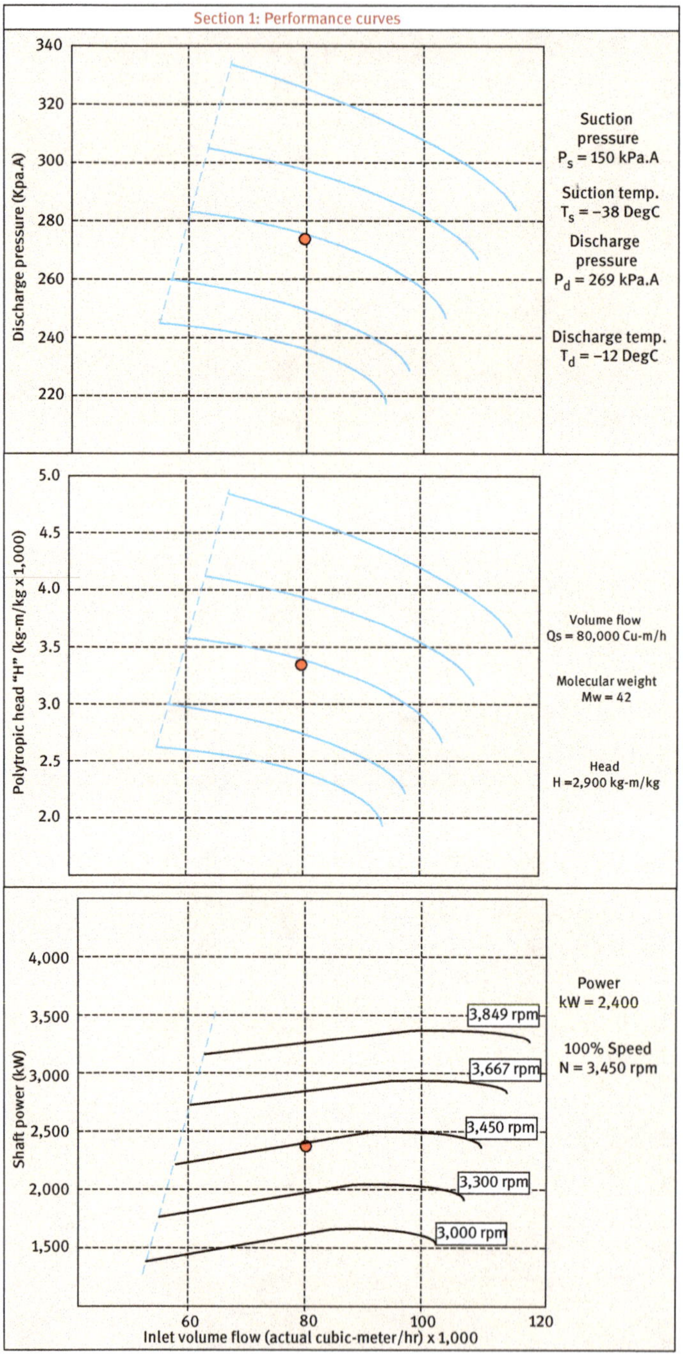

Figure B.9: Top: *Discharge pressure versus flow*; middle: polytropic head versus Flow; bottom: shaft power versus flow.

Neglecting Reynolds number and Mach number effects, we can also use dimensionless numbers to evaluate off-design operating conditions for a compressor stage or section operating on a specific gas:

Head coefficient
$$\mu = \frac{H_1}{N_1^2 D^2} = \frac{H_2}{N_2^2 D^2}$$
$$= \frac{H_1}{N_1^2} = \frac{H_2}{N_2^2}$$

Flow coefficient
$$\varphi = \frac{Q_1}{N_1 D^3} = \frac{Q_2}{N_2 D^3}$$
$$= \frac{Q_1}{N_1} = \frac{Q_2}{N_2}$$

Power/work coefficient
$$\psi = \frac{P_1}{N_1^3 D^5} = \frac{P_2}{N_2^3 D^5}$$
$$= \frac{P_1}{N_1^3} = \frac{P_2}{N_2^3}$$

The above relations are referred to as the fan laws; they are commonly used in turbomachinery to assess off-design situations. These laws are not entirely accurate but can be highly accurate if used with experience-based correction factors.

Specific speed is a parameter that has been used in the past to select compressor stages. Specific speed is defined as the speed (in rpm) at which geometrically similar impellers would have to rotate and deliver one unit volume of gas. Specific speed makes use of the relationship between speed, volume flow and developed head:

$$\text{Specific speed } N_S = \frac{NQ^{1/2}}{H^{3/4}}$$

The specific speed of a centrifugal compressor impeller relates efficiency to the type of impeller. Ranging from radial flow impellers to axial flow impellers with radial flow impellers being the least efficient along the range of impellers. Maximum efficiency is obtained midway between radial flow and axial flow impellers; efficiency declines somewhat as the geometry approaches purely axial flow [1].

By definition, specific speed N_s should be a dimensionless parameter; however, in practice Ns has been used in various forms that are not dimensionless. Because this practice can be confusing, the more current approach to selecting efficient impellers is based on the flow coefficient.

Figure B.10 shows the most recent trend in efficiency comparisons. The comparison of different impellers makes use of their respective coefficients at conditions of suction flow.

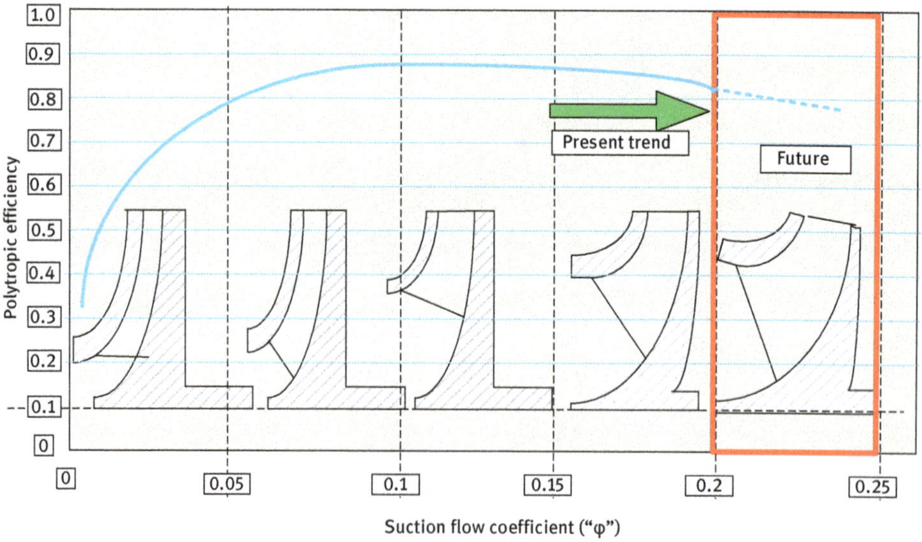

Figure B.10: Recent trend in the design of centrifugal impellers.

B.1.2 Axial compressors

Axial compressors shares much with centrifugal compressors. Both belong to the dynamic family of compressors; they are dynamic in the sense that each continuously compresses the gas. The two machine types have the same limitations – surging and choking – at the end of their operating curves. However, axial machines have a narrower operating range than centrifugal compressors.

Axial compressors are considered rotating airfoil machines because the airfoils on the rotor impart energy to the fluid. They also have stationary airfoil blades (called vanes) that help increase the gas pressure by diffusion. These rotating and stationary airfoils continuously increase the pressure by compound staging along the axis of the rotor.

To maintain uniform mean axial velocity of the gas through the compressor, the blade height must decrease toward the exhaust end of the machine. Decreasing blade height allows the designer to use similar blade contours for each stage.

Velocity triangles are developed as shown earlier for radial flow impellers in centrifugal compressors. However, in axial compressors the rotating blades interact with a stationary blade instead of a volute that would be utilized in centrifugal compressors.

Applying Euler's equation from the velocity triangles shown in Figure B.11, we can derive the work done per unit mass as

$$W = (UCw_2 - UCw_1)$$

where U is the peripheral/tangential velocity at the mean blade height, C_a is the axial component, C_w is the whirl or tangential component, β_1 is the inlet blade angle, β_2 is the outlet blade angle, α_1 is the fluid angle at inlet to the rotating blade, α_2 is the fluid angle at inlet to stationary blade.

Note:

$$C_{a1} = C_{a2} = C_a \text{ for axial compressors}$$

$$\begin{aligned} W &= U\,(Cw_2 - Cw_1) \\ &= UCa\,(\tan\alpha_2 - \tan\alpha_1) \end{aligned} \tag{B.8}$$

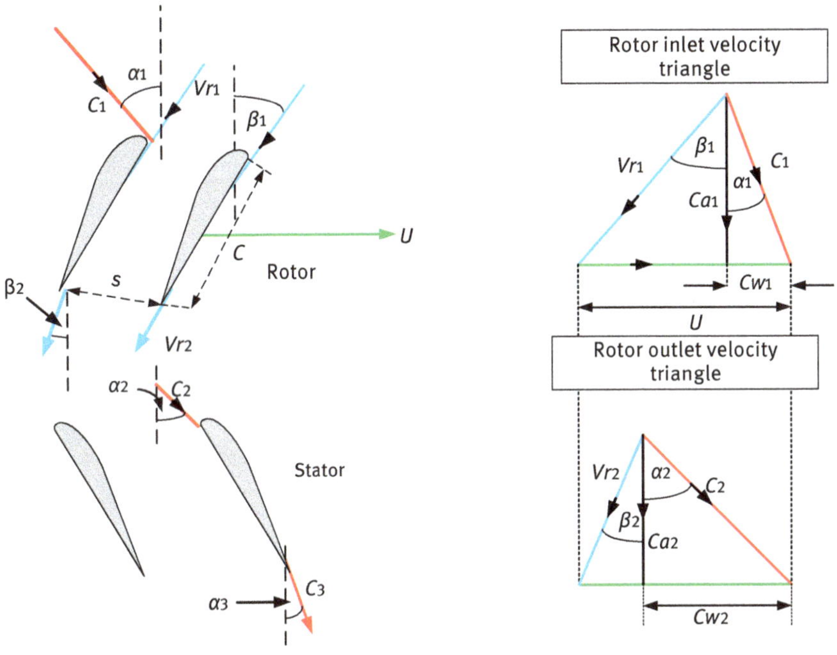

Figure B.11: Velocity triangle for a single stage axial compressor.

Generally, the fluid angles entering both the rotating and stationary blades are not known; therefore, eq. (B.8) is rewritten in terms of blade angles that are known:

$$W = UC_a(\tan\beta_1 - \tan\beta_2) \tag{B.9}$$

W is the actual change in enthalpy = $\Delta h_{12} = h_{t2} - h_{t1}$ from the Mollier chart depiction shown in Figure B.12.

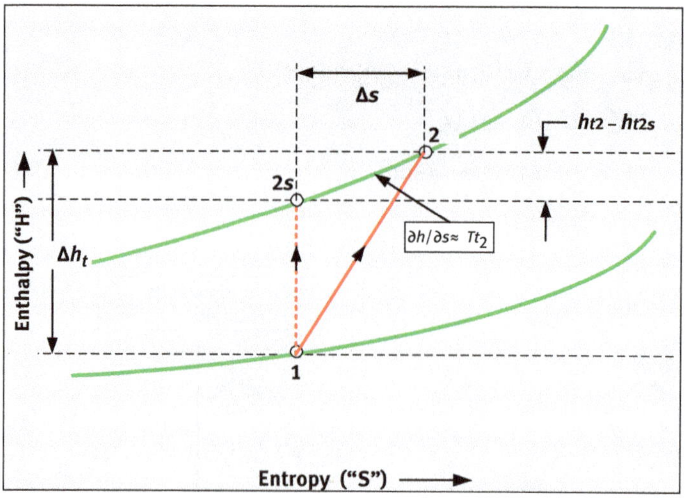

Figure B.12: Mollier chart depiction on how to calculate stage efficiency.

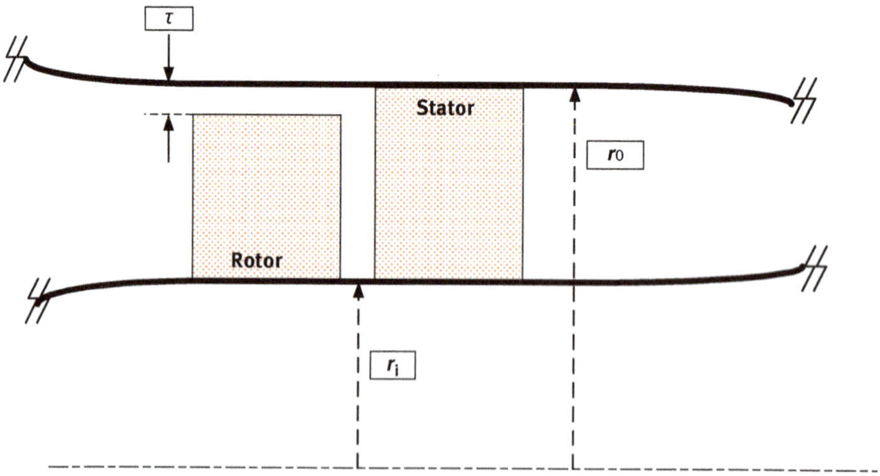

Figure B.13: Rotor and stator geometrical parameters.

The relationship between entropy and efficiency for the small pressure rise across a stage is of importance:

$$\text{Stage efficiency } \eta_s = \frac{\text{Insentropic change in enthalpy}}{\text{Actual change in enthalpy}} = \frac{h_{t2s} - h_{t1}}{h_{t2} - h_{t1}}$$

$$\approx 1 - \frac{T_{t2}\,\Delta s}{\Delta h_t} \qquad \text{(B.10)}$$

where T_{t2} is the stagnation temperature at state point 2, Δs is the change in entropy.

Equation (B.10) shows that entropy increases through the stage as a result of irreversibility. The losses from irreversibility are directly linked to decreased efficiency.

The main sources for losses are:

- Viscous losses from boundary layers created near the blade, vane, and end wall surfaces
- Mixing of wakes that are created downstream of each row of blades
- Rotor tip leakage away from the main flow through the machine
- Losses from shock etc., exist; however, these are extremely low and are thus not being considered.

Figure B.13 represents a typical stage of an axial compressor showing key geometical dimensions.

Notable parameters that are usually considered when designing to optimize stage efficiency include:

Flow coefficient	φ	$= C_a/U$
		$= C_a/\omega r$ (r = the mean of the blade)
	C_a	$= m/\rho A$ (ρ = density and A = area)
Stage loading coefficient	ψ	$= \Delta h_t/(\omega r)^2$ [same as work coefficient]
Reynolds Number	R_e	$= C_a c/v$ (v = dynamic viscosity)
Blade aspect ratio	AR	$= (r_o - r_i)/c$
Hub-to-tip ratio		$= r_i/r_o$
Gap-to-height ratio		$= \tau/(r_o - r_i)$
Solidity σ		$= c/s$

$$\psi = \varphi\ (\tan\alpha_2 - \tan\alpha_1) \qquad \text{(B.11)}$$

Researchers Dickens and Day [2] concluded that increasing stage-loading coefficients above the conventional industrial values (0.2 to 0.4) leads to decreasing stage efficiencies. We find it important that equipment manufacturers share their experience regarding these parameters with the purchaser during design audits of the machines they have purchased.

In one actual example case, the purchaser did not pay much attention to the design of a machine that was being built as a first-off model. This then resulted in several failures during shop testing. We think the issues could have been avoided if the purchaser had demonstrated more diligence and experience in performing such audits. Having knowledge of the various design parameters previously mentioned in this appendix would have equipped the purchaser to ask the right questions. The vendor–manufacturer's answers would have shown considerable deviations from good practices and might have allowed to redirect a risk-prone design into safer regions.

B.1.2.1 Static temperature rise across the stage

If we make the reasonable assumption that the velocity entering the stage is also equal to the velocity leaving the stage, it follows that all the useful energy is transferred to the fluid. A small portion of energy will have been consumed as frictional losses. Based on this reasoning, the static temperature of the fluid will increase.

If ΔT_s is the static temperature rise across the stage, then

$$W_1 = Cp\Delta Ts = UCa\ (\tan\beta_1 - \tan\beta_2)$$

$$\Delta Ts = \frac{UC_a}{C_p}\ (\tan\beta_1 - \tan\beta_2) \tag{B.12}$$

However, the actual work done across the stage is less than W_1 because of the non-uniform fluid velocity distribution across the annulus. A *work-done factor Ω* is introduced to compensate.

Rewriting eq. (B.12),

$$\Delta TS = \Omega\frac{UC_a}{C_p}\ (\tan\beta_1 - \tan\beta_2) \tag{B.13}$$

B.1.2.2 Work-done factor

The work-done factor is the ratio of the actual work absorbing capacity of the stage to the ideal work done as calculated from the Euler equation. It usually ranges from 0.96 at the first stage to 0.85 at the fourth and ensuing stages.

The pressure ratio across a stage can be written as

$$r_p = \left[1 + \eta_s\frac{\Delta T_s}{T_{0i}}\right]^{\frac{k-1}{k}} \tag{B.14}$$

where η_s is the isentropic efficiency, k is the specific heat ratio, T_{0i} is the inlet stagnation gas temperature.

Typical single stage compression ratio ranges from 1.15 to 1.25 for industrial-type applications, with efficiencies ranging from 88% to 92%. If ΔT_s can be calculated from eq. (B.10), then the pressure ratio should be calculated and checked against typical ranges for industrial machines. By knowing the overall pressure ratio and calculating the individual stage pressure ratio, the amount of stages required can be estimated.

B.1.2.3 Degree of reaction "R "

$$R = \frac{\text{Work transfer by pressure difference across rotor}}{\text{Total work transfer across the stage}}$$

$$= \frac{\text{Enthalpy drop across the rotor}}{\text{Enthalpy drop across the stage}} = \frac{C_P(\Delta T_a)}{C_p(\Delta T_a + \Delta T_b)}$$

where ΔT_a is the static temperature rise in the rotor and ΔT_b is the static temperature rise in stator

$$W = Cp(\Delta Ta + \Delta Tb) = Cp\Delta Ts$$

Therefore, it can be shown that

$$R = 1 - \frac{1}{2}\frac{C_a}{U}(\tan\beta_1 + \tan\beta_2)$$

The degree-of-reaction "*R*" lies between 0 and 1. If all the pressure or enthalpy change occurs in the stator of the machine (or in the nozzles in the case of turbines), then $R = 0$, as is the case with impulse turbines. However, if there is change in pressure or enthalpy in both rotor and stator, the machine is considered a reaction machine.

Ideally, axial compressor designs approach $R = 0.5$ (50% reaction), which makes the rotor and stator blades symmetrical.

H–Q curves for axial compressors are developed in a similar manner to what was done previously with centrifugal compressors. We start with Euler equations and utilize experience-based slip factors to arrive at actual performance curves. These *H–Q* performance curves are much steeper for axial compressors than for centrifugal compressors. The difference between surge and stonewall can range from 2% to 10%, making the performance range in axials much narrower than in centrifugal compressor.

The narrowing of the performance range of axial compressors can be mitigated with variable speed drives, using adjustable stators, or interstage bleed valves.

Although traditional fixed stator blades are used in older compressor designs, this arrangement offers less operational flexibility. Selecting variable speed drives makes up for not using adjustable stator blades. Modern axial compressors incorporate adjustable vanes; these are sometimes designed to be automatically adjusted.

Figures B.14 (a) and (b) show axial compressor maps with speed control and stator vane control, respectively.

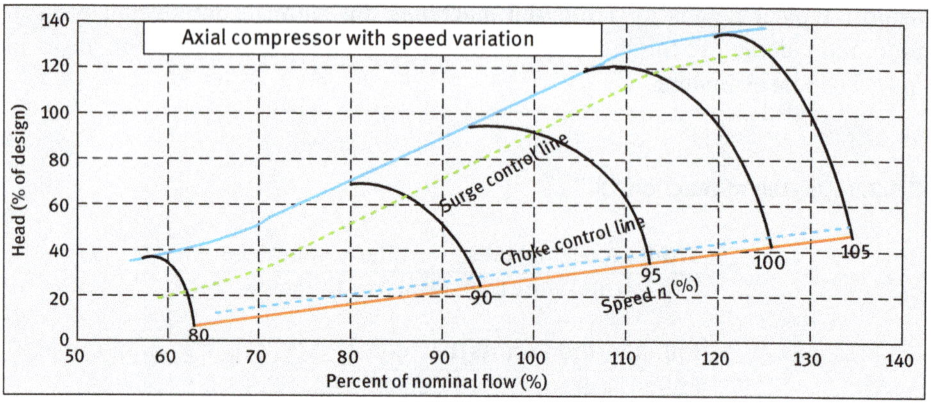

Figure B.14a: Axial compressor performance map with speed variation control.

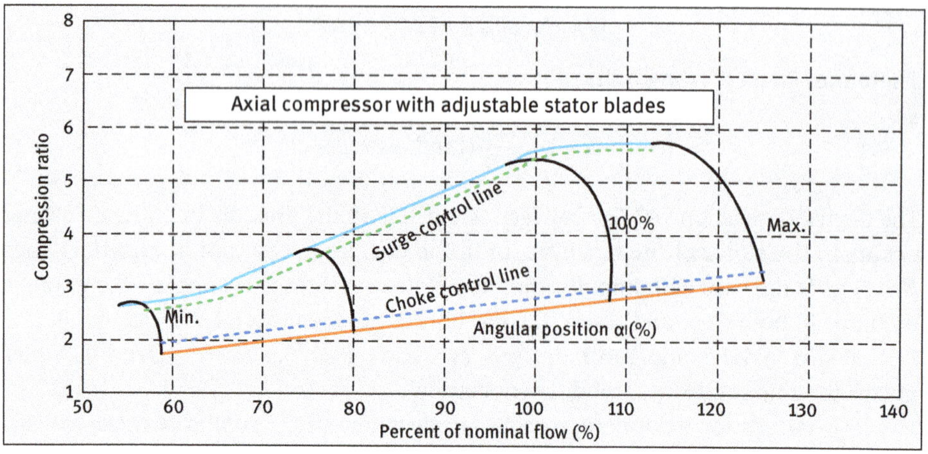

Figure B.14b: Axial compressor performance map with adjustable stator vane control.

B.1.3 Reciprocating compressors

Reciprocating compressors were described earlier in Chapter 2. "Recips," as they are often called, belong to the family of positive displacement machines. Gas is drawn inside the cylinder and trapped, while the piston compresses the gas into a smaller volume. The gas is then discharged from the cylinder when the pressure inside the cylinder is high enough to overcome the delta-p of the discharge valve.

The objective of this section is to review the following as it pertains to reciprocating compressors:

- Performance fundamentals of reciprocating compressors
- Examination of relevant basic thermodynamics
- Reviewing practical limitations of a few key components

We will start with some basic theoretical concepts by using Figure B.15. It depicts a pressure–volume (*P–V*) diagram for a single-acting reciprocating compressor without clearance. As the piston moves back and forth, the cycles of operation or "paths" are:

- **0 to 1:** Suction valve opens and aspirates gas to the cylinder at constant pressure
- **1 to 2:** Both suction and discharge valves are closed. The gas is compressed polytropically until the pressures in cylinder and discharge line are equal. The discharge valve opens, and gas leaves the cylinder at constant pressure.
- **2 to 3:** Gas continues to leave the cylinder at constant pressure.
- **3 to 0:** Since theoretically there is no clearance, no residual gas is left inside the cylinder. Therefore, as the piston starts moving in the opposite direction, it creates a vacuum and the suction valve opens. The cycle then repeats.

The cycle work done per unit mass on the gas is the negative area under the *P–V* diagram. Therefore, from Figure B.15, the cycle work done by an isothermal process will be less than work done by the more realistic polytropic cycle.

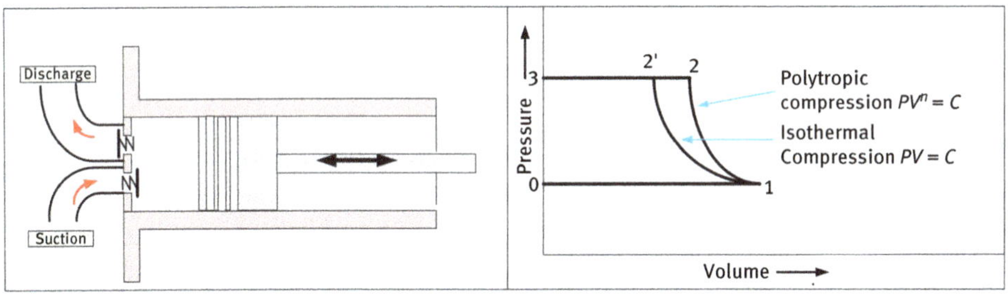

Figure B.15: Theoretical *P–V* diagram without internal clearance.

B.1.3.1 Cycle work/power

Using ideal gas laws to calculate the cycle work done "W" on the gas:

$$W = -\int_1^2 v dp$$

$$= \frac{n}{n-1}(p_1 v_1 - p_2 v_2) \tag{B.15}$$

The cycle work done can be further arranged by eliminating V_2.

Assuming the compression process from state 1 to state 2 is polytropic:

Then,

$$\frac{V_2}{V_1} = \left(\frac{P_2}{P_1}\right)^{\frac{-1}{n}}$$

And

$$\frac{P_2 V_2}{P_1 V_1} = \left(\frac{P_2}{P_1}\right)^{\frac{n-1}{n}} \tag{B.16}$$

Substituting eq. (B.16) into eq. (B.15):

$$\text{Cycle work } W = \frac{n}{n-1} P_1 V_1 \left[1 - \left(\frac{P_2}{P_1}\right)^{\frac{n-1}{n}}\right]$$

Alternatively,

$$W = \frac{n}{n-1} RT_1 \left[1 - \left(\frac{P_2}{P_1}\right)^{\frac{n-1}{n}}\right] \tag{B.17}$$

$$= \frac{n}{n-1} R\, [T_1 - T_2] \tag{B.18}$$

If the flow is treated as a steady flow process, then from first law of thermodynamics

$$W = h_1 - h_2 [\text{neglecting losses, kinetic energy, and heat transfer}]$$

$$= C_p (T_1 - T_2)$$

Cycle work for the isothermal case:

$$W = -p_1 V_1 \ln\left(\frac{P_2}{P_1}\right) \tag{B.19}$$

In practice, industrial reciprocating compressors tend to operate somewhere between isothermal and polytropic, and the compression process fits better as an adiabatic process. The polytropic index in eq. (B.17) will now become "*k*," the specific heat ratio of the gas.

The reciprocating compressor is considered an open system as defined in Figure 1.7.

Therefore, from the *first law of thermodynamics*

$$Q - W = \frac{k}{k-1} RT_1 \left[1 - \left(\frac{P_2}{P_1}\right)^{\frac{k-1}{k}}\right]$$

But since the cylinders are almost always cooled, there is no heat transferred and

$$Q = 0$$

$$0 - W = \frac{k}{k-1} RT_1 \left[1 - \left(\frac{P_2}{P_1}\right)^{\frac{k-1}{k}}\right]$$

or

$$W = \frac{k}{k-1} RT_1 \left[\left(\frac{P_2}{P_1}\right)^{\frac{k-1}{k}} - 1\right] \tag{B.20}$$

Rewriting eq. (B.20) in terms of an adiabatic process

$$\text{Cycle work } W = \frac{k}{k-1} RT_1 \left[r_p^{\frac{k-1}{k}} - 1\right] \text{ (where } r_p \text{ is the compression ratio)} \tag{B.21}$$

We can now extend our theoretical knowledge gained from reciprocating compressors without clearance to actual compressors with clearance.

B.1.3.2 Clearance volume

Figure B.16 illustrates the cycle of operation for a *single-acting*, single-cylinder reciprocating compressor with clearance. As mentioned earlier, in these compressors, the piston moves to the top of the cylinder, and at the end of the compression stroke a space remains in the cylinder; the primary location of this "dead space" is often near the valves. The space is called *clearance volume*.

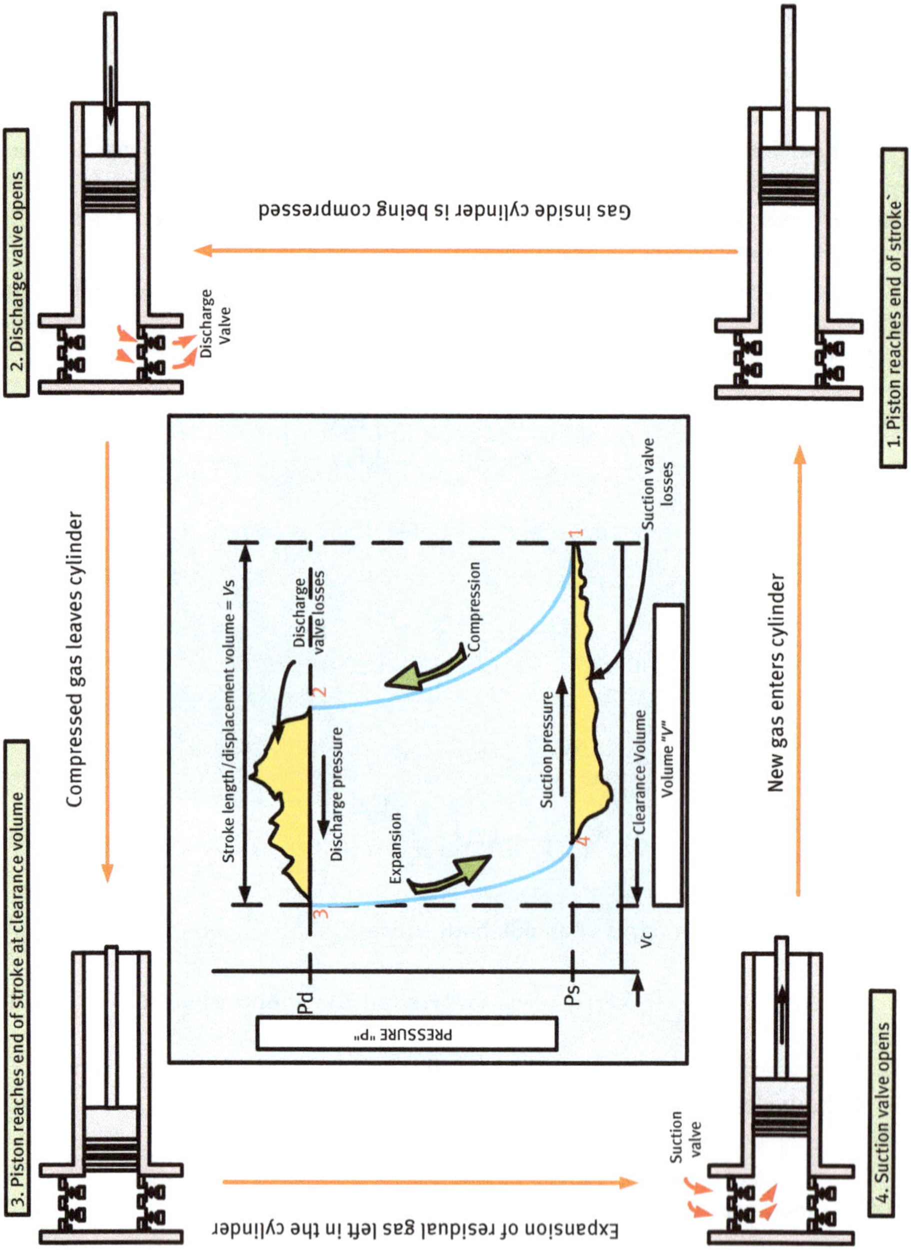

Figure B.16: Single acting reciprocating compressor work cycle with clearance volume.

The clearance volume is usually expressed as a percentage of the displacement volume. It typically ranges between 3% and 15% for industrial machines. A good average number is 10%; the volumetric efficiency "VE" decreases as clearance volume increases.

We occasionally see the notation:

$$VE = 1 - \frac{\text{Clearance volume}}{\text{Displacement volume}} = 1 - \frac{V_c}{V_s} \quad \text{[rough approximation]}$$

By definition, $V_c = cV_s$; where c is the clearance volume fraction

If m_1 is the mass of gas being compressed at beginning of the compression process, m_2 is the mass of gas remaining at the end of compression

Then,

$$\text{Mass of gas delivered } m = m_1 - m_2$$

Since the clearance mass is small, the work done during the expansion process is also small. Therefore, even if we assume that the expansion process follows $PV^n = C$ ("C" meaning constant), the error of adding the work done in the expansion process to the overall work cycle will be small.

Neglecting the valve losses shown in Figure B.16

$$W = \frac{n}{n-1} m_1 R\ (T_1 - T_2) - \frac{n}{n-1} m_2 R\ (T_4 - T_3)$$

Assuming $T_2 = T_3$ and $T_4 = T_1$

$$W = \frac{n}{n-1} (m_1 - m_2)\ R\ (T_1 - T_2)$$

$$= \frac{n}{n-1} mR\ (T_1 - T_2) \tag{B.22}$$

Or

$$W_{cyl} = \frac{n}{n-1} mRT_1 \left[\left(\frac{P_2}{P_1} \right)^{\frac{n-1}{n}} - 1 \right] \tag{B.23}$$

Substituting P_1Q_1 for mRT_1 into eq. (B.23)

$$W_{cyl} = \frac{n}{n-1} P_1 Q_1 \left[\left(\frac{P_2}{P_1} \right)^{\frac{n-1}{n}} - 1 \right] \quad (Q_1 \text{ is the theoretical capacity of the compressor})$$

B.1.3.3 Theoretical work/power per stage

When $n = k$.

$$W_{cyl} = \frac{k}{k-1} P_1 Q_1 \left[\left(\frac{P_2}{P_1} \right)^{\frac{k-1}{k}} - 1 \right]$$

For real gases:

$$W_{cyl} = \frac{k}{k-1} P_1 Q_1 \left[\left(\frac{P_2}{P_1} \right)^{\frac{k-1}{k}} - 1 \right] \times \frac{Z_s + Z_d}{2} \tag{B.24}$$

where Z_s is the suction compressibility factor and Z_d is the discharge compressibility factor.

B.1.3.4 Volumetric efficiency

Volumetric efficiency "VE" is a term used to describe how efficiently the gas is drawn into the compressor.

It is expressed as

$$\text{VE} = \frac{\text{Actual intake volume}}{\text{Displacement volume}} = \frac{V_1 - V_4}{V_s} \tag{B.25}$$

Let V_c be the clearance volume, V_1 the total cylinder volume, V_4 the volume after expansion, and V_s the displacement volume.

We previously showed $V_c = cV_s$

Referring to Figure B.16 $\quad \mathrm{V_1 = V_S + cV_S}$

And $P_4 V_4{}^n = P_3 V_3{}^n = P_3 Vc^n$ or $P_1 V_4{}^n = \mathrm{P}_2 Vc^n$

Substituting V_1 and V_4 into eq. (B.25)

$$\begin{aligned} \text{VE} &= 1 + c - c\left(\frac{P_2}{P_1}\right)^{\frac{1}{n}} \\ &= 1 - c\left(r_p^{\frac{1}{k}} - 1\right) \end{aligned} \tag{B.26}$$

$n = k$ for an adiabatic compression.

Note that the volumetric efficiency decreases as the clearance "c" increases.

Equation (B.26) is the theoretical formula for volumetric efficiency; however, as is often the case, theoretical formulae must be modified to represent real-world situations.

There are many different modifications to the theoretical volumetric efficiency that are used in industry; however, the one most used is shown in eq. (B.27)

$$\mathrm{VE_{act}} = 0.97 - c\left(r_p{}^{\frac{1}{k}} - 1\right) - L \tag{B.27}$$

In eq. (B.26), the constant 1.00 becomes 0.97 to correct for inefficiencies and unavoidable losses during the suction stroke. "L" is a correction for internal leakage; it is based on the particular process application and is equipment manufacturer specific.

While "L" is very difficult to generalize, an average value of 3% can be used for moderate pressure oil-lubricated cylinders. A higher value, say, 7% is more appropriate for light gas in nonlubricated cylinders.

We can also calculate "L" as shown below during initial sizing calculations for most applications, and determine if it ranges within the guidelines mentioned:

$$L = (\mathrm{LF}) \times r_p \times \left(\frac{P_s}{P_a}\right) \times 0.02$$

where P_s is the suction pressure, psia (barA); P_a is the atmospheric pressure, psia (barA); LF = 0.18 for molecular weights ranging from 2 to 7, LF = 0.13 for molecular weights ranging from 7.1 to 14.99, and LF = 0.008 for molecular weights > 15.0.

Sometimes, temperature and pressure corrections are made to the volumetric efficiency, most times relative to standard conditions:

$$\mathrm{VE_a ctual} = \mathrm{VE} \times \left(\frac{P_1}{P_0}\right)\left(\frac{T_0}{T_1}\right)$$

where T_0 and P_0 are the temperature and pressure of the surroundings, T_1 and P_1 are the temperature and pressure, respectively, of state 1 of the gas.

B.1.3.5 Capacity

Once the volumetric efficiency of a compressor is known, then the compressor's capacity can be calculated by multiplying the volumetric efficiency by the piston displacement.

The theoretical maximum capacity of a reciprocating compressor is given as follows:

$$Q = 0.0509 \times \frac{P_s}{T_s} \times \frac{Z_{std}}{Z_s} \times \mathrm{DISP} \times \left[1 - c\left(r_p{}^{\frac{1}{k}} - 1\right)\right] \tag{B.28}$$

where Q is the capacity, million standard cubic feet per day [ref. 14.7 psia and 520 °R]; P_s is the suction pressure – psia [at compressor flange]; Z_{std} is the compressibility factor at standard conditions; Z_s is the compressibility factor at suction conditions; DISP is the displacement rate in cubic feet per minute; c is the cylinder clearance volume as a decimal fraction of the displaced volume; r_p is the pressure ratio across the cylinder [flange to flange]; k is the specific heat ratio for an adiabatic process using ideal gas laws.

One of the notable variables in eq. (B.28) is the theoretical volumetric efficiency, which shows that by increasing the volumetric clearance, the capacity decreases. This must be properly accounted for during sizing of reciprocating compressors. Therefore, we recommend using the modified volumetric efficiency shown in eq. (B.27).

B.1.3.6 Valve losses

The shaded areas in Figure B.16 give us an indication of the fluid frictional losses as gas leaves or enters the compressor. Valve losses are among the major factors in reciprocating compressor losses. This was discussed in more detail in Chapter 4 (Section 4.2.2).

The valve losses vary as the square of the gas velocity through the valves. In most cases, losses are directly proportional to the specific gravity of the gas being compressed. The engineer can therefore get a relative indication of likely losses by noting the valve velocity for the valve being proposed.

API average valve velocity calculations are the ones most frequently used:

In SI units,

$$\text{Valve velocity } W = F \times \frac{C_{\mathrm{m}}}{f} \tag{B.29}$$

where W is the average gas velocity in m/s, F is the effective piston area of the cylinder end or ends at issue. For a double-acting cylinder, this area is the area of the crank-end of the cylinder less the piston rod plus the area of the outer end of the piston in cm^2; f is the product of the actual lift, the valve opening periphery, and the number of inlet or discharge valves in cm^2; C_{m} is the average piston velocity in m/s.

Please note that API treats the results calculated from eq. (B.29) only as an indication of valve performance; it does not necessarily represent anticipated power loss. However, it can be used to compare valve styles, types, or brands. It should be noted that modular reed valves ("Zahroof Valves"), introduced in 2015, often defy the old rules; they deserve to be looked at in terms of efficiency gains and greatly extended operating life.

In US customary units,

$$\text{Valve velocity } V = 288 \times \frac{D}{A}$$

where V is the average gas velocity; D is the piston displacement per cylinder, in ft^3/min; A is the product of the actual lift, the valve opening periphery, and the number of inlet or discharge valves per cylinder in in^2.

B.1.3.7 Piston speed

Industry defines average piston speed as the product of compressor stroke times operating speed, and is given as follows:

$$\text{Average piston speed (ft/min)} = \frac{\text{Stroke (inch)} \times \text{rpm} \times 2}{12 \text{ (in per ft)}}$$

$$\text{Average piston speed (m/s)} = \frac{\text{Stroke (mm)} \times \text{rpm} \times 2}{60 \left(\frac{\text{s}}{\text{min}}\right) \times 1{,}000 \left(\frac{\text{mm}}{\text{m}}\right)}$$

From Figure B.17 the piston speed reaches a maximum near the middle of the compression stroke, but not at the 90° crank angle. However, the average/mean piston speed used by the industry is approximately 40% less than the maximum piston speed.

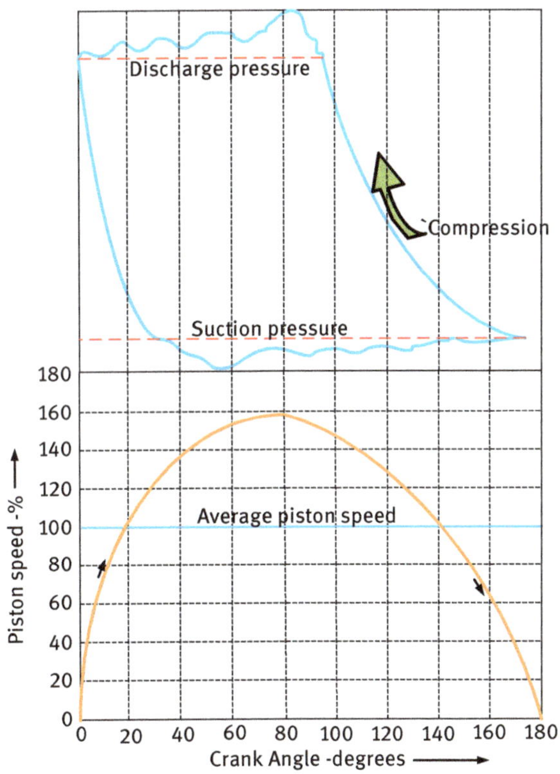

Figure B.17: Piston speed relative to crank angle and compression process.

The main reason for the evaluation of piston speed by end users is based on the concept that lower piston speeds offer less wear and thus greater reliability. This is essentially correct because higher piston speeds promote increase rates of wear on rider bands, piston rings, and packing. This often puts the end user and equipment manufacturer at odds, since the manufacturer favors higher speeds in order to keep costs and compressor size on the low side of the ledger. Thorough studies of life cycle costs will include maintenance, downtime expense for periods of overhaul and other factors. Good machines are rarely cheap machines, and the results of in-depth life cycle cost studies may surprise even the seasoned investigator.

Historically, our view of lower piston speeds was endorsed by API 618 when the standard was first published in 1964. These early design reciprocating compressors were limited to around 750 ft/min (3.8 m/s) for lubricated compressors, and ≤ 700 ft/min for nonlubricated compressors to achieve approximately 3 years between overhauls. Operating speeds ranging from 200 to 600 rpm became the standard at that time, with stroke lengths having upper limits between 14 and 15 in (381–356 mm). These limits were based on the materials used for rider-bands, piston rings, driver technology and general experience at the time.

However, since the early 1990s, there have been significant improvements in materials, driver technology, and so on. These improvements have allowed reciprocating compressor piston speeds to trend upwards. As of 2020, nonlubricated compressor piston speeds are as high as 750 ft/min (3.81 m/s) and lubricated pistons at a high of 850 ft/min (4.32 m/s) [3]. Clearly, these higher piston speed machines can be designed with almost endless combinations of shorter stroke and higher rotating speeds, or vice versa.

API 618, in their effort to acknowledge speed differences in reciprocating compressors, now brackets these machines into two categories, namely low-speed and moderate-speed compressors. However, the standard does not define the terms low speed, and moderate speed. Because of this, there is confusion among end users and purchasers. While some relate these two categories to rotating speed, others relate it to piston speed.

In actuality, low-speed and moderate-speed compressors are categorized by their rotating speeds. Relative to the historical definition of low speed in early editions of API 618, we can define moderate speed as speeds > 600 rpm. Table B.1 shows a popular distinction between the categories.

Table B.1: Low- and moderate-speed compressors [3].

	Low speed	Moderate speed
RPM	200–600	600–1,200

B.1.3.8 Allowable discharge temperature

There are few performance parameters which challenge the skills and experience of compressor designers more than the prediction of discharge temperature for specific machines.

The adiabatic temperature is generally used to calculate discharge temperature as an initial attempt to specifying heat load on intercoolers and aftercoolers. This is the normal calculation routine even though the actual discharge temperature might be less or sometimes more than the calculated adiabatic temperature.

API 618, fifth edition, under Section 6.5, mentions that the maximum predicted discharge temperature shall not exceed 150 °C (300 °F). For high-pressure applications with molar mass <12, predicted discharge shall be limited to 135 °C (275 °F).

Predicted discharge temperature can be calculated as follows:

$$T_2 = T_1 \left(r_{\mathrm{p}} \right)^{\frac{k-1}{k}}$$

where T_1 is the absolute inlet temperature and T_2 is the absolute discharge temperature.

B.1.3.9 Staging and interstage cooling

Compressing gases to very high pressures will necessitate that compression be done in stages.

Multistage compression gives the following advantages:

- Allowable discharge temperature is more easily achieved.
- Improvement in overall volumetric efficiency will result. If all the compression were done in one cylinder, the gas in the clearance volume would expand to a large volume before the new intake begins. By cooling the gas between stages, a much higher efficiency can be achieved.
- Cycle work is reduced.
- Size and strength of cylinders can be adjusted to suit volume and gas pressure; and
- Allows for much better mechanical balance.

Figure B.18 illustrates a two-stage compressor with interstage cooling between the first and second stages.

Total cycle work $W = W_{\mathrm{LP}} + W_{\mathrm{HP}}$

$$= \frac{n}{n-1} mRT_1 \left[\left(\frac{P_2}{P_1} \right)^{\frac{n-1}{n}} - 1 \right] + \frac{n}{n-1} mRT_3 \left[\left(\frac{P_4}{P_3} \right)^{\frac{n-1}{n}} - 1 \right] \quad \text{(B.30)}$$

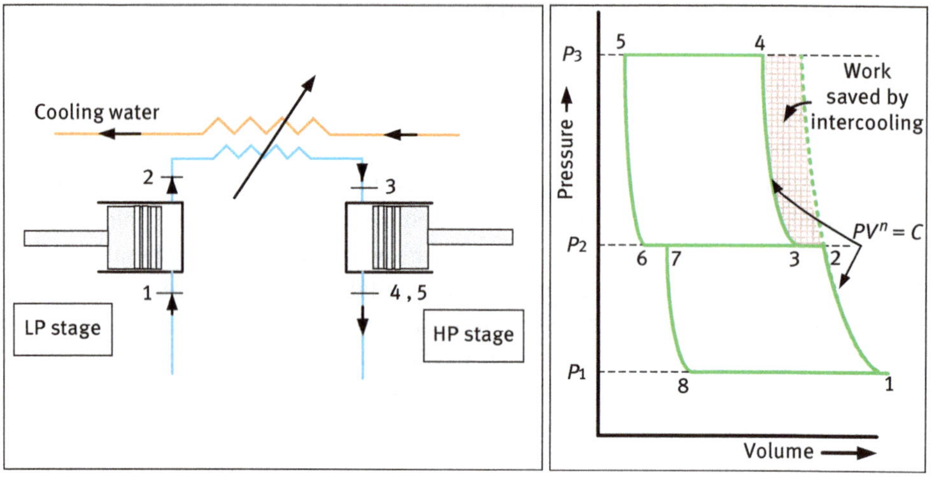

Figure B.18: Two-stage compressor with interstage cooling.

Ideally, if intercooling is complete then $T_1 = T_3$ and $P_3 = P_2$. Also, $P_1V_1 = P_2V_2 = P_3V_3 = P_4V_4$

$$W = \frac{n}{n-1} mRT_1 \left[\left(\frac{P_2}{P_1} \right)^{\frac{n-1}{n}} + \left(\frac{P_4}{P_2} \right)^{\frac{n-1}{n}} - 2 \right]$$

If we let $n/(n-1) = a$ and

$$X = \left[\left(\frac{P_2}{P_1} \right)^{\frac{n-1}{n}} + \left(\frac{P_4}{P_2} \right)^{\frac{n-1}{n}} - 2 \right]$$

Then, for maximum efficiency, there will be some intermediate pressure, P_2 at which the work will be minimized.

Let $dX/dP_2 = 0$.
That is, $P_4 / P_2 = P_2 / P_1$ and $P_2 = [P_1P_4]^{1/2}$
Therefore,

$$\frac{P_2}{P_1} = \left[\frac{P_4}{P_1} \right]^{\frac{1}{2}} \text{(for a two-stage compressor)} \tag{B.31}$$

Or

$r_{\text{p-stage}} = r_{\text{p-overall}}{}^{1/S}$
where R_p is the compression ratio and S is the number of stages

$$W_{\text{Total}} = \frac{Sn}{n-1} mRT_1 \left[\left(\frac{P_{S+1}}{P_1} \right)^{\frac{n-1}{Sn}} - 1 \right] \tag{B.32}$$

B.1.3.10 Rod and gas loading

Overloading of the piston rod can cause much damage to the compressor. Reciprocating compressor design must ensure that there is a complete stress reversal of the piston rod during each revolution for all operating conditions. This includes all cylinder unloading combinations.

Ensuring stress reversal during loading of the piston rod allows lubricating oil to flow around the crosshead pin.

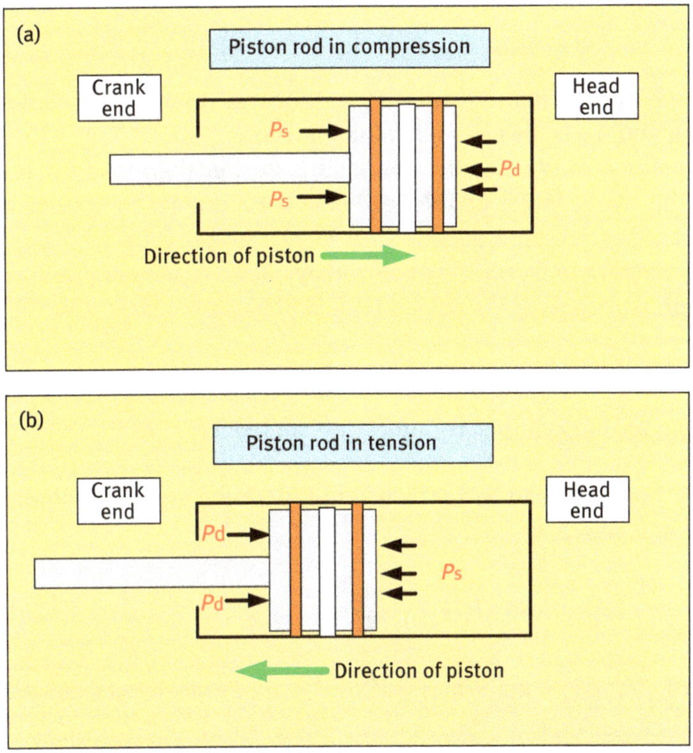

Figure B.19: (a) Piston rod in compression and (b) piston rod in tension.

The following are the three types of loads acting on the piston rod:
- Gas load
- Inertia load
- Combined rod load (gas plus inertia)

Gas loads

Gas load is the net force transferred to the piston rod due to the pressure difference that acts on opposing sides of the piston. The force acting on a piston rod will alternate between *compression* and *tension* as the piston moves back and forth.

Compression load: From Figure B.19a, visualize the net force as the piston moves toward the head-end of the cylinder:

$$\mathrm{RL_c} = P_\mathrm{d} \times A_\mathrm{he} - P_\mathrm{s} \times A_\mathrm{ce}$$

Tension load: From Figure B.19b, as the piston moves toward the crank-end of the cylinder, the net force will be

$$\mathrm{RLt} = Ps \times A_{he} - P_d \times A_{ce}$$

where RL is the rod load, P_s is suction pressure, P_d is discharge pressure, A_he is the cylinder head-end area, and A_ce is the cylinder crank-end area.

Normally, the magnitude of the compression load is greater than load in tension, mainly because the area is larger on the head-end. Also, the net compression load will increase as suction pressure decreases or discharge pressure increases. Because of this, operators have to be careful on how they decrease suction pressure or increase discharge pressure during loading and unloading of the compressor.

Inertia load

Inertia load is the force on the cylinder reciprocating components resulting from the change in velocity.

Let the location of the crosshead in the *x*-direction with respect to the crank angle be "*x*" as shown in Figure B.20.

$$X = r\mathrm{Cos}\theta + l\sqrt{1 - \frac{r^2\mathrm{Sin}^2\theta}{l^2}}$$

The reciprocating inertia force $\mathrm{F_i} = \mathrm{ma} = \mathrm{m_{recip}}\ \mathrm{d^2X/dt^2}$ can be calculated from the following equation:

$$F_\mathrm{i} = m_\mathrm{recip} r\omega^2 \left[\mathrm{Cos}t(\omega t) + \frac{r}{l}\mathrm{Cos}(2\omega t)\right]$$

Combined rod loads

The combined rod load is the algebraic sum of the gas load and the inertia force acting on the crosshead pin. Referring to Figure B.21, load reversals take place at approximately 92° and 320°.

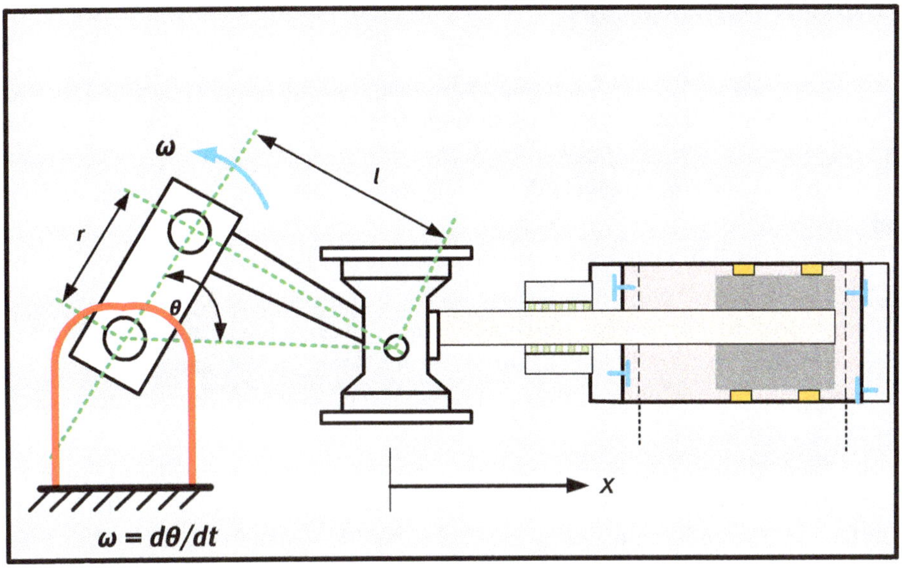

Figure B.20: Reciprocating compressor slider-crank geometry.

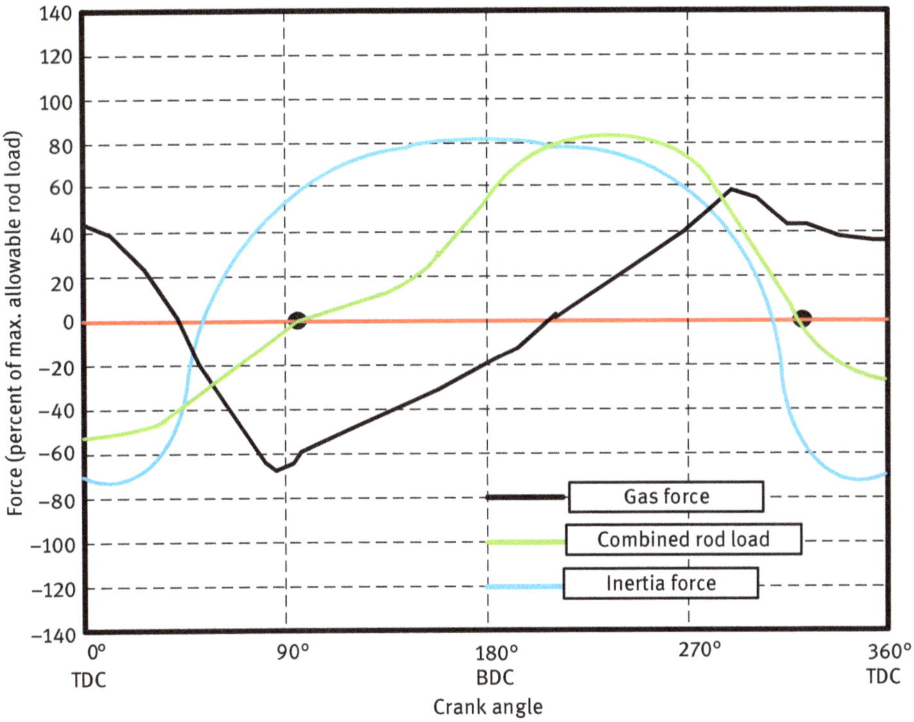

Figure B.21: Illustrated example of piston rod loads versus Crank angle.

B.1.4 Rotary screw compressors

As previously mentioned in Chapter 2, screw machines are classified as rotary positive displacement compressors. This designation is based on the way in which these compressors raise the pressure of a gas by trapping a fixed volume of gas on the suction side and progressively decreasing the volume of the gas until it reaches the discharge port.

The idealized *P–V* diagram for a screw compressor is shown in Figure B.22; the diagram is quite similar to the *P–V* diagram for a reciprocating compressor without clearance. However, unlike reciprocating compressors, screw compressors do not have valves that open and close for intake and discharge. The exchange of gas is done through fixed openings in the casing that are related to definite angular positions of the rotor scrolls.

The fixed opening are the spaces formed between the cylinder wall and the interlocking convolutions of the two helical rotors. The opening position of the edge of the outlet port determines the design parameter *built-in volumetric ratio*, V_i. This ratio is defined as the ratio of the maximum volume V_1 at inlet to the chamber to the volume V_2 at the beginning of the discharge opening.

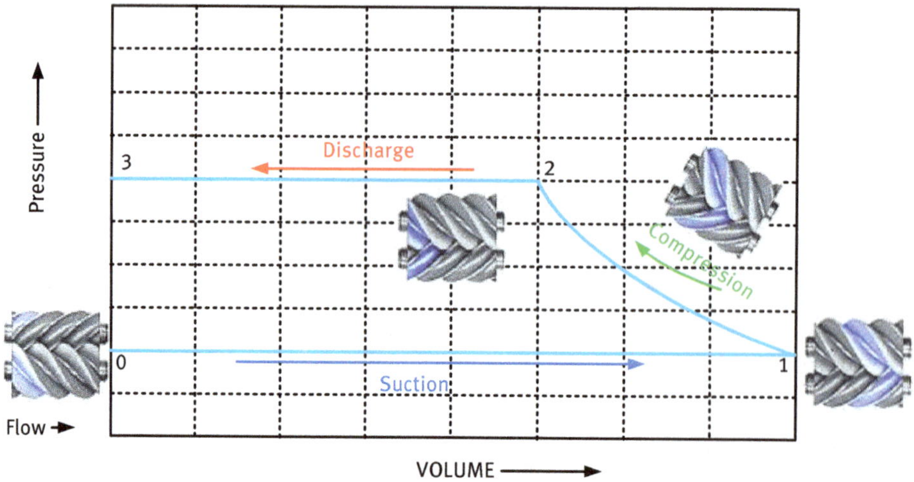

Figure B.22: Screw compressor compression cycle.

Built-in volume ratio $Vi = (V_1/V_2)$

Built-in volume ratio is a specific value designed into different frame size compressors; it is determined by rotor and casing geometry. This parameter is related to the "*built-in pressure ratio*" Π_i.

Built-in pressure ratio $\prod_i = V_i^k$, where k is the adiabatic exponent or specific heat ratio of the gas being compressed. Note that, unlike V_i, $\prod_I$ is not a fixed value for a particular frame size compressor since it depends on the k-value of the gas being compressed.

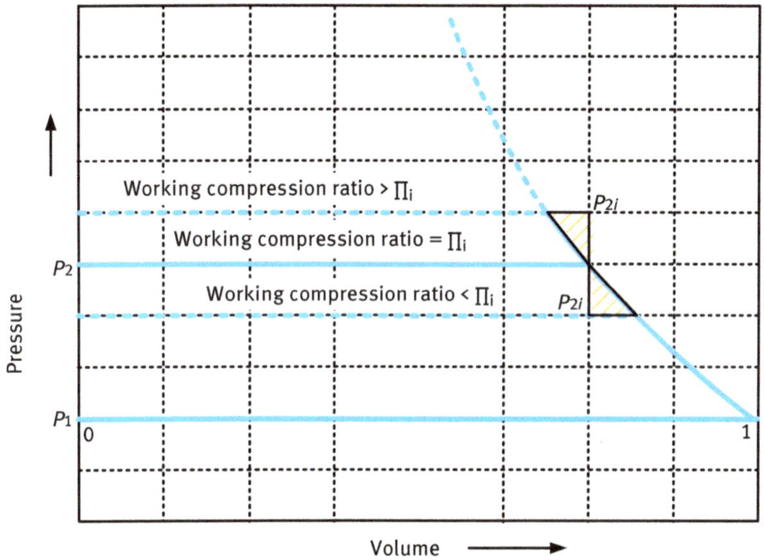

Figure B.23: Idealized *P–V* diagram showing a screw compressor with under- and overcompression.

Modern rotary screw compressors are designed for a specific anticipated compression ratio $\prod_i$ as illustrated in Figure B.23. However, if the compressor discharges into a system with a compression ratio in excess of $\prod_i$, the compressor end-wall will be exposed to higher discharge pressure. Subject only to component strength and input power, the compressor will be able to produce this higher discharge pressure P_{2i}.

Less than as-designed built-in compression ratios can also be accommodated by screw compressors. This is also illustrated in Figure B.22 although, in this situation, some of the compressor's efficiency will be sacrificed.

P_{2i} is related to suction pressure P_1 by

$$P_{2i} = P_1 \prod{}_i = P_1 V_i^k \tag{B.33}$$

The induced volume of a screw compressor is a function of the interlobe volume and speed.

Displacement volume per hour at suction condition is given by

$$Q_{th} = C_n C_{th} d^3 \frac{L}{D} N \times 60 \text{ Ref. [4]} \tag{B.34}$$

where C_{th} represent constants depending on rotor lobe profile (total wrap angle of male rotor), d is the rotor diameter, L/D is the length-to-diameter ratio of the rotor, N is the speed in rpm, and C_n is the geometric constant relating to rotor exposed area.

Equation (B.34) can be simplified as

$$q_0 = \frac{d^3(L/D)}{C} \tag{B.35}$$

where q_0 is the displacement per revolution (liters/rev), C is the constant for different profile geometries. For example a 4 + 6 rotor lobe assembly, $C = 2.231$ for circular profiles [5] and $C = 2.10$ for asymmetric profiles [5].

Theoretical induced volume

$$Q_0 = \frac{N q_0}{1{,}000} \left(\text{m}^3/\text{min}\right) \text{ [Ref.6]} \tag{B.36}$$

Actual induced volume Q_1 is lowered by the amount of gas Q_v flowing back through the very small clearances:

$$Q_1 = Q_0 - Q_v \left[\text{m}^3/\text{min}\right] \text{ [Ref.5]}$$

Volumetric efficiency is expressed as

$$\eta_v = \frac{Q_1}{Q_0} = 1 - \frac{Q_v}{Q_0} \tag{B.37}$$

The theoretical power input required to compress the induced flow volume Q_1 is

$$P_{th} = \frac{10^{-3}}{60} \times \delta_1 Q_0 h_{1ad} [\text{kW}] \tag{B.38}$$

where δ_1 is the gas density at inlet conditions (kg/m^3), h_{1ad} is the energy required for adiabatic compression of 1.0 kg from P_1 to P_2.

Alternatively

$$P_{th} = \frac{k}{k-1} P_1 Q_1 \left[\left(\frac{P_2}{P_1}\right)^{\frac{k-1}{k}} - 1\right] \times \frac{10^4}{6{,}000} \text{kW} \tag{B.39}$$

where $Q = \text{m}^3/\text{min}$ and $P_1 = \text{bara}$

$$\text{Total transmitted power } P_{total} = (P_{th} + P_{dyn} + P_v)\text{kW} \tag{B.40}$$

where P_{dyn} is the dynamic flow loss and P_v is the mechanical losses.

B.2 Understanding Mach numbers as applied to centrifugal compressors

The Mach number is a dimensionless quantity. It represents the ratio of the speed of an impeller component through-gas and the speed with which sound travels through the gas at that point. Vendors use Mach numbers to evaluate the aerodynamic performance of centrifugal compressor designs. Understanding how vendors or manufacturers use these numbers will assist owner/end users and their engineering staff. Knowing acceptable ranges of Mach numbers will enhance or diminish how confident we can be that a gas compressor will meet or even exceed predicted performance. Some vendors are a bit too optimistic in this regard and the desire to sell may win out over the desire to be accurate.

In any event, understanding Mach numbers as a basic predictor of reasonable performance expectations will be of value. Mach numbers can be examined at the proposal stage for either a compressor revamp or procurement of a completely new machine. That said, Mach-related parameters are defined below and somewhat simplified guidelines are helpful in spotting extrapolated or limiting criteria.

B.2.1 Inlet relative Mach number

For impellers to perform at design efficiency, the impeller relative velocity between the impeller blade inlet edges and the gas going into the eye area of the impeller must be at its lowest possible value. By keeping this number low, a design avoids propagation of a disturbance ("shock") and the almost explosive diffusion of such shocks within the impeller. The relative inlet velocity is computed as the vector between the absolute gas velocity entering the impeller eye, and the peripheral velocity of the blade tip at the impeller eye.

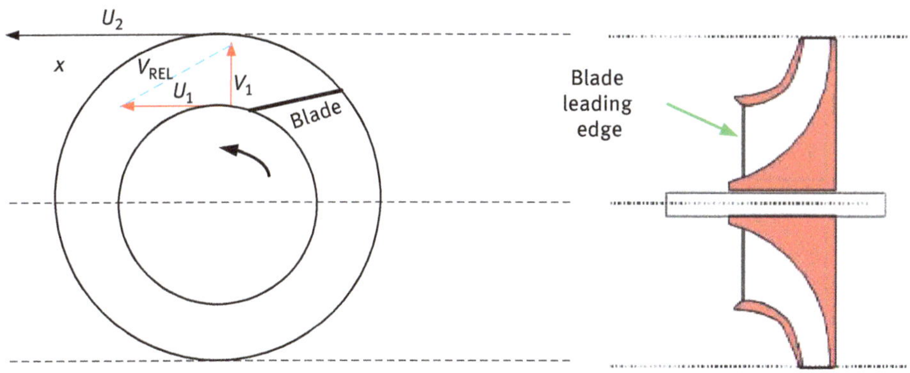

Figure B.24: Velocity triangle at the leading edge of the impeller blade.

The magnitude of V_{rel} compared to the speed of sound at the inlet is called the relative inlet Mach number. It is the magnitude of this ratio which largely influences "stonewall" (the abrupt loss of pressure, sometimes called "choke limit") in a conventional stage. Although true stonewall should, theoretically, not be reached until the relative inlet Mach number is unity, it is conventional practice not to exceed 0.85 or 0.90 at design flow.

From Figure B.24 it can be discerned that for a given rpm, the magnitude of V_{rel} will diminish with decreasing flow, since V_1 is proportional to flow. If V_{rel} decreases, the relative inlet Mach number decreases. Accordingly, the stonewall effect is normally not a factor at flows below design. It is also evident that at low flows, the direction of V_{rel} is such that the gas impinges on the leading side of the blade (positive incidence), a factor that is not very detrimental to performance until very high values of positive incidence are reached.

Suppose now that flow is being increased beyond the design point. As V_1 increases, so do V_{rel} and relative inlet Mach number. In addition, V_{rel} now impinges on the *trailing* side of the blade, a condition known as *negative incidence*. It has been observed that high degrees of negative incidence tend to contribute to stonewall problem as Mach 1 is approached, presumably because of boundary layer separation and reduction of effective flow area in the space from blade to blade.

B.2.2 Impeller tip Mach number

Impeller tip Mach number, often called "machine number," is sometimes confused with "inlet relative Mach number" by engineers. The impeller tip Mach number is defined as the impeller's tangential velocity at exit/tip speed compared to the sonic velocity of the gas. In contrast with "inlet relative Mach number" discussed earlier, satisfactory performance of impellers can be obtained at levels 1.0 and slightly above. This parameter is used to qualitatively judge the amount of stability, turndown, and rise to surge for an impeller. The higher the tip Mach number, the lower the stability, turndown, and/or rise to surge.

B.2.3 Nozzle Mach number

Nozzle Mach number is defined as the velocity entering or leaving a compressor nozzle compared to the sonic velocity of the gas. Designers and end users pay more attention to nozzle Mach number at the inlet nozzles rather than that of discharge nozzles. An emphasis on inlet conditions is linked to the desire of having proper flow distribution as gas enters the suction chamber; then into the impeller eye without affecting performance. Higher velocities translate into higher pressure drops through the nozzles, resulting in lower overall compressor efficiency/performance, and thus higher power requirements.

It follows that the guidelines discussed here are referring to suction nozzle Mach number. Most compressor manufacturers use 140 ft/sec (42 m/s) as the upper limit for inlet nozzle velocity. This number is traditionally based on air at 80 °F and, consequently, a better approach is to ratio the nozzle velocity to the sonic velocity at the inlet conditions. This then will take into account the particular molecular weight of the gas.

Conservative guidelines for actual inlet Mach number are ≤0.11 for new machines and 0.11–0.15 for machines being rerated. When a compressor design occasionally pushes the 1.35 limit previously mentioned in section 3.2.3.2, the actual performance may significantly deviate from the user/purchaser's expectations.

B.3 Compressor selection

Compressor selection sometimes starts as early as the project's conception phase; early selection may determine if a project is feasible at all. Process designers often ask for several cases to be evaluated at that phase in the project. In some instances, the responsible machinery engineer must investigate several different compressor types to find the best fit.

Once the feasibility is completed, the selection process is again repeated during Front-End-Enginnering-Design (FEED) to get better definitions of equipment including compressors. And remember that, thermodynamic feasibility is just the very first of the many ingredients of life cycle cost evaluation.

The process design engineer provides the following basic information:

- Gas mixture
- Flows
- Suction and discharge pressures
- Suction and discharge temperatures

Factors to be considered by the machinery engineer for selection of compressors include:

1. Understanding the *duty cycle* of the compressor application. How will the compressors be operated? Will it be a continuous duty? Will it be a batch process application? What is the planned turnaround time for the compressor? What will be the turnaround interval? Will it be 4–5 years or less?
2. Type of *gas mixture* being compressed. Is the mixture hazardous? Fouling type service?
3. *Flow rate* at normal and rated point. Are there other cases to be considered?
4. *Suction and discharge pressure.* Are there process limitations on pressure?
5. *Suction and discharge temperature.* Are there process limitations on temperature?
6. **Head** – To be determined from calculations
7. **Power** – To be determined from calculations

8. **Speed** – To be determined from calculations
9. **Mach number** – To be determined from calculations and to assess other potential machine limitations.

The type of compressor can be selected once all of the above are known, and then evaluated.

The preliminary selection process begins by examining where the design operating limits fit on Figure 2.2 or by interrogating Tables B.2 and B.3.

Several compressor types can sometimes fit an application; their performance may overlap as well. Therefore, advantages and disadvantages of each type must be weighed against capital expenditure, safety, and reliability.

Following preliminary selection, other factors to be considered are important support systems or auxiliaries. These will generally include

- Seal and seal system
- Lubrication skid

We will assume that the type of compressor is selected based on the factors previously discussed. Therefore, a few detailed examples will be shown to reinforce the helpfulness of some basic calculations. Viewing the examples may also pave the way to competently estimate compressor size selection.

Centrifugal compressors

Example 1: A process application consists of a gas mixture whose molar composition is 81% propane, 15% isobutane, and 4% ethane. Calculate the molecular weight, compressibility factor, and specific heat of the gas if the condition of the mixture is 30 psia and 60 °F.

Table B.4 is generated similar to what was done in Table A.8, and the compressibility factor and specific heat calculated as follows:

Molecular weight 45.56

Specific heat ratio $k = \frac{MC_p}{MC_p - 1.99} = \frac{17.7446}{17.7446 - 1.99} = 1.126$

Reduced pressure $Pr = \frac{30}{607.8} = 0.049$

Reduced temperature $Tr = \frac{(60 + 460)}{671.56} = 0.774$

Compressibility factor $Z_1 = 0.955$ [from Chart A – 1]

Table B.2: Typical operating range of compressors (metric units) [7].

	Inlet capacity (acmh m^3/h)	Maximum discharge pressure (bar)	Adiabatic efficiency (%)	Operating speed (rpm)	Maximum power (MW)	Application
Dynamic Compressors						
Centrifugal	170–850,000	690	70–87	1,800–50,000	38	Process gas and air
Integrally geared centrifugal	500–500,000	350	80	7,000–50,000	60	Process gas and air
Axial	50,000–850,000	17	87–90 +	1,500–10,000	75	Mainly air
Positive displacement compressors						
Reciprocating (piston)	20–34,000	4,150	80–90	200–900	15	Air and process gas
Diaphragm	0–250	1,400	60–70	300–500	1.5	Corrosive and hazardous process gas
Rotary screw (wet)	100–12,000	24	65–70	1,500–3,600	1.5	Air, refrigeration, and process gas
Rotary screw (dry)	200–100,000	1–50	55–70	1,000–20,000	6	Air and dirty process gas
Rotary lobe	25–50,000	0.3–1.7	55–65	300–4,000	0.4	Pneumatic conveying, process gas and vacuum
Sliding vane	15–5,000	10	40–70	400–1,800	0.35	Vacuum service and process gas
Liquid ring	10–17,000	5.5–10.5	25–50	200–3,600	0.3	Vacuum and corrosive process gas

Table B.3: Typical operating range of compressors (US customary units) [7].

	Inlet capacity (acfm)	Maximum discharge pressure (psig)	Adiabatic efficiency (%)		Maximum power (hp)	Application
Dynamic compressors						
Centrifugal	100–200,000	10,000	70–87	1800–50,000	50,000 +	Process gas and air
Integrally geared centrifugal	300–30,000	4,000	80	7000–50,000	80,000	Process gas and air
Axial	30,000–500,000	250	87–90 +	1,500–10,000	100,000	Mainly air
Positive displacement compressors						
Reciprocating (piston)	10–20,000	60,000	80–95	200–900	20,000	Air and process gas
Diaphragm	0.5–150	20,000	60–70	300–500	2,000	Corrosive and hazardous process gas
Rotary screw (wet)	50–7,000	350	65–70	1,500–3,600	2,000	Air, refrigeration and process gas
Rotary screw (dry)	120–58,000	15–700	55–70	1,000–20,000	8,000	Air and dirty Process gas
Rotary lobe	15–30,000	5–25	55–65	300–4,000	500	Pneumatic conveying, process gas and vacuum
Sliding vane	10–3,000	150	40–70	400–1,800	450	Vacuum service and process gas
Liquid ring	5–10,000	80–150	25–50	200–3,600	400	Vacuum and corrosive process gas

Table B.4: Gas mixture at 60 °F and 30 psia.

Gas composition	Mole fraction Y	Mol. weight of weight of individual M (1)	Mol. weight of weight of individual mixture $\sum Y \times$	Individual molal specific heat MCp (2)	Molal specific heat of mixture $\sum Y \times$ MCp	Component critical pressure (psia) P_c (3)	Critical pressure of mixture × P_c	Component critical temperature (°R) T_c (3)	Critical temperature of mixture $\sum Y \times T_c$
Propane	0.810	44.096	35.71776	17.13	13.8753	617.4	500.094	666	539.46
Isobutane	0.150	58.124	8.7186	22.51	3.3765	529.2	79.38	734	110.1
Ethane	0.040	28.052	1.12208	12.32	0.4928	708.0	28.32	550	22
Mixture properties	**1.000**		**45.55844**		**17.7446**		**607.794**		**671.56**

Notes:

(1) Taken from Table A1

(2) Taken from Table A5

(3) Taken from Table A2

Example 2: If 5,100 lb/min of the gas in Example 1 is to be compressed to 120 psia, determine the compressor frame size, number of stages (number of impellers), rotational speed, power requirement, and discharge temperature.

Table B.5: Centrifugal compressor selection table.

Frame	Nominal flow range (acfm)	Nominal flow range (m^3/ h)	Avg. polytropic efficiency (%)	Type of casing (VS = Vertical split, HS = Horizontal split)	Straight through flow or double flow (H-W = Double flow, H = Straight through flow)	Speed required to Develop 10,000 ft-lb/ lb or 3,048 m (rpm)
1	410,000	700,000	85	HS	H-W	4,000
2	350,000	600,000	85	HS	H	4,000
3	150,000	250,000	84	HS or VS	H	5,500
4	44,000	75,000	83	HS or VS	H	7,000
5	30,000	50,000	81	HS or VS	H	10,000
6	20,000	35,000	79	HS or VS	H	12,000
7	12,000	21,000	72	VS	H	16,000

Source: *Mitsubishi Heavy Industries.*

1. Determine the inlet volume Q_1:

$$Q_1 = m_1\left(\frac{Z_1RT_1}{144P_1}\right) = 5,100\left[\frac{0.955\times 1545\times(60+460)}{45.56\times 144\times 30}\right] = 19,881\,\text{icfm (acfm)}$$

   Choose a compressor frame with minimum uprateability of 15%; *QFrame* is 1.15 x 19,881 = 22,863 ACFM
2. Select a compressor from Table B.5. Frame 5 is selected based on the volume flow calculated:
   Nominal head Hp_{nom} = 10,000 ft-lb/lb
   η_p = 81%
   Nominal speed N_{nom} = 10,000 rpm
3. Determine the overall pressure ratio:

$$r_p = \frac{P_2}{P_1} = \frac{120}{30} = 4.0$$

4. Determine the estimated discharge temperature:
   Since $k_1 = 1.126$

As shown in Chapter 1:

$$\frac{n}{n-1} = \left(\frac{k}{k-1}\right)\left(\eta_p\right) = \frac{1.126}{0.126} \times 0.81 = 7.2386$$

Therefore,

$$T_2 = T_1 r_p^{\frac{n-1}{n}} = 520 \times 4^{0.1381} = 629.72\ ^\circ\mathrm{R}\ \text{or}\ 169.73\ ^\circ\mathrm{F}$$

5. Determine the average compressibility:
Reduced pressure

$$Pr_2 = \frac{P_2}{P_c} = \frac{120}{607.8} = 0.197$$

Reduced temperature

$$T_{r_2} = \frac{T_2}{T_c} = \frac{629.72}{671.56} = 0.938$$

$$Z_{avg} = \frac{Z_1 + Z_2}{2} = \frac{0.955 + 0.93}{2} = 0.9425$$

6. Determine the average k value:
$MC_{p2} = 20.464$

$$K_2 = \frac{MC_p}{MC_p - 1.99} = \frac{20.464}{20.464 - 1.99} = 1.1077$$

$$K_{avg} = \frac{k_1 + k_2}{2} = \frac{1.126 + 1.1077}{2} = 1.117$$

7. Determine the average $n/n-1$ from K_{avg}:

$$\frac{n}{n-1} = \left(\frac{k}{k-1}\right)\left(\eta_p\right) = \frac{1.117}{0.117} \times 0.81 = 7.733$$

8. Determine the polytropic head:

$$H_p = Z_{avg} R\ T1 \frac{n-1}{n}\left[r_p^{\frac{n-1}{n}} - 1\right] = 0.9425 \times \left(\frac{1,545}{45.6}\right) \times 520 \times 7.733 \times \left[\ 4^{0.1293} - 1\right]$$

$$= 25,209\ ft - lb/lb$$

9. Determine the number of compressor stages
Max. head/stage h_{max} = 10,000 ft-lb/lb [from Table B.5].
Therefore, number of stages:

$$S_t = \frac{H_p}{H_{max}} = \frac{25,209}{10,000} = 2.52\ [\text{Make it three stages}]$$

10. Determine the rotational speed

$$N = N_{nom}\sqrt{\frac{H_p}{H_{max} \times \text{no.of stages}}} = 10,000 \times \sqrt{\frac{25,209}{10,000 \times 3}}\ [N_{nom}\ \text{is taken from Table B5}]$$

$$= 9,167\ \text{rpm}$$

Note: To lower the speed more stages would be required

11. Determine the required shaft power:

$$P_{gas} = \dot{m} \times \frac{H_p}{33,000\eta_p} = \frac{5,100 \times 25,209}{33,000 \times 0.81} = 4,809.8 \text{ hp}$$

Assume mechanical losses to be approximately 2.5%
Mechanical losses M_L = 0.025 × 4,809.8 = 120.2 hp
Shaft power = P_{gas} + M_L = 4,809.8 + 120.2 = 4,930 hp

12. Determine the actual discharge temperature:

$$T_2 = T_1 (r_p)^{\frac{n-1}{n}} = 520 \times 4^{0.1293} \text{ °R}$$
$$= 622 \text{ °R or } 162 \text{ °F}$$

Example 3: This example deals with selecting the number of compressor sections in a process compressor with intercooling between sections. For simplicity, we use dry air as the gas being compressed from 250 kPa-a and suction temperature 34 °C. The mass is given as 70,000 kg/h and discharge pressure 1200 kPa-a. Discharge temperature should not exceed 190 °C. Assume a pressure drop of 2% across heat exchanger and a gas reentry temperature of approximately 38 °C.

Since the specific heat ratio and compressibility for air does not vary much, we can assume the following for this example:

$Z = 1.0$
$K = 1.4$
MW = 28.96

1. Determine the inlet volume

$$Q_1 = \dot{m}\left[\frac{Z_1 R T_1}{P_1 \times 10^3}\right] = 70,000\left[\frac{1.0\text{x}8314 \times (34+273)}{28.96 \times (250 \times 10^3)}\right] = 24,678 \text{ m}^3/\text{h}$$

Choose a compressor frame with minimum uprateability of 15%; QFrame = 1.15 x 24, 678 m3/h = 28,379.7 m3/h

Select the smallest compressor frame from Table B.5 capable of delivering the calculated volume flow. In this case, the selected frame is Frame 6:

Hp_{nom} = 3,038 m
= 3,048/101 kN-m/kg
= 30.0 kN-m/kg
Hp_{nom} = 79%
N_{nom} = 12,000 rpm

2. Determine the overall pressure ratio:

$$rp = \frac{P_2}{P_1} = \frac{1,200}{250} = 4.8$$

3. Determine the approximate discharge temperature and the number of compressor sections: Assuming one section

$$k = 1.4$$

$$\frac{n}{n-1} = \left(\frac{k}{k-1}\right)\left(\eta_p\right) = \frac{1.4}{0.4} \times 0.79 = 2.765$$

Assuming one section to start:

$$T_2 = T_1 r_p^{\frac{n-1}{n}} = 307 \times 4.8^{0.362} = 541.7\ ^\circ K$$
$$= 268.7\ ^\circ C$$

This discharge temperature exceeds the discharge temperature limit of 190 °C; therefore, intercooling is necessary, and more than one section is required.

$$r_{\text{p-section}} = r_{\text{p-overall}}^{1/S} \text{ [Previously derived for calculating number of stages]}$$

Assuming two section

$$r_{\text{p-section}} = \sqrt[2]{r_{\text{p-overall}}} = \sqrt[2]{4.8} = 2.19 \text{ sections}$$
$$T_2 = 307 \times 2.19^{0.362} = 407.74 \text{ K}$$
$$= 134.74\ ^\circ\text{C} \text{ [This temperature is well below the imposed limit of 190 °C.]}$$

Figure B.25 represents a two section centrifugal compressors with interstage cooling. Let P2' be the discharge pressure for the 1st section, and P1' the inlet pressure to the 2nd section.

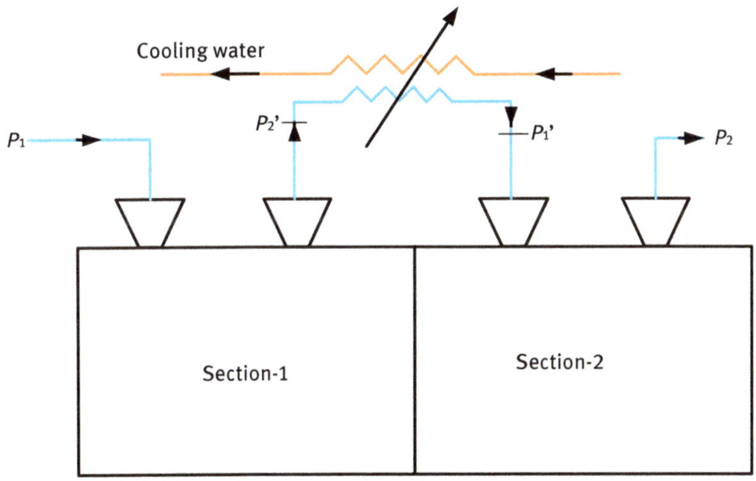

Figure B.25: Two-stage centrifugal compressor with interstage cooling.

4. Determine the average compressibility, average *k*-value, and average $n/n{-}1$ as was done previously before. However, since we are dealing with dry air, we will assume the following.
   Average *k*-value = 1.4
   Average $Z = 1$
   Average $n/n - 1 = 2.765$

5. Determine the polytropic head
   $P_1 = 250$ kPa-a

$$P_{2'} = r_{\text{p-section}} \times P_1 \times 1.02 = 2.19 \times 250 \times 1.02 = 558.45 \text{ kPa-a}$$

$$P_1{}' = \frac{P'_2}{1.02} = \frac{558.45}{1.02} = 547.5 \text{ kPa} - \text{a}$$

$$P_2 = 1{,}200 \text{ kPa-a}$$

$$r_{\text{p}_1} = \frac{P'_2}{P_1} = \frac{558.45}{250} = 2.23$$

$$r_{\text{p}_2} = \frac{P_2}{P'_1} = \frac{1,200}{547.5} = 2.19$$

Total polytropic head $H_{\text{p}} = H_{\text{p1}} + H_{\text{p2}}$

$$Hp_1 = \frac{1}{1,000} \times 1 \times (\frac{8,314}{28.96}) \times (34 + 273)\ (2.765)\ \left[2.23^{0.362} - 1\right]$$

$$= 82.09 \text{ kN-m/kg}$$

$$Hp_2 = \frac{1}{1,000} \times 1 \times \left(\frac{8,314}{28.96}\right) \times (38 + 273)\ (2.765)\ \left[2.19^{0.362} - 1\right]$$

$$= 81.00 \text{ kN-m/kg}$$

Total head = 82.09 + 81.00

$$= 163.1 \text{ kN-m/kg}$$

6. Determine the number of stages:
Section 1:

$$\text{No. of stages } S_{\text{t}} = Hp_1 \backslash Hp_{\text{nom}} = \frac{82.09}{30} = 2.73 \text{ or approximately three stages}$$

Section 2:

$$\text{No. of stages } S_{\text{t}} = Hp_2 \backslash Hp_{\text{nom}} = \frac{81.00}{30} = 2.70 \text{ or approximately three stages}$$

Total number of impellers in the casing will be six. This is well within the maximum number of impellers of 10–12 previously recommended as the limit per casing.

7. Determine the rotational speed:
Section 1:

$$N_1 = N_{\text{nom}} \sqrt{\frac{H_{\text{p1}}}{H_{\text{nom}} \times \text{no. of stages}}} = 12,000 \times \sqrt{\frac{82.09}{30 \times 3}}\ [N_{\text{nom}} \text{ is taken from Table B.5}]$$

$$= 11,460 \text{ rpm}$$

Section 2:

$$N_2 = N_{\text{nom}} \sqrt{\frac{H_{\text{p2}}}{H_{\text{nom}} \times \text{no. of stages}}} = 12,000 \times \sqrt{\frac{81.00}{30 \times 3}}\ [N_{\text{nom}} \text{ is taken from Table B.5}]$$

$$= 11,384 \text{ rpm}$$

Therefore, rotational speed selected is 11,460 rpm. We recommend adding more impellers (stages) in each section to lower the speed.

8. Determine the actual shaft power:

Section 1:

$$P_{gas1} = \dot{m} \times \frac{H_{p1}}{3,600\eta_p} = 70,000\frac{82.09}{3,600 \times 0.79} = 2,020 \text{ kW}$$

Section 2:

$$P_{gas2} = \dot{m} \times \frac{H_{p2}}{3,600\eta_p} = 70,000\frac{81.00}{3,600 \times 0.79} = 1,993.67 \text{ kW}$$

Total power

$$P_{gas} = (2,020 + 1,993.67) \text{ kW} = 4,013.67 \text{ kW}$$

Mechanical losses

$$M_L = 0.025 \times 4,013.67 \text{ kW} = 100.3 \text{ kW}$$

$$\text{Shaft power} = (4,013.67 + 100.3) \text{ kW}$$

$$= 4,114 \text{ kW, approximately}$$

9. Determine the actual discharge temperature

Section 1:

$$T_{2'} = T_1(r_{p1})^{n\text{-}1/n} = 307 \times 2.23^{0.362} \text{ K}$$

$$= 410.42 \text{ °K or } 137 \text{ °C approximately}$$

Section 2:

$$T_2 = T_{1'}(r_{p2})^{n\text{-}1/n} = 311 \times 2.19^{0.362} \text{ K}$$

$$= 413.04 \text{ K or } 140 \text{ °C approximately.}$$

Once a decision is made on the type of compressor – based on the factors presented earlier – a similar process in selecting and sizing the compressor can be carried out. We could follow the approach described in Examples 1–3.

Using the "aerothermodynamic fundamentals" covered earlier in this text and the appended material is perfect for screening studies. Of course, for years computers have been even more precise and most certainly should be used by compressor manufacturers. Nevertheless, a screening study is immensely useful to reliability professionals engaged in machinery quality assessment (see indexed words).

Hand calculations should be welcomed as a test of the human brain and a verification of what we know as engineers. The human brain will never be replaced in the context of assessing if numbers make sense. Only knowledgeable humans will become wise through the application of what they have learned. Wisdom is the application of knowledge and, finally, humans alone can know the true meaning of common sense.

As to compressors: Using the proper manual (as opposed to automated) approach for different types of compressors, along with equipment manufacturers' charts and tables, will be more than adequate for selecting and sizing different types of compressors.

References

[1] Bloch, H.P. A Practical Guide to Compressor Technology, 2nd Edition, John Wiley & Sons, Inc., Hoboken, New Jersey, USA, 1998.

[2] Dickens, T., Day, I. The Design of Highly Loaded Axial Compressors, Journal of Turbomachinery, July 2011, 133.

[3] Kopsick, G.M. Application and Operating History of Moderate-Speed API Reciprocating Compressors, Proceedings of the 33rd Turbomachinery Symposium, Texas A & M University, College Station, Texas, 2004, 103.

[4] Conversion Procedure of Shop Performance Test Data – Kobe Steel, Compressor Department, Japan

[5] Brown, R.N. Compressor Selection and Sizing, 2nd Edition, Gulf Publishing Company, Houston, Texas, USA, 1986.

[6] Bloch, H.P., Noack, P.W. Recent Experience with Large Liquid Injected Rotary Screw Process Compressors, Proceedings of the 20th Turbomachinery Symposium 1991, Texas A & M University, Dallas, Texas

[7] Process Industry Practices (PIP) – Machinery, PIP REEC001, Compressor Selection Guidelines, April 2013.

Appendix C Novel mitigation techniques for solving rotor stability problems

C.1 Combining an arc spring with a squeeze film damper bearing to mitigate a rotor stability problem

Rotor-bearing stability problems, although not as common today as in the past, are problems that occasionally occur whenever one or more of the following shortcuts are taken:

1. Buying new machines and not doing either a stability verification test or an ASME PTC-10 type 1 test because of (unwise) project cost constraints
2. Not following the proper guidelines outlined in API 617 and API 684 during design audits for new or rerate compressor trains
3. End user doing no design audit validation due to (unwise) project budget constraints

Since the American Petroleum Institute (API) 684 guidelines for designing rotor-bearing systems were first issued in 2002, rotor-bearing stability problems are no longer as prevalent today as in the past. However, when stability issues arise, they can become one of the most damaging problem causes. The consequential damage to critically important machines is often quite extreme [1]; it can result in extended downtime in operating plants. In view of this and to thoroughly address the potential issues, turbomachinery manufacturers have invested in modern computer tools. The actions taken by competent manufacturers are fully aligned with up-to-date versions of the relevant guidelines found in API 684.

The most frequent step to mitigate rotor-bearing stability problem is to install a squeeze film damper (SFD) bearing. A typical SFD is shown in Figure C.1; its O-ring components are noteworthy because they perform three important functions:

1. They act as oil seals.
2. They provide a centering element for the bearing shell.
3. They give needed stiffness to the bearing assembly.

SFD bearings are frequently adopted as the main solution to rotor-bearing instability; they are sometimes favored by compressor manufacturers and their end users or customers. However, the lack of longevity of the installed O-rings in these SFDs in many cases makes them not as reliable as most end user would like. The less-than-desired O-ring life in SFDs can be attributed to the following:

– The inherent life expectancy of an elastomeric O-ring is approximately 4–5 years.

https://doi.org/10.1515/9783110678765-016

- The O-rings function as a centering component for the bearing shell, and therefore bear the full weight of the rotor-bearing assembly. If the rotor is relatively light, the O-ring will remain flexible. However, heavier rotors will cause the O-rings to develop a set (permanent deformation) over time; this elastomeric component will, therefore, lose some of its mechanical properties.

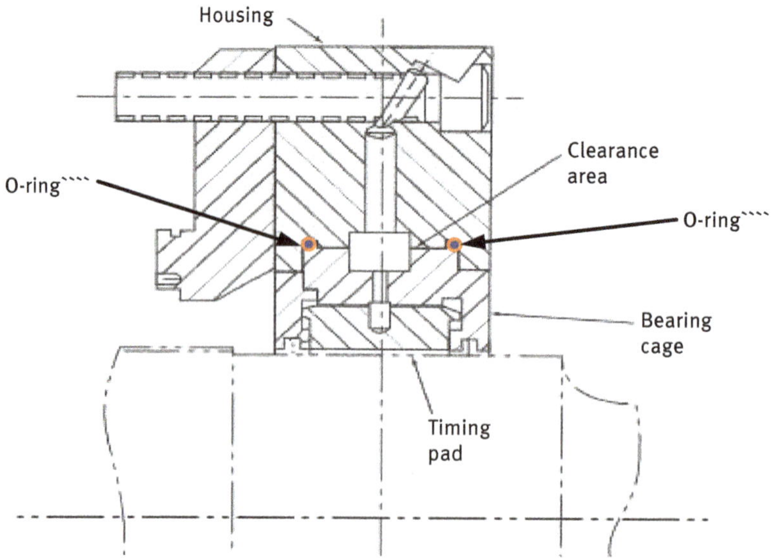

Figure C.1: Typical squeeze film damper bearing assembly [2].

An arc-spring damper proved advantageous to get around, or at least minimize, the shortcoming of O-rings in the application at issue here. In the designers' and manufacturers' efforts to optimize the performance of SFD bearings, fitting the SFD with an arc-spring damper has proved advantageous when extending run lengths, warding off, and solving rotor-bearing stability problems [3, 4]. A case study will be used to highlight this fact.

In 2001, the high-pressure casing of a three-casing compressor train was replaced to increase plant throughput and to accommodate new process conditions. Major changes to the train included a power increase from previously 16,500 kW to now 27,000 kW, and a substantial molecular weight change from 16.4 to 26.8. The new casing had similar construction features to the old casing; these features included identical oil seals, five-pad (tilting pad) radial bearings, and balance piston/center labyrinth seal with a bypass (shunt). The machine was of the back-to-back impeller arrangement as shown in Figure C.2. Table C.1 shows some additional rotor-bearing design information for the new compressor.

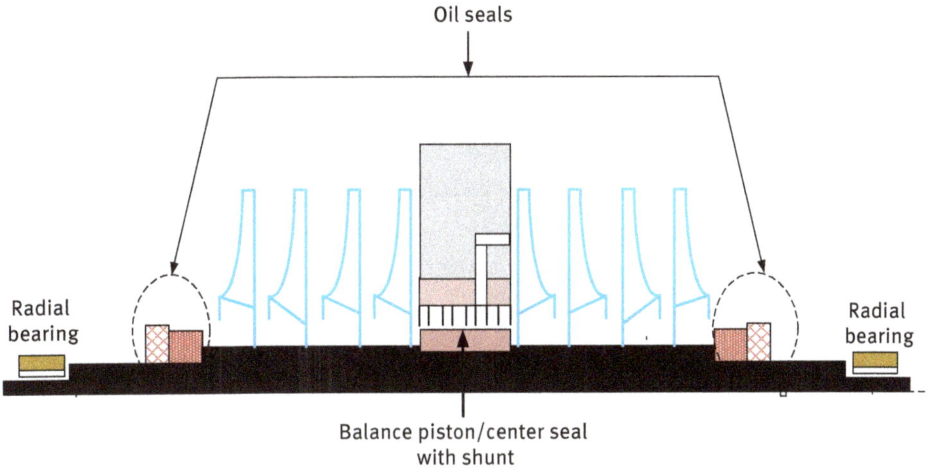

Figure C.2: Casing configuration showing back-to-back impeller arrangement, oil seals, and balance piston with shunt.

Table C.1: Rotor-bearing design information.

Design parameters	Values	Comments
Rotor L_b/D_i L_b = Bearing span D_i = Impeller bore diameter	14.8	Good stable designs are generally <10.0. Values between 10 and 12 should be evaluated based on present level 1 API 684 stability guidelines. Values over 12 should most likely have a level 2 check [5].
Five tilting pad radial bearing, load on pad, with center pivot		These bearing parameters are well within typical industry values for minimizing the risk of rotor instability.
Radial bearings L/D L = Length of bearing D = Journal diameter	0.4	
radial bearing preload load on pad	0.04–0.364	

The new compressor casing underwent high speed balancing and an API 617 mechanical run test without any observed vibration problems. The original equipment manufacturer had performed a lateral stability analysis during the design stage; they used their own in-house computational tools and found no problems. We should note that the stability analysis done by the vendor using their in-house software codes was done just prior to release of the API 617 rotor stability guidelines (available since late 2002). That meant that API 617 guidelines were not followed in this instance. Also, because of (risk-inducing) project budget constraints, no PTC-10 performance test was done.

During startup of the plant, the compressor train surged and initiated an uncontrollable instability of the high-pressure casing rotor, which resulted in a wreck. Following the wreck, vibration data showed a predominant subsynchronous vibration at approximately 0.5 × running speed (see Figures C.3 and C.4).

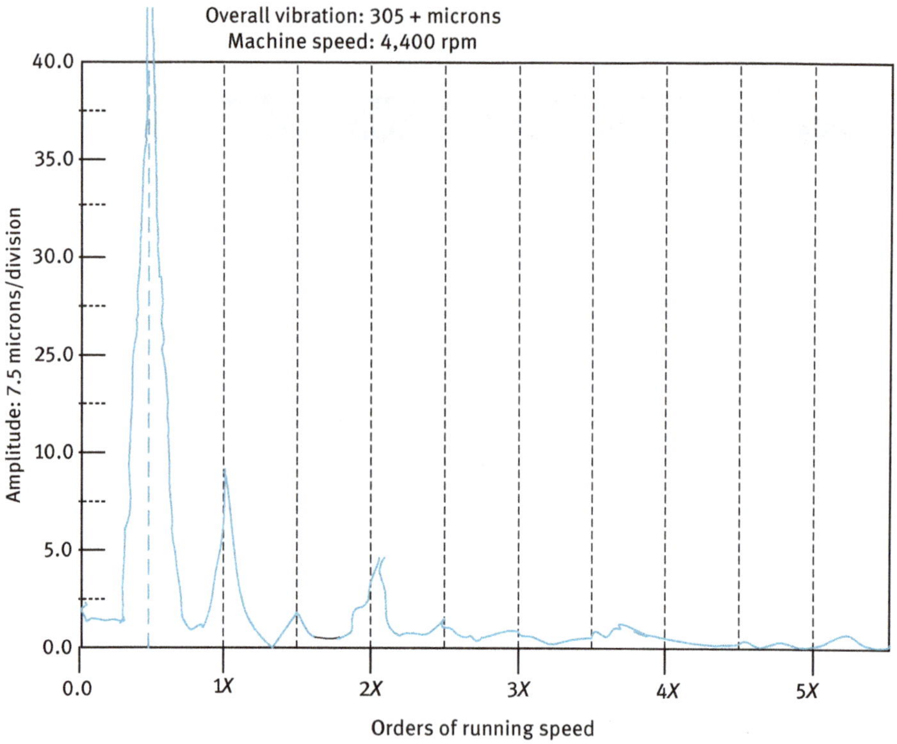

Figure C.3: Vibration spectral data showing predominant subsynchronous vibration.

Upon shutdown and subsequent disassembly of the compressor, all bearings and seals were found damaged. The vibration excursion had exceeded the internal clearances and had reached as high as 300 µm. The compressor was rebuilt with the spare rotor, bearings, and seals, but without any further in-depth vibration analysis. The compressor train was restarted with new operating procedures to prevent surge. Unfortunately, the new procedures did not help, and the compressor was again severely damaged again during startup. There were now no spare parts and with longer than foreseen downtime, the plant was compelled to stop operating by trial and error. The proper resources – finally – had to be mobilized to analyze causes and implement lasting solutions to a very obvious problem.

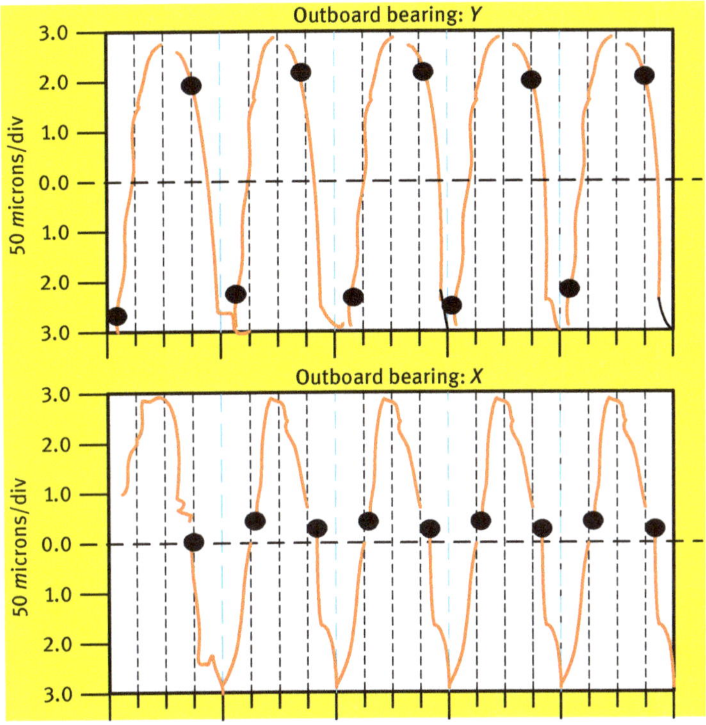

Figure C.4: Vibration waveform showing two key phasor dots per cycle indicating subsynch vibration.

The compressor manufacturer's and end user's engineers rather than concentrating only on startup procedures, formed a team and began to analyze the failures from a rotordynamic point of view. Attention was focused on a rotor-bearing stability analysis. Both failures were analyzed using advanced rotordynamic software; the intent, of course, was to determine the rotor's sensitivity to prevailing destabilizing forces. After running these analyses, the team concluded that the compressor was indeed sensitive to destabilizing forces, as can be seen from the results shown in Figures C.5 and C.6. Figure C.5 was generated using the "simplified Alford" equation mentioned earlier (see Chapter 7, Section 7.3).

The team had to find possible sources of the instability and to firmly recommend steps to eliminate the problem. In so doing, the team decided that since the journal bearings were already optimized for stability, they would look at the other potential sources where destabilizing forces can originate. Thus, there was a clear need to examine three previously overlooked locations: oil seals, impeller eye seals, and the balance piston labyrinth (or rotor center) seal.

Using modern in-house software tools the team concluded that, although the machine was sensitive to destabilizing forces, the overall destabilizing force was

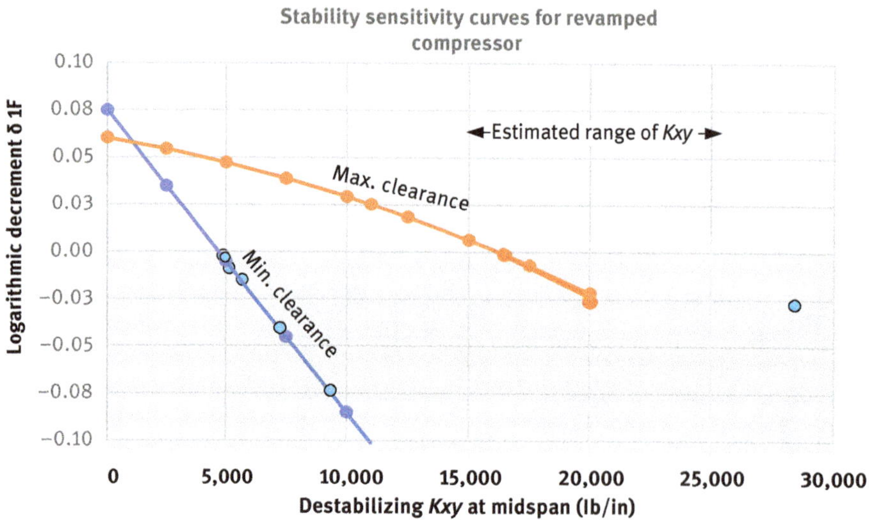

Figure C.5: Stability sensitivity analysis results of failed compressor (*results courtesy of BRG. Machinery Consulting LLC*).

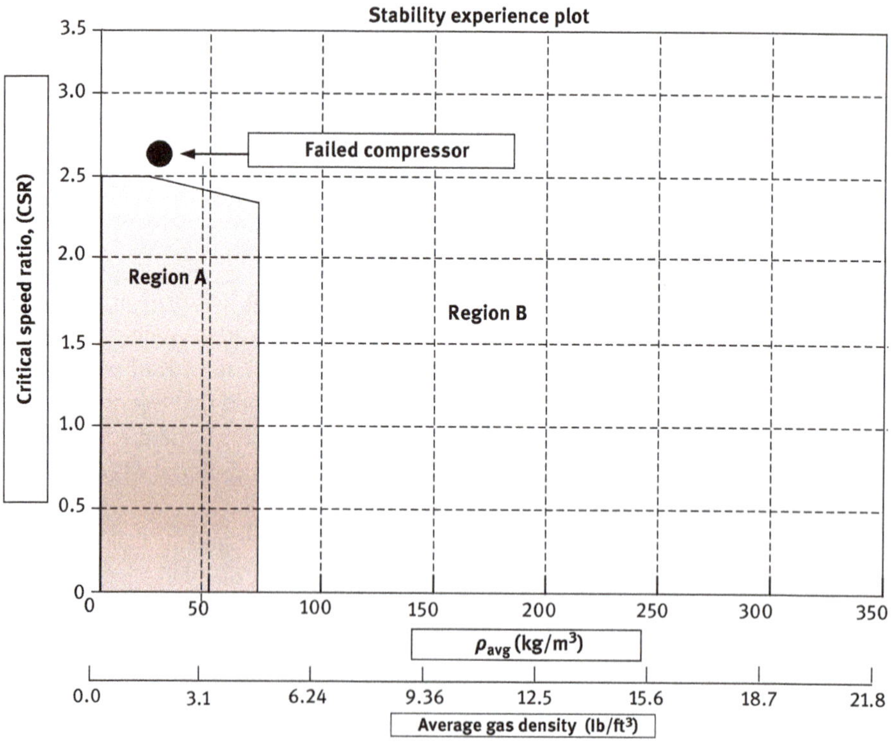

Figure C.6: Level 1 stability experience plot (*results courtesy of BRG. Machinery Consulting LLC*).

small after analyzing the two oil seals, and the balance piston seal. Therefore, the rotor-bearing system was in need of more damping, and the SFD-bearing solution was the quickest available.

An SFD bearing was designed, tested, and tuned to the rotor-bearing system. This design used O-rings to help keep the bearing cage centered. Figure C.7 shows a sizeable improvement in the logarithmic decrement. The machine started up successfully with the SFD bearing installed. There were no vibration problems, and the machine survived several subsequent surge events without the initiation of any instability.

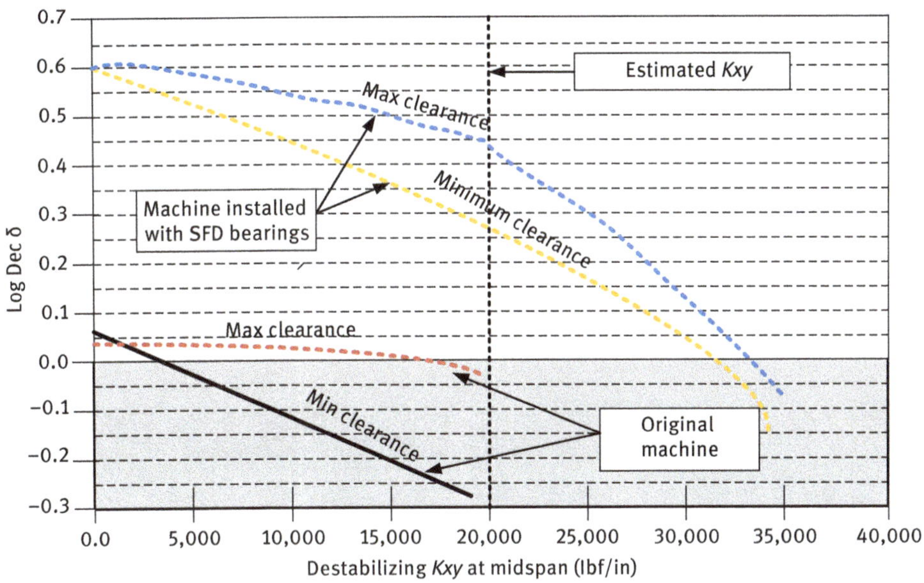

Figure C.7: Improvement in rotor-bearing stability after O-ring squeeze-film-damper installed (*results courtesy of BRG. Machinery Consulting LLC*).

Although the SFD bearing proved successful, there was reasonable concern regarding the reliability of the O-rings installed in the SFD's. The plant engineers voiced concern that the O-rings would degrade over time due to the load imposed by the weight of the rotor-bearing assembly. Such degradation would reduce their effectiveness, possibly resulting in an unplanned outage of the plant. With this in mind, a plan was put in place to improve the longevity of the SFD. It was decided to install an arc spring damper inside each bearing that would help support and center the SFD. The plan called for this to be carried out during the next planned plant outage.

Figure C.8a–d shows the sequence in which the arc spring was installed to become part of the overall SFD bearing assembly.

The new arc spring damper bearings were designed by the compressor manufacturer and would eliminate the load-bearing feature on the O-rings, thus improving the overall reliability of the system. There were three major design objectives of the SFD supplemental geometry:

- The basic SFD oil film design should be kept the same.
- The arc spring damper should be designed soft enough to allow motion, yet strong enough to center the so-called bearing cage (labeled in Figure C.1).
- The arc spring should be sufficiently robust to not experience fatigue failure.

Figure C.8a

Figure C.8b

Figure C.8c

Figure C.8d

Figure C.8: (a)–(d) Installation of the arc spring damper.

This new damper bearing system was thoroughly tested at the compressor manufacturer's facility; it was a complete success.

A design audit study was also done by the end user to determine the effectiveness of the new damper bearings. The study confirmed the stability of the rotor-bearing system as compared to the existing system. The auditors concluded that the overall reliability of the machine would be increased, since the new stability margin exceeded the stability threshold of 20,000 lb/in by a factor of 2, whereas it had been only 1.5 in the existing SFD design.

Achieving long-term reliable operation by combining the traditional SFD and arc spring does come with a minor disadvantage. The overall stiffness of the combined centering support increased because the stiffness of the arc spring adds stiffness to that preexisting in the O-rings. This combined centering stiffness is therefore higher than that of the O-rings alone; the combination reduces the effective damping of the squeeze film and causes the logarithmic decrement to undergo the directional decrease shown in Figure C.9.

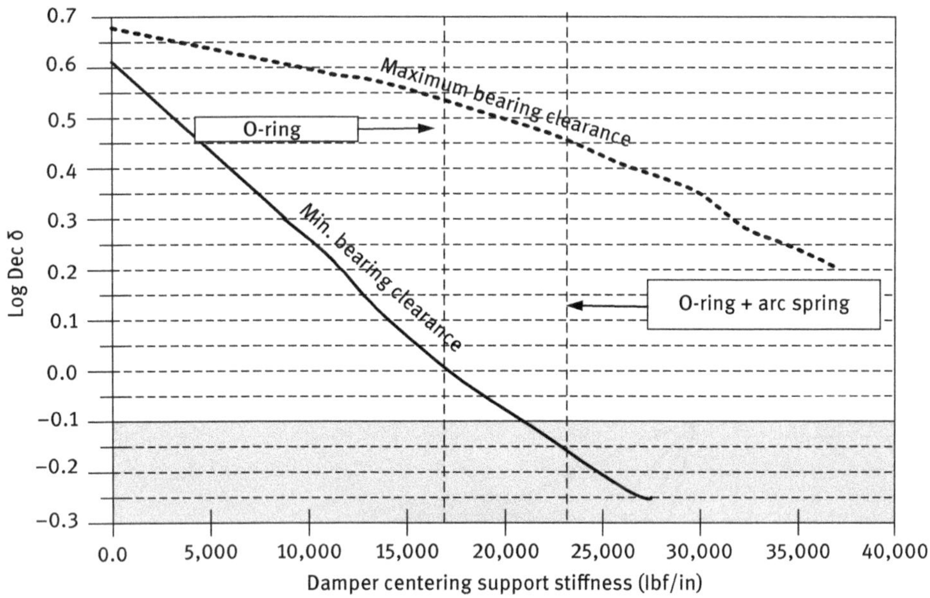

Figure C.9: Impact of centering support stiffness on stability (*results courtesy of BRG. Machinery Consulting LLC*).

The stability sensitivity curves for the three different bearing configurations are depicted in Figure C.10:
- Original bearings
- SFD bearings with O-rings
- SFD bearings with O-rings and added arc spring damper

After installing the arc spring along with the SFD bearing, the machine ran for many years well beyond the normally scheduled turnaround intervals.

Because it did so without any instability problems, the next iteration would be a fully integrated, laser-machined one-piece design. Figure C.11 introduces a true state-of-the-art design.

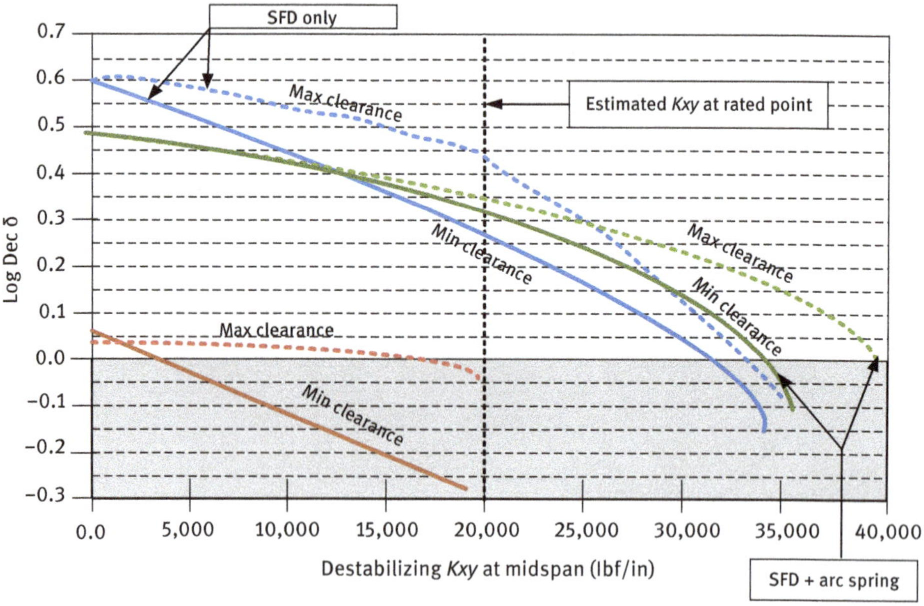

Figure C.10: Stability curves for different bearing configurations (*results courtesy of BRG. Machinery Consulting LLC*).

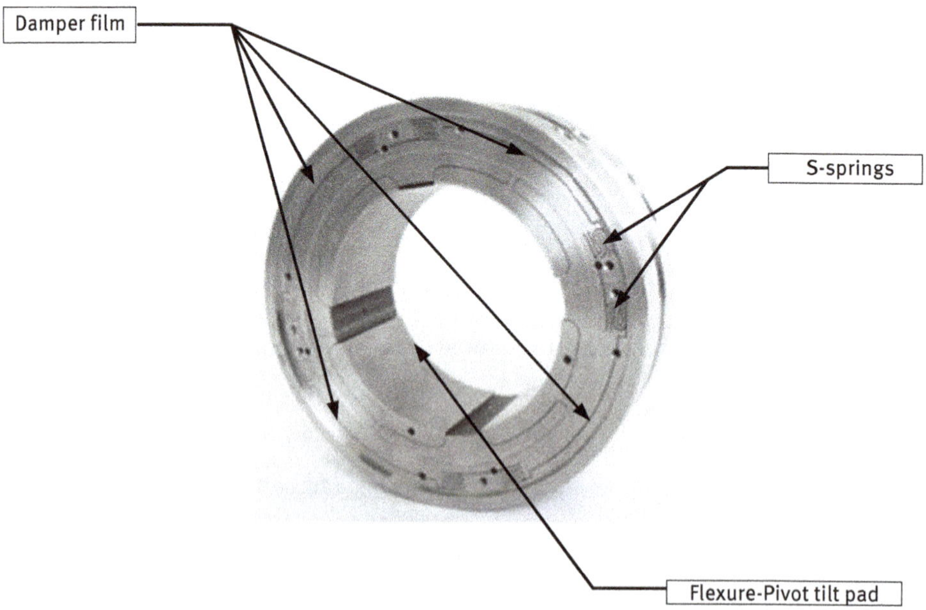

Figure C.11: The ISFD bearing with Flexure–Pivot tilt pads (*courtesy of Waukesha Bearings*).

Recall that the approach of retrofitting the add-on arc spring damper represents an actual field example of immense value to the compressor owner at the time, in 2001. Today, in 2020, an even more modern approach for solving the rotor-bearing stability problem previously described would be to replace the arc spring/SFD bearing with an integral squeeze film damper (ISFD) bearing shown in Figure C.11. The ISFD design eliminates the shortcomings of the typical SFD, while having the capabilities of the arc spring/SFD bearing combination previously discussed. The solution is to use the ISFD bearing. It integrates both SFD and Flexure–Pivot tilt pad technologies without using O-rings for centering or stiffening of the bearing shell. Damping and stiffness of the bearing are accurately determined by incorporating s-type springs within the bearing shell. Specific damper radial clearances are obtained in the shell by using electrical discharge machining or laser-based machining technologies.

C.2 Torsional vibration problem solved using a new design viscous rotational damper

Couplings are the most flexible components used to transmit torque from one shaft to another in turbomachinery trains. Compared to the connecting shafts, they are the torsionally least-rigid component; they act as torsional springs. Therefore, during design of the turbomachinery train engineers usually quantify the influence which the coupling has on torsional modes. The information is then used to torsionally tune the turbomachinery trains.

Previously, in Section 7.6.1, we explained that the lower torsional frequencies of turbomachinery shaft strings cause twisting of couplings. This, of course, sets up high stress amplitudes within the couplings. Excessively high stresses can result in fatigue failure of the coupling and, in severe cases, can cause ejection of the coupling from the turbomachinery train.

The API 617 has given design guidelines for equipment manufacturers to avoid turbomachinery failures caused by torsional excitation. API 617 states that torsional natural frequencies shall preferably be 10% below or 10% above any possible excitation frequency within the specified operating speed range; minimum to maximum continuous speed.

If the separation margin recommended by API 617 cannot be achieved, then suitable experience-based remedial techniques should be executed in the following order:

1. Perform a fatigue analysis to assess the risk. Determine if the resulting stresses are within the endurance limit of the coupling's material
2. Establish if adjustment can be made to the torsional spring rate of the rotor-coupling assembly to improve the separation margin, and verify that frequency interferences are avoided.

3. Determine if the rotor-coupling assembly has sufficient damping to minimize the torsional vibratory amplitudes
4. Determine if changes can be made to any of the inertias within the train

If, after doing a fatigue analysis, we find that the risk of failure is high, then the next mitigating step is to detune the coupling. Detuning the coupling is done to either reduce or increase the torsional stiffness of the coupling. The intent is to move the torsional natural frequencies away from any interference zones and to thus obtain a proper separation margin. However, there are instances where tuning does not change the torsional natural frequencies of the machinery train. Figure C.12 shows a particular case where the three torsional modes were insensitive to changes in torsional stiffness. Therefore, damping would be the next step to ensure lower vibratory amplitudes and reduced stress levels.

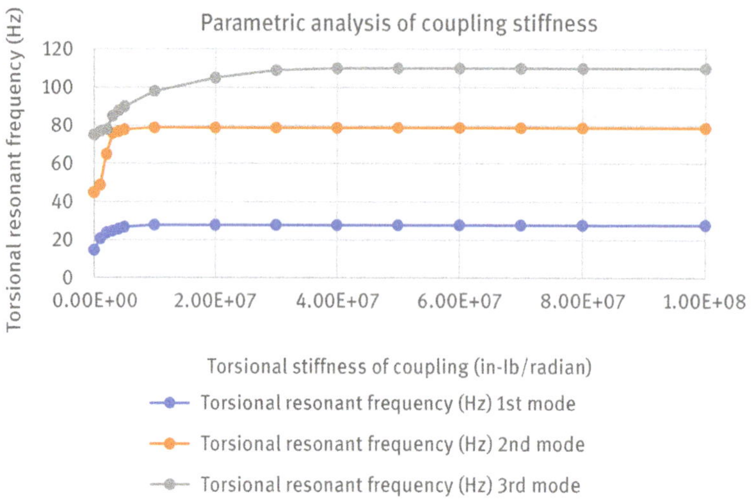

Figure C.12: Sensitivity of torsional resonant frequencies to change in stiffness [6].

In Section 7.6.2, we explained that, while elastomeric couplings can be used to add damping in a shaft system, they possess nonlinearities that are hard to quantify or monitor while in service. This makes them unpopular with equipment manufacturers and end users of turbomachinery. In addition, these elastomeric couplings tend to be bulky in size and weight, have large residual unbalance, speed limitations, and limited life expectancies. These properties make them unsuitable for most special-purpose machinery applications. Note, however, that large elastomer block couplings are indeed used in some reciprocating compressors and in a number of small general-purpose fluid machines.

It is accepted by equipment manufacturers that turbomachinery trains not having elastomeric couplings, gear boxes or gear couplings, tend to have only a small amount of damping. The manufacturers assume that only a small damping ratio ($\zeta \le 2\%$) exists and use the 2% number for torsional modeling. Because some manufacturers and analysts use only a small damping ratio in their models, it would be preferred to make *avoidance* of interference zones our first step in the design process. In other words, avoiding the issue by design is better than trying to remedy the problem after it manifests itself during or after commissioning. This is why best-in-class users perform design audits and allocate 5% of the asset's value in their original budget. Reliability pays back handsomely, but achieving reliability requires effort, not complacency.

Coupling vendors are now alerting users to new technologies by manufacturing special-purpose couplings with high damping ratios. These couplings can be designed with very accurate damping ratios and can be customized to resolve specific torsional vibratory problems.

Contrary to the present industrial practice where avoidance of interference zones should be made the first step in the design process, arguments against that notion can now be made with regard to torsional issues. As is the case with lateral vibration and advanced bearing designs, low amplification factors (AF) for torsional vibration can now be achieved with special-purpose couplings designed with high damping ratios. In essence, up-front designs with low AFs give reasonable assurance that, even in cases of touching an interference zone, there will be virtually no increase in vibration amplitude.

The viscous rotational damper (VRD) coupling is a product designed for special-purpose turbomachinery requiring flexible couplings with large damping ratios. The VRD concept has been specifically designed and optimized to address the exact needs of these critical machines. VRD couplings are designed to a required damping ratio; this ratio can be specified either by the equipment manufacturer or by end user. Damping ratios can be verified during shop testing. A VRD safely modulates unwanted torsional vibrations without sacrificing the typical requirements of low residual unbalance and weight.

The VRD in Figure C.13 is a quill-type coupling spacer with a damping element wrapped around its core. Damper designs can be tube-shaped, as shown; they can also be manufactured as a plate configuration whenever high damping is required. High damping values are possible with "fluid shear dampers."

Referring to Figure C.13, useful or steady-state power enters the drive flange and is transmitted to the load flange via the central shaft. The shaft torsional stiffness is tunable as needed for the application. The tunable steel central shaft is one of the key elements of the VRD design.

During torsional vibration, dynamic power associated with a torsional excursion enters the VRD through one of the flanges. Note that torsional vibration can enter from either the driver or driven equipment. The dynamic power is then directed into the damper tubes where fluid shear forces between the outer and inner tubes are

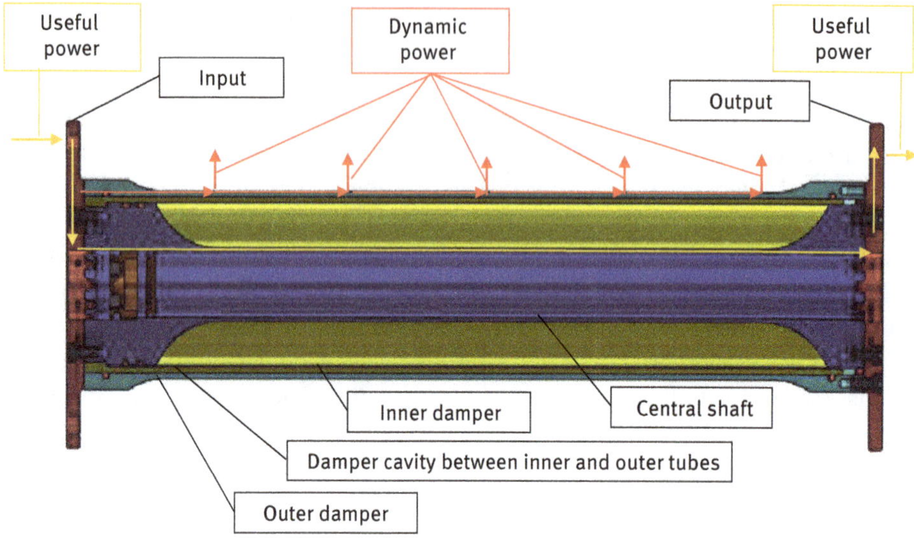

Figure C.13: Cross-section of the VRD tube-style coupling spacer.

absorbed. The absorbed power or waste energy is converted to heat and is efficiently conducted out to the atmosphere. A short conductive heat path is very important for effective dissipation of this unwanted energy. Optimized effective heat dissipation gives VRD the ability to manage greater amounts of dynamic power than most competing technologies.

Unlike common damping devices, the VRD is designed and built to meet the infinite dynamic life requirements of today's special-purpose couplings. Infinite life surpasses the design life of prior damping technologies. The VRD is designed using forged steel to utilize well-documented endurance properties that few other materials possess. It is also designed using smooth transitions to minimize the effects of stress concentration.

The design of the "quill" center shaft is based on the modified Goodman approach, normally used to rate API 671 flexible coupling elements. This approach brings the VRD coupling into the realm of full API compliance.

The effectiveness of torsional damping is demonstrated in the following two examples:

Example 1: Consider a simple induction motor-driven centrifugal compressor rated at 4,000 hp (2,983 kW) and 3,570 rpm. The steady-state torque "*T*" is calculated to be 70,616 in-lb (7978.5 N-m). A 1% amplitude torque pulse of 706 in-lb (79.8 N-m) is expected to excite the shaft string at the first torsional natural frequency "N1" for extended periods of time.

Generally, a 1% torque amplitude would not cause much concern. However, in this case interaction will be considered based on how the damping ratio influences the AF.

In Section 7.6.2 we observed that the damping ratios ζ are typically ≤2% for turbomachinery systems. Vance [7] suggests values ranging from 1.5% to 2.0%. This example will attempt to show how varying the damping ratio can affect the amplification of a shaft string. Notice how not knowing the true damping ratio of a system leaves us in the dark regarding the true AF; the integrity of an entire system will be uncertain.

amplification for steady-state systems can be expressed as

$$A = \frac{1}{2 \times \zeta}$$

where A is the amplification factor (AF) and ζ is the damping ratio.

The exercise will assume a range of $\zeta = 1.5\%$ to 2%
For damping of 1.5%, A = 33.33
Torque amplitude becomes $T_{1.5}$ = 706 × 33.33 = 23,876 in-lb (2,697.6 N-m)
For damping of 2.0%, A = 25.0
Torque amplitude becomes $T_{2.0}$ = 706 × 25.0 = 17,650 in-lb (1994 N-m)
Comparing these values, $T_{1.5} / T_{2.0}$ = 1.333
Torque variation between ζ1.5% and ζ2%
= 100% × $(T_{1.5} - T_{2.0})/T$ = 8.8% of steady-state torque

This variation is solely attributable to the difference in the allowable assumption of damping values within the train.

Amplitude variability without added damping is 33.3% or 6,217 in-lb (702.4 N-m). Relative to the application requirement, this represents 8.8% uncertainty. Such uncertainty must be accommodated by an increase in safety factor. A conscientious reliability professional cannot venture to guess that the uncertainty can be ignored.

Let us now look at how torque amplitude varies using the same assumptions but with a *verifiable* damping element inserted into the coupling and having a damping ratio of 10%. The two damping ratios can be combined by simple addition, 10% + 1.5% = 11.5%. The total damping ratio now becomes 11.5% and A = 4.35

Torque amplitude becomes $T_{1.5}$ = 706 × 4.35 = 3,071 in-lb (347 N-m)
For shaft damping of 2.0%, total damping becomes 12.0% and A = 4.17
Torque amplitude becomes $T_{2.0}$ = 706 × 4.17 = 2,944 in-lb (332.6 N-m)
Comparing these new values, $T_{1.5} / T_{2.0}$ = 1.043
Torque variation between ζ1.5% and ζ2%
= 100% × $(T_{1.5} - T_{2.0})/T$ = 0.18% of full-load torque.

Comparing the results of the traditional train to that of a train damped with a *verifiable* damper, it can be seen that the torque amplitude drastically decreases; variability drops even further.

With added *verifiable* damping, amplitude variability drops to 4.3% or a torque of 127 in-lb (4.35 N-m). Relative to the steady-state torque this is only 0.18%.

How these assumptions compare with respect to an actual coupling selection is explored next for the type of coupling often used in large compressor trains.

Suppose that here we require a steady-state torque, as per API 671, $T_s = 1.5 \times 70{,}616 = 105{,}925$ in-lb (11,968 N-m). The number 1.5 is the service factor of such couplings.

A disk coupling is selected that meets this requirement with the following published properties:
Maximum continuous torque, 126,750 in-lb (14,320 N-m)
Peak torque capacity, 253,500 in-lb (28,642 N-m)

Coupling manufacturers rarely publish *dynamic torque* capacity. However, experience has shown that a *dynamic torque* above 15% of the maximum continuous torque will require a stronger coupling. We, therefore, justify erring on the safe side and will stipulate that the *dynamic torque* should not go above 15% of the maximum continuous torque. Therefore, the *maximum dynamic torque* capacity for the selected coupling becomes .15 × 126,750 = 19,013 in-lb.

Table C.2 shows that the selected coupling will not meet the required reliability for a train which has only 1.5% damping. Nevertheless, the selected coupling will work for the case of 2.0% damping, although the margin relative to the *maximum dynamic torque* is low and will leave little room for deviations from actual damping.

Table C.2: Dynamic loading of coupling.

Damping case	Torque amplitude in-lb	% of maximum continuous torque	Acceptability
1.5% shaft damping	23,876	18.8	Not acceptable
2.0% shaft damping	17,650	13.9	Acceptable
1.5% + 10%	3,071	2.42	Acceptable
2.0% + 10%	2,944	2.32	Acceptable

By adding a *verifiable* damping element, not only does torque amplitude drop significantly, but dynamic torque amplitudes stay well within the maximum allowed.

Understanding that unexpected conditions can arise over time, it makes sense to add extra margin for damping. This will reduce the variability in the train's response to torsional excursions.

Example 2: How would the synchronous motor train that was discussed in Section 7.6 respond with the addition of a *verifiable* damper coupling?

Figure 7.15 showed a stress versus time plot of the across-the-line start of a synchronous motor. Since stress is proportional to torque, we can analyze the system using the information provided in units of stress instead of torque. Of interest in this figure is the high amplitude region where motor pulse rate crosses over the N1 train torsional natural frequency. Rough scaling of the graph yields the following:

Positive peak ~ 55,000 psi (379.2 MPa)
Negative peak ~ 50,000 psi (344.74 MPa)
Mean stress = (55,000–50,000)/2 = 2,500 psi (17.23 MPa)

From Figure 7.15, speed oscillations commenced at approximately 1,100 rpm (91.7% of synchronous speed). The response stress at this speed caused by the excitation is approximately 52,500 psi (358.5 MPa):

$$\text{Response stress amplitude} = 52{,}500\ \text{psi}\ (358.5\ \text{MPa})$$
$$= (\text{AF}) \times \text{Excitation stress amplitude}$$
$$\text{Excitation stress amplitude} = \frac{\text{Response stress amplitude}}{\text{AF}}$$

Assuming that a maximum damping ratio of 2% is used for this model, we compute an AF of 25, which is similar to that shown in Example 1.

Therefore, the Excitation stress amplitude $= \frac{52{,}500}{25}\ \text{psi} = 2{,}100\text{psi}(14.5\text{MPa})$

A quick review of Figure 7.13 shows that the oscillating torque amplitude continues to increase steeply even after 91.7% of synchronous speed is reached, although the motor's average torque drops. Therefore, this calculation should be considered a rough approximation of the excitation stress amplitude.

Borrowing again from Example 1: Adding a *verifiable* damper set at 10% of critical damping will result in a total damping factor of 12.0%. This now yields an AF of 4.17.

The response stress amplitude for the damped train becomes 2,100 × 4.17 = 9,135 psi (63 MPa) compared to 52,500 MPa (358. 5 MPa). This reduction in peak response stress is an improvement of about 5.7 times.

The figure below from a similar evaluation is offered here for comparison. It shows results from an analysis of a synchronous motor driven multistage compressor train start.

Figure C.14 is the before versus after response obtained by adding a VRD at the low-speed coupling end of a gearbox-driven train. For this case, a thorough analysis was done, yielding a reduction in the response torque amplitude of approximately five times.

(a) (b)

Figure C.14: Comparison of dynamic torque response before and after application of damping at low-speed coupling. (a) Before adding VRD and (b) after adding VRD.

C.3 Conclusion

Our examples show that dramatic reductions in startup stress can be gained by adding damping at the low-speed coupling. All trains of this sort can be customized by design, requiring detailed analysis. Reductions in dynamic torque amplitudes of 5× can be gained; these quite obviously offer the potential of significant improvements in equipment life. VRDs add definable benefits in both availability and reliability when the damping value is known, can be verified, and remains consistent over long periods of operation.

References

[1] Cloud, C.H. Case History: Violent Compressor Instability, Fundamentals of Rotordynamics for End-Users & Manufacturers Short Course, BRG Machinery Consulting, LLC, Charlottesville, VA, October 2016.

[2] Memmott, E.A. Stability of Centrifugal Compressors by Applications of Tilt Pad Seals, Damper Bearings, and Shunt Holes, IMechE, 5th International Conference on Vibrations in Rotating Machinery, Bath, 1992, 99–106.

[3] Edney, S.L., Nicholas, J.C. Retrofitting a Large Steam Turbine with a Mechanically Centered Squeeze Film Damper, Proceedings of the 28th Turbomachinery Symposium, Texas A&M University, 1999, 29–40.
[4] Kuzdzal, M.J., Hustak, J.F. Squeeze Film Damper Bearing Experimental versus Analytical Results for Various Damper Configurations, Proceedings of the 25th Turbomachinery Symposium, Texas A&M University, 1996, 57–70.
[5] Sandberg, M. Centrifugal Compressor Configuration, Selection and Arrangement: A User's Perspective. Proceedings of the 45thTexas A&M University Turbomachinery Symposium, Houston, TX, 2016.
[6] Engineering Dynamics Inc. June 20, 1992: Technical Report 20983-1 Torsional Analysis of Vent Gas Screw Compressor Failure Saudi Petrochemical Company (SADAF)
[7] Vance, J.M. Rotordynamics of Turbomachinery 1st Edition, Wiley-Interscience, Hoboken, NJ, 1988, ISBN 978-0471-802587.

Appendix D Machinery quality assessment (MQA)

Design basics of tabletop foundations for machinery engineers

D.1 Introduction

As readers perhaps refer back to Chapter 11, they may recall that the detailed design of large concrete foundations for centrifugal compressor trains can be very complex and is well beyond the scope of this book. For that reason, the actual design of these foundations remains within the purview of structural and geotechnical specialists. These specialists must accurately determine the static and dynamic responses of these structures.

Nevertheless, a project machinery specialist or subject matter expert (SME) is expected to take responsibility for every system or subsystem that affects machine reliability and uptime achieved. With that in mind, Chapter 11 served as an overview aimed at allowing the SME to understand how to engage with the team that designs the foundation and structures for major machines.

We know of more than one career-limiting experiences where unsatisfactory machine performance was blamed on the machinery expert. His explanations of not being an expert on foundation design were uniformly resented and rejected by the client's managers as an inappropriate defense. They pointed out that although a plant manager has no formal training in safety management, he or she will at all times be held accountable for the safety performance of their plant.

While encouraging rotating equipment engineers to ask questions whenever they are not comfortable with design results, we steered clear of inundating the SME with too much detail. But what if the SME wanted to learn more about these details? What if he or she wishes to go line for line over the report or printout a civil engineer or structural geotechnical specialist had submitted? That may be necessary in some cases to complete the SME's task of ascertaining, and at least understanding, that the design is sound.

We compiled this appendix to make that SME's job easier and call it the model machinery quality assessment (MQA) approach. MQA is a review and verification activity used by best-in-class companies since the early 1960s [1]. The activity can be defined as a structured up-front (prepurchase and preconstruction) process that uses the input of experienced SMEs to determine which equipment, lubricant application strategy, foundation configuration, future maintenance pursuit, or potential upgrading measures should be designed-in, or built-in, to ensure high reliability and profitability of the assets being acquired by the SME's employer.

MQA activities are budgeted by the owner–operator as part of capital expenditures. The allocated amount for MQA activities is 5% of the total cost of the assets being assessed. MQA progresses in several time-related phases. These phases consist

https://doi.org/10.1515/9783110678765-017

of prepurchase (which includes identifying acceptable bidders), issuance of comprehensive purchase specifications, resolution of exceptions to our specifications taken by bidders in their bid or offer, thorough review of a proposed design, assessment of future upgrade opportunities, proposed execution of routine maintenance, definition of spare parts, and many other reviews [2]. As mentioned earlier, MQA for tabletop structures is often overlooked by machinery SMEs whose natural inclination is to focus on the machines. They must, however, include the elements described in Chapter 11 of this text and will find supplementary information and considerably more detail in this appendix. The appendix starts by describing typical models.

D.1.1 Typical models

Historically, mathematical modeling of tabletop foundations is done using different techniques, depending on the software and computer power available. Before today's massive computer power became available, relevant literatures showed the feasibility of modeling foundations as *spring–mass–damper* systems.

For example, using a three degrees-of-freedom (3DoF) system with the vertical, lateral, and rocking modes as per Figure D.1, the linear equations associated with the model can be derived [3].

Figure D.1 shows the tabletop foundation system modeled into two spring–mass systems. In Figure D.1, "A" represents the top floor plus the supporting columns. The equations associated with this part of the model are as follows:

$$m_u\ddot{X}_2 + K_h X_2 = F_x(t) \tag{D.1}$$

$$m_u\ddot{Z}_2 + K_y Z_2 = F_z(t) \tag{D.2}$$

Similarly, "B" represents the total structure, rotating equipment, and bearing soil

$$m_x\ddot{X}_1 + C_X\dot{X}_1 + K_x X_1 = F_x(t) \tag{D.3}$$

$$m_x\ddot{Z}_{1+} C_z\dot{Z}_1 + K_z Z_1 = Fz(t) \tag{D.4}$$

$$I_\varphi\ddot{\varphi} + C_\varphi\dot{\varphi} + K_\varphi\varphi = Fx(t)H \tag{D.5}$$

where I_φ is the moment of inertia for the combined masses in part "B," and H is the height between the center of gravity (CG) of the combined masses in "B" and the CG of the mass in part "A."

By not incorporating the interaction of structure and soil, only its three predominant motions (i.e., lateral, vertical, and rotational) are predictable at the CG of the masses. The model assumes the two subsystems to have independent dynamic responses, and that the column bases are fixed.

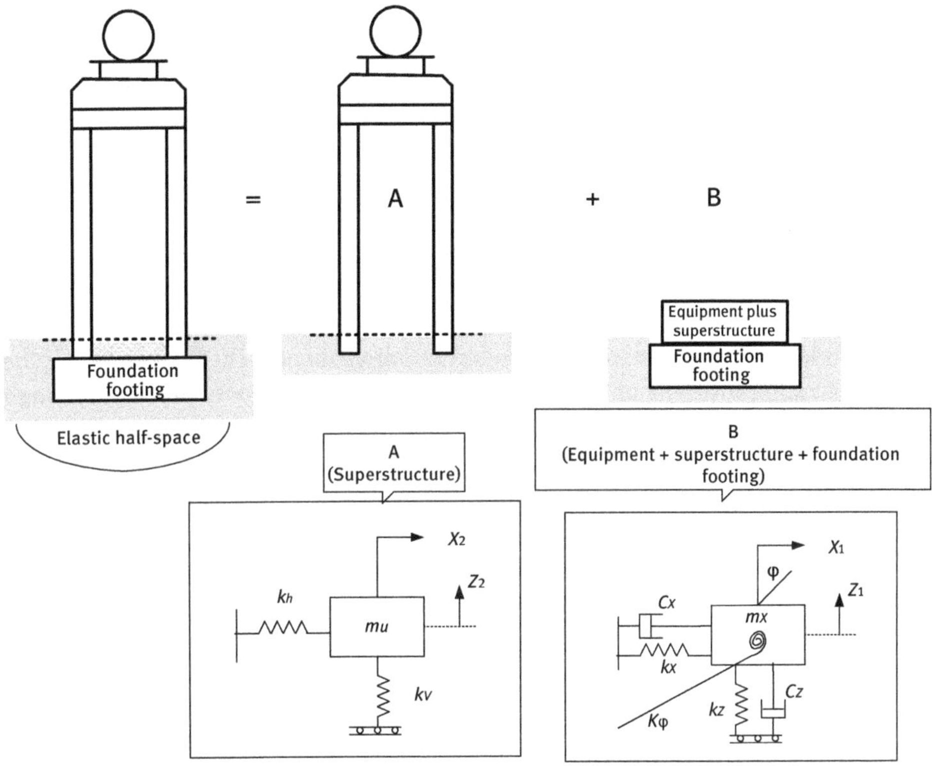

Figure D.1: Three degrees-of-freedom (3DoF) model of a tabletop foundation [3].

More refinement of the model can be achieved by incorporating interaction between A and B, as was done in [3]. The equations can be solved manually using linear algebra.

However, available finite element analysis (FEA) software in combination with powerful computers represents better structural models. Several design studies by Bhatia [4] proved of interest here. Bhatia found that variation in the natural frequencies of frame structures modeled using *manual computational methods* versus FEA models is within perhaps 10–20% accuracy. Therefore, the *manual computational method* is recommended as a useful initial trial tool to evaluate free–free vibration as well as forced vibration responses. The manual method represents a close initial assessment of the frame's response frequencies. Refinements are done later using FEA models.

Soil or piles are modeled as isolated springs and dampers to represent the soil's impedance. However, the soil or piles may also be modeled using FEA to obtain more accurate results; of course, they will require more computer power.

The structural FEA model output permits adjustments to the pile spacing by changing applicable dimensions. The most common verification of the results is to compare vibration frequencies and amplitudes for the normal operating speed range with the desired separation margin. It is also necessary to study the behavior of the structure during start-up of the equipment. The rotating equipment engineer must verify that no potentially destructive structural vibrating modes exist during start-up and normal operation.

Commercial FEA software permits realistic modeling using cubic elements for the slab or pile cap, columns, beams, and tabletop. The optimum size of each cube should be found by letting the analysis converge. Secondary elements may be modeled with less precision than the main structural elements. These models may be time consuming for dynamic analyses; therefore, use of a computer with enough RAM, graphic processor capability, and a sizable hard disk is recommended.

A simplified *spring–mass–damper* model using linear elements can be used for rapid verification of the preliminary dimensions; these could later be adjusted to reflect the dimensions for optimum performance of the structure. This analysis is rapid and occupies less computer memory. Once the response is within the allowable range, the cubic FEA model may be used for a final analysis. For large tabletops, the cubic analysis may be slow, requiring a large memory to save both input data and results.

D.1.2 Required information for structural analysis and design

Figure D.2 schematically represents the modeling process as well as the necessary input. The geotechnical engineer provides the dynamic properties of the soil and decides whether to use piles. A geotechnical report must include the mechanical properties of each soil layer, such as:
- Thickness
- Soil classification (saturated clay, dense sand, gravel, silt, etc.)
- Modulus of elasticity
- Poisson's ratio
- Soil density
- Soil damping

Using this geotechnical information and following the preliminary design rules, the structural engineer selects the size of the FEA elements and develops the structural model. The equipment manufacturer provides the following information for use in the static and dynamic models:
- Equipment outline dimensional drawing
- Static weight of the equipment
- Operating speed range

- Allowable peak-to-peak vibration at top of foundation
- Unbalanced load
- Dynamic loading for the foundation in all three axes (x, y, and z)

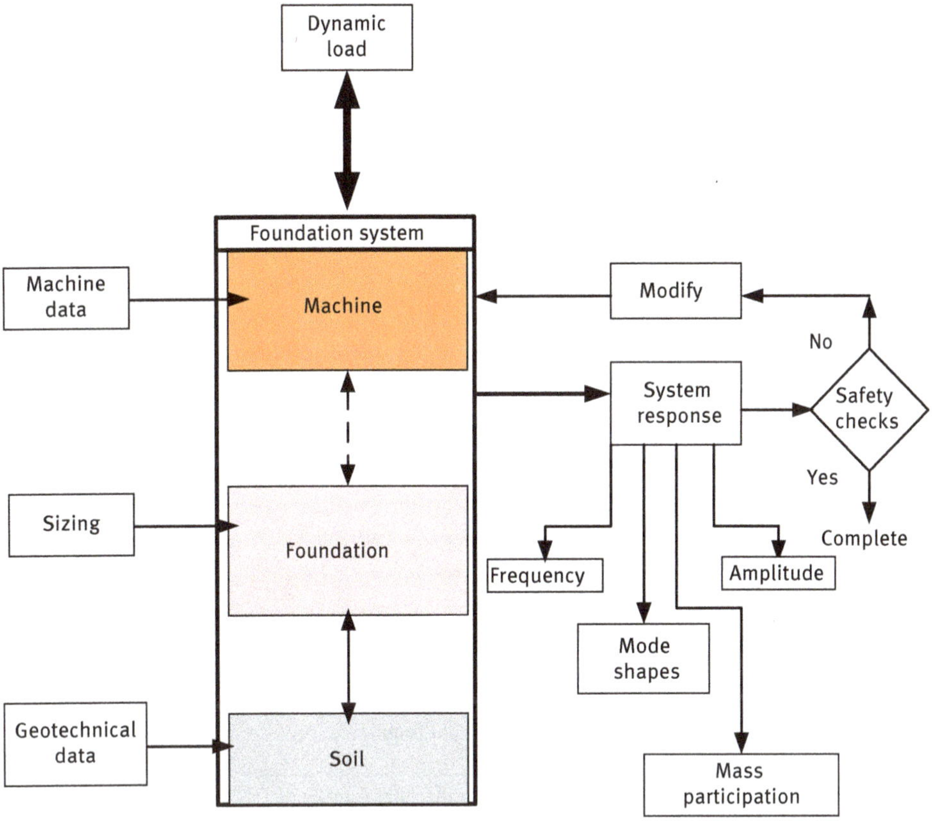

Figure D.2: Schematic of a typical dynamic model.

Other forces that must be considered are live loads, piping loads, nearby equipment, earthquake, and wind forces. Depending on the weight of tabletop and machine, earthquake loads may be very important and need to be studied closely.

The following information must be obtained to perform the structural modeling:

a) *Soil.* Typically, the soil is modeled as springs and dampers. For static loads, the geotechnical engineer provides the coefficient of subgrade reaction, which permits the calculation of the spring constant and associated damping ratios.

Spring constants for an elastic half-space model are shown in Table D.1. For clarity and basic understanding of how these constants are derived and used, only three of the six DoF in Figure D.3 will be represented in this appendix.

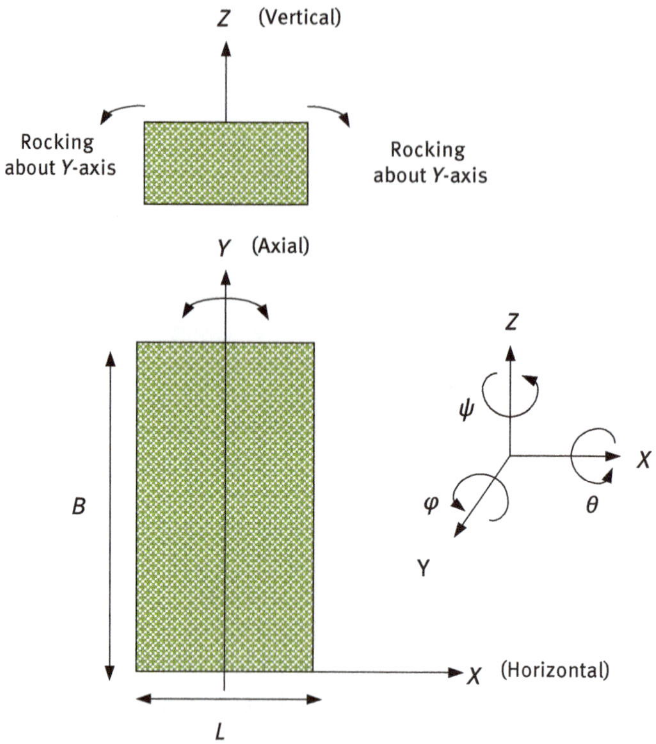

Figure D.3: Six DoF for a typical footing of a tabletop foundation.

Table D.1: Equivalent spring constants for rectangular footings [3, 5].

Mode of vibration	Rectangular footing
Vertical (*z*-direction)	$Kz = \frac{G}{1-v}\beta z \,.\, \sqrt{BL} \,.\, \eta_z$
Horizontal (*x*-direction)	$Kx = 2(1+v)\, G\beta x \,.\, \sqrt{BL} \,.\, \eta_x$
Rocking (around the axis of machine rotation)	$K\varphi = \frac{G}{1-v}\, \beta_\varphi \,.\, BL^2\eta_\varphi$

Where G is the shear modulus of the soil, v is the Poisson ratio of the soil, β is the geometry factors of the footing particulars defined in Figure D.4, B is the dimension of the footing along y-axis, L is the dimension of the footing along x-axis, η_z is the embedment coefficient in the vertical direction = $1 + 0.6(1 - v).\ (h/r_0)$, η_x is the embedment coefficient in the horizontal direction = $1 + 0.55(2 - v).\ (h/r_0)$, η_φ is the embedment coefficient for rocking = $1 + 1.2(2 - v)\ .\ (h/r_0) + 0.2(2 - v)\ (h/r_0)$, h is the depth of the embedment, r_0 is the equivalent radius of the rectangular footing = $(\sqrt{BL})/\pi$ for translatory motion and $((B/3\pi)L^3)^{0.25}$ for rocking mode.

Note: Embedment coefficients are from [4, 6].

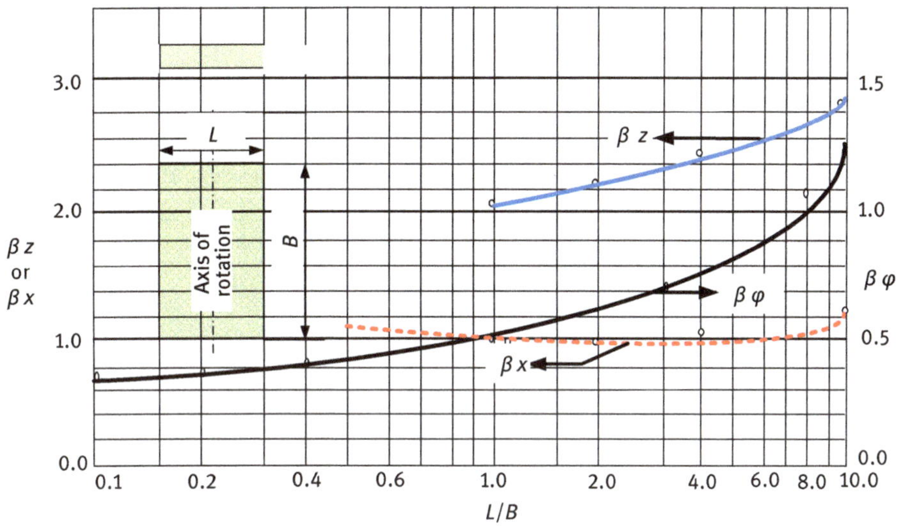

Figure D.4: Geometry factors βz, βx, and $\beta\varphi$ for rectangular footings [5].

The damping ratio of the soil foundation system consists of two components, namely:

1. **Geometric damping ratio D_G:** This derives its value from the weight, inertia of the foundation, and the geometric spreading of energy on the soil away from the foundation. Mathematical expressions to evaluate Geometric damping ratios for the three primary vibratory modes are shown in Table D.2.
2. **Material damping ratio D_m:** This is the inherent internal damping characteristic of the soil caused by energy dissipation due to friction in the soil. It is a measure of the energy lost because of the hysteresis effects of the soil.

Table D.2: Geometric damping ratio D_G for rectangular footings [3, 5].

Mode of vibration	Mass/inertia ratio	Damping ratio
Vertical (z-direction)	$Bz = \frac{(1-v)}{4} .m/(\rho r_0)^3$	$Dz = \frac{0.425}{\sqrt{Bz}} \alpha_z$
Horizontal (x-direction)	$Bx = \frac{7-8v}{32(1-v)} .m/(\rho r_0)^3$	$Dx = \frac{0.288}{\sqrt{Bx}} \alpha_x$
Rocking (around the axis of machine rotation)	$B_\varphi = \frac{3(1-v)}{8} . I_\varphi/\rho r_0^5$	$D_\varphi = \alpha_\varphi \frac{0.15}{\left(1+n_\varphi B_\varphi\right)\sqrt{n_\varphi B_\varphi}}$

where, Dz, Dx, and D_φ are damping ratios as a fraction of critical damping. Bz, Bx are mass ratios for vibration in the z- and x-directions, B_φ is the inertia ratio for vibration rocking around the y-axis (axis of machine rotation), m is the total mass of machine and foundation, ρ is the soil mass density, I_φ is the mass moment of inertia of foundation and machine, α_z, α_x, α_φ are correction factors for embedment (see Table D.3).

Table D.3: Damping correction factors showing effect of depth of embedment [3, 7].

Mode of vibration	Damping ratio embedment correction factor
Vertical (z-direction)	$\alpha_z = \frac{1+1.9(1-v)h/r_0}{\sqrt{\eta z}}$
Horizontal (x-direction)	$\alpha_x = \frac{1+1.9(2-v)h/r_0}{\sqrt{\eta x}}$
Rocking (around the axis of machine rotation)	$\alpha_\varphi = \frac{1+\frac{0.7(1-v)h}{r_0}+0.6(2-v)\left(\frac{h}{r_0}\right)^3}{\sqrt{n_\varphi}}$

D.1.2.1 Material damping ratio

Usually obtained from resonant column testing of the soil. Vibration of the soil takes place using an exciter. After the exciter is stopped, the displacement signal is measured as a sinusoidal time-based wave, with amplitudes decreasing with time. By measuring successive amplitudes, the material damping ratio is calculated as $D_m = (\ln[a_1/a_2])\ [4\pi^2 + (\ln[a_1/a_2])^2]^{-0.5}$ [3]. Typical range for machine foundations is 0.03–0.05 depending on the soil.

The summation of geometric and material damping ratios gives the total damping ratio D_t is as follows:

$$D_t = D_m + D_G$$

Using derivations from classical vibration theory [3]

Natural frequency of a spring–mass system $f_n = \frac{1}{2\pi}\sqrt{\frac{\text{Stiffness}}{\text{Mass}}}$

Frequency for a damped spring–mass system $f_d = \frac{f_n}{\sqrt{1-2D_t^2}}$

Damping ratio $D_t = C/Cc$

where C is the actual damping of the system and $Cc = 2mfn$ (critical damping).

Bhatia [4] recommends, in the absence of damping values at a site, using total damping coefficients in the range of $D_t = 0.08$–0.1.

If piles are used, the spring constant can be estimated by the slope of the red line in Figure D.5. The pile stiffness should be obtained using the service load, which is

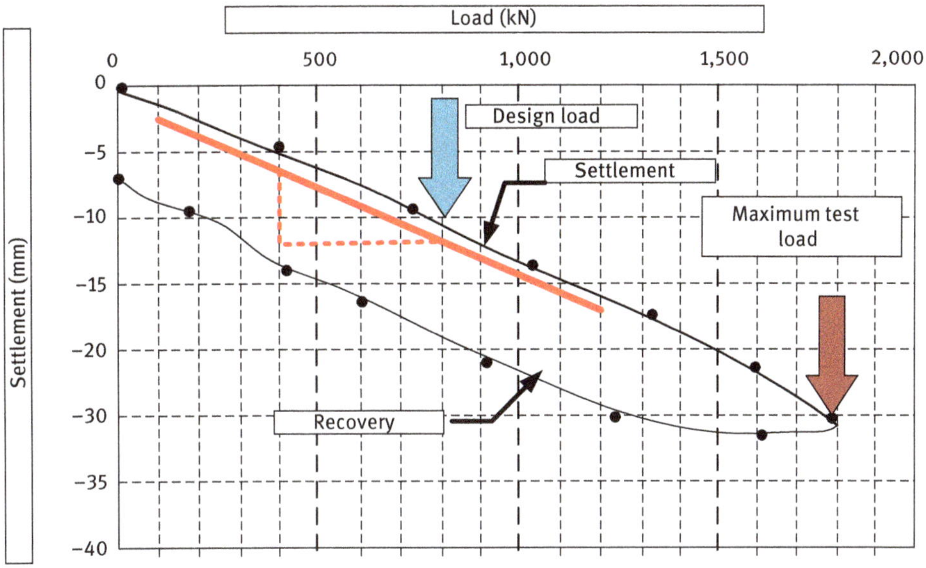

Figure D.5: Load settlement curve for a single pile.

the ultimate load divided by the safety factor, generally the curve is nonlinear and the initial slope is higher than the secant slope at ultimate capacity.

For the dynamic analysis, it is necessary to determine the soil or pile impedance, consisting of vertical and horizontal springs and dampers. The American Concrete Institute (ACI) specification ACI 351-3R (2018) [6] presents a methodology commonly used to obtain these parameters. The engineer may use cubic finite elements for the dynamic soil model, including the piles if necessary. The geotechnical engineer provides the soil dynamic properties, like the shear modulus, G, the Poisson's ratio, ν, and the material damping, D_m.

Considering the nature of soils, it is a sound recommendation to work with a range of values of G; a lower boundary G_{LB} and an upper boundary G_{UB}. ACI 351.3R-18 provides empirical correlations of G with other soil properties. Better values of G may be obtained using laboratory or field test data.

Poisson's ratio may be estimated by the geotechnical engineer or obtained from field tests using wave velocities. Values in the range from 0.25 to 0.40 have no significant effect on the dynamic response of the foundation; however, the dynamic response is sensitive to variation for values between 0.40 and 0.50, which are common for saturated soils. The typical ranges of ν values are:

- For cohesionless granular soils (free draining), ν values range from 0.25 to 0.40
- For cohesive soils (impervious), ν values range from 0.33 to 0.50

For saturated soils under the groundwater table: ν values are close to 0.50, for analysis one uses $\nu = 0.48$ to avoid numerical instability.

The material damping may be obtained in the laboratory using the torsional shear test or a resonant column test; a conservative decision would be to use an available minimum damping number. When the experimental data are not available, the minimum damping ratio value can be set to 1%.

b) *When piles are necessary, the following information is needed:*
   - Size, shape, and material
   - Type and construction method
   - Static allowable capacity in compression, shear, and tension
   - Load versus settlement curves
   - Properties of the backfill
   - Pile arrangement
   - The pile impedance (dynamic stiffness and damping) may be obtained following the methodology given by ACI 351.3R (2018)

c) *Reinforced concrete*:
   - The properties of steel and concrete are necessary for the structural model. For both materials, the most important properties are rebar yield strength and their respective moduli of elasticity.
   - The modulus of elasticity of the concrete, E_c, is given by ACI 318 as a function of the 28 days strength (f'_c), and the concrete density (w_c)

$E_c = 33\, w_c^{1.5}\, \mathrm{sqrt}(f'_c)$ lb/in^2 – in US customary units
where (w_c in lb/ft^3 and f'_c in psi)
$E_c = 0.04\, w_c^{1.5}\, \mathrm{sqrt}(f'_c)$ – in SI units
where (w_c in kg/m^3 and f'_c in MPa)

For normal weight concrete (NWC) ($w_c = 145$ lb/ft^3 or 2,400 kg/m^3), the equation reduces to

$E_c = 57{,}000\, \mathrm{sqrt}(f'_c)$ – in US customary units (f'c in psi)
$E_c = 4{,}700\, \mathrm{sqrt}(f'_c)$ – in SI units (f'_c in MPa)

The modulus of elasticity provided by ACI 318 corresponds to correlations of the secant of the stress versus strain curve at 45% of f'_c.

However, research literature shows that the modulus of elasticity of concrete is sensitive to other factors such as:

- Type of aggregate and additions used. The modulus of elasticity is greater when crushed limestone (+20%) and fly ash (+10%) are used.
- Age of concrete. The modulus of elasticity increases with the age of the concrete; it is about 8% more for 1-year-old concretes, as shown by ACI 209.R (1992).
- Static or dynamic loads. The dynamic modulus of elasticity of concrete may be greater than its static value (+20%) (ACI-CRC, 2008).

- Effect of steel reinforcement in the total area for the elastic range of loads (+4%).
- For better results, static and dynamic tests are valuable; they are used to find the true modulus of elasticity. The values from statistical relationships may underestimate the modulus of elasticity of the concrete. The static and dynamic analysis of the structure is sensitive to the assumed modulus of elasticity of the concrete.
- Another important value is the tensile capacity of the concrete in flexion or modulus of rupture (f_r), which may be obtained with statistical relationships or by using laboratory tests. ACI 318 (2018) provides the following equation to find f_r for NWC:

$$f_r = 7.5 \text{ sqrt}(f'_c)$$

This tensile strength is sensitive to the aggregates and additions used. It is recommended that laboratory tests are used to find the best value for design.

- The confined structural elements shall be designed to resist the stresses produced by the differential of temperature or shrinkage. For this analysis it is necessary that the thermal expansion coefficient, the creep, and shrinkage properties of the concrete required by ACI 207.2R (2007) are used.

D.1.3 Refining and interpreting the system response (results)

D.1.3.1 Static analysis

The static analysis provides information for verification of the stiffness of the structure by:

- Verifying the foundation capacity (soil and/or piles)
- Reinforced concrete design of the structural elements

The stiffness of the structure is limited by the equipment manufacturer's recommendation to use the equivalent static loads previously described.

For the reinforced concrete design, consider the results from the different load combinations and verify the shear, flexural, and flexo-compression capacity of each structural element. Typically, the elements are massive. The reinforcement design is governed by the minimum required by codes, taking care of the steel distribution to facilitate the construction procedure, and avoiding concentration of steel in small areas.

Additionally, the changes in volume due to thermal effects, shrinkage, and creep produce stresses in the elements confined by massive columns and slabs. The engineer must verify that the concrete resists these tensions. The correct reinforcement distribution must be verified to avoid tension cracks that would change the stiffness of the structure.

D.1.3.2 Dynamic analysis

The dynamic analysis provides information on how the structure behaves under dynamic load. It yields the following information:
- Natural frequency
- Mode shapes
- Displacement amplitudes
- Velocity or acceleration
- Mass participation factor

Each natural frequency is associated with a mode shape. The mode shape is a function or property of the structure and is independent of the dynamic loads. The pre-dimensioning guidelines ensure that the principal natural modes are transverse, longitudinal, and vertical with low torsional effects. A characteristic of the finite elements analysis is the presentation of several natural modes with different mass participation factors.

The *mass participation factor* provides a measure of the energy contained within each resonant mode; it represents the amount of mass participation in the system for the particular vibratory mode. To obtain acceptable results for the analysis of dynamic forces, the engineer must verify that these factors are greater than 90% in each direction (*X*, *Y*, and *Z*). Table D.4 is an example showing the results from the first six modes of a dynamic analysis.

Table D.4: Example of percentage mass participation for each mode.

Mode	Frequency (Hz)	Period (s)	Mass participation – X (%)	Mass Participation – Y (%)	Mass Participation – Z (%)
1	6.92	0.145	0	0	70.70
2	10.25	0.098	18.87	30.45	0
3	11.89	0.084	22.45	60.45	0
4	13.67	0.073	0	3.10	28.65
5	15.89	0.063	50.23	0	0
6	21.35	0.047	7.98	5.43	0
Motor-operating speed: 3,600 rpm (60 Hz)			99.53	99.43	99.35

The example in Table D.4 shows that the summation of the mass participation is greater than 90% in each direction after the first six modes. It is generally recommended to terminate the modal analysis after summing 90% or more mass participation in each direction being analyzed.

It is also necessary to verify that the natural frequency of the structure is not in resonance with the equipment. The natural frequency of modes with mass participation factors greater than 5% cannot be in the range of 0.8–1.2 × the operating speed.

A cursory check of the transient phase during machine start-up must be performed. One uses a typical Campbell diagram to determine interference frequencies, and to see if displacements associated with these modes exceed allowables.

D.1.4 Amplitude

The dynamic force depends on the residual unbalance of the rotor. The following equation may be used if the manufacturer does not specify the dynamic force:

$$\text{Dynamic force}, Fn = \frac{(\text{rotor weight}) \times (\text{rotor speed, rpm})}{6{,}000}$$

This force is applied using a sinusoidal function with the angle varying according to the angular frequency of the rotor (ω). The force components are:

$$F_{vert} = F_n.\text{Sin}(\omega t) \qquad \text{Vertical component}$$
$$F_{tr} = F_n.\text{Cos}(\omega t) \qquad \text{Transverse component}$$

The dynamic force of each rotor must be applied at the CG of the respective rotor.

The maximum vibration amplitude of the machine foundation system is given by the equipment manufacturer. However, the engineer can also compare the amplitude results with tables and charts in ACI 351.3R (2018).

Collaborating with the structural engineer, a cautious rotating equipment engineer will sometimes ask that other possible sources of vibration be fed as data input to the analysis software. These might include:
- Short circuit effects and faulty synchronizing conditions from motor drivers
- Oil whirl frequency
- Rotor critical speeds
- Shaft bow effects when starting large steam turbines

Absent permissible vibration amplitudes given by the equipment manufacturer, the values in Table D.5 and Figure D.6 can be used. However, the hatched limit lines in Figure D.6 indicate safety limits and not limits of satisfactory machinery operation.

Table D.5: Maximum allowable peak-to-peak amplitude (source: [7]).

Machine type	Maximum allowable peak-to-peak amplitude (mm)
Low-speed machinery (500 rpm)	0.2–0.25
High-speed machinery:	
(a) 3,000 rpm	
i. Vertical vibration	0.02–0.03
ii. Horizontal vibration	0.04–0.05
(a) 1,500 rpm	
i. Vertical vibration	0.04–0.06
ii. Horizontal vibration	0.07–0.09

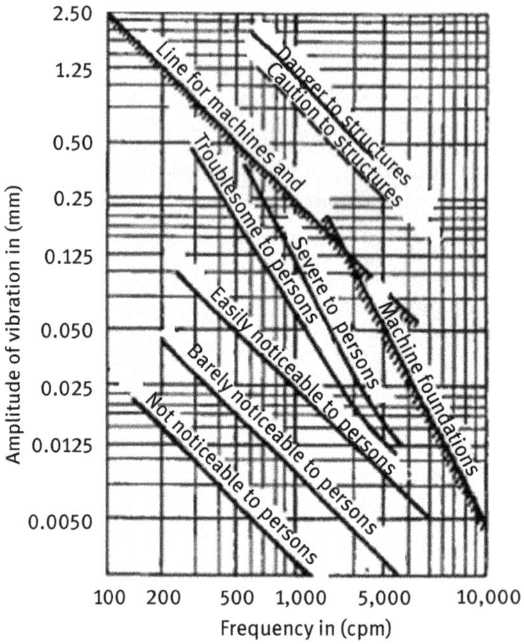

Figure D.6: General limits of vibration with respect to frequency.

D.2 Example of a tabletop design

Figures D.7 and D.8 show the dimensions of a tabletop foundation previously studied; the EPC (Engineering-Procurement-Construction) contractor and one of the authors were assigned to the project team and collaborated on the foregoing analysis.

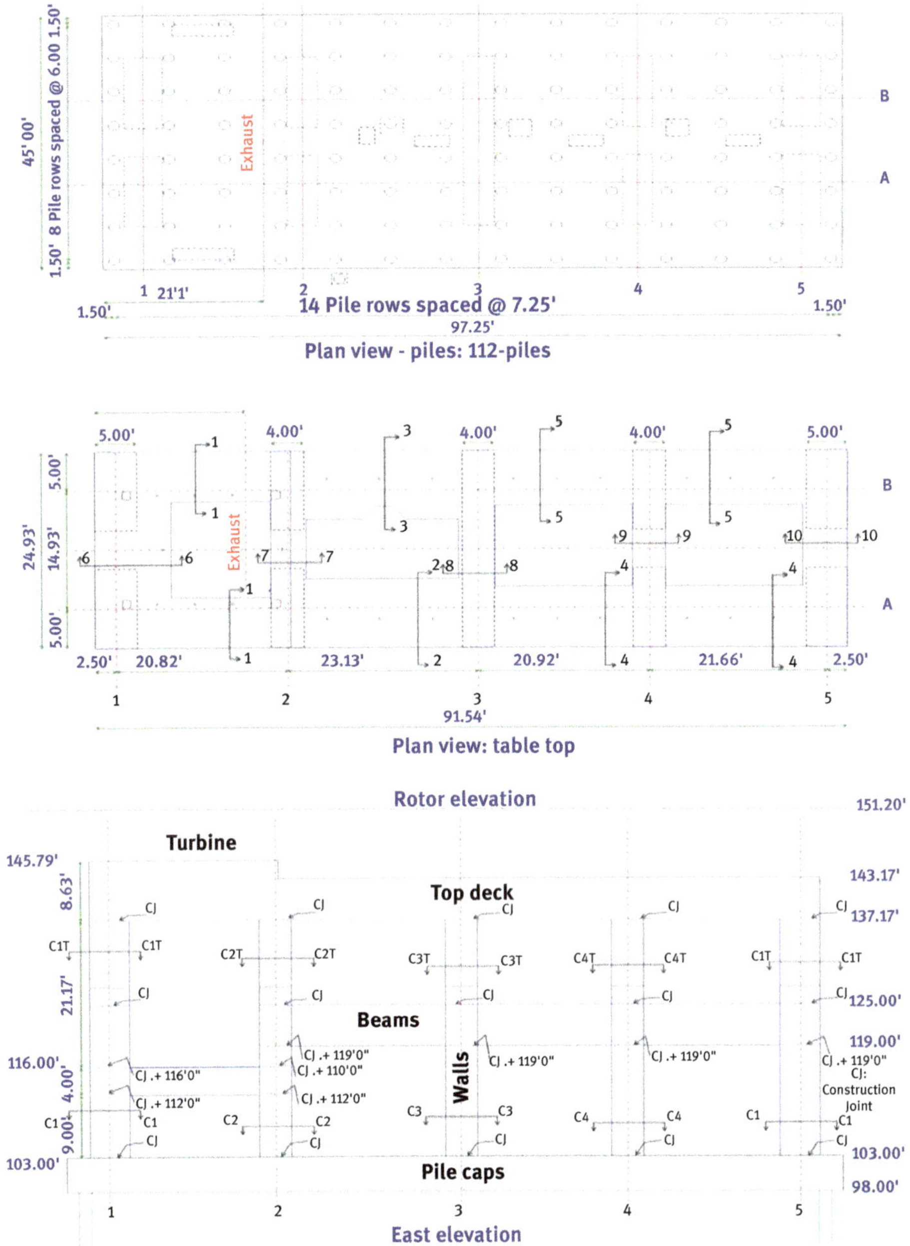

Figure D.7: Tabletop on piles.

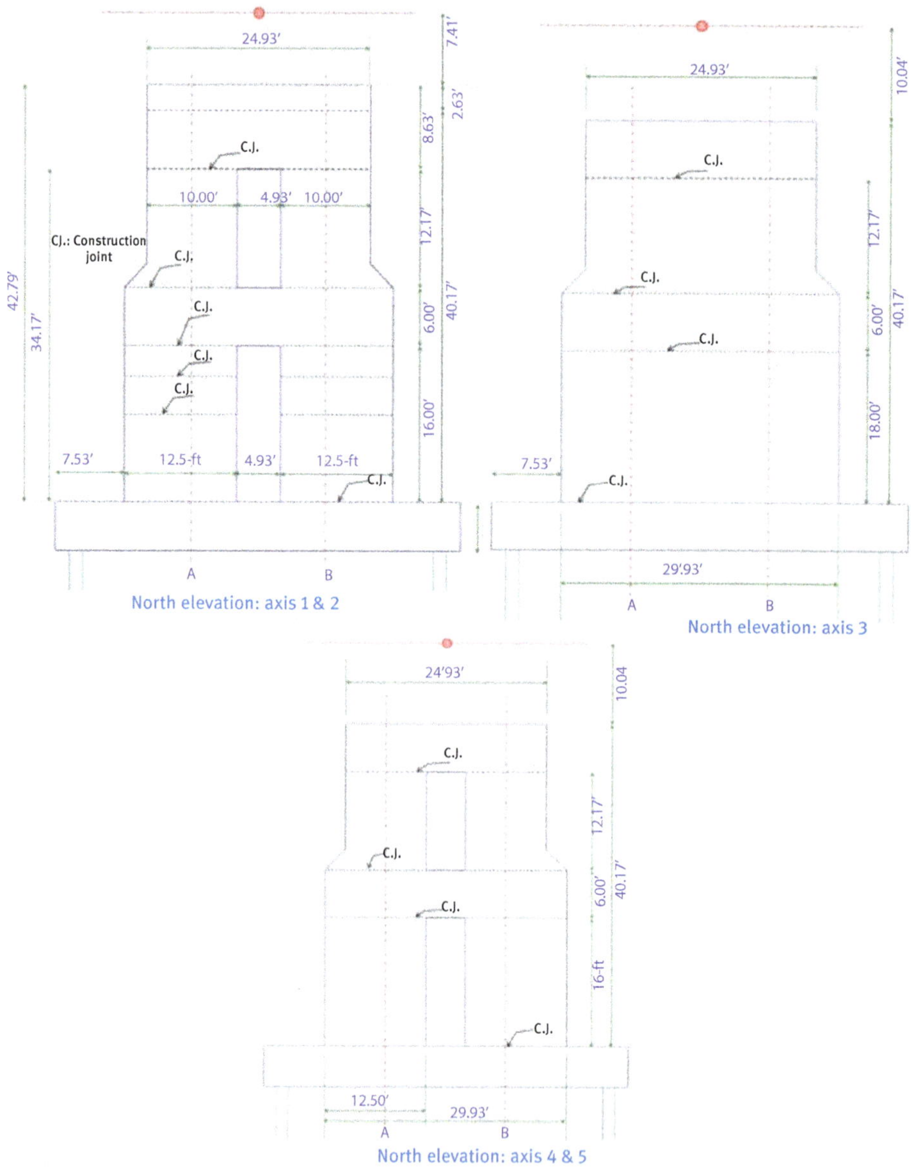

Figure D.8: Tabletop walls on piles.

The machine train involved a 39 MW centrifugal compressor; it consisted of three compressor bodies driven in tandem by a steam turbine. The equipment manufacturer's guidelines for the design of the tabletop foundation are as follows:

1. Equipment foundation design shall have strength and rigidity based on the following criteria:
   - *Have sufficient strength for a short term to undergo dynamic loads during transients, that is, start-up and overspeed conditions.*
   - *Peak-to-peak vibration velocity at the top of the foundation shall not exceed 1.8 mm/s rms for steady-state long-term basis.*
   - *Design loads*
     - *Vertical load (Z-direction) = 1.5 × Operating weight* W_T
     - *Horizontal load (X-direction) = 0.25 × Operating weight* W_T
     - *Axial load (Y-direction) = 0.1 × Operating weight* W_T
   - *Maximum bending of columns and horizontal beams shall not exceed 0.008 in (0.2 mm) under any operating load.*

D.2.1 Description of the structural components

D.2.1.1 Concrete and rebar

The pile cap and the structure mentioned above had concrete with an f'_c of 4 ksi (28 MPa). The water-to-cement ratio was $w/c = 0.40$, with fly ash class F, and aggregates from limestone rock. The pouring followed the recommendations of ACI for massive concrete [6]. The rebar complied with ASTM A-615, yield strength of 60 ksi (414 MPa), and modulus of elasticity of 29,000 ksi (200,000 MPa).

D.2.1.2 Piles

This structure included 112 auger-cast piles, 75 ft (22.9 m) long, 18 in (457.2 mm) diameter. The piles had an allowable compression capacity of 260 kip (1,793 MPa), no-tension capacity, and 8 kips (55 MPa) of allowable shear strength. The concrete was rendered with strength $f'_c = 4$ ksi (28 MPa), and the reinforcement consisted of six no. 8 rebars with no. 3 tie-bars spaced 12 in (305 mm) apart. The piles were 4 in (102 mm) embedded in the pile cap.

D.2.1.3 Pile cap

The pile cap was a massive concrete block. The top of the concrete was 6 in (152 mm) above the pavement. The top surface was sloped to permit drainage; its thickness varied from a minimum 5 ft 2 in (1.58 m), to a maximum of 5 ft 5 in (1.65 m). However, for the analysis the thickness was considered as 5 ft (1.52 m). The plan dimensions are 97 ft 3 in (29.64 m) long and 45 ft (13.72 m) wide, as shown in Figure D.7.

The pile cap had reinforcements to satisfy the estimated minimum requirement differential temperature during curing and flexural strength. The reinforcements were no. 10 rebars spaced 6 in (152.4 mm) center-to-center each way – top and bottom layers; three longitudinal no. 6 rebar and no. 6 u-bars spaced 6 in (152.4 mm) all around the edges, producing a *ratio of rebar weight to concrete volume of 7.2 lb/ft^3 (115.33 kg/m^3).*

The tension stresses due to flexure were designed to be less than the tension strength of the concrete; the intent was to prevent external forces from producing cracks. The shear stresses due to external and internal forces were taken by the concrete.

ACI 207.2R-95 was followed to verify that the reinforcement was enough to resist the tensions due to differential temperatures normally experienced during concrete curing.

D.2.1.4 Walls

There were five lines of transversal walls, defined as axes 1 to 5. The end walls (axes 1 and 5) were designed with 5 ft (1.52 m) thickness, and the central walls (axes 2, 3, and 4) had 4 ft (1.22 m) thickness. The walls were 29 ft 11 in (9.12 m) long at the base and 24 ft 11 in (7.6 m) at the top. They all had a central opening of 4 ft 11 in (1.5 m) to permit the installation of piping, except the wall at axis 3. The height from the top of the pile cap to the bottom of the top slab was 34 ft 2 in (10.42 m). Figures D.7 and D.8 show the dimensions of the structure.

The walls of axes 1 and 5 had a steel ratio of 1.06% at their base section, and 1.03% at their top section. The steel ratio is defined as the area of rebar with respect to the cross-sectional area.

The walls of axes 2 and 4 had a steel ratio of 0.93% at the base, and 0.94% at the top.

The wall of axis 3 had a steel ratio of 0.73% at the base, and 0.71% at the top.

The interaction diagram curves of axial loading versus bending moment capacity showed that the actual loads were well within the allowable region, as shown in Figure D.9, therefore requiring minimum steel only. For these walls, *the ratio of rebar weight to concrete volume was 6.5 lb/ft^3(104.2 kg/m^3).*

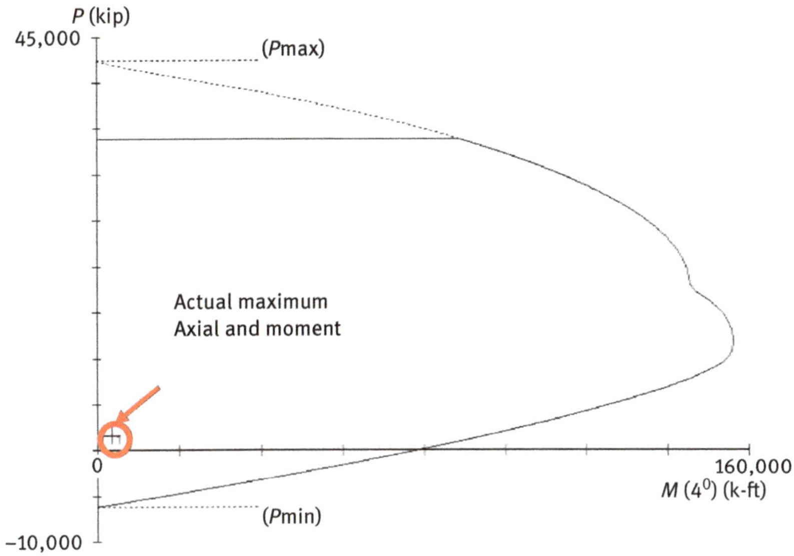

Figure D.9: Interaction diagram of axial versus moment capacity for wall of axis 3.

D.2.1.5 Beams

Longitudinal beams were needed to support pipes and a condenser. The beams also provided stiffness to the structure. The thickness of the beams was 4 ft (1.22 m). The height for the beams between axes 1 and 2 was 4 ft (1.22 m), and the height for the beams between axes 2 and 5 was 6 ft (1.83 m). The reinforcement steel was the minimum required by codes considering flexion and thermal movement. For these beams, *the ratio of rebar weight to concrete volume was 9.0 lb/ft^3(144.2 kg/m^3).*

D.2.1.6 Top deck

The top deck was dimensioned according to the equipment manufacturer's general assembly (GA) drawing. The GA drawing included the compressor train, auxiliary piping and other equipment attached to the machine train. The slab/footing was designed not to be in contact with the adjacent platforms, and with a joint designed to avoid contact when the platform moves.

The top deck was 8 ft 7½ in (2.63 m) between axes 1 and 2, and 6 ft (1.83 m) between other axes; 24 ft 11 in (7.6 m) wide and 91 ft 6½ in (28 m) long.

For the reinforced concrete design: The deck was divided in strips, and the concrete in each strip reinforced like beams, taking into consideration the external forces and temperature effects on each beam. The *ratio of rebar weight to concrete volume was approximately 6.2 lb/ft^3 (99.3 kg/m^3).*

D.2.2 Structural model

The preliminary design was done following the guidelines in section D.3; the equipment manufacturer's guidelines were adhered to. The equipment manufacturer's guidelines were considered stringent; especially so the criteria limiting the lateral stiffness of the structure. The lateral drift of the columns and vertical deflection of the beams were restricted to 0.008 in (0.2 mm) under the design loads. Design loads consisted of the independent combination of the vertical dead load (DL), along with the equivalent static load. These loads governed the structural design for the walls and the top deck. Fortunately, the dimensions in the drawings did satisfy the design loads when they were first used to perform the corresponding static analysis.

The concrete used had a strength $f'_c = 4$ ksi (28 MPa) and the rebar conformed to ASTM A-615 grade 60, or $Fy = 60$ ksi (420 MPa).

Considering the low level of stresses in the structural elements and because crushed limestone and fly ash class F were being used, the concrete modulus of elasticity was set 20% higher than the modulus of elasticity given by ACI-318, resulting in a value of

$$E_c = 4,400 \text{ ksi } (30 \text{ GPa})$$

The damping for the structure was set to 5%, which is a conservative value considering that the interaction of the soil and the foundation may produce a larger damping.

A commercial FEA software package was used for the structural modeling and analysis of the tabletop structure. The piles were modeled as springs with the stiffnesses computed following the methodology provided by Prakash and Puri [7]. The pile cap, walls, and top deck were modeled using cubic elements with an approximate maximum side of 2 ft (0.61 m).

The machines were modeled as beams using rigid elements. The weight and dynamic forces of each were located along their respective centerlines and center of gravities. The software used the weight inputs as DLs and masses for the dynamic analysis.

Figure D.10 shows a finite element 3D view of the structural model using cubic finite elements. Figure D.11 is a finite element 2D view of the compressor train arrangement. Figure D.12 is a finite element 3D view of the dynamic forces acting on the foundation.

D.2.3 Static loads

Additional to the loads specified by the construction code ASCE 7, the following equivalent static loads were also applied:

Vertical: 50% of the equipment operation weight
Transversal: 25% of the equipment operation weight
Longitudinal: 10% of the equipment operation weight

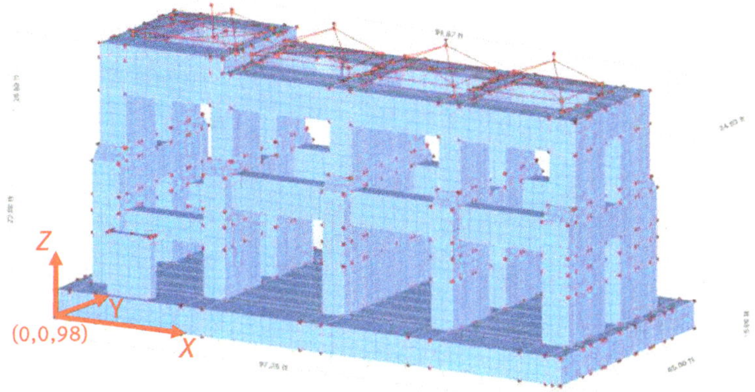

Figure D.10: Finite element 3D view of structural model.

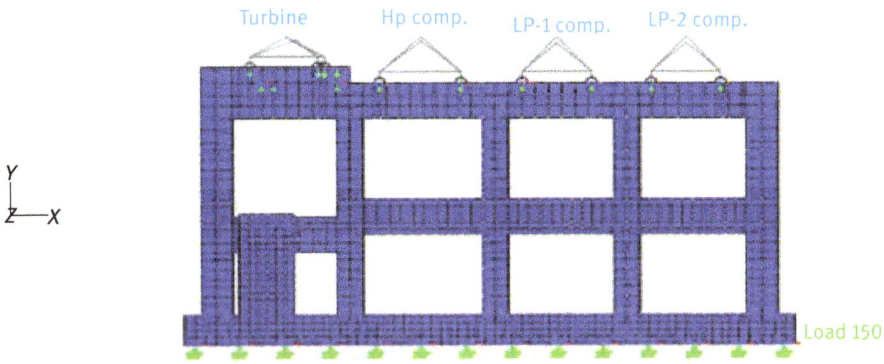

Figure D.11: Finite element 2D view showing arrangement of compressor train.

Each one of these equivalent static loads were combined with the DL, and then used in the analysis to verify that the deflection for the reinforced concrete design was less than the equipment manufacturer value of 0.008 in (0.2 mm) along X,Y, and Z directions. The machine weights are summarized in Table D.6. The equipment weight and the equivalent static loads are applied at the CG of each machine.

D.2.4 Dynamic loads

The machine train consisted of four machine bodies: A steam turbine, and three centrifugal compressors. The compressors were designated as high pressure (HP), low pressure 1 (LP1), and low pressure 2 (LP2). The centerline of the train was located at 5.42-ft (1650 mm) above the top deck, as shown earlier in Figure D.7.

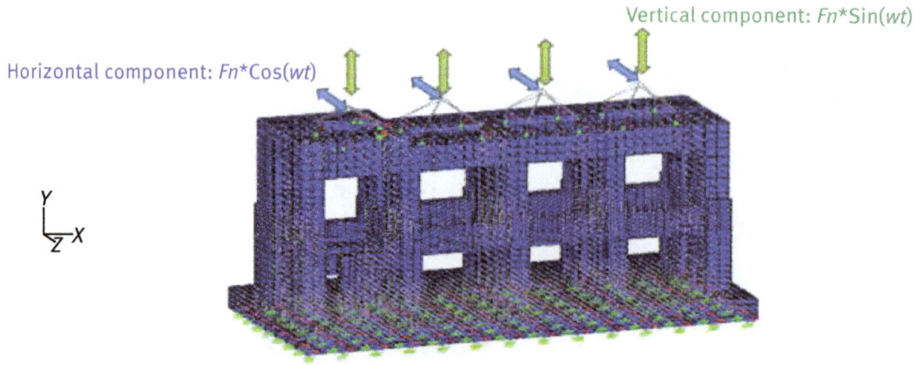

Figure D.12: Finite element 3D view showing dynamic forces acting on the foundation.

Table D.6: Static loads – machine and piping.

Load	Location of CG (*X*, *Y*, *Z*) ft	Weight (DL) (kip)	Equiv. *Fx* (longitudinal) (10% equipment) (kip)	Equiv. *Fy* (transversal) (25% equipment) (kip)	Equiv. *Fz* (vertical) (50% equipment) (kip)
Steam turbine	(21.08, 22.50, 145.79)	67.15	6.72	16.79	33.58
Steam turbine	(17.54, 22.37, 151.21)	152.34	15.23	38.09	76.17
HP compressor	(37.41, 22.50, 151.21)	396.83	39.68	99.21	198.42
LP1 compressor	(58.54, 22.50, 151.21)	282.19	28.22	70.55	141.1
LP2 compressor	(79.56, 22.50, 151.21)	282.19	28.22	70.55	141.1
Base plate for turbine	(16.49, 22.50, 147.17)	22.71	2.27	5.68	11.36
Base plate for HP comp.	(38.44, 22.50, 145.69)	65.26	6.53	16.32	32.63

Table D.6 (continued)

Load	Location of CG (*X*, *Y*, *Z*) ft	Weight (DL) (kip)	Equiv. *Fx* (longitudinal) (10% equipment) (kip)	Equiv. *Fy* (transversal) (25% equipment) (kip)	Equiv. *Fz* (vertical) (50% equipment) (kip)
Base plate for LP1 comp.	(59.67, 22.50, 145.50)	66.15	6.62	16.54	33.08
Base plate for LP2 comp.	(80.73, 22.50, 145.50)	66.15	6.62	16.54	33.08
Trip throttle valve	(17.57, 21.80, 151.75)	5.29	0.53	1.32	2.65
Condenser	(13.18, 22.50, 119.84)	197.5	n/a	n/a	n/a
Pipe 1	(32.99, 49.50, 123.08)	42.0	4.2	10.5	21
Pipe 2	(34.68, 23.48, 123.08)	12.5	1.25	3.13	6.25
Pipe 3	(37.93, 25.29, 123.08)	45.0	4.5	11.25	22.5
Pipe 4	(43.22, 22.50, 123.08)	97.0	9.7	24.25	48.5
Pipe 5	(54.79, 24.79, 123.08)	52.5	5.25	13.13	26.25
Pipe 6	(63.45, 22.50, 123.08)	155.0	15.5	38.75	77.5
Pipe 7	(75.52, 24.79, 123.08)	52.5	5.25	13.13	26.25

Table D.6 (continued)

Load	Location of CG (*X*, *Y*, *Z*) ft	Weight (DL) (kip)	Equiv. *Fx* (longitudinal) (10% equipment) (kip)	Equiv. *Fy* (transversal) (25% equipment) (kip)	Equiv. *Fz* (vertical) (50% equipment) (kip)
Pipe 8	(84.18, 22.50, 123.08)	155.0	15.5	38.75	77.5
	Total	**2215**	**201.79**	**504.48**	**1008.92**

The rotors were coupled in tandem; hence, they all had the same frequency. The machines were skid mounted and installed on baseplates on the top deck. The manufacturer provided anchor bolts to be installed using ducts through the top deck. These special anchor bolts had end plates that ware designed to interact with the top deck soffit and the top of the skid.

The dynamic loads were obtained using the following equation given by the equipment manufacturer:

$$F_n = \frac{W}{2g} \times \frac{\varepsilon}{2} \times 10^6 \times \left(\frac{2\pi N}{60}\right)^2$$

where W is the weight of the rotor, g is the acceleration of gravity = 9.81 m/s^2, N is the rotor frequency (revolutions per minute, rpm), $\mathcal{E}$ is the vibration amplitude at alarm = $1.5 \times 25.4 \times \sqrt{12,000/N}$

The dynamic forces were modeled by applying the forces at the center of each rotor using these equations:

$F_{vert} = F_n \,.\, \mathrm{Sin}(\omega t)$ vertical component
$F_{tr} = F_n \,.\, \mathrm{Cos}(\omega t)$ transversal component

where $\omega = f \times (2\pi)$ rad/s and $f = N/60$ cycles per second or Hz, s represents seconds, and t represents time in seconds.

The equipment manufacturer limited the vibration at the top of the tabletop to 1.8 mm/s RMS for all operating speeds including start-up:

$$\mathrm{RMS} = \frac{\text{peak vibration}}{\sqrt{2}} <= 0.07\,\mathrm{in/s}(1.8\,\mathrm{mm/s})$$

The dynamic analysis was done for these cases:
- Case I. Operation at maximum continuous speed: 3,849 rpm (64.15 Hz).
- Case II. Operation at minimum operation speed: 2,792 rpm (45.53 Hz).
- Case III. Start-up/slow roll warmup condition: Analyze at 1,000 rpm (16.67 Hz).
- Case IV. Start-up/slow roll warmup condition: Analyze at 500 rpm (8.33 Hz).

- Case V. Start-up and slow roll turbine with bowed rotor: Analyze for 1,000 rpm (16.67 Hz). The eccentricity of the turbine rotor was set at 0.008 in.

Cases III, IV, and V were special cases given to the team by the rotating equipment engineer to evaluate. The objective was to determine if during slow roll of the turbine (500–1,000 rpm) whether or not the modal frequencies could excite resonance of the structure. The rotating equipment engineer was curious to know what effect a bowed turbine rotor would have on the foundation structure if a resonant frequency was encountered during slow roll. A "Campbell interference" diagram (see Figure D.17, later) was used to determine if the modal frequencies could excite the structure during slow roll. Table D.7 shows the dynamic forces for the different cases studied.

Table D.7: Dynamic forces applied at center of each rotor.

Cases studied	Rotor	F_n (kip)	F_n (kN)
Case I			
Normal operation 3,849 rpm $f = 64$ Hz, $\omega = 403$ rad/s $e = 67 \times 10^{-6}$ mm (2.65×10^{-6} in)	Turbine	6.26	27.9
	HP comp.	8.82	39.2
	LP1 comp.	8.66	38.5
	LP2 comp.	8.66	38.5
Case II			
Minimum speed operation 2,792 rpm $f = 47$ Hz, $\omega = 292$ rad/s $e = 79 \times 10^{-6}$ mm (3.11×10^{-6} in)	Turbine	3.87	17.2
	HP comp.	5.45	24.2
	LP1 comp.	5.35	23.8
	LP2 comp.	5.35	23.8
Case III			
Start-up – phase 3 1,000 rpm $f = 17$ Hz, $\omega = 105$ rad/s $e = 132 \times 10^{-6}$ mm (5.2×10^{-6} in)	Turbine	0.83	3.69
	HP comp.	1.17	5.19
	LP1 comp.	1.15	5.10
	LP2 comp.	1.15	5.10
Case IV			
Start-up – phase 2 500 rpm $f = 8.3$ Hz, $\omega = 52.4$ rad/s $e = 187 \times 10^{-6}$ mm (7.35×10^{-6} in)	Turbine	0.29	1.30
	HP comp.	0.41	1.84
	LP1 comp.	0.41	1.80
	LP2 comp.	0.41	1.80

Table D.7 (continued)

Cases studied	Rotor	F_n (kip)	F_n (kN)
Case V			
Start-up – phase 3 with "bowed rotor/banana effect" at turbine	Turbine	101.38	450.9
1,000 rpm	HP comp.	1.17	5.2
$f = 17$ Hz, $\omega = 105$ rad/s	LP1 comp.	1.15	5.1
$e = 132 \times 10^{-6}$ mm (5.2×10^{-6} in)	LP2 comp.	1.15	5.1
Eccentricity of turbine rotor: Add $e = 0.008$ in (0.203 mm)			

D.2.5 Results

Figure D.13 confirms that the maximum drift or deflection due to the equivalent transverse static load was 0.008 in (0.2 mm), which is in the range required by the manufacturer. Longitudinal and vertical drift were also in the required (or allowable) range.

The structural analysis shows that internal forces were relatively low, therefore, the reinforced concrete design is governed by the minimum reinforcement required by codes. The structural elements were verified against changes of temperature, which could have generated forces due to restriction of movement.

Figure D.14 shows the principal modes of vibration of the structure. The dynamic analysis was done using the first 10 modes of vibration with a mass participation of 99% for the transverse, longitudinal and vertical modes. The software uses these modes for calculating displacement, velocity and acceleration for the first three seconds.

Figures D.15 and D.16 depict curves of velocity versus time. The equipment manufacturer restricted the maximum RMS velocity to 0.07 in/s (1.8 mm/s). Cases I and II are steady state conditions and represent normal operation. In these cases, the velocity is lower than required. However, the cases III, IV, and V are transient special situations during warm up/slow roll of the turbine. Case V was done to determine how a bowed rotor (banana-shaped) would affect the foundation's design. This was a requirement stipulated by the rotating equipment engineer. The vibration limit for case V was well outside the allowable limit specified by the equipment manufacturer. Based on the results of case V, the rotating equipment engineer recommended to keep the machine longer at 500 rpm to reduce the rotor bow before going up in speed to 1,000 rpm during warmup.

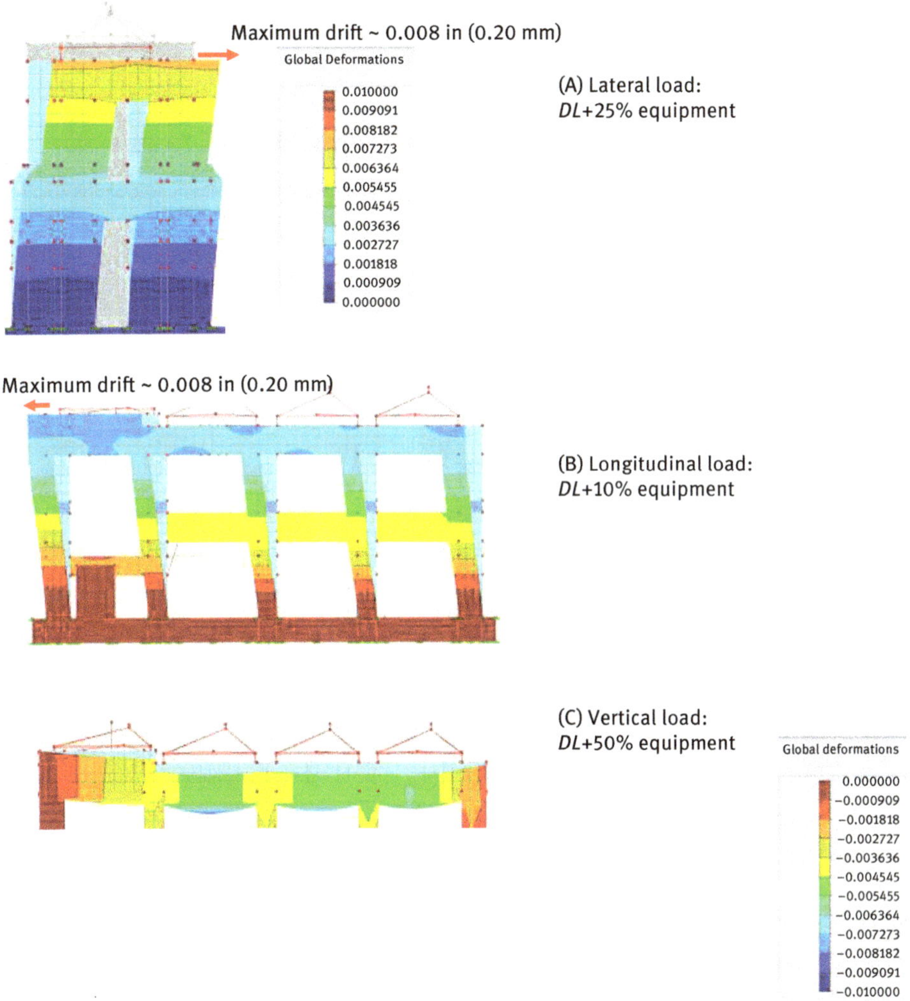

Figure D.13: Drift/deflection due to equivalent static loads.

To gain additional confidence in the results, a Campbell interference diagram (Figure D.17) was plotted as an elementary check to:

- Determine whether or not modal frequencies with large mass participation could vibrate the foundation during warmup.
- Ensure there was enough separation margin during steady state operation between operating speed range and the primary sources of excitation, namely rotor unbalance (1× running speed) and rotor misalignment (2× running speed).

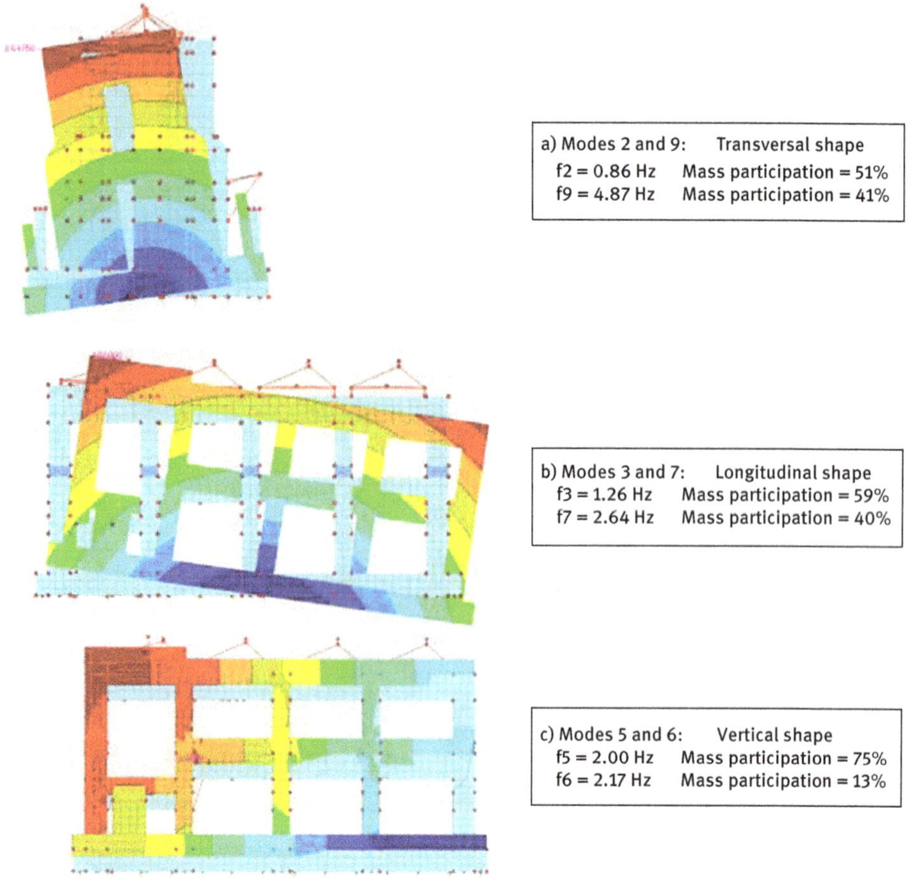

Figure D.14: Modes of vibration with higher mass participation.

Figure D.17 shows no interference of the modal frequencies having high mass participations with the primary sources of machine vibration (1× running speed and 2x running speed). This was the case for both steady state operation and warmup of the turbine. Figure D.18 shows the start-up diagram (or dwell speed versus time plot) for warming up the turbine. Unless speeds are carefully ramped up in timed intervals, the steam turbine rotor may thermally expand more rapidly than the massive casing that surrounds it. The very instant a rotating element contacts a stationary part of the turbine frictional heat will be generated and multimillion-dollar damage will be the likely result. The industrial world's safety and reliability literature attest to the potentially grave consequences.

The curves shown correspond to the velocity at node 1081 located at center of the top deck.

Maximum allowed RMS peak-to-peak Velocity = 0.071 in/s (1.8 mm/s)

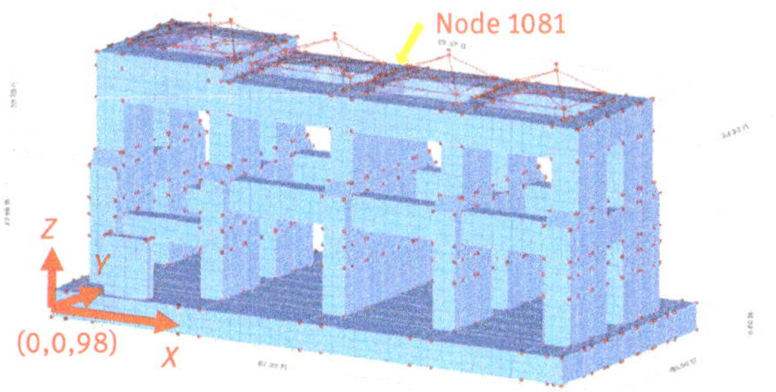

Case I: Normal operation
speed = 3,849 rpm
$\varepsilon = 2.65 \times 10^{-6}$ in (67×10^{-6} mm)
Transverse (Y-Y)
RMS peak-to-peak velocity = 0.020 in/s (0.52 mm/s)

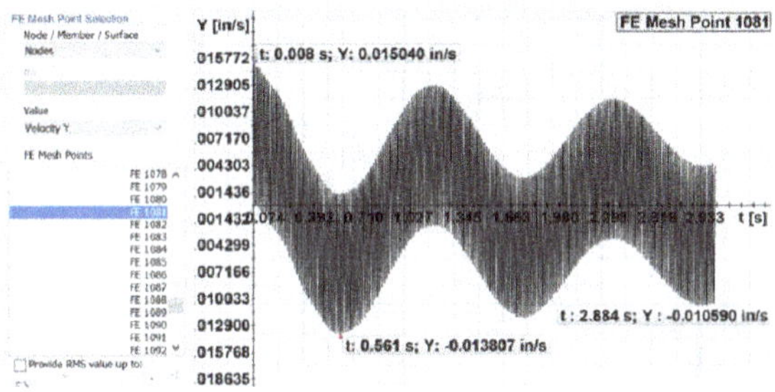

Case II: Minimum speed operation

Speed = 2,792 rpm
$\varepsilon = 3.11 \times 10^{-6}$ in (79×10^{-6} mm)
Transverse (Y-Y)
RMS peak-to-peak velocity = 0.017 in/s (0.44 mm/s)

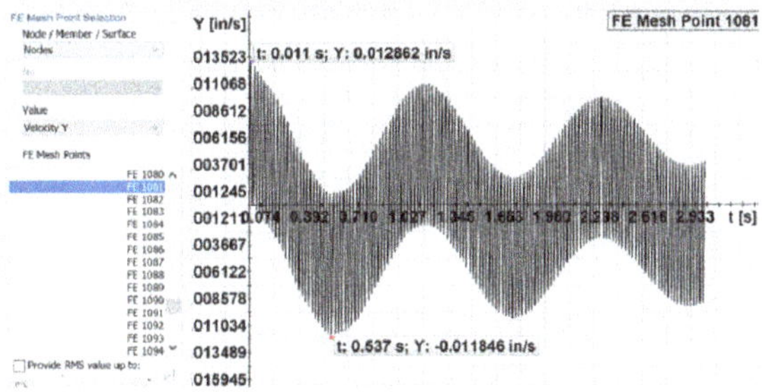

Case III: Start-up – Phase 3

Speed = 1000 RPM
$\varepsilon = 5.2 \times 10^{-6}$ in (132×10^{-6} mm)
Transversal (Y-Y)
RMS peak-to-peak velocity = 0.011 in/s (0.27 mm/s)

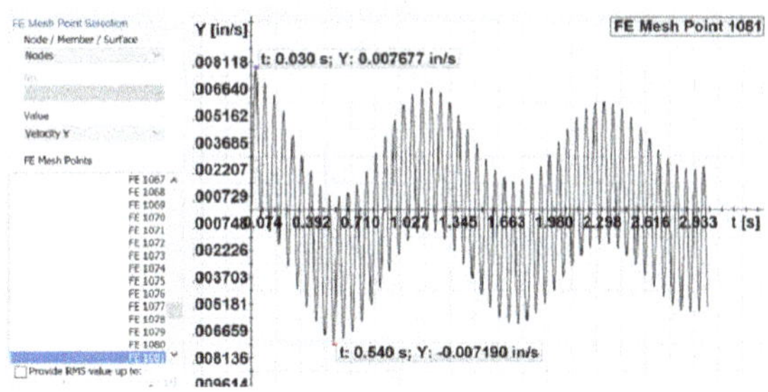

Figure D.15: Curves of velocity versus time at the center of top deck.

Case IV: Start-up –phase 2

Speed = 500 rpm
ε = 7.35x10^{-6} in (187x10^{-6} mm)
Transversal (Y-Y)
RMS peak-to-peak velocity = 0.007 in/s (0.19 mm/s)

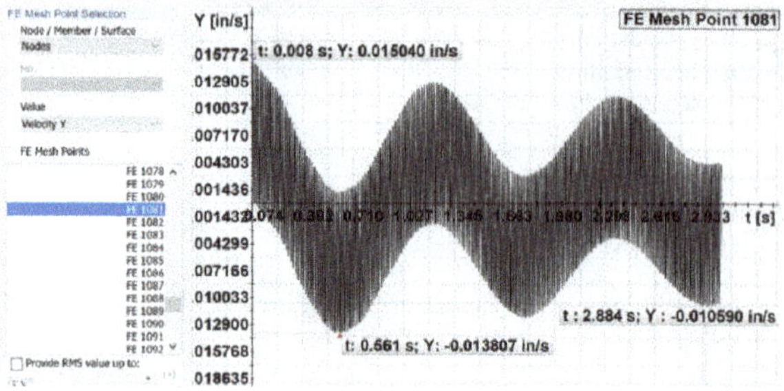

Case V: Start-up –phase 3 with "Bowed Rotor/Banana Effect"

Speed = 1,000 rpm
ε = 132x10^{-6} mm (5.2x10^{-6} in)
For turbine only, add:
e = 0.0080 in (0.203 mm)
Transversal (Y-Y)
RMS peak-to-peak velocity = 0.238 in/s (6.05 mm/s)

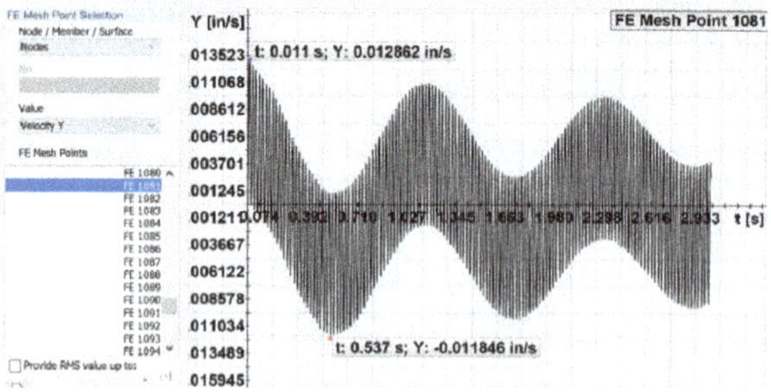

Figure D.16: Curves of velocity versus time at the center of top deck.

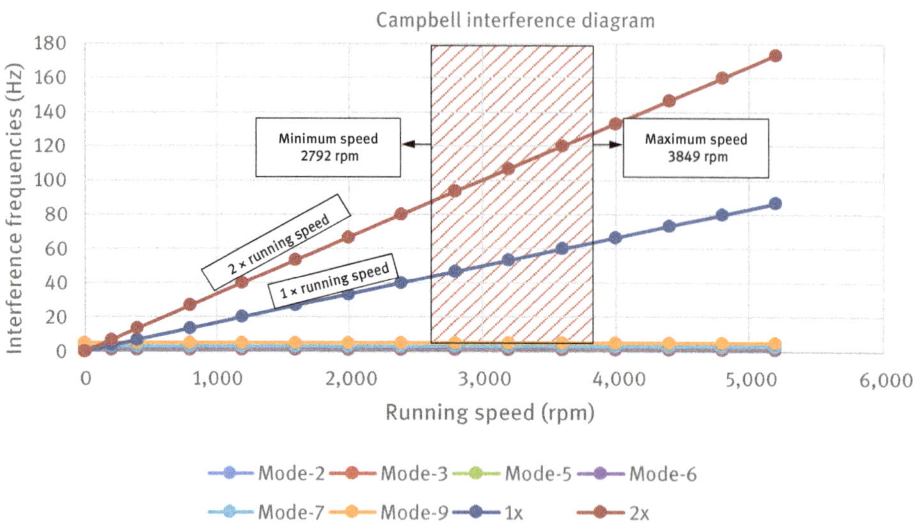

Figure D.17: Campbell interference diagram plotted to check potential resonance of the foundation.

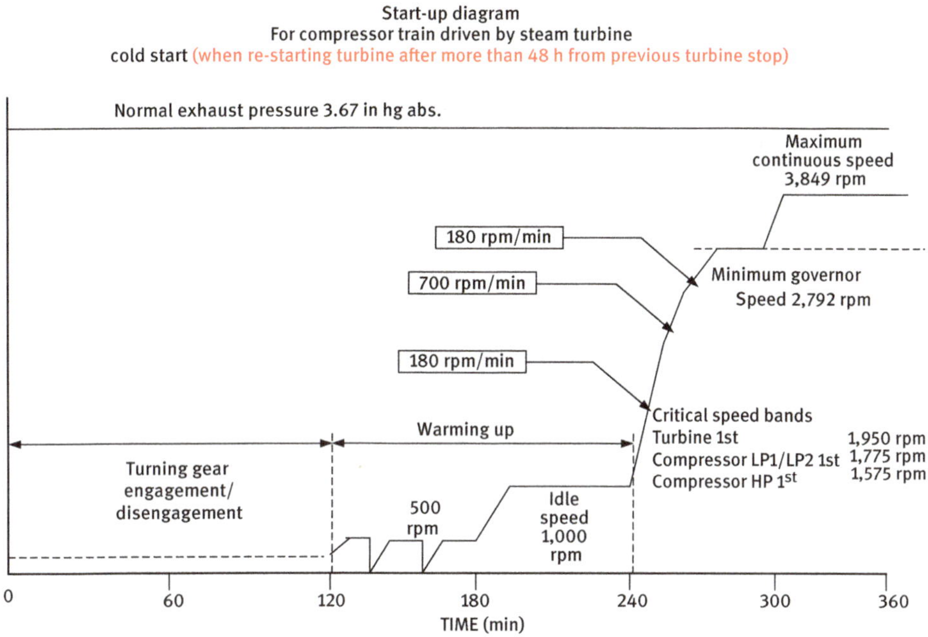

Figure D.18: Start-up diagram for steam turbine-driven compressor train.

D.2.6 Construction and installation of the equipment described in the example

Finally, some construction images: Figure D.19 illustrates piles or pile caps. The as-installed surface condenser is depicted in Figure D.20, followed by six site photos (comprising Figure D.21) for the actual field example narrated in this appendix. The reader can also track the construction sequence for tabletop foundations by following the arrows in Figure D.22.

Figure D.19: Piles/pile caps.

Figure D.20: Surface condenser installed.

Installation of rebars for top-deck (Looking North)

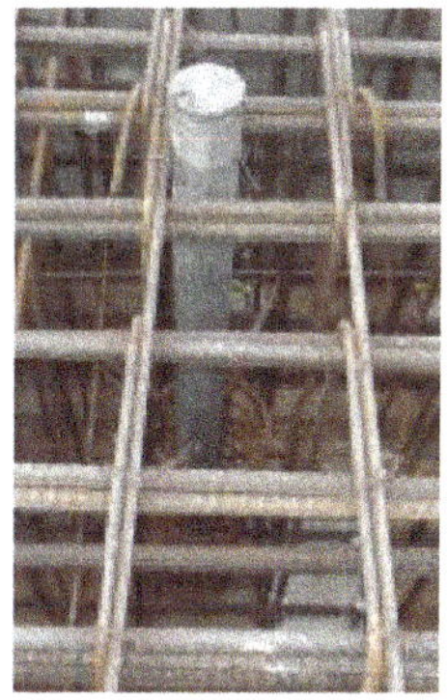

Duct for installation of
compressor anchor bolts

Compressors and steam turbine driver
installed on top-deck (Looking from the north side)

Compressors and steam turbine driver installed on top-deck (Looking from the west side)

Figure D.21: Construction site pictures.

Figure D.22: Typical construction sequence of a tabletop foundation.

References

[1] Bloch, H.P. Petrochemical Machinery Insights, Elsevier Publishing, Oxford UK, and Cambridge, MA, 2017. ISBN978-0-12-809272-9.

[2] Bloch, H.P. Fluid Machinery Innovations, DeGruyter Publishing, Berlin, Germany, 2020. ISBN 978-3-11-067413-2.

[3] Arya, S., O'Neill, M., Pincus, G. Design of Structures and Foundations for Vibrating Machines, Gulf Publishing Co, Houston, TX, 1979.

[4] Bhatia, K.G. March-June 2008, ISET Journal of Earthquake Technology, Paper 495, 45, 13–29.

[5] Whitman, R.V., Richart, F.E. Jr. Design Procedures for Dynamically Loaded Foundations, Journal of Soil Mechanics and Foundation Division, A.S.C.E., November 1967, 93(SM 6).

[6] American Concrete Institute, ACI 207.2R-07. 2007. Report on Thermal and Volume Change Effects on Cracking of Mass Concrete.

[7] Prakash, S., Puri, V. Foundation for Machines: Analysis and Design, John Wiley & Sons, New York, 1988.

Index

https://doi.org/10.1515/9783110678765-018

www.ingramcontent.com/pod-product-compliance
Lightning Source LLC
Chambersburg PA
CBHW080931240926
51849CB00003BA/431